Fundamentals of Quantum Mechanics

The book discusses fundamental concepts of quantum mechanics, including the state of a quantum mechanical system, operators, superposition principle and measurement postulate. The notion of an operator and the algebra of operators is introduced with the help of elementary concepts of mathematical analysis. The mathematical tools developed here will help resolve the difficulties encountered in classical physics while trying to explain the experimental results involving atomic spectra and other phenomena. The differential equations that arise while solving eigenvalue problems are solved rigorously, to make the text self-sufficient. The solutions are then physically interpreted and explained.

The book covers modern algebraic language of quantum mechanics, wherein the fundamental concepts and methods of solutions are translated into the algebraic formalism and compared with the earlier simpler approach. The text offers solved examples and homework problems to help students in solving practical problems of physics requiring quantum mechanical treatment.

Ajit Kumar is a Professor at the Department of Physics, Indian Institute of Technology, New Delhi. He is a Fellow of the Alexander von Humboldt Foundation, Bonn, Germany since 1987. A recipient of the Teaching Excellence Award from I. I. T. Delhi, he has been teaching the core subjects of theoretical physics including quantum field theory, group theory and its applications, and general theory of relativity for the past 34 years. His current research is related to the problems of nonlinear optics, solution of nonlinear Schrödinger equation in nonlinear optical media, fiber-optic solitons and their switching dynamics and electromagnetic wave propagation in metamaterials.

Fundamentals of Quantum Mechanics

Ajit Kumar

CAMBRIDGE
UNIVERSITY PRESS

University Printing House, Cambridge CB2 8BS, United Kingdom

One Liberty Plaza, 20th Floor, New York, NY 10006, USA

477 Williamstown Road, Port Melbourne, VIC 3207, Australia

314 to 321, 3rd Floor, Plot No.3, Splendor Forum, Jasola District Centre, New Delhi 110025, India

79 Anson Road, #06–04/06, Singapore 079906

Cambridge University Press is part of the University of Cambridge.

It furthers the University's mission by disseminating knowledge in the pursuit of education, learning and research at the highest international levels of excellence.

www.cambridge.org
Information on this title: www.cambridge.org/9781107185586

First published 2018

Printed in India by Rajkamal Electric Press, Kundli, Haryana.

A catalogue record for this publication is available from the British Library

ISBN 978-1-107-18558-6 Hardback
ISBN 978-1-108-46593-9 Paperback

Additional resources for this publication at www.cambridge.org/9781107185586

Dedicated to my parents

Contents

Figures

Tables

Preface

The given book presents an introduction to the basic concepts and mathematical tools of quantum mechanics. It is based on the material that I have been using in teaching the first course on quantum mechanics to the undergraduate and M.Sc. students at I. I. T. Delhi. The last chapter on relativistic generalization of quantum mechanics does not constitute a part of the usual course and has been added for those who wish to have some basic ideas of relativistic quantum mechanics.

In presenting the material, I have taken into account the feedback of the students about the conceptual as well as the mathematical difficulties faced by them during the course. As a result, I have tried to be as simple as possible. Therefore, I might appear to be too simple and repetitive at times and I hope the knowledgeable reader will pardon me for that.

The book starts with the basics of quantum mechanics in the traditional way by using the fundamental tools of mathematical analysis with an emphasis on the physical explanation for the mathematical treatment of the topics. This part includes the introduction to the concept of the state of a quantum mechanical system, operators and their algebra, the basic postulates of quantum mechanics and the solution of the Schrödinger equation for important one-dimensional systems. The algebraic formalism in the traditional language of Dirac is then introduced and the entire earlier material is reformulated in this language so as to make the reader comfortable with the modern language of quantum mechanics. In the later chapters of the book, I deal with the three-dimensional problems, hydrogen atom, quantum mechanical theory of orbital as well as spin angular momentum, and many particle systems. Simple effects related to the quantum mechanical treatment of the motion of a charged particle in the presence of a magnetic field are also presented. The basic concepts related to the symmetries of a system and the corresponding laws of conservation are then introduced and developed. In particular, the relationship between the fundamental quantum mechanical operators and the generators of the continuous groups of symmetries of spacetime are established and discussed. The book ends with an introduction to relativistic quantum mechanics.

I would like to thank my students whose active participation in the discussions, inside as well as outside the class rooms, has worked as the main motivation for writing this book. Finally, I would like to thank V. Ravishankar for going through some parts of the manuscript and making valuable suggestions.

Chapter 1

Introduction

Before the advent of quantum mechanics, classical physics studied the universe as a system consisting of matter and radiation. Matter was supposed to be made up of tiny building blocks called particles whose motion in space and time was assumed to be governed solely by the laws of mechanics formulated by Newton. The state of a classical system at a given instant of time, t_0, was completely defined by prescribing its position vector, $\vec{r}_0$, and momentum, $\vec{p}_0$. For any $t > t_0$, this state was uniquely determined by $\vec{r}(t)$ and $\vec{p}(t)$, the solutions of Newton's equations of motion with the initial conditions $\vec{r}(t_0) = \vec{r}_0$ and $\vec{p}(t_0) = \vec{p}_0$.

Radiation, on the other hand, was assumed to be governed by the laws of electromagnetism formulated in a unified manner by James Clark Maxwell. The dynamical variables of the radiation field were the components of the electric field, $\vec{E}$, and the magnetic field, $\vec{B}$, at every point in space. The spatio-temporal evolution of these fields was governed by Maxwell's equations. Unlike matter, the radiation field consisted of waves, with their characteristic properties exhibited in the phenomena of interference and diffraction.

Until the end of the nineteenth century, both these theories were on firm footings. The results of the kinetic theory of gases and statistical thermodynamics made it possible to verify qualitatively and also quantitatively the basic predictions of this corpuscular theory of matter. Besides that, all the contradictions that arose between the corpuscular theory and the wave theory of radiation were overcome by Huygens and Fresnel whose wave theory of light enabled one to explain all the phenomena involving light, including geometrical optics. At this stage, it was firmly believed that all phenomena in physics could be satisfactorily explained in the general framework of matter–radiation theory. If, in some cases, the explanation was not possible, the blame was put on the mathematical difficulties involved in the solution of the problems.

However, towards the end of the nineteenth century and the beginning of the twentieth century, a number of experiments, which were aimed at probing the atomic and sub-atomic structures of matter, were carried out. The results of these experiments strongly suggested a non-classical behaviour of matter at the microscopic level. Precisely speaking, two major groups of phenomena emerged that could not be comprehended by classical physics. The first was related to the existence of discrete energy levels for atoms (*discrete physical characteristic up against a continuous one in classical physics*) and the second was related to the so-called wave–particle dualism which appeared both in the behaviour of light as well as in the behaviour of the then-known elementary particles of matter. The physical phenomena underlying these experimental observations could not be explained on the basis of the classical framework of matter–radiation theory. As a result, it was felt that a new physical insight, radically different from the traditional one, was needed to explain the physical phenomena behind those, seemingly unusual, experimental results. As we know now, the result was a new physical theory called *quantum mechanics*.

In the rest of this Chapter, we shall dwell upon some of the key phenomena that gave a decisive jolt to the foundations of classical physics and played the most crucial role in the development of quantum mechanics.

1.1 The Blackbody Radiation

The radiation emitted by a body due to its temperature is called thermal radiation. In general, the spectral distribution in the thermal radiation emitted by a hot body depends on its composition. However, experiments show that there is a class of hot bodies whose spectra of thermal radiation have a universal character. Such hot bodies are known as blackbodies. Blackbodies are bodies with surfaces that absorb all of thermal radiation incident on them.

An important example of a blackbody is a cavity with a small hole on the surface. Any radiation incident upon the hole enters the cavity and undergoes a very large number of reflections off the walls of the cavity. In this process, it gets trapped and eventually absorbed by the walls before it can escape through the hole. Thus, the hole is equivalent to a surface that is a perfect absorber of radiation, like the surface of a blackbody.

Let us now assume that the walls of a cavity with a small hole is heated uniformly to a temperature T. Clearly, the interior of the cavity will be filled by the thermal radiation from the walls of the cavity. A small amount of radiation from the interior will fall on the hole and escape outside. The hole now acts as an emitter of thermal radiation. As discussed earlier, the hole has all the properties of the surface of a blackbody and hence the radiation from the hole of the cavity is called blackbody radiation.

The experimental data about blackbody radiation for various objects show that, at equilibrium, the radiation emitted has a well defined, continuous energy distribution and for each frequency, it is characterized by a quantity $u(\nu,T)$, which is called the energy density. It is defined as the energy content per unit volume per unit frequency interval of a cavity at temperature T. $u(\nu,T)$ does not depend on the chemical composition of the object nor does it depend on its shape. It depends only on the temperature of the walls of the cavity. Apart from that, the energy density shows a pronounced maximum at a given frequency, which increases with temperature. The experimentally observed dependence of $u(\nu,T)$ on frequency, ν, is shown in Figure 1.1 for three different temperatures.

In 1879, J. Stefan, on the basis of his experimental data, established the following empirical expression for the total power per unit area emitted by a hot body at temperature T:

$$P = \sigma T^4, \tag{1.1.1}$$

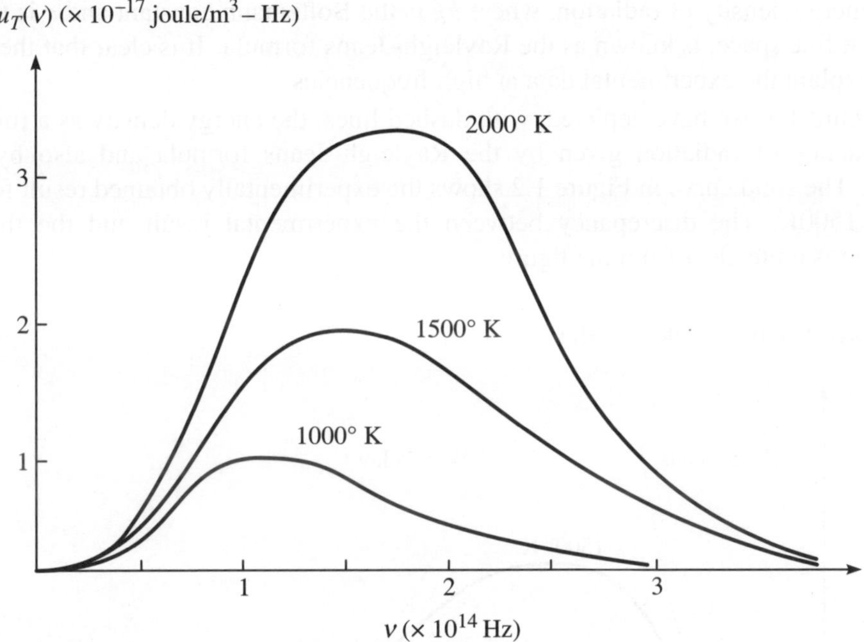

Figure 1.1 The energy density of a blackbody radiator as a function of frequency ν for three different temperatures $T = 1000K, 1500K$ and $2000K$.

where $\sigma = \pi^2 k_B^4/(15\hbar^3 c^3) = 5.67 \times 10^{-8}$ W m^{-2}K^{-4} is the Stefan–Boltzmann constant. The theoretical derivation of Stefan's law, given by (1.1.1), was provided by Boltzmann in 1884 by combining the thermodynamical calculations with the principles of Maxwell's electrodynamics.

There were various attempts to explain the continuous nature of the blackbody radiation. In 1894, Wien, using the Stefan–Boltzmann law (1.1.1), obtained the following formula for the energy density distribution $u(\nu, T)$

$$u(\nu, T) = \alpha \nu^3 e^{-\beta\nu/T}, \tag{1.1.2}$$

where α and β are constants that can be adjusted to fit the experimental data. Wien's formula explains the experimental data pretty well at high frequencies but fails miserably at low frequencies.

Later, Rayleigh and Jean attempted the theoretical modelling of the blackbody radiation on the basis of the classical radiation theory and the principle of equipartition of energy among the degrees of freedom. The resultant formula

$$u(\nu,T)\,d\nu = \frac{8\pi k_B T}{c^3}\,\nu^2, \tag{1.1.3}$$

for the energy density of radiation, where k_B is the Boltzmann constant and c is the speed of light in free space, is known as the Rayleigh–Jeans formula. It is clear that the formula fails to explain the experimental data at high frequencies.

In Figure 1.2, we have depicted, with dashed lines, the energy density as a function of the frequency of radiation given by the Rayleigh–Jeans formula and also by Wien's formula. The solid curve in Figure 1.2 shows the experimentally obtained result for $u_T(\nu)$ at $T = 1500$K. The discrepancy between the experimental result and the theoretical prediction is quite clear from the figure.

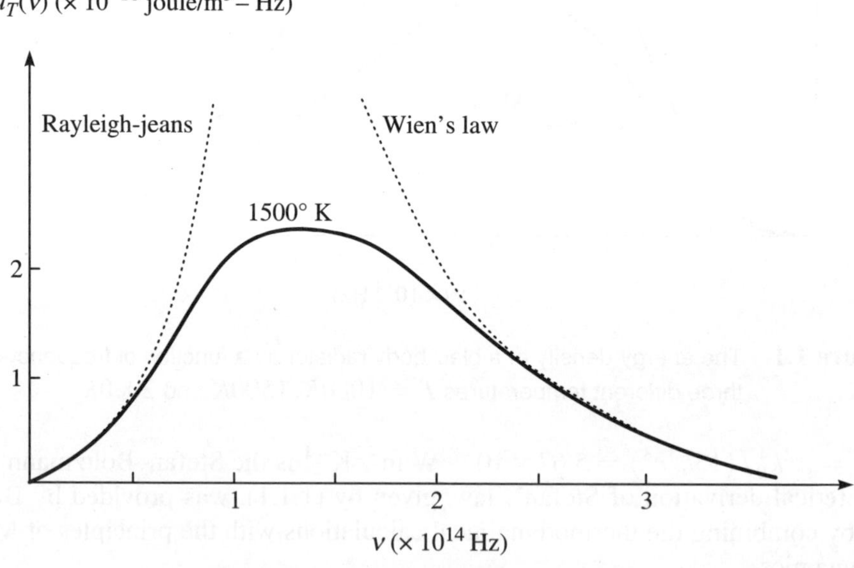

Figure 1.2 The energy density of a blackbody radiator as a function of frequency ν. The solid line represents the experimental curve, while the dashed lines correspond to the Rayleigh–Jeans and Wien's formulae, respectively.

All attempts, based on statistical thermodynamics and electromagnetic theory, to explain the continuous nature of the spectral distribution of thermal radiation over the entire range of frequencies resulted into utter failure until 1901 when Max Planck formulated his celebrated theory of blackbody radiation. He showed that the experimental curve could be explained only if one postulated that the energy of the radiation, emitted by the walls of the cavity, was quantized, that is, it was emitted only in multiples of the quantity $h\nu$, where $h = 1.63 \times 10^{-34}$ J s. is a universal constant. On the basis of his hypothesis of quantized radiation, Planck derived the spectral distribution function

$$u(\nu,T) = \frac{8\pi\nu^2}{c^3}\,\frac{h\nu}{e^{h\nu/k_BT-1}}, \tag{1.1.4}$$

valid for the entire range of frequency. In terms of the angular frequency, $\omega = 2\pi\nu$, Planck's formula can be written as

$$u(\omega,T) = \frac{\omega^2}{\pi^2c^3}\,\frac{\hbar\omega}{e^{\hbar\omega/k_BT-1}}. \tag{1.1.5}$$

At a given temperature T, the quantity $u(\omega,T)d\omega$ is the energy density in the frequency interval $[\omega, \omega + d\omega]$.

In the low frequency limit, when $h\nu \ll k_BT$, we have $e^{\hbar\omega/k_BT} \approx 1 + \hbar\omega/k_BT$ and we get

$$u(\omega,T) = \frac{\omega^2}{\pi^2c^3},\, k_BT \quad \text{or} \quad u(\nu,T)\,d\nu = \frac{8\pi\,\nu^2}{c^3}\,k_BT, \tag{1.1.6}$$

which is nothing but the Rayleigh–Jeans formula (1.1.3). On the other hand, in the high frequency limit, when $h\nu \gg k_BT$, Planck's formula reduces to Wien's formula:

$$u(\omega,T) = \frac{\hbar\omega^3}{\pi^2c^3}\,e^{-\hbar\omega/k_BT}. \tag{1.1.7}$$

Moreover, if we integrate equation (1.1.3) over all frequencies, then using the well known result

$$\int_0^\infty \frac{x^3}{e^x - 1}dx = \frac{\pi^5}{15}, \tag{1.1.8}$$

we recover Stefan's formula.

Thus, Planck's hypothesis of quantized radiation was a complete departure from the notions of classical physics and marked the beginning of a new era in physics, the era of quantum physics.

1.2 The Photoelectric Effect

It was experimentally established that irradiation of metallic surfaces with light led to the ejection of electrons. *This phenomenon of ejection of electrons from a metallic surface under the action of light is known as photoelectric effect.* The phenomenon had some peculiar properties: (a) when the frequency of the irradiating light was less than the so-called threshold frequency (ν_0), irrespective of the intensity of the irradiating light, no electrons could be ejected, (b) irrespective of the intensity of radiation (high or low), electrons were ejected instantaneously provided the frequency of radiation was greater than the threshold frequency and (c) the kinetic energy of the ejected electrons depended on the frequency but not on the intensity of the irradiating light.

Note that the threshold frequency is a characteristic of a given metal and is defined as $\nu_0 = \Phi/h$, where Φ is the work function of the metal and h Planck's constant. The work function is the minimum energy required for an electron to overcome the attractive forces that bind it to the metal surface.

The dependence of the photoelectric effect on the frequency of the falling radiation, which must be greater than the threshold frequency, could not be explained in the framework of classical physics. According to classical wave theory of light, the intensity of light is proportional to the square of the amplitude of the oscillating electric field. Hence, light of any frequency with sufficient intensity should be able to supply the required amount of energy for the electrons to overcome the potential barrier (work function) and become free. However, this was not the case in reality.

The second important point was the instantaneous ejection of electrons from the surface. According to classical physics, even if the falling radiation is weak, the electron would continuously absorb energy from it and in the process would be able to accumulate enough energy to leave the surface of the metal. Hence, according to classical physics, if the radiation is weak, the photoelectric effect should take some time before it shows up. However, no detectable time lag has ever been measured.

In 1905, Einstein used Planck's concept of quantized radiation and formulated the theory of photoelectric effect. He assumed light to be consisting of discrete quanta (photons), each of energy $h\nu$, ν being the frequency of light. When a metal is irradiated by light, an electron absorbs a photon (a quantum of energy) and gains an amount of energy equal to $h\nu$ irrespective of the intensity of the falling light. If this amount of energy is greater than the work function of the metal, the electron will be ejected. If the amount of energy is not greater than the work function of the metal, the electron will not be able to overcome the potential barrier. Therefore, the photoelectric effect can take place only if $h\nu \geq \Phi$. As a result, Einstein's fundamental equation for photoelectric effect reads:

$$h\nu = \Phi + K, \tag{1.2.1}$$

where K is the kinetic energy of the ejected electron. Einstein's theory of photoelectric effect explained the experimental data completely, including the linear increase in the kinetic energy of the electrons with the increase in the frequency of the irradiating light. The latter is evident from the following equation

$$K = h(\nu - \nu_0), \tag{1.2.2}$$

representing the dependence of the kinetic energy K on frequency ν. It is clear from (1.2.2) that $K = 0$ for $\nu = \nu_0$. This shows that a photon of frequency ν_0 has just enough energy to make an electron free, that is, to make the electron overcome the work function barrier. For an electron to leave the metallic surface with some non-zero kinetic energy, the frequency ν of the irradiating light must be greater than ν_0. This explains why the frequency of the light falling on a metal surface should be greater than the threshold frequency for photoelectric effect to take place. It also explains the dependence of the kinetic energy of the ejected electrons on the frequency of light rather than on its intensity. So far as the discrepancy regarding the time lag is concerned, this is explained by the fact that the energy is provided in discrete quanta rather than continuously. Therefore, if the quantum of energy $h\nu$ is greater than the work function, it will be absorbed by an electron and the electron will be ejected immediately; no time lag should be observed.

In summary, Einstein's theory of photoelectric effect and its subsequent experimental verification gave strong evidence in favour of Planck's hypothesis of quantized radiation and showed that light, in general, consists of discrete particles called photons (corpuscles) of finite energy. This was contrary to the classical description, according to which light of a given frequency was nothing but a continuous electromagnetic wave. A very important point to be noted here is that, besides explaining the peculiar properties of photoelectric effect, Einstein's theory clearly demonstrated the characteristic wave–particle duality of the micro-world.

1.3 The Bohr Model of an Atom

Rutherford's nuclear model of an atom consisting of a positively charged nucleus with electrons spread around it underwent rigorous experimental verification and left little doubt about its validity except that it did not guarantee the stability of the atom. For instance, if the electrons are supposed to be in a stationary arrangement, there is no stable configuration that would prevent the electrons from being pulled into the nucleus by the Coulomb force of attraction. On the other hand, if we assume the electrons to be orbiting the nucleus, in a similar way as the planets revolve around the sun in our solar system, they would be in an accelerated motion and being charged particles, they would emit energy in the form of electromagnetic radiation. As a result, the electrons would spiral into the nucleus.

To overcome these difficulties associated with the stability of atoms in the Rutherford model, Niels Bohr developed his model of an atom on the basis of a set of postulates. He assumed the electrons to be moving in circular orbits around the positively charged nucleus under the attractive Coulomb force. He, contrary to the principles of classical mechanics that allowed an electron to have a infinite set of circular orbits, postulated that electrons could move only in a selected discrete set of orbits for which its angular momentum was quantized. He assumed the angular momentum, L of an electron in such orbits to be an integral multiple of Planck's constant h, that is, $L = n\hbar$, where $\hbar = h/2\pi$ and n is a positive integer. Bohr also postulated that an electron in any of these allowed discrete orbits does not radiate and hence, its total energy E is conserved. These Bohr orbits are called stationary orbits.

Finally, in an attempt to explain the peculiarities involved in the discrete emission and absorption of radiation by atoms, Bohr postulated that if an electron made a discontinuous transition from one stationary orbit with energy E_i to another stationary orbit with energy

E_f such that $E_i > E_f$, it must emit energy in the form of electromagnetic radiation (photon) whose frequency, ν, is given by

$$\nu = \frac{E_i - E_f}{h}. \tag{1.3.1}$$

Using these postulates, Bohr derived his celebrated formula for the total energy of an electron of mass m_e and charge $-e$ in an atom with a positively charged nucleus of charge $+Ze$ and mass M. It is given by

$$E_n = -\frac{m_e Z^2 e^4}{32\pi^2 \varepsilon_0^2 \hbar^2} \frac{1}{n^2}, \tag{1.3.2}$$

where ε_0 is the permittivity of free space. We thus see that, according to Bohr, the energy of an electron in an atom is quantized and its various stationary states are characterized by the integer n, which is there in the condition for quantized angular momentum. The integer n, as we shall see later, is called the principal quantum number.

As we know today, Bohr theory of atom was not a consistent quantum theory in the sense that it imposed an ad hoc quantization condition on otherwise continuous classical variables. However, it did establish the discrete nature of physical characteristics of microscopic entities and strengthened the belief that the new microscopic theory of matter must inevitably incorporate this discreteness in its formalism.

1.4 The Compton Effect

The increase in the wavelength of X-rays after scattering off a free electron is known as Compton effect. The experiment was first conducted by Compton in 1923, who found that the wavelength of the scattered radiation was larger than the wavelength of the incident radiation. This was contrary to the classical theory of radiation according to which the incident and the scattered radiations should have the same wavelength. In order to explain the experimental results, Compton assumed that the X-ray beam was not a wave of frequency ν but a bunch of photons, each with energy $E = h\nu$. He treated the phenomenon as an elastic scattering of photons off a free electron. Qualitatively, according to Compton, the recoil photons emerging from the target constituted the scattered radiation and, since a photon transfers some of its initial energy, E_i, to the electron it bounces off, the scattered photon must have a lower final energy E_f. This means that it must have a lower final frequency ν_f and consequently a larger final wavelength λ_f. Compton applied energy and momentum conservation to obtain the required theoretical expression for the shift in the wavelength of the scattered photons.

Consider a photon of energy $E^\gamma = h\nu$ and momentum $p^\gamma = E^\gamma / c = h\nu / c$, where c is the speed of light in vacuum, undergoing an elastic collision with an electron which is initially at rest. Let the momentum of the photon, after scattering at an angle θ, be $\vec{p}\,'^\gamma$.

If the recoil momentum of the electron after scattering is $\vec{p}_e$, the conservation of momentum yields:

$$\vec{p}^{\gamma} = \vec{p}'^{\gamma} + \vec{p}_e. \tag{1.4.1}$$

On the other hand, if $E'^{\gamma} = h\nu'$ is the energy of the scattered photon and m_e the rest mass of the electron, then energy conservation leads to

$$h\nu + E_{e0} = E'^{\gamma} + E_e, \tag{1.4.2}$$

where $E_{e0} = m_e c^2$ and $E_e = \sqrt{p_e^2 c^2 + m_e^2 c^4}$ are the energy of the electron before and after the collision, respectively. After a bit of algebra, involving equations (1.4.1) and (1.4.2), we arrive at

$$h\nu + m_e c^2 = h\nu' + h\sqrt{\nu^2 + \nu'^2 - 2\nu\nu'\cos\theta + \frac{m_e^2 c^4}{h^2}}. \tag{1.4.3}$$

Dividing (1.4.3) throughout by h, we obtain

$$\nu - \nu' + \frac{m_e c^2}{h} = \sqrt{\nu^2 + \nu'^2 - 2\nu\nu'\cos\theta + \frac{m_e^2 c^4}{h^2}} \tag{1.4.4}$$

Squaring both sides of (1.4.4) and simplifying, we end up with

$$\frac{1}{\nu'} - \frac{1}{\nu} = \frac{1}{m_e c^2}(1 - \cos\theta) = \frac{2h}{m_e c^2}\sin\left(\frac{\theta^2}{2}\right). \tag{1.4.5}$$

Finally, converting to wavelength, we have

$$\Delta\lambda = \lambda' - \lambda = 4\pi\lambda_c \sin\left(\frac{\theta^2}{2}\right), \tag{1.4.6}$$

where $\lambda_c = h/2\pi m_e c = 3.86 \times 10^{-13}$ m is the Compton wavelength of the electron. This formula, which explains the experimental results, shows that the change in wavelength depends only on the angle of scattering θ and not on the wavelength of the incident radiation. The Compton effect confirmed the corpuscular behaviour of light that had surfaced as a surprise after Einstein formulated his theory of photoelectric effect.

The final and decisive impetus was provided by Louis in 1924, who put forward a courageous hypothesis that every material particle, besides its usual corpuscular properties described by Newton's equations of motion, also possesses properties that we assign to a wave. According to him, a free particle with momentum $\vec{p}$ can be represented by a plane wave

$$\psi(\vec{r}, t) = A\, e^{i(\vec{k}\cdot\vec{r} - \omega t)}, \tag{1.4.7}$$

where A is a constant. The wave vector $\vec{k}$ and the angular frequency ω of the wave are related to the momentum $\vec{p}$ and the energy E of the particle by the following expressions

$$\vec{k} = \frac{\vec{p}}{\hbar}, \quad \omega = \frac{E}{\hbar}, \tag{1.4.8}$$

where $\hbar$ is equal to Planck's constant divided by 2π. de Broglie's hypothesis was experimentally confirmed by Davisson–Germer in 1927.

All these peculiar and scientifically new developments at the end of the nineteenth and the beginning of the twentieth centuries inspired the discovery of a new physical theory capable of describing the phenomena occurring at the microscopic scales. It was formulated in two different forms: *Matrix mechanics* (Heisenberg 1925) and *wave mechanics* (Schrödinger 1926). Later, they were shown to be equivalent. This new theory was universally called *quantum mechanics*.

Example 1.4.1: A light source of wavelength λ_1 illuminates a metal and ejects photoelectrons with a maximum kinetic energy of 2 eV. If the same metal is irradiated with another source that emits light at a wavelength half that of the first source, what will be the maximum kinetic energy of the ejected electrons? The work function of the metal is 1 eV.

Solution: According to Einstein's equation of photoelectric effect

$$K_{\max} = \frac{hc}{\lambda} - \Phi, \tag{1.4.9}$$

where K stands for the kinetic energy of the ejected electrons, $h \approx 6.62 \times 10^{-34}$ J s is Planck's constant, c is the speed of light in free space, λ is the wavelength of the light incident on the metal and Φ is the work function of the metal. In the given case, we have

$$K_{1\max} = \frac{hc}{\lambda_1} - \Phi, \tag{1.4.10}$$

$$K_{2\max} = \frac{hc}{\lambda_2} - \Phi = \frac{2hc}{\lambda_1} - \Phi. \tag{1.4.11}$$

From (1.4.10), we have $hc/\lambda_1 = K_{1\max} + \Phi$, and hence

$$K_{2\max} = 2K_{1\max} + \Phi. \tag{1.4.12}$$

Thus, $K_{2\max} = 4 + 1 = 5$ eV.

Example 1.4.2: A 500 MeV photon undergoes a head-on collision with a proton $(Mc^2 = 936 \text{ MeV})$ at rest. Find the maximum loss of energy in the process.

Solution: Let the initial and the final momenta of the photon be $+p$ and p' respectively. Since the proton is initially at rest, momentum conservation requires its final momentum to be $(p-p')$. The relativistic formula for energy conservation gives

$$pc+Mc^2 = |p'|c+\sqrt{(Mc^2)^2+(p-p')^2c^2}, \tag{1.4.13}$$

where we have taken into account that the final momentum p' may be positive as well as negative. If $p'>0$, then we get

$$(pc-p'c+Mc^2)^2 = (Mc^2)^2+(p-p')^2c^2, \quad \Rightarrow \quad 2M(p-p')c^3 = 0. \tag{1.4.14}$$

This means that $p = p'$ and the proton remains at rest. Therefore, there is no energy loss. However, if $p'<0$, then the final proton momentum is $(p+|p'|)$ and we have

$$(pc-|p'|c+Mc^2)^2 = (Mc^2)^2+(p+p')^2c^2, \quad \Rightarrow \quad 2Mc(p-|p'|) = 4p|p'|. \tag{1.4.15}$$

This gives the magnitude of the final momentum of the photon to be

$$|p'| = \frac{Mcp}{Mc+2p} = \frac{p}{1+2p/Mc}. \tag{1.4.16}$$

As a consequence, the energy loss is given by

$$\Delta E = c(p-|p'|) = \frac{2(cp)^2}{Mc^2}\cdot\frac{1}{1+\dfrac{2cp}{Mc^2}}. \tag{1.4.17}$$

Putting the values $cp = 500$ MeV and $Mc^2 = 936$ MeV, we obtain $\Delta E \cong 258$ MeV.

Example 1.4.3: Muonic hydrogen consists of a muon (a particle just like an electron but with a mass $m_\mu = 105.7$ MeV/c^2) bound to a proton. Use the Bohr model to find the energy and radius of the lowest orbit, and the wavelength of the $n=2 \to n=1$ transition.

Solution: For the energy of the lowest orbit, we use the Bohr formula

$$E = -\left(\frac{e^2}{4\pi\varepsilon_0}\right)^2\frac{m}{\hbar^2} = -\left(\frac{1}{4\pi\varepsilon_0}\frac{e^2}{\hbar c}\right)^2 mc^2. \tag{1.4.18}$$

Taking into account that, in general, the nucleus is also moving, we use the reduced mass for the muon-proton atom

$$\mu = \frac{m_\mu m_p}{m_\mu+m_p} = \frac{105.7\times 638.28}{105.7+638.28}\,\mathrm{MeV/c^2} = 94.998\,\mathrm{MeV/c^2}. \tag{1.4.19}$$

Hence, the energy is given by

$$E = -\frac{1}{2}\left(\frac{1}{137}\right)^2 94.998\,\mathrm{MeV}/\mathrm{c}^2 = -2531\,\mathrm{eV}. \tag{1.4.20}$$

The radius of the orbit is

$$r = \left(\frac{4\pi\varepsilon_0}{e^2}\right)^2 \frac{\hbar^2 c^2}{\mu c^2} = \left(\frac{1}{1.44\,\mathrm{eV\,nm}}\frac{[1240\,\mathrm{eV\,nm}/2\pi]^2}{94.998\,\mathrm{MeV}}\right) = 2.85\,\mathrm{nm}. \tag{1.4.21}$$

For $n = 2 \to n = 1$ transition, the energy of the photon is given by

$$E_\gamma = E_2 - E_1 = -\left(\frac{1}{4} - 1\right)E_1 = -\frac{3}{4}E_1 = \frac{3}{4}2531\,\mathrm{eV} = 1898\,\mathrm{eV}. \tag{1.4.22}$$

The wavelength is therefore

$$\lambda = \frac{\hbar c}{E} = \frac{1240\,\mathrm{eV\,nm}}{1898\,\mathrm{eV}} = 0.653\,\mathrm{nm}. \tag{1.4.23}$$

Homework Problems

1. The work function for photo-effect in potassium is 2.25 eV. When a light of wavelength 3.6×10^{-7} m falls on the potassium atom, calculate: (a) the stopping potential $V_{\max}$ of the photoelectrons and (b) the kinetic energy and the velocity of the fastest of the ejected electrons.
2. A photon with 10^4 eV energy collides with a free electron at rest and is scattered through an angle of 60°. Calculate (a) the change in energy, frequency and wavelength of the photon, and (b) the kinetic energy, momentum and direction of the electron after the collision.
3. Calculate the de Broglie wavelength for an electron with a kinetic energy of 20 eV and 200 keV.
4. Consider an atom with a nucleus consisting of 2 protons, and instead of an electron, a muon (207 times as heavy as an electron, or $m_\mu = 207 \times m_e$). Calculate the energy of the ground state and the first excited state, and the wavelength of light associated with the transition.
5. A 100 keV photon undergoes Compton scattering at an angle of 40°. Find the energy of the scattered photon, and the energy and angle of the recoil electron.

Chapter 2

The Postulates of Quantum Mechanics

The basic questions that quantum mechanics ventures to answer are the following:

I. How can one define the state of a quantum mechanical system at a given time t?

II. How can one represent or describe a measurable physical quantity (position, momentum, energy, angular momentum etc) in quantum mechanics?

III. How can one calculate the values of various measurable physical quantities in a given quantum state?

IV. Do all possible measurements of an observable in a given state lead to the same numerical value? If not, how do we define the value of an observable in a given state?

V. How, knowing the state of the system at any given instant of time t_0, can one determine the state at any later instant $t > t_0$?

In what follows, we shall discuss these questions and the possible answers to them. This will eventually lead us to the basic theory of quantum mechanics. Let us start with the definition of a state of a microscopic system.

As in classical mechanics, we have to first decide how we are going to define the state of a quantum mechanical system (particle) at a given instant of time and then determine the equation that governs the time-evolution of the state. In classical mechanics, the state of a particle at a given instant of time is completely defined by specifying its position and momentum at that instant. In quantum mechanics, however, due to some specific properties (which will be discussed later), it is impossible to determine both the position and momentum of a particle simultaneously and accurately at a given instant of time. Therefore, for a quantum system we need a new, completely different from the classical point of view, definition of state. It is done, as we shall see, by introducing a mathematical object called the wave function.

Note that in this entire book, we shall use the word particle to denote not only a particle, but also a given quantum mechanical system. That is, the word 'particle' is synonymous to system and vice versa.

2.1 Specification of State. Statistical Interpretation

Postulate 1: The state of a quantum system at a given instant of time, t, is completely defined by a function, $\psi(\vec{r},t)$, of position $\vec{r} = \{x,y,z\}$ and time, t.

It is called the wave function of the particle. The wave function is a complex-valued function and contains information about the position of the particle at time t.

Statistical Interpretation: The wave function $\psi(\vec{r},t)$ does not have any physical meaning of its own. However, it has been accepted to consider it as the probability amplitude in the sense that, if a measurement of the position of the particle is carried out, the probability that at a given instant of time, t, the particle will be found in an infinitesimal volume element, $\Delta V = dx\,dy\,dz$, is given by $|\psi(\vec{r},t)|^2\,dV$. This is the so-called statistical interpretation of the wave function proposed by Max Born. It is then obvious that the quantity $|\psi(\vec{r})|^2$ plays the role of the probability density for locating the particle in space at a given instant of time.

There are two important points to be noted here. Firstly, if we multiply ψ by a complex number $e^{i\alpha}$, where α is a real constant, its physical meaning does not change because $|\psi|^2$ remains unchanged. Consequently, the probability, P, of locating the particle in a given volume V,

$$P = \int_V |\psi(\vec{r},t)|^2\,dV, \tag{2.1.1}$$

also remains unchanged.

Now, since the probability of finding the particle at some point in space at a given instant of time is definitely equal to 1, if the volume of integration, V, in the above formula is replaced by all space (i.e., the entire universe), we arrive at:

$$\int_{\text{all space}} |\psi(\vec{r},t)|^2\,dV = \int_{-\infty}^{+\infty} dx \int_{-\infty}^{+\infty} dy \int_{-\infty}^{+\infty} dz\,|\psi(\vec{r},t)|^2 = 1. \tag{2.1.2}$$

Equation (2.1.2) is known as the normalization of the wave function. This condition of normalizability requires the wave function to be square-integrable. As a particular case, the square–integrability requires the wave function to vanish at spatial infinity:

$$\lim_{(x,y,z)\to\pm\infty} \psi(x,y,z,t) = 0. \tag{2.1.3}$$

In general, an acceptable wave function must satisfy the following conditions:

(a) The wave function must be single-valued.
(b) It must be continuous in the entire region of its arguments (that is, of the independent variables).
(c) It must be finite everywhere.
(d) The wave function must also be square-integrable.

We shall call these as standard conditions. If a function does not satisfy even one of these conditions, it cannot be an acceptable quantum mechanical wave function.

Before we proceed further, let us note that the set of all possible complex wave functions, $\psi(\vec{r},t), \phi(\vec{r},t), \chi(\vec{r},t), \ldots$, of a quantum mechanical system forms a linear vector space which is equipped with a scalar product. This scalar product between any pair of functions, ψ and ϕ, is defined as

$$\int_{-\infty}^{+\infty} \psi^* \phi \, d^3x \equiv (\psi, \phi), \tag{2.1.4}$$

where $d^3x = dx\,dy\,dz$. According to the standard conditions, mentioned earlier, this scalar product has to exist, that is, $(\psi, \phi) < \infty$.

Example 2.1.1: Which of the wave functions $\psi(x,t) = Axe^{-i\omega t}$ and $\phi(x,t) = Axe^{-\beta x^2} e^{-i\omega t}$, where A, ω and β are arbitrary constants, can be an acceptable quantum mechanical wave function?

Solution: Both the functions are single-valued and continuous. However, $\psi(x,t)$ is not finite everywhere and is not square-integrable:

$$\int_{-\infty}^{+\infty} |\psi_1(x,t)|^2 \, dx = |A|^2 \int_{-\infty}^{+\infty} x^2 \, dx \rightarrow \infty. \tag{2.1.5}$$

Hence, it cannot be an acceptable wave function. On the other hand, $\phi(x,t)$ is finite everywhere and is also square-integrable as shown below:

$$\int_{-\infty}^{+\infty} |\psi_2(x,t)|^2 \, dx = |A|^2 \int_{-\infty}^{+\infty} x^2 \, e^{-2\beta x^2} \, dx. \tag{2.1.6}$$

The integral can be evaluated with the help of the Gaussian integral

$$I(\alpha) = \int_{-\infty}^{+\infty} e^{-\alpha x^2} \, dx = \sqrt{\frac{\pi}{\alpha}}. \tag{2.1.7}$$

If we differentiate $I(\alpha)$ with respect to α, we get

$$\int_{-\infty}^{+\infty} x^2 \, e^{-\alpha x^2} \, dx = \sqrt{\frac{\pi}{4\alpha^3}}. \tag{2.1.8}$$

In our case, $\alpha = 2\beta$. Therefore,

$$\int_{-\infty}^{+\infty} |\psi_2(x,t)|^2 \, dx = \sqrt{\frac{\pi}{32\beta^3}}. \tag{2.1.9}$$

We see that $\phi(x,t)$ satisfies the standard conditions and, hence, is an acceptable quantum mechanical wave function.

Example 2.1.2: Does the function,

$$\phi(x) = \begin{cases} \sin\left(\frac{\pi x}{3}\right), & 0 < x < 3 \\ 0, & \text{Otherwise} \end{cases}$$

represent an acceptable wave function?

Solution: The first three standard conditions are clearly satisfied by the given function. Let us check the last condition of square-integrability. We have

$$I = \int_{-\infty}^{+\infty} |\psi(x,t)|^2 \, dx = \int_0^3 \sin^2\left(\frac{\pi x}{3}\right) dx. \tag{2.1.10}$$

Using the trigonometrical formula $\cos(2x) = 1 - 2\sin^2 x$, we get

$$I = \frac{1}{2}\int_0^3 dx - \frac{1}{2}\int_0^3 \cos\left(\frac{2\pi x}{3}\right) dx = \frac{3}{2} - \frac{3}{4}\sin\left(\frac{2\pi x}{3}\right)\Big|_0^3 = \frac{3}{2} < \infty. \tag{2.1.11}$$

Hence, the given function satisfies all the standard conditions and is an acceptable quantum mechanical wave function.

Example 2.1.3: The wave function of a particle is

$$\psi(x,t) = A\,e^{-\alpha(x-\beta)^2} e^{-i\omega t},$$

where A is an arbitrary constant. Here, α, β and ω are known real constants. (a) Find A. (b) Where is the particle most likely to be found?

Solution: The normalization condition for the wave function reads

$$\begin{aligned} \int_{-\infty}^{+\infty} |\psi(x,t)|^2 \, dx = & \ |A|^2 \int_{-\infty}^{+\infty} e^{-\alpha(x-\beta)^2} e^{-i\omega t} e^{-\alpha(x-\beta)^2} e^{+i\omega t} \, dx \\ = & \ |A|^2 \int_{-\infty}^{+\infty} e^{-2\alpha(x-\beta)^2} \, dx = 1. \end{aligned} \tag{2.1.12}$$

Changing the variable $x - \beta$ to ξ, we get

$$|A|^2 \int_{-\infty}^{+\infty} e^{-2\alpha\xi^2} \, d\xi = 1. \tag{2.1.13}$$

Taking the integral, we obtain

$$\sqrt{\frac{\pi}{2\alpha}}\;|A|^2 = 1. \quad \Rightarrow \quad A = \frac{(2\alpha)^{1/4}}{\pi^{1/4}}. \tag{2.1.14}$$

For the extremum of the probability density, $\rho = |\psi|^2$, we must have

$$\frac{d\rho}{dx} = \frac{(2\alpha)^{1/2}}{\pi^{1/2}}\frac{d}{dx}\left(e^{-2\alpha(x-\beta)^2}\right) = -\frac{(2\alpha)^{1/2}}{\pi^{1/2}}4\alpha(x-\beta)e^{-2\alpha(x-\beta)^2} = 0. \tag{2.1.15}$$

Equation (2.1.15) shows that $x = \beta$ is the point of extremum for ρ. Since

$$\left.\frac{d^2\rho}{dx^2}\right|_{x=\beta} = \frac{(2\alpha)^{1/2}}{\pi^{1/2}}\left[\left(-4\alpha + 16\alpha^2(x-\beta)^2\right)e^{-2\alpha(x-\beta)^2}\right]_{x=\beta}$$

$$= -4\frac{(2\alpha)^{1/2}}{\pi^{1/2}}\alpha < 0, \tag{2.1.16}$$

the probability density reaches its maximal value at $x = \beta$. Therefore, the particle is most likely to be found at $x = \beta$.

Example 2.1.4: A particle of mass m is confined to move in the region $0 < x < L$. It is in a state described by the wave function

$$\psi(x,0) = A\left[\sin\left(\frac{\pi x}{L}\right) + \sin\left(\frac{2\pi x}{L}\right)\right],$$

where A and L are arbitrary real constants. (a) Find A. (b) Find the probability of locating the particle in the interval $L/4 \le x \le 3L/4$.

Solution: The normalization condition leads to

$$I = |A|^2\int_0^L\left(\sin^2\left[\frac{\pi x}{L}\right] + \sin^2\left[\frac{2\pi x}{L}\right] + 2\sin\left[\frac{2\pi x}{L}\right]\sin\left[\frac{\pi x}{L}\right]\right)dx = 1. \tag{2.1.17}$$

Using the formulae $\cos(2x) = 1 - 2\sin^2 x$ and $\cos(\alpha-\beta) - \cos(\alpha+\beta) = 2\sin(\alpha)\sin(\beta)$, we get

$$I = \frac{|A|^2}{2}\int_0^L\left(2 - \cos\left[\frac{2\pi x}{L}\right] - \cos\left[\frac{4\pi x}{L}\right]\right)dx$$

$$+\int_0^L |A|^2\left(\cos\left[\frac{\pi x}{L}\right] - \cos\left[\frac{3\pi x}{L}\right]\right)dx. \tag{2.1.18}$$

Integrating and using the identity $\sin\left(\frac{n\pi x}{L}\right)\Big|_0^a = 0$, we get

$$A^2 L = 1. \quad \Rightarrow \quad A = \frac{1}{\sqrt{L}}. \tag{2.1.19}$$

The probability of finding the particle in the interval $L/4 \leq x \leq 3L/4$ is given by

$$\begin{aligned} P =& \frac{1}{2L}\int_{L/4}^{3L/4}\left(2-\cos\left[\frac{2\pi x}{L}\right]-\cos\left[\frac{4\pi x}{L}\right]\right)dx \\ &+\frac{1}{L}\int_{L/4}^{3L/4}\left(\cos\left[\frac{\pi x}{L}\right]-\cos\left[\frac{3\pi x}{L}\right]\right)dx. \end{aligned} \tag{2.1.20}$$

Taking the integrals, we finally obtain

$$\begin{aligned} P =& \frac{1}{2L}\left(-\frac{L}{2\pi}(-1-1)-\frac{L}{4\pi}(0-0)\right) \\ &+\frac{1}{L}\left(\frac{L}{\pi}\left[\frac{1}{\sqrt{2}}-\frac{1}{\sqrt{2}}\right]+\frac{L}{3\pi}\left[\frac{1}{\sqrt{2}}-\frac{1}{\sqrt{2}}\right]\right) = \frac{1}{2\pi}. \end{aligned} \tag{2.1.21}$$

2.2 Observables and Operators

An observable, A, is a dynamical variable of a particle that can be measured. Position, momentum, angular momentum and energy are examples of observables that we encounter in classical mechanics. Naturally, the question arises about how to represent these in the framework of quantum mechanics. Here, and in the following couple of subsections, we shall discuss this problem.

Postulate 2: An observable, A, is represented by a linear and hermitian operator that is written as $\hat{A}$.

An operator, $\hat{A}$, is a mathematical instruction, which when applied to a mathematical object, say ψ, gives another mathematical object ϕ of the same nature. It is symbolically written as $\hat{A}\psi = \phi$.

Let ψ be a usual function of one or several variables $x_i, i = 1,2,3,\ldots$: $\psi(x_1,x_2,x_3,\ldots)$. Then, in the given context, defining an operator means defining the mathematical instruction(s) that must be applied to the function $\psi(x_1,x_2,x_3,\ldots)$ to obtain a new function $\phi(x_1,x_2,x_3,\ldots)$ of the same number of independent variables.

A simple operator, for instance, may contain the instruction for just taking the derivative of a function. Suppose we call this operator $\hat{D}$. Then, for a given function $\psi(x)$:

$$\hat{D}\psi(x) = \frac{d\psi}{dx} = \phi(x). \tag{2.2.1}$$

Or, in shorthand notation, we just write $\hat{D}\psi(x) = \phi(x)$. The function ϕ belongs to the same functional space to which ψ belongs. Another operator could be the instruction to obtain a new function $\phi(x)$ by multiplying a given function $\psi(x)$ by its independent variable x, that is, $\phi(x) = x\psi(x)$. In other words, the independent variable x can also be looked upon as an operator. It is usually denoted by $\hat{x}$. Thus, $\phi(x) = \hat{x}\psi(x) = x\psi(x)$.

Linear operators: If, for the given scalars α and β and functions $\psi(x)$ and $\phi(x)$, the operator $\hat{A}$ satisfies the relation

$$\hat{A}\left(\alpha\psi(x) + \beta\phi(x)\right) = \alpha\hat{A}\psi(x) + \beta\hat{A}\phi(x), \tag{2.2.2}$$

it is said to be a linear operator. If the action of an operator $\hat{A}$ on a function $\phi(x)$ is to multiply that function by some constant a:

$$\hat{A}\phi(x) = a\,\phi(x), \tag{2.2.3}$$

we say that the constant a is an *eigenvalue* of the operator $\hat{A}$, and we call $\phi(x)$ an *eigenfunction* of $\hat{A}$. An operator can have more than one eigenvalues. The set of all possible eigenvalues of an operator constitutes the so-called *eigenvalue spectrum (or, simply spectrum)* of the operator. If for every eigenvalue, there is a single eigenfunction, the spectrum of the operator is called non-degenerate. If for a given eigenvalue, a, there are more than one eigenfunctions, the eigenvalue a is said to be *degenerate*. If, for instance, for a given eigenvalue, a_k, there exist m linearly independent eigenfunctions, then the eigenvalue a_k is said to be *m-fold degenerate*.

Operators in quantum mechanics are, in general, complex and act in the space of complex functions. Consequently, their eigenvalues are also, in general, complex.

2.3 Hermitian Operators

Given an operator $\hat{A}$, let us define an operator $\hat{A}^\dagger$ by

$$\int_{-\infty}^{+\infty} \phi^*(\vec{r})\left(\hat{A}\psi(\vec{r})\right) d^3x = \int_{-\infty}^{+\infty} \left(\hat{A}^\dagger\phi(\vec{r})\right)^* \psi(\vec{r})\, d^3x. \tag{2.3.1}$$

The operator $\hat{A}^\dagger$ is called the operator *hermitian conjugate (adjoint)* to the operator $\hat{A}$.

If an operator $\hat{A}$ is equal to its hermitian conjugate operator, that is, $\hat{A}^\dagger = \hat{A}$, it is called a hermitian operator. On the other hand, if $\hat{A}^\dagger = -\hat{A}$, then the operator $\hat{A}$ is called *anti-hermitian*.

If an operator, $\hat{A}$, is represented by a matrix A of appropriate dimensions, then the hermitian conjugate operator is given by the matrix, denoted as $A^\dagger$, which is hermitian conjugate to the matrix A. $A^\dagger$ is obtained by first transposing the matrix A and then performing the complex conjugation, that is,

$$A^\dagger = \left[(A)^T\right]^*, \tag{2.3.2}$$

where $*$ stands for complex conjugation.

We may ask why do we use hermitian operators to represent dynamical variables in quantum mechanics. As we shall see later, in quantum mechanics, it is postulated that if we measure a dynamical variable A of a system in a given quantum state ψ, the result will be one of the eigenvalues of the operator $\hat{A}$ that represents the dynamical variable A. Since the results of measurement are real numbers (in appropriate units), the eigenvalues of $\hat{A}$ must be real. That is why we have to represent a physical characteristic of a system by a hermitian operator.

Theorem 2.3.1: The eigenvalues of a hermitian operator are real.

Proof: Consider a hermitian operator $\hat{A}$ and its eigenvalue equation

$$\hat{A}\psi_n(\vec{r}) = \lambda_n \psi_n(\vec{r}), \;\; \vec{r} = (x,y,z). \tag{2.3.3}$$

As stated earlier, λ_n are in general complex. Now, since $\hat{A}$ is hermitian, we have

$$\int_{-\infty}^{+\infty} \psi_n^*(\hat{A}\psi_n)d^3x = \int_{-\infty}^{+\infty} (\hat{A}\psi_n)^* \psi_n \, d^3x. \tag{2.3.4}$$

Using the eigenvalue equation for $\hat{A}$, we have

$$(\lambda_n - \lambda_n^*) \int_{-\infty}^{+\infty} \psi_n^* \psi_n \, d^3x = 0. \tag{2.3.5}$$

For the non-trivial solutions to the eigenvalue equation, $\int_{-\infty}^{+\infty} \psi_n^* \psi_n \, d^3x \neq 0$ and we get: $\lambda_n = \lambda_n^*$. Hence, λ_n are real.

Theorem 2.3.2: The eigenfunctions of a hermitian operator, corresponding to distinct eigenvalues, are orthogonal.

Proof: Let us assume that the eigenvalues are non-degenerate. Consider the eigenvalue equations for the operator $\hat{A}$:

$$\hat{A}\psi_m = \lambda_m \psi_m, \tag{2.3.6}$$

$$\hat{A}|\psi_n\rangle = \lambda_n \psi_n, \tag{2.3.7}$$

where $\lambda_m \neq \lambda_n$. Using the hermiticity of $\hat{A}$, we have

$$\int_{-\infty}^{+\infty} \psi_n^*(\hat{A}\psi_m)d^3x = \int_{-\infty}^{+\infty} (\hat{A}\psi_n)^* \psi_m \, d^3x. \tag{2.3.8}$$

If we now use the eigenvalue equation for $\hat{A}$ and the fact that the eigenvalues are real, we have

$$(\lambda_m - \lambda_n) \int_{-\infty}^{+\infty} \psi_n^* \psi_m d^3x = 0. \tag{2.3.9}$$

Since $\lambda_m \neq \lambda_n$,

$$\int_{-\infty}^{+\infty} \psi_n^* \psi_m d^3x = 0. \tag{2.3.10}$$

The last equation shows that the eigenfunctions ψ_m and ψ_n are orthogonal. We can normalize these eigenfunctions and write the content of Theorem 2.3.2 symbolically as

$$\int_{-\infty}^{+\infty} \phi_n^* \phi_m d^3x = \delta_{mn} \tag{2.3.11}$$

where δ_{mn} is the Kronecker delta symbol and

$$\phi_i = \frac{\psi_i}{\sqrt{\int_{-\infty}^{+\infty} |\psi_i|^2 d^3x}}. \tag{2.3.12}$$

Let us now see whether Theorem 2.3.2 holds or not for the case when an eigenvalue of $\hat{A}$ is degenerate. For simplicity, consider the case of two-fold degeneracy, that is, assume that a given eigenvalue λ is two-fold degenerate. Let ψ_1 and ψ_2 be the two distinct eigenfunctions corresponding to λ. Clearly, two (or more) eigenfunctions corresponding to the same eigenvalue will be distinct, if they are linearly independent. If they are not linearly independent, one of them will be proportional to the other and the constant of proportionality can be absorbed in the normalization of the wave function. As a result, we will be left with only one eigenfunction. Therefore, we shall assume ψ_1 and ψ_2 to be linearly independent. Since, they are linearly independent, the linear combination

$$\psi = c_1\psi_1 + c_2\psi_2 \tag{2.3.13}$$

where c_1 and c_2 are arbitrary non-zero constants, must be nonzero. We can use this fact and the eigenfunctions ψ_1 and ψ_2 to construct two new mutually orthogonal eigenfunctions corresponding to λ. Firstly, we normalize ψ_1 and take it as the first eigenfunction of $\hat{A}$ with eigenvalue λ:

$$\phi_1 = \frac{\psi_1}{\sqrt{\int_{-\infty}^{+\infty} \psi_1^* \psi_1 d^3x}}. \tag{2.3.14}$$

Consider now the linear combination

$$u = c_{21}\phi_1 + \psi_2, \tag{2.3.15}$$

where c_{21} is an arbitrary constant, and let it be orthogonal to ϕ_1. This allows us to determine c_{21}:

$$c_{21} = -\int_{-\infty}^{+\infty} \phi_1^* \psi_2 d^3x. \tag{2.3.16}$$

Thus, the function u, with c_{21} given by (2.3.16), is orthogonal to ϕ_1. Finally, we normalize u and take it as the second eigenfunction with the same eigenvalue λ:

$$\phi_2 = \frac{u}{\sqrt{\int_{-\infty}^{+\infty} u^* u d^3x}}. \tag{2.3.17}$$

Thus, from the linearly independent eigenfunctions ψ_1 and ψ_2 of the two-fold degenerate eigenvalue λ of $\hat{A}$, we were able to construct two new orthonormal eigenfunctions ϕ_1 and ϕ_2 of $\hat{A}$ with the same eigenvalue λ. Clearly, this procedure can easily be generalized to the case of m-fold degeneracy ($m > 2$).

In view of the two theorems proved here, we shall always assume that the eigenfunctions of a hermitian operator satisfy the orthonormality condition (Equation (2.3.11)).

So, we have a very important and general result: The eigenfunctions of a hermitian operator form a complete set of mutually orthonormal functions and, hence, can be taken to be a basis. This set is unique if the operator has no degenerate eigenvalues; it is not unique if there is even one degenerate eigenvalue.

In particular, the normalized eigenfunctions, $\{\phi_n\}, n = 1, 2, 3, ...$ of the Hamiltonian of a particle (system) constitute an orthonormal basis. Therefore, an arbitrary function ψ, representing an arbitrary energy state of a particle, can be expanded into a series with respect to $\{\phi_n\}$:

$$\psi = \sum_n c_n \phi_n, \tag{2.3.18}$$

where the expansion coefficients c_n are determined as

$$c_n = \int_{-\infty}^{+\infty} \phi_n^* \psi d^3x. \tag{2.3.19}$$

This kind of expansion is frequently used to solve various problems in quantum mechanics. *In fact, this is another reason why representing a dynamical variable with a hermitian operator is advantageous.*

Before we conclude, let us have the list of fundamental operators that are used in quantum mechanics:

Dynamical variables in classical mechanics	***Corresponding operators in quantum mechanics***
Coordinates: $\begin{cases} \vec{r} \\ x, \quad y, \quad z \end{cases}$	$\begin{cases} \hat{\vec{r}} \\ x, \quad y, \quad z \end{cases}$
Momentum: $\begin{cases} \vec{p} \\ p_x, \quad p_y, \quad p_z \end{cases}$	$\begin{cases} -i\hbar\vec{\nabla} \\ -i\hbar\frac{\partial}{\partial x}, \quad -i\hbar\frac{\partial}{\partial y}, \quad -i\hbar\frac{\partial}{\partial z} \end{cases}$
Angular momentum: $\begin{cases} \vec{L} \\ L_x = yp_z - zp_y \\ L_y = zp_x - xp_z \\ L_z = xp_y - yp_x \end{cases}$	$\begin{cases} \hat{\vec{r}} \times \hat{\vec{p}} \\ -i\hbar\left(y\frac{\partial}{\partial z} - z\frac{\partial}{\partial y}\right) \\ -i\hbar\left(z\frac{\partial}{\partial x} - x\frac{\partial}{\partial z}\right) \\ -i\hbar\left(x\frac{\partial}{\partial y} - y\frac{\partial}{\partial x}\right) \end{cases}$
Energy: $H = \frac{\vec{p}^2}{2m} + V(\vec{r})$	$-\frac{\hbar^2}{2m}\vec{\nabla}^2 + \hat{V}(\vec{r})$

It is not difficult to check that all the listed operators are hermitian.

Example 2.3.1: Check whether an operator $\hat{F}$, acting in the space of square-integrable functions and defined by the relation $\hat{F}\psi(x) = -(d^2\psi(x)/dx^2)$, is hermitian or not.

Solution: The operator hermitian conjugate to the operator $\hat{F}$ is defined by following relation

$$\int_{-\infty}^{+\infty} \phi^*(x)\left(\hat{F}\psi(x)\right) dx = \int_{-\infty}^{+\infty} \left(\hat{F}^\dagger\phi(x)\right)^* \psi(x)\, dx, \tag{2.3.20}$$

where $\phi(x)$ is an arbitrary square-integrable function. Replacing $\hat{F}\psi(x)$ by $-(d^2\psi(x)/dx^2)$ on the left-hand side of (2.3.20) and integrating by parts, we have

$$\int_{-\infty}^{+\infty} \phi^*(x)\left(\hat{F}\psi(x)\right) dx = -\left.\phi^*(x)\frac{d\psi(x)}{dx}\right|_{-\infty}^{+\infty} + \int_{-\infty}^{+\infty} \frac{d\phi^*(x)}{dx}\frac{d\psi(x)}{dx}\, dx. \tag{2.3.21}$$

Since the wave functions and their first derivatives must be zero at $x = \pm\infty$, the first term on the right-hand side of (2.3.21) vanishes. Integrating once more by parts and using the same boundary conditions, we arrive at

$$\int_{-\infty}^{+\infty} \phi^*(x)\,(\hat{F}\psi(x))\,dx = -\int_{-\infty}^{+\infty} \frac{d^2\phi^*(x)}{dx^2}\,\psi(x)\,dx = \int_{-\infty}^{+\infty} \left(-\frac{d^2}{dx^2}\phi(x)\right)^*\psi(x)\,dx. \tag{2.3.22}$$

From the equations (2.3.20) and (2.3.22), we obtain

$$\hat{F}^\dagger = -\frac{d^2}{dx^2} = \hat{F}. \tag{2.3.23}$$

Hence, the operator $\hat{F} = -d^2/dx^2$ is hermitian.

Example 2.3.2: Using the definition of a hermitian conjugate operator (through the integral expression), find the operator hermitian conjugate to the operator

$$\hat{A} = x\frac{d}{dx} + \alpha\cos(x),$$

where α is a real constant.

Solution: From the definition of the hermitian conjugate operator, we have

$$\int_{-\infty}^{+\infty} \phi^*(x)\,(\hat{A}\psi(x))\,dx = \int_{-\infty}^{+\infty} (\hat{A}^\dagger\phi(x))^*\,\psi(x)\,dx, \tag{2.3.24}$$

where ϕ and ψ are arbitrary square-integrable functions. Let us simplify the right-hand side (RHS) of (2.3.24),

$$\text{RHS} = \int_{-\infty}^{+\infty} \phi(x)^* x\frac{d\psi}{dx}\,dx + \int_{-\infty}^{+\infty} \alpha\cos(x)\,\phi^*(x)\,\psi(x)\,dx \equiv I_1 + I_2. \tag{2.3.25}$$

Integrating the first term by parts and using the fact that the functions ϕ and ψ tend to zero at spatial infinities, we obtain

$$\begin{aligned} I_1 &= [x\phi^*\psi]_{-\infty}^{+\infty} - \int_{-\infty}^{+\infty} \frac{d(x\phi^*)}{dx}\,\psi\,dx = -\int_{-\infty}^{+\infty} \frac{d(x\phi^*)}{dx}\,\psi\,dx \\ &= -\int_{-\infty}^{+\infty} \left(\phi^* + x\frac{d\phi^*}{dx}\right)\psi\,dx. \end{aligned} \tag{2.3.26}$$

As a consequence, we arrive at

$$\int_{-\infty}^{+\infty} \phi^*(x)\,(\hat{A}\psi(x))\,dx = \int_{-\infty}^{+\infty} \left[\left(-1 - x\frac{d}{dx} + \alpha\cos(x)\right)\phi(x)\right]^*\psi(x)\,dx. \tag{2.3.27}$$

Comparing the equations (2.3.24) and (2.3.27), we see that the hermitian conjugate operator is given by

$$\hat{A}^\dagger = \left(-1 - \frac{d}{dx} + \alpha \cos(x)\right). \tag{2.3.28}$$

Theorem 2.3.3: The eigenvalues of an anti-hermitian operator are either purely imaginary or equal to zero.

Proof: Consider an anti-hermitian operator $\hat{A}$ and its eigenvalue equation

$$\hat{A}\psi_n = \lambda_n \psi_n. \tag{2.3.29}$$

As stated earlier, λ_n are in general complex. Now, since $\hat{A}$ is anti-hermitian, we have

$$\int_{-\infty}^{+\infty} \psi_n^* (\hat{A}\psi_n) d^3x = \int_{-\infty}^{+\infty} (-\hat{A}\psi_n)^* \psi_n \, d^3x, \tag{2.3.30}$$

Using the eigenvalue equation for $\hat{A}$, we have

$$(\lambda_n + \lambda_n^*) \int_{-\infty}^{+\infty} \psi_n^* \psi_n \, d^3x = 0. \tag{2.3.31}$$

For the non-trivial solutions of the eigenvalue equation, $\int_{-\infty}^{+\infty} \psi_n^* \psi_n \, d^3x \neq 0$ and we get: $\lambda_n + \lambda_n^* = 0$. This means that $\mathrm{Re}\{\lambda\} = 0$. Hence, λ_n is purely imaginary. Clearly, if $\lambda = 0$, the equation $\lambda_n + \lambda_n^* = 0$ is automatically satisfied.

Inverse of an operator: The operator $\hat{A}^{-1}$, inverse of an operator $\hat{A}$, is defined by the relation: $\hat{A}\hat{A}^{-1} = \hat{A}^{-1}\hat{A} = \hat{I}$, where $\hat{I}$ is the unit operator, i.e., the operator that leaves any state unchanged: $\hat{I}\psi = \psi$.

A matrix operator $\hat{A}$ has an inverse only if the matrix $A = (a_{ij})$, representing the operator $\hat{A}$, is a square matrix and its determinant ($\det(A)$) is non-zero. A matrix that has an inverse is called a non-singular matrix, while the one with no inverse is called a singular matrix. The inverse of an operator $\hat{A}$ is given by the matrix A^{-1}, which is the inverse of the matrix A.

Given a matrix A, we can find A^{-1} as follows:

$$A^{-1} = \frac{1}{det(A)} adj(A), \tag{2.3.32}$$

where $adj(A) = C^T$ is the matrix adjoint to A and $det(A)$ stands for the determinant of A. The elements c_{ij} of the matrix C are given by

$$c_{ij} = (-1)^{i+j} M_{ij}, \tag{2.3.33}$$

where M_{ij} is the determinant of the submatrix that remains after the ith row and j^{th} column are deleted from A. Obviously, if the determinant of the matrix A equals zero, the corresponding operator $\hat{A}$ does not possess an inverse.

The inverse of the product of any two operators $\hat{A}$ and $\hat{B}$ is determined by the formula: $(\hat{A}\hat{B})^{-1} = \hat{B}^{-1}\hat{A}^{-1}$. Similarly, we have

$$(\hat{A}\hat{B}\hat{C})^{-1} = \hat{C}^{-1}\hat{B}^{-1}\hat{A}^{-1}, \quad (\hat{A}\hat{B}\hat{C}\hat{D})^{-1} = \hat{D}^{-1}\hat{C}^{-1}\hat{B}^{-1}\hat{A}^{-1}, \tag{2.3.34}$$

and so on and so forth.

Quotient of two operators: Dividing an operator $\hat{A}$ by another operator $\hat{B}$ (provided that the inverse $\hat{B}^{-1}$ exists) is equivalent to multiplying $\hat{A}$ by $\hat{B}^{-1}$:

$$\frac{\hat{A}}{\hat{B}} = \hat{A}\hat{B}^{-1}. \tag{2.3.35}$$

Note that the quotients $\hat{A}\hat{B}^{-1}$ and $\hat{B}^{-1}\hat{A}$ differ, that is, in general $\hat{A}\hat{B}^{-1} \neq \hat{B}^{-1}\hat{A}$.

Unitary operator: A linear operator $\hat{U}$ is said to be unitary, if it satisfies

$$\hat{U}^{\dagger}\hat{U} = \hat{U}\hat{U}^{\dagger} = \hat{I}, \tag{2.3.36}$$

that is, its inverse operator coincides with its hermitian conjugate.

The product of any two unitary operators is also unitary. Let $\hat{U}$ and $\hat{V}$ be two unitary operators. Then, we have

$$(\hat{U}\hat{V})^{\dagger}(\hat{U}\hat{V}) = \hat{V}^{\dagger}\hat{U}^{\dagger}\hat{U}\hat{V} = \hat{V}^{\dagger}\hat{I}\hat{V} = \hat{V}^{\dagger}\hat{V} = \hat{I}. \tag{2.3.37}$$

It means that $(\hat{U}\hat{V})^{\dagger} = (\hat{U}\hat{V})^{-1}$ and hence, the product operator is unitary. This result can be generalized to any number of unitary operators:

$$(\hat{A}\hat{B}\hat{C}\hat{D}...)^{\dagger}(\hat{A}\hat{B}\hat{C}\hat{D}...) = (\hat{A}\hat{B}\hat{C}\hat{D}...)(\hat{A}\hat{B}\hat{C}\hat{D}...)^{\dagger} = \hat{I}. \tag{2.3.38}$$

Theorem 2.3.4: The eigenvalues of a unitary operator are complex numbers of moduli equal to one and the eigenfunctions of a unitary operator, that does not have any degenerate eigenvalue, are mutually orthogonal.

Proof: Let ψ_m and ψ_n be the eigenfunctions of a unitary operator $\hat{U}$ corresponding to non-degenerate eigenvalues λ_m and λ_n, respectively. Using the unitarity of $\hat{U}$ and the eigenvalue equations, we have

$$\begin{aligned}\int_{-\infty}^{+\infty} \psi_m^* \psi_n d^3x &= \int_{-\infty}^{+\infty} (\psi_m^* \hat{U}^{\dagger})(\hat{U}\psi_n)\, d^3x \\ &= \int_{-\infty}^{+\infty} (\hat{U}\psi_m)^{\dagger}(\hat{U}\psi_n)\, d^3x = \lambda_m^*\lambda_n \int_{-\infty}^{+\infty} \psi_m^*\psi_n\, d^3x = 0. \end{aligned} \tag{2.3.39}$$

We can rewrite (2.3.38) as

$$(\lambda_m^* \lambda_n - 1) \int_{-\infty}^{+\infty} \psi_m^* \psi_n \, d^3x = 0. \tag{2.3.40}$$

If $m = n$, we have $\lambda_m^* \lambda_m = |\lambda_m|^2 = 1$. Therefore, $|\lambda_m| = 1$. If $n \neq m$, then the only possibility for the equation (2.3.39) to be satisfied, is that

$$\int_{-\infty}^{+\infty} \psi_m^* \psi_n \, d^3x = (\psi_m, \psi_n) = 0, \tag{2.3.41}$$

that is, the eigenfunctions ψ_m and ψ_n are orthogonal.

2.4 Algebra of Operators

(a) The sum of two operators $\hat{A}$ and $\hat{B}$, given by $\hat{C} = \hat{A} + \hat{B}$, is defined through the relation

$$\hat{C}\psi(x) = (\hat{A} + \hat{B})\psi(x) = \hat{A}\psi(x) + \hat{B}\psi(x). \tag{2.4.1}$$

According to this definition, we act on ψ with $\hat{A}$ and $\hat{B}$ one by one and then add up the results. The new function, thus obtained, is the function that would result, if we act on ψ directly with $\hat{C}$.

(b) The product of an operator $\hat{A}$ with a complex number c, that is, the operator $c\hat{A}$ is defined by the relation

$$(c\hat{A})\psi = c(\hat{A}\psi). \tag{2.4.2}$$

(c) The product of two operators $\hat{A}$ and $\hat{B}$ is an operator $\hat{C} = \hat{A}\hat{B}$, which by acting on the function ψ transforms it into another function ϕ. The function ϕ is obtained by first applying $\hat{B}$ on ψ (which yields a function, say, χ) and then acting on the resulting function by $\hat{A}$:

$$\hat{C}\psi = (\hat{A}\hat{B})\psi = \hat{A}(\hat{B}\psi) = \hat{A}\chi = \phi. \tag{2.4.3}$$

For instance, if $\hat{A} = x$ and $\hat{B} = \frac{d}{dx}$, then

$$\left(x\frac{d}{dx}\right)\psi = x\left(\frac{d\psi}{dx}\right) = x\frac{d\psi}{dx}. \tag{2.4.4}$$

(d) Successive application of an operator, $\hat{A}$, n times on a function, ψ, is written as the power of that operator:

$$\underbrace{\hat{A}\hat{A}\hat{A}...\hat{A}}_{n-1}(\hat{A}\psi) = \hat{A}^n\psi. \tag{2.4.5}$$

(e) In general, the product of any two operators, $\hat{A}$ and $\hat{B}$, is not commutative. That is,

$$\hat{A}\hat{B} \neq \hat{B}\hat{A}. \tag{2.4.6}$$

The difference $\hat{A}\hat{B} - \hat{B}\hat{A}$ is an operator $\hat{C}$, which is called the commutator of $\hat{A}$ and $\hat{B}$ and is written as

$$\hat{C} = [\hat{A}, \hat{B}]. \tag{2.4.7}$$

Similarly, $\hat{A}\hat{B} + \hat{B}\hat{A} \neq 2\hat{A}\hat{B}$ and the operator $[\hat{A}, \hat{B}]_+ = \hat{A}\hat{B} + \hat{B}\hat{A}$ is called the anti-commutator of the operators $\hat{A}$ and $\hat{B}$. If $[\hat{A}, \hat{B}] = 0$, the operators are said to be commuting with each other, while, if $[\hat{A}, \hat{B}]_+ = 0$, the operators anti-commute.

(f) *Rules for hermitian conjugation*: Given some algebraic combinations of operators, we need to consider the rules that allow us to get the corresponding hermitian conjugate operators. To obtain the hermitian conjugate of any expression, we must cyclically reverse the order of the factors and make the following replacements:

(i) Replace constants, $\alpha, \beta, \gamma, ...$, by their complex conjugates: $\alpha \to \alpha^*$, $\beta \to \beta^*$, $\gamma \to \gamma^*, \ldots$.

(ii) Replace operators by their hermitian conjugates: $\hat{A} \to \hat{A}^\dagger$, $\hat{B} \to \hat{B}^\dagger$, $\hat{C} \to \hat{C}^\dagger$, etc.

Following these rules, we can write

(i) $(\hat{A}^\dagger)^\dagger = \hat{A}$, (ii) $(\alpha\hat{A})^\dagger = \alpha^*\hat{A}^\dagger$, (iii) $(\hat{A}^n)^\dagger = (\hat{A}^\dagger)^n$,

(iv) $(\hat{A} + \hat{B} + \hat{C} + \hat{D})^\dagger = \hat{A}^\dagger + \hat{B}^\dagger + \hat{C}^\dagger + \hat{D}^\dagger$, (v) $(\hat{A}\hat{B}\hat{C}\hat{D})^\dagger = \hat{D}^\dagger\hat{C}^\dagger\hat{B}^\dagger\hat{A}^\dagger$, etc.

(g) The commutator of two hermitian operators is anti-hermitian.

Proof: Consider two hermitian operators $\hat{A} = \hat{A}^\dagger$ and $\hat{B} = \hat{B}^\dagger$. Then, using the rules of hermitian conjugation, we obtain

$$\begin{aligned}[\hat{A}, \hat{B}]^\dagger &= (\hat{A}\hat{B} - \hat{B}\hat{A})^\dagger = (\hat{A}\hat{B})^\dagger - (\hat{B}\hat{A})^\dagger = \hat{B}^\dagger\hat{A}^\dagger - \hat{A}^\dagger\hat{B}^\dagger \\ &= -(\hat{A}\hat{B} - \hat{B}\hat{A}) = -[\hat{A}, \hat{B}].\end{aligned} \tag{2.4.8}$$

Hence, $[\hat{A}, \hat{B}]$ is anti-hermitian.

(h) The anti-commutator of two hermitian operators is hermitian.

Proof: Consider two hermitian operators $\hat{A} = \hat{A}^\dagger$ and $\hat{B} = \hat{B}^\dagger$. Then, using the rules of hermitian conjugation, we obtain

$$\begin{aligned}[\hat{A}, \hat{B}]_+^\dagger &= (\hat{A}\hat{B} + \hat{B}\hat{A})^\dagger = (\hat{A}\hat{B})^\dagger + (\hat{B}\hat{A})^\dagger = \hat{B}^\dagger\hat{A}^\dagger + \hat{A}^\dagger\hat{B}^\dagger \\ &= (\hat{A}\hat{B} + \hat{B}\hat{A}) = [\hat{A}, \hat{B}]_+.\end{aligned} \tag{2.4.9}$$

Hence, $[\hat{A}, \hat{B}]_+$ is hermitian.

(i) **Some other important properties of the commutators are:**

$$[\hat{A},\hat{B}] = -[\hat{B},\hat{A}], \qquad \text{(Anti-symmetry)} \qquad (2.4.10)$$

$$[\hat{A},\hat{B}+\hat{C}] = [\hat{A},\hat{B}] + [\hat{A},\hat{C}], \qquad \text{(Linearity)} \qquad (2.4.11)$$

$$[\hat{A},\hat{B}\hat{C}] = \hat{B}[\hat{A},\hat{C}] + [\hat{A},\hat{B}]\hat{C}, \qquad \text{(Distributivity)} \qquad (2.4.12)$$

$$[\hat{A},\hat{B}]^\dagger = [\hat{B}^\dagger,\hat{A}^\dagger], \qquad \text{(Hermitian conjugation)} \qquad (2.4.13)$$

$$[\hat{A},[\hat{B},\hat{C}]] + [\hat{B},[\hat{C},\hat{A}]] + [\hat{C},[\hat{A},\hat{B}]] = 0. \qquad \text{(Jacobi Identity)} \qquad (2.4.14)$$

In addition, by repeatedly using (2.4.12), we can prove that

$$[\hat{A},\hat{B}^n] = \sum_{k=0}^{n-1} \hat{B}^k\,[\hat{A},\hat{B}]\,\hat{B}^{n-k-1}, \qquad (2.4.15)$$

$$[\hat{A}^n,\hat{B}] = \sum_{k=0}^{n-1} \hat{A}^{n-k-1}\hat{B}^k\,[\hat{A},\hat{B}]\,\hat{A}^k \qquad (2.4.16)$$

Example 2.4.1: Show that the operators $\hat{A} = \hat{x}$ (defined by $\hat{A}\psi(x) = x\psi(x)$) and the operator $\hat{B} = \hat{p}_x$ (defined by $\hat{B}\psi(x) = -i\hbar(d\psi(x)/dx)$) satisfy the commutation relation $[\hat{x},\hat{p}_x] = \hat{x}\hat{p}_x - \hat{p}_x\hat{x} = i\hbar$.

Solution: We have

$$[\hat{x},\hat{p}_x]\psi(x) = -i\hbar\left(x\frac{d}{dx} - \frac{d}{dx}x\right)\psi = (-i\hbar)\left[x\frac{d\psi}{dx} - \frac{d}{dx}(x\psi)\right]$$

$$= i\hbar\left[\psi + x\frac{d\psi}{dx} - x\frac{d\psi}{dx}\right] = i\hbar\psi(x) \qquad (2.4.17)$$

Since $\psi(x)$ is an arbitrary function, we get

$$[\hat{x},\hat{p}_x] = i\hbar \qquad (2.4.18)$$

In general, the commutation relations between various Cartesian components of the position operator, $\hat{\vec{r}} = \{\hat{x},\hat{y},\hat{z}\} \equiv \{\hat{x}_1,\hat{x}_2,\hat{x}_3\}$, and the momentum operator, $\hat{\vec{p}} = \{\hat{p}_x,\hat{p}_y,\hat{p}_z\} \equiv \{\hat{p}_1,\hat{p}_2,\hat{p}_3\}$, are

$$[\hat{x}_j,\hat{p}_k] = i\hbar\delta_{jk}, \qquad (2.4.19)$$

where $j,k = 1,2,3$ and δ_{jk} is the Kronecker delta. These are the fundamental commutation relations. Using these, we can calculate any commutation relation between any two observables in quantum mechanics.

Function of an operator: A function, $f(\hat{A})$, of an operator, $\hat{A}$, is defined through the corresponding Taylor expansion of the function $f(x)$. That is, if

$$f(x) = \sum_n \frac{1}{n!} c_n x^n, \quad c_n = \left. \frac{\partial f}{\partial x^n} \right|_{x=0}, \tag{2.4.20}$$

exists, then

$$f(\hat{A}) = \sum_n \frac{1}{n!} c_n \hat{A}^n, \quad c_n = \left. \frac{\partial f}{\partial \hat{A}^n} \right|_{\hat{A}=0}. \tag{2.4.21}$$

In the above expression, the matrix raised to the power n is defined recursively in terms of the products $\hat{A}^n = \hat{A}\left(\hat{A}^{n-1}\right)$ (see Eq.(2.4.5)).

For instance, consider $f(\hat{A}) = \exp\left[a\hat{A}\right]$, where a is a constant. Using the above definition, we obtain

$$f(\hat{A}) = \sum_n \frac{1}{n!} (a\hat{A})^n = \hat{I} + a\hat{A} + \frac{1}{2!} a^2 \hat{A}^2 + \frac{1}{3!} a^3 \hat{A}^3 + \ldots. \tag{2.4.22}$$

Parity operator: Consider the operation of space inversion in which we change the space variables from $\vec{r} = \{x, y, z\}$ to $-\vec{r} = \{-x, -y, -z\}$. As a result, a function $\psi(\vec{r})$ goes into $\psi(-\vec{r})$. If $\psi(-\vec{r}) = \psi(\vec{r})$, the function $\psi(\vec{r})$ is said to be symmetric (even) or, equivalently, a function with even parity. On the other hand, if $\psi(-\vec{r}) = -\psi(\vec{r})$, the function $\psi(\vec{r})$ is said to be anti-symmetric (odd) or, equivalently, a function with odd parity.

The transformation of a function $\psi(\vec{r})$ under space inversion can be written in operator form as

$$\psi(-\vec{r}) = \hat{\mathscr{P}} \psi(\vec{r}), \tag{2.4.23}$$

where $\hat{\mathscr{P}}$ is the parity operator or space inversion operator.

Theorem 2.4.1: The parity operator is hermitian, that is $\hat{\mathscr{P}}^\dagger = \hat{\mathscr{P}}$.

Proof:

$$\begin{aligned} \int_{-\infty}^{+\infty} \phi^*(\vec{r}) \left[\hat{\mathscr{P}} \psi(\vec{r})\right] d^3x &= \int_{-\infty}^{+\infty} \phi^*(\vec{r})\, \psi(-\vec{r})\, d^3x \\ = -\int_{\infty}^{+\infty} \phi^*(-\vec{r})\, \psi(\vec{r})\, d^3x &= \int_{-\infty}^{+\infty} \left[\hat{\mathscr{P}} \phi(\vec{r})\right]^* \psi(\vec{r})\, d^3x. \end{aligned} \tag{2.4.24}$$

From here we get that $\hat{\mathscr{P}}^\dagger = \hat{\mathscr{P}}$. The theorem is proved.

Note that the parity operator is also a unitary operator. This can be seen as follows. We have

$$\hat{\mathscr{P}}^2\psi(\vec{r}) = \hat{\mathscr{P}}(\hat{\mathscr{P}}\psi(\vec{r})) = \hat{\mathscr{P}}\psi(-\vec{r}) = \psi(\vec{r}). \tag{2.4.25}$$

It follows from the last equation that $\hat{\mathscr{P}}^2 = \hat{I}$, which means $\hat{\mathscr{P}} = \hat{\mathscr{P}}^{-1}$. The parity operator is, therefore, unitary, since its hermitian conjugate is equal to its inverse: $\hat{\mathscr{P}}^\dagger = \hat{\mathscr{P}}^{-1}$.

Further, since $\hat{\mathscr{P}}^2 = \hat{I}$, the eigenvalues of the parity operator are ± 1. Let the eigenfunction corresponding to the eigenvalue $+1$ be ψ_+. Then, on one hand we have $\hat{\mathscr{P}}\psi_+(\vec{r}) = \psi_+(\vec{r})$, while on the other $\hat{\mathscr{P}}\psi_+(\vec{r}) = \psi_+(-\vec{r})$. Hence, $\psi_+(\vec{r}) = \psi_+(-\vec{r})$. Therefore, $\psi_+(\vec{r})$, is an even function. Similarly, the eigenfunction, ψ_-, corresponding to the eigenvalue -1 is an odd function:: $\psi_-(\vec{r}) = -\psi_-(\vec{r})$. It then follows that the *eigenfunctions of the parity operator have definite parity: they are either even or odd.* In addition, the eigenfunctions $\psi_+(\vec{r})$ and $\psi_-(\vec{r})$ are orthogonal. In fact, the scalar product of these eigenfunctions satisfies

$$\begin{aligned}(\psi_+, \psi_-) &= \int_{-\infty}^{+\infty} \psi_+^*(\vec{r})\psi_-(\vec{r})\, d^3x = \int_{-\infty}^{+\infty} \psi_+^*(-\vec{r})\psi_-(-\vec{r})\, d^3x = \int_{-\infty}^{+\infty} \psi_+^*(\vec{r})\psi_-(-\vec{r})\, d^3x \\ &= -\int_{-\infty}^{+\infty} \psi_+^*(\vec{r})\psi_-(\vec{r})\, d^3x = -(\psi_+, \psi_-).\end{aligned} \tag{2.4.26}$$

Hence it vanishes. Therefore, the functions $\psi_+(\vec{r})$ and $\psi_-(\vec{r})$ are orthogonal. It turns out that, since any function $\phi(\vec{r})$ can be written as $\phi = \psi_+(\vec{r}) + \psi_-(\vec{r})$ with

$$\psi_+ = \frac{1}{2}[\phi(\vec{r}) + \phi(-\vec{r})], \quad \psi_- = \frac{1}{2}[\phi(\vec{r}) - \phi(-\vec{r})], \tag{2.4.27}$$

the eigenfunctions of the parity operator form a complete set of functions.

Definition: An operator $\hat{A}$ is said to an even operator, if $\hat{\mathscr{P}}\hat{A}\hat{\mathscr{P}} = \hat{A}$. On the other hand, if $\hat{\mathscr{P}}\hat{A}\hat{\mathscr{P}} = -\hat{A}$ the operator is said to be odd.
It is easy to show that even operators commute with the parity operator $\hat{\mathscr{P}}$, while odd operators anticommute with $\hat{\mathscr{P}}$:

$$[\hat{A}, \hat{\mathscr{P}}] = 0, \quad [\hat{B}, \hat{\mathscr{P}}]_+ = 0, \tag{2.4.28}$$

where $\hat{A}$ is an even operator and $\hat{B}$ is an odd operator.

Since even operators commute with the parity operator and the eigenfunctions of the parity operator are either even or odd, the eigenfunctions of even operators must have definite parity, provided they possess non-degenerate eigenvalues. If an even operator has a degenerate eigenspectrum, its eigenvectors do not necessarily have a definite parity.

Example 2.4.2: Find the value of the commutator

$$\hat{A} = [\hat{p}_x^2, (\hat{x}\hat{p}_y - \hat{y}\hat{p}_x)],$$

where, $\hat{\vec{r}} = (\hat{x}, \hat{y}, \hat{z})$ and $\hat{\vec{p}} = (\hat{p}_x, \hat{p}_y, \hat{p}_z)$ are the position and momentum operators of a particle, respectively.

Solution: Using the properties of the commutator of operators

$$[\hat{A}, \hat{B} + \hat{C}] = [\hat{A}, \hat{B}] + [\hat{A}, \hat{C}], \tag{2.4.29}$$

$$[\hat{A}\hat{B}, \hat{C}] = \hat{A}[\hat{B}, \hat{C}] + [\hat{A}, \hat{C}]\hat{B}, \tag{2.4.30}$$

$$[\hat{A}, \hat{B}\hat{C}] = \hat{B}[\hat{A}, \hat{C}] + [\hat{A}, \hat{B}]\hat{C}, \tag{2.4.31}$$

we get

$$\begin{aligned}
[\hat{p}_x^2, (\hat{x}\hat{p}_y - \hat{y}\hat{p}_x)] &= [\hat{p}_x^2, \hat{x}\hat{p}_y] - [\hat{p}_x^2, \hat{y}\hat{p}_x] \\
&= \hat{p}_x[\hat{p}_x, \hat{x}\hat{p}_y] + [\hat{p}_x, \hat{x}\hat{p}_y]\hat{p}_x - \hat{p}_x[\hat{p}_x, \hat{y}\hat{p}_x] - [\hat{p}_x, \hat{y}\hat{p}_y]\hat{p}_x \\
&= \hat{p}_x\hat{x}[\hat{p}_x, \hat{p}_y] + \hat{p}_x[\hat{p}_x, \hat{x}]\hat{p}_y + \hat{x}[\hat{p}_x, \hat{p}_y]\hat{p}_x + [\hat{p}_x, \hat{x}]\hat{p}_y\hat{p}_x \\
&\quad - \hat{p}_x\hat{y}[\hat{p}_x, \hat{p}_x] - \hat{p}_x[\hat{p}_x, \hat{y}]\hat{p}_x - \hat{y}[\hat{p}_x, \hat{p}_y]\hat{p}_x - [\hat{p}_x, \hat{y}]\hat{p}_y\hat{p}_x \\
&= -i\hbar(\hat{p}_x\hat{p}_y + \hat{p}_y\hat{p}_x) = -2i\,\hat{p}_x\hat{p}_y,
\end{aligned} \tag{2.4.32}$$

where we have used the fundamental commutators $[\hat{x}_j, \hat{p}_k] = i\hbar\delta_{jk},\ j,k = 1,2,3.$

2.5 The Schrödinger Equation

Postulate 3: The time evolution of the wave function, $\psi(\vec{r},t)$, representing the state of a quantum mechanical system is governed by the following partial differential equation:

$$i\hbar\frac{\partial\psi(\vec{r},t)}{\partial t} = -\frac{\hbar^2}{2m}\vec{\nabla}^2\psi(\vec{r},t) + V(\vec{r})\psi(\vec{r},t), \tag{2.5.1}$$

where $\vec{\nabla}^2 = \frac{\partial^2}{\partial x^2} + \frac{\partial^2}{\partial y^2} + \frac{\partial^2}{\partial z^2}$ is the Laplacian or Laplace operator and V is the potential energy function. This is the well-known time-dependent Schrödinger equation.

In one spatial dimension, equation (2.5.1) reduces to:

$$i\hbar\frac{\partial\psi}{\partial t} = -\frac{\hbar^2}{2m}\frac{\partial^2\psi}{\partial x^2} + V(x)\psi, \quad \psi = \psi(x,t). \tag{2.5.2}$$

Note that the potential V, in principle, can depend on time. However, we shall not deal with such problems in this book.

Important properties of the Schrödinger equation:

(a) Consider a collection of a large number, N, of non-interacting particles in an infinitesimal volume element $d\tau$. Assume also that at some instant t, they are all in the same given state $\psi(\vec{r},t)$. Since $\psi^*\psi\, d\tau$ is the probability of finding the particle in an infinitesimal volume element, $d\tau$, centered around a given point $\vec{r}$, the quantity $N\psi^*\psi\, d\tau \equiv \langle N\rangle$ will be the average number of particles in $d\tau$. By calculating $\langle N\rangle$ for a large number of points in space, we can construct the probability density distribution function at a given instant of time t. *The significance of the time-dependent Schrödinger equation lies in the fact that it allows us to determine the time evolution of the wave function ψ, enabling thereby the prediction of the time evolution of the density distribution function. This in turn allows us to follow the changes that take place in the system as it evolves in time.*

(b) *Linearity and homogeneity:* Since the Schrödinger equation contains only the first power of ψ, it is linear in ψ. The homogeneity is related to the fact that for a given solution ψ of the Schrödinger equation, the function $\alpha\psi$, where α is a constant, is also a solution of the Schrödinger equation.

(c) The equation is of first order with respect to time and, therefore, the knowledge of the wave function at some initial time, t_0, allows us to determine the wave function at any later time, t, uniquely.

An important consequence of the first property is that the superposition principle holds. This means that if $\psi_1(\vec{r},t), \psi_2(\vec{r},t), \psi_3(\vec{r},t), ..., \psi_n(\vec{r},t)$ are solutions of the Schrödinger equation, then the linear combination of these functions:

$$\psi(\vec{r},t) = \sum_{j=1}^{n} c_j \psi_j(\vec{r},t), \tag{2.5.3}$$

where $c_j, j = 1,2,3,...,n$ are arbitrary complex constants, is also a solution.

It is important to note here that the quantum mechanical superposition principle is radically different from its counterpart in classical physics. Firstly, it reflects the wave-like properties of the micro-particles and hence represents the wave–particle duality that is a characteristic of the micro-world. Secondly, suppose that a particle is in a state ψ, which is a superposition of two other possible states ψ_1 and ψ_2 of the particle, that is, $\psi = \psi_1 + \psi_2$. If we measure a physical characteristic A of the particle in the state ψ_1, we shall get some value A_1. Similarly, we shall get a value A_2 for A in the state ψ_2. Let us now assume that we measure A in the superposition state ψ. Classically, the measurement of A in this state will yield a value between A_1 and A_2. As we shall see later, quantum mechanics says that if you measure A in the state ψ, you will get either A_1 or A_2 with certain probabilities but nothing in between A_1 and A_2. In other words, the process of measurement induces a sudden transition of the particle from the superposition state ψ to one of the states ψ_1 and ψ_2. This phenomenon is known as the collapse of the wave function and does not have a classical analogue. This is one of the conceptual difficulties in quantum mechanics that is yet to be resolved.

Thus, the quantum mechanical superposition has far-reaching consequences including the conceptual difficulties that we encounter when we try to draw a naive parallel between a classical system and a quantum system. However, on the positive side, the validity of the superposition principle enormously helps in solving various quantum mechanical problems.

2.6 Time-independent Potentials and the Stationary States

Solutions to the Schrödinger equation with time-independent potentials, $V(\vec{r})$, can be found by employing the method of separation of variables; well known from the theory of differential equations. This is done by writing the wave function in the form:

$$\psi(\vec{r},t) = \phi(\vec{r})\, f(t). \tag{2.6.1}$$

The Schrödinger equation (2.5.1) then leads to

$$i\hbar \frac{1}{f}\frac{df}{dt} = -\frac{\hbar^2}{2m}\frac{1}{\phi(\vec{r})}\vec{\nabla}^2\phi(\vec{r}) + V(\vec{r}). \tag{2.6.2}$$

The left-hand side of (2.6.2) is a function of time, whereas the right-hand side depends only on spatial variables, x, y and z. Therefore, for this equality to hold, both the left-hand side and the right-hand side must be equal to a constant (same for both the sides). Let us call it E. As a consequence, we get a system of two ordinary differential equations:

$$i\hbar \frac{1}{f}\frac{df}{dt} = E \quad \Rightarrow \quad \frac{df}{dt} = -\frac{i}{\hbar}E\, f(t), \tag{2.6.3}$$

and

$$-\frac{\hbar^2}{2m}\frac{1}{\phi(\vec{r})}\vec{\nabla}^2\phi(\vec{r}) + V(\vec{r}) = E \;\Rightarrow\; -\frac{\hbar^2}{2m}\vec{\nabla}^2\phi(\vec{r}) + V(\vec{r})\phi(\vec{r}) = E\phi(\vec{r}). \tag{2.6.4}$$

The first of these equations, (see (2.6.3)), can be readily integrated to yield

$$f(t) = e^{-\frac{i}{\hbar}Et}. \tag{2.6.5}$$

The differential equation (2.6.4), satisfied by $\phi(\vec{r})$ is called the *time-independent Schrödinger equation* (TISE) and its solution depends on the form of the potential $V(\vec{r})$. In view of the standard conditions (to be satisfied by the overall wave function $\psi(\vec{r},t)$), a given specific form of $V(\vec{r})$ imposes specific boundary conditions on $\phi(\vec{r})$, which we shall discuss later.

Note that the equation (2.6.4) can be written as

$$\hat{H}\phi(\vec{r}) = E\phi(\vec{r}), \tag{2.6.6}$$

where

$$\hat{H} = -\frac{\hbar^2}{2m}\vec{\nabla}^2 + V(\vec{r}). \tag{2.6.7}$$

As we shall discuss later, $\hat{H}$ is nothing but a differential operator representing the classical Hamiltonian, $H = \frac{p^2}{2m} + V(\vec{r})$, equal to the total energy of the particle. The solutions of this equation are called the *eigenfunctions* of the Hamiltonian and the corresponding values of E (for which the solutions of this differential equation exist), are called the *energy eigenvalues*. Therefore, the separation constant E represents the total energy of the particle in a given state.

It is now clear that, for time-independent potentials, a full solution to the Schrödinger equation can be written as

$$\psi(\vec{r},t) = \phi(\vec{r})\, e^{-\frac{i}{\hbar}Et}. \tag{2.6.8}$$

If we calculate the probability density $|\psi|^2$ for such states, we obtain

$$|\psi(x,t)|^2 = \psi^*(\vec{r},t)\psi(\vec{r},t) = \phi(\vec{r})^*\, e^{\frac{i}{\hbar}Et}\,\phi(\vec{r})\, e^{-\frac{i}{\hbar}Et} = |\phi(\vec{r})|^2. \tag{2.6.9}$$

That is, for such states, the probability density is independent of time.

Definition: *The states of a quantum mechanical system (particle) for which the probability density does not depend on time (i.e., remains constant at all times) are called stationary states.*

Depending on the potential, there could be more than one solution to the equation (2.6.4). Let there be m such solutions: $\phi_1(\vec{r}), \phi_2(\vec{r}), \phi_3(\vec{r}), \ldots, \phi_m(\vec{r})$, with energies $E_1, E_2, E_3, \ldots, E_m$, respectively. In view of the superposition principle, the general stationary state solution is then given by

$$\psi(\vec{r},t) = \sum_{j=1}^{m} c_j\phi_j(\vec{r})\, e^{-\frac{i}{\hbar}E_jt} = c_1\phi_1(\vec{r})\, e^{-\frac{i}{\hbar}E_1t} + c_2\phi_2(\vec{r})\, e^{-\frac{i}{\hbar}E_2t} + \ldots + c_m\phi_m(\vec{r})\, e^{-\frac{i}{\hbar}E_mt}, \tag{2.6.10}$$

where c_js are arbitrary complex (in general) constants.

2.7 Measurement and Compatible Operators

Up till now we have discussed three postulates of quantum mechanics regarding the characterization of state of a quantum system, representation of the observables and the time-evolution of the state. Let us now discuss how quantum mechanics describes the process of measurement of an observable and what is its prescription for comparing the theoretically calculated values of an observable with its experimentally measured ones.

In quantum mechanics, the act of measurement of an observable A, at a given instant of time in a given state, is theoretically represented as the action of the corresponding operator, $\hat{A}$, on the wave function, $\psi(x,t)$, representing the state of the particle at that instant of time: $\hat{A}\psi(x,t)$. So far as the result of measurement is concerned, quantum mechanics puts forward the following postulate.

Postulate 4: The only possible result of measurement of an observable A, at a given instant of time, t, in a given state $\psi(\vec{r},t)$, is one of the eigenvalues of the corresponding operator $\hat{A}$.

(a) If the operator, $\hat{A}$, has discrete and non-degenerate eigenvalues, $\{a_n\} : \hat{A}\phi_n(\vec{r}) = a_n\phi_n(\vec{r})$, the probability that the measurement of A will yield the eigenvalue a_j is given by

$$P(a_j) = \frac{|(\phi_j,\psi)|^2}{(\psi,\psi)}, \tag{2.7.1}$$

where $\phi_j(\vec{r})$ is the eigenfunction of $\hat{A}$ with eigenvalue a_j and

$$(\phi_j,\psi) \equiv \int_{-\infty}^{+\infty} \phi^*(\vec{r})\,\psi(\vec{r},t)\,d^3x, \tag{2.7.2}$$

$$(\psi,\psi) \equiv \int_{-\infty}^{+\infty} |\psi(\vec{r},t)|^2\,d^3x. \tag{2.7.3}$$

If the wave function $\psi(\vec{r},t)$ is normalized to unity, the above mentioned probability is simply written as

$$P(a_j) = \left|\int_{-\infty}^{+\infty} \phi^*(\vec{r})\,\psi(\vec{r},t)\,d^3x,\right|^2. \tag{2.7.4}$$

As we see here, the act of measurement changes the state of the system. The state of the system, immediately after the measurement, changes from $\psi(\vec{r},t)$ to ϕ_j, the jth eigenstate of $\hat{A}$: $\psi_{\text{after}} = \phi_j(\vec{r})$. However, if the particle is in one of the eigenstates, say, $\phi_k(\vec{r})$ of the operator $\hat{A}$, then the result of measurement will with certainty give the value a_k and the state of the particle will remain unchanged.

(b) If the eigenvalue a_j is m-fold degenerate (i.e., there are m linearly independent eigenfunctions $\phi_j^{(m)}$ with the same eigenvalue a_j), the probability of getting the value a_j is given by

$$P(a_j) = \frac{\sum_{k=1}^{m} \left|(\phi_j^k, \psi)\right|^2}{(\psi,\psi)} = \frac{\sum_{k=1}^{m} \left|\int_{-\infty}^{+\infty} \phi_j^{k\,*}(\vec{r})\,\psi(\vec{r},t)\,d^3x\right|^2}{\int_{-\infty}^{+\infty} |\psi(\vec{r},t)|^2\,d^3x}. \tag{2.7.5}$$

(c) If the operator $\hat{A}$ possesses a continuous eigenspectrum $\{a\}$, the probability that the result of measurement will yield a value between a and $a + da$ is given by

$$dP(a) = \frac{|\psi(a)|^2}{\int_{-\infty}^{+\infty} |\psi(a')|^2\,da'}\,da. \tag{2.7.6}$$

For instance, for a free particle, there is no restriction on its position in space. Hence, the eigenvalues x of the position operator, $\hat{x}$, can take continuous values. Therefore, when the position of a particle is measured, the probability of obtaining a value between x and $x + dx$ is

$$dP(x) = \frac{|\psi(x)|^2}{\int_{-\infty}^{+\infty} |\psi(x')|^2\,dx'}\,dx. \tag{2.7.7}$$

On the basis of the discussions so far, we conclude that, in its essence, quantum mechanics is a statistical theory in which the results of measurement of a given physical quantity cannot be predicted exactly, They can only be predicted with certain probabilities. Therefore, the question as to which result of quantum mechanical calculations should be compared with the experimentally measured value of a physical quantity must be answered. We shall talk about this problem and try to come to a certain conclusion.

In classical mechanics, every dynamical variable takes on a definite numerical value. This number is nothing but the value that we obtain when we measure it. The justification for assigning a definite value to the result of measurement in classical physics is the fact that all possible measurements of a dynamical variable on the system in a given state, yield the same numerical value.

The situation is radically different in quantum mechanics. Before we show the difference, let us recollect that the superposition principle holds for the Schrödinger equation. That is, if ψ_1 and ψ_2 are two distinct solutions then an arbitrary linear combination of these wave functions is also a solution of the Schrödinger equation.

Let a system be prepared in a superposition state $\phi = a_1\phi_1 + a_2\phi_2$, where a_1 and a_2 are arbitrary constants and ϕ_1 and ϕ_2 are the eigenstates of a given observable $\hat{A}$ corresponding to the eigenvalues λ_1 and λ_2, respectively. Clearly, had the system been in the state ψ_1, or in the state ψ_2, the measurement of $\hat{A}$ would have given a value equal to λ_1, or λ_2, respectively. Let us pose the question: If the the dynamical variable is measured in the superposition state ϕ, what values will be obtained and with what probabilities?

Note that, had the system been a classical system, a definite value λ, intermediate between λ_1 and λ_2 would have resulted. In quantum mechanics, however, measurements will not produce a value intermediate between λ_1 and λ_2 but one of the values λ_1 or λ_2 (sometimes λ_1 and sometimes λ_2) and no other value except λ_1 or λ_2. Besides that, it is not possible to predict which measurement will yield λ_1 and which would give λ_2. One or the other result is obtained with a well-defined probability. The system, which was in the state ψ, before the measurement, makes an abrupt transition either into the state ψ_1 or into the state ψ_2.

This discussion leads to the conclusion that, at a given instant of time, it is not possible to assign a definite value to an observable in quantum mechanics. It is, however, always possible to assign a definite probability to the occurrence of one of the possible values. This implies that the formalism of quantum mechanics allows us to compute only the probabilities of occurrence of various possible values of an observable. *Hence, only the average value (or, as it is called, the expectation value) of a dynamical variable, computed in accordance with the theory of probability, should be compared with the experimental result.*

Average value of a dynamical variable: The average value, $\langle A\rangle$, of a dynamical variable A, in a given state ψ of the system, is defined as

$$\langle A\rangle = \int_{-\infty}^{+\infty} \psi^*(\vec{r})\,[\hat{A}\psi(\vec{r})]\,d^3x \Big/ \int_{-\infty}^{+\infty} \psi^*(\vec{r})\psi(\vec{r})\,d^3x, \tag{2.7.8}$$

where the integration is over the entire region of variation of the independent variables, x, y, and z. The asterisk stands for complex conjugation. If the wave function is normalized to unity, the required average value is given by

$$\langle A\rangle = \int_{-\infty}^{+\infty} \psi^*(\vec{r})(\hat{A}\psi(\vec{r}))\,d^3x. \tag{2.7.9}$$

For instance, the average value of the position operator, $\hat{x}$, in one spatial dimension in the normalized state ψ,

$$\langle x\rangle = \int_{-\infty}^{+\infty} \psi^*(x)(\hat{x}\psi)\,dx = \int_{-\infty}^{+\infty} \psi^*(x)x\psi(x)\,dx. \tag{2.7.10}$$

Similarly, the expectation value of the x component of momentum, $\langle p_x\rangle$, is given by

$$\langle p_x\rangle = \int_{-\infty}^{+\infty} \psi^*(x)(\hat{p}_x\psi(x))\,dx = -i\hbar\int_{-\infty}^{+\infty} \psi^*(x)\frac{d\psi(x)}{dx}\,dx. \tag{2.7.11}$$

Let $\psi_1(\vec{r})$, $\psi_2(\vec{r})$, $\psi_3(\vec{r})$,..., be the normalized eigenfunctions of a hermitian operator $\hat{A}$ with discrete eigenvalues λ_1, λ_2, λ_3,..., respectively. Let the particle (system), on which the measurement of $\hat{A}$ is done, be in a state $\psi(\vec{r})$, which is not an eigenstate of $\hat{A}$. Since the eigenfunctions of a hermitian operator form a complete set, we can expand ψ as

$$\psi(\vec{r}) = \sum_{k=1} c_k \psi_k(\vec{r}), \tag{2.7.12}$$

where c_ks are arbitrary complex coefficients. Using the eigenvalue equation, $\hat{A}\psi_k(\vec{r}) = \lambda_k \psi_k(\vec{r})$ and the orthonormality of the eigenfunctions $\{\psi_n(\vec{r})\}$, the average value $\bar{A}$ can be written as

$$\begin{aligned}\langle A\rangle &= \sum_\ell \sum_k \lambda_k c_\ell^* c_k \int_{-\infty}^{+\infty} \psi_\ell^*(\vec{r})\psi_k(\vec{r})\, d^3x = \sum_\ell \sum_k \lambda_k c_\ell^* c_k\, \delta_{\ell k} = \sum_k \lambda_k |c_k|^2 \\ &= \lambda_1 |c_1|^2 + \lambda_2 |c_2|^2 + \lambda_3 |c_3|^2 + \ldots \end{aligned} \tag{2.7.13}$$

In view of the orthonormality of the eigenfunctions of $\hat{A}$, we have

$$\begin{aligned}\sum_m \sum_n c_m^* c_n \int_{-\infty}^{+\infty} \psi_m^* \psi_n\, d^3x &= \sum_m \sum_n c_m^* c_n\, \delta_{mn} = \sum_n |c_n|^2 \\ &= |c_1|^2 + |c_2|^2 + |c_3|^2 + \ldots = 1. \end{aligned} \tag{2.7.14}$$

If we now recall the expression for the average value of a random variable y from the theory of probability ($\langle y\rangle = \sum_j y_j w_j$, where w_j is the probability of getting the value y_j of the random variable y and $\sum_j w_k = 1$), then the equation (2.7.13) suggests that the number $|c_k|^2$ represents the probability, P_k, of obtaining a value λ_k for A in the state ψ. Therefore, the average value of $\hat{A}$ takes the form

$$\langle A\rangle = \lambda_1\, P_1 + \lambda_2\, P_2 + \lambda_3\, P_3 + \ldots, \tag{2.7.15}$$

where, in view of (2.7.14), the sum total of all the probabilities $P_k, k = 1, 2, 3, \ldots$ is equal to 1:

$$\sum_k P_k = 1. \tag{2.7.16}$$

On the basis of our discussions, we can now conclude:

1. If an observable, $\hat{A}$, is measured on a system in a state ψ, then it can have a definite value if and only if ψ happens to be an eigenstate of the operator $\hat{A}$. In such a case, the result will yield the corresponding eigenvalue of the operator $\hat{A}$ in the state ψ.
2. On the other hand, if ψ is not an eigenfunction of $\hat{A}$, the state ψ has to be expanded into a series with respect to the complete set of eigenfunctions of $\hat{A}$ (see equation (2.7.12)). The result of a measurement of $\hat{A}$ will then be one of the eigenvalues a_k of $\hat{A}$ with a probability $|c_k|^2$.

3. The act of measurement, in general, changes the state of the system from a given initial state to one of the eigenstates of the observable being measured.
4. For the measurement of an observable on a system in a given state, ψ, a large number of identical copies of the system are to be prepared and the observable has to be measured on each of them independently. The average of the obtained results, calculated according to the prescription of quantum mechanics, will be the value of the observable in the state ψ..

Example 2.7.1: Consider a system, in one spatial dimension, which is in a state with a wave function

$$\psi(x) = A\,e^{-\alpha(x-b)^2}, \tag{2.7.17}$$

where α and b are real constants and A, in general, is a complex constant. Calculate the expectation values of the operators $\hat{x}$, $\hat{x}^2$ and $\hat{p}_x$.

Solution: Using (2.7.8), we get

$$\langle\hat{x}\rangle = \int_{-\infty}^{+\infty} \psi^*(x)\,x\psi(x)\,dx \Big/ \int_{-\infty}^{+\infty} \psi^*(x)\,\psi(x)\,dx. \tag{2.7.18}$$

Let us compute the integrals. We have

$$\int_{-\infty}^{+\infty} \psi^*(x)\,\psi(x)\,dx = |A|^2 \int_{-\infty}^{+\infty} e^{-2\alpha(x-b)^2}\,dx = |A|^2\sqrt{\frac{\pi}{2\alpha}}. \tag{2.7.19}$$

The numerator is

$$\int_{-\infty}^{+\infty} \psi^*(x)\,x\,\psi(x)\,dx = |A|^2 \int_{-\infty}^{+\infty} xe^{-2\alpha(x-b)^2}\,\psi(x)\,dx = 0, \tag{2.7.20}$$

because the integrand is an odd function of x and it is integrated over the region that is symmetric with respect to the origin. Therefore, $\langle\hat{x}\rangle = 0$.

Since

$$\begin{aligned}\int_{-\infty}^{+\infty} \psi^*(x)\,x^2\,\psi(x)\,dx &= |A|^2 \int_{-\infty}^{+\infty} x^2 e^{-2\alpha(x-b)^2}\,dx \\ &= |A|^2 \frac{\sqrt{\pi}}{2\sqrt{8\alpha^3}},\end{aligned} \tag{2.7.21}$$

the average value of $\hat{x}^2$ is given by

$$\langle\hat{x}^2\rangle = \left(|A|^2\frac{\sqrt{\pi}}{4\alpha\sqrt{2\alpha}}\right) \Big/ \left(|A|^2\frac{\sqrt{\pi}}{\sqrt{2\alpha}}\right) = 1/4\alpha. \tag{2.7.22}$$

The momentum operator $\hat{p}_x = -i\hbar(\partial/\partial x)$, and we have

$$\langle \hat{p}_x \rangle = \int_{-\infty}^{+\infty} \psi^*(x)\,(\hat{p}_x \psi(x))\,dx = -i\hbar|A|^2 \int_{-\infty}^{+\infty} \psi^*(x)\,\frac{d\psi}{dx}\,dx$$

$$= 2i\alpha\hbar|A|^2 \int_{-\infty}^{+\infty} (x-b)\,e^{-2\alpha(x-b)^2}\,dx = 2i\alpha\hbar|A|^2 \int_{-\infty}^{+\infty} y e^{-2\alpha y^2}\,dy = 0, \tag{2.7.23}$$

where we have changed the variable of integration from $(x-b)$ to $y = x-b$ and taken into account the fact that the integrand is an odd function of y and hence, the integral is zero. As a result, the expectation value of $\hat{p}_x$ is equal to zero: $\langle \hat{p}_x \rangle = 0$.

Further, we calculate

$$\hat{p}_x^2 \psi(x) = 2\alpha\hbar^2 \left(e^{-2\alpha(x-b)^2} - 2\alpha\, e^{-2\alpha(x-b)^2}\right). \tag{2.7.24}$$

$$\langle \hat{p}_x^2 \rangle = \int_{-\infty}^{+\infty} \psi^*(x)\,(\hat{p}_x\,\psi(x))\,dx = 2\alpha\hbar^2|A|^2 \int_{-\infty}^{+\infty} \left[e^{-2\alpha(x-b)^2} - 2\alpha(x-b)^2\,e^{-2\alpha(x-b)^2}\right] dx$$

$$= 2|A|^2\alpha\hbar^2 \left[\sqrt{\frac{\pi}{2\alpha}} - \frac{1}{2}\sqrt{\frac{\pi}{2\alpha}}\right] = \frac{|A|^2\sqrt{\pi}\sqrt{2\alpha}\hbar^2}{2}. \tag{2.7.25}$$

The expectation value of $\hat{p}_x^2$ is then given by

$$\langle \hat{p}_x^2 \rangle = \frac{|A|^2\sqrt{\pi}\sqrt{2\alpha}\hbar^2}{2} \times \frac{\sqrt{2\alpha}}{|A|^2\sqrt{\pi}} = \alpha\hbar^2. \tag{2.7.26}$$

Example 2.7.2: Consider a particle of mass m confined to move in one spatial dimension in the region $0 < x < a$. Let the particle be in a state described by the wave function $\psi_1(x,t) = \sin(\pi a/x)\exp(-i\omega t)$, where ω is a constant. Find the average values of the position and momentum operators in this state.

Solution: First, let us check whether the wave function of the particle is normalized or not. We have

$$\int_0^a |\psi_1(x,t)|^2\,dx = \int_0^a \sin^2(\pi a/x)\,dx$$

$$= \int_0^a \frac{((1-\cos(2\pi a/x))}{2}\,dx = \frac{a}{2}. \tag{2.7.27}$$

Therefore, the normalized wave function is

$$\psi(x,t) = \sqrt{\frac{2}{a}} \sin(\pi a/x) \exp(-i\omega t). \tag{2.7.28}$$

The average value of the position operator $\hat{x}$ will be given by

$$\begin{aligned}\langle \hat{x} \rangle &= \int_0^a \hat{x}\, |\psi(x,t)|^2\, dx = \frac{2}{a}\int_0^a x \sin^2(\pi a/x)\, dx \\ &= \frac{2}{a}\int_0^a \frac{((x - x\cos(2\pi a/x))}{2}\, dx. \end{aligned} \tag{2.7.29}$$

Integrating the second term by parts, we arrive at $\langle \hat{x} \rangle = a/2$. The average value of the momentum operator $\hat{p}_x$ will be

$$\langle \hat{p}_x \rangle = \int_0^a \psi^*(x,t)\left(-i\hbar\frac{d}{dx}\right)\psi(x,t)\, dx = -\frac{i\hbar}{\pi}\int_0^a \sin(2\pi a/x)\, dx. \tag{2.7.30}$$

Taking the integral, we get

$$\langle \hat{p}_x \rangle = -\frac{i\hbar}{\pi}\int_0^a \sin(2\pi a/x)\, dx = 0. \tag{2.7.31}$$

Example 2.7.3: Consider a particle of mass m confined to move in a one-dimensional infinite potential well of width a. Let, at $t = 0$, the particle be in a state described by the wave function $\psi(x,t) = \sin^3(\pi a/x)$. If the energy of the particle is measured, what values will be obtained and with what probabilities? What will be the average value of energy in this state?

Solution: We shall show in Chapter 3 that the eigenfunctions and the corresponding eigenvalues of the Hamiltonian, for a particle of mass m moving in a 1D infinite potential well of width a, are given by

$$\psi_n(x) = \sqrt{\frac{2}{a}} \sin(n\pi a/x), \quad E_n = \frac{n^2\pi^2\hbar^2}{2ma^2}, \quad n = 1,2,3,... \tag{2.7.32}$$

The wave function of the particle at $t = 0$ can be written as

$$\psi(x) = \frac{3}{4}\sin(\pi a/x) - \frac{1}{4}\sin(3\pi a/x) = \frac{3\sqrt{a}}{4\sqrt{2}}\phi_1(x) - \frac{\sqrt{a}}{4\sqrt{2}}\phi_3(x), \tag{2.7.33}$$

where ϕ_1 and ϕ_3 are the ground state and the second excited state wave functions of the particle in the infinite potential well. Let us check whether the wave function (2.7.33) is normalized or not. We have

$$\int_0^a |\psi(x)|^2\,dx = \frac{9a}{32}\int_0^a |\phi_1(x)|^2\,dx + \frac{a}{32}\int_0^a |\phi_3(x)|^2\,dx$$
$$- \frac{6a}{32}\int_0^a \phi_1(x)\phi_3(x)\,dx = \frac{9a}{32} + \frac{a}{32} = \frac{5a}{16}, \tag{2.7.34}$$

where we have used the fact that the eigenfunctions of the Hamiltonian are orthonormal:

$$\int_0^a \phi_i^*(x)\phi_j(x)\,dx = \delta_{ij} = \begin{cases} 1, & \text{if} \quad i = j \\ 0, & \text{if} \quad i \neq j. \end{cases} \tag{2.7.35}$$

As a result, the normalized wave function at $t = 0$ is

$$\phi(x) = \frac{4}{\sqrt{5a}}\frac{3\sqrt{a}}{4\sqrt{2}}\phi_1(x) - \frac{4}{\sqrt{5a}}\frac{\sqrt{a}}{4\sqrt{2}}\phi_3(x) = \frac{3}{\sqrt{10}}\phi_1(x) - \frac{1}{\sqrt{10}}\phi_3(x). \tag{2.7.36}$$

Therefore, when energy is measured on the system, the values that can result are

$$E_1 = \frac{\pi^2\hbar^2}{2ma^2} \qquad \text{and} \qquad E_3 = \frac{9\pi^2\hbar^2}{2ma^2}. \tag{2.7.37}$$

Now the probability of getting E_1 is

$$P_1 = |\langle \phi_1 \mid \phi \rangle|^2 = \frac{9}{10}, \tag{2.7.38}$$

while the probability of getting E_3 is

$$P_3 = |\langle \phi_3 \mid \phi \rangle|^2 = \frac{1}{10}. \tag{2.7.39}$$

The average value of energy in the state $\psi(x)$ is

$$\langle E \rangle = P_1E_1 + P_3E_3 = \frac{9}{10} \times \frac{\pi^2\hbar^2}{2ma^2} + \frac{1}{10} \times \frac{9\pi^2\hbar^2}{2ma^2} = \frac{9\pi^2\hbar^2}{10ma^2}. \tag{2.7.40}$$

***Compatible operators*:** In physics, we want to have the maximal information about the system under study. The same applies to a quantum mechanical system. Obviously, this is possible only if we are able to determine and measure accurately the maximal number of relevant physical characteristics of the system. Therefore, it is important to discuss the possibilities that quantum mechanics offers and the restrictions it puts on achieving this goal. Keeping this in mind, we are now going to prove some important theorems that will tell us about the extent to which we can accomplish this task.

Theorem 2.7.1: If two observables A and B are compatible, they possess a common set of eigenfunctions (this theorem holds for both degenerate and non-degenerate eigenstates).

Proof: Since the two observables are compatible, the corresponding operators $\hat{A}$ and $\hat{B}$ commute, that is, $\hat{A}\hat{B} = \hat{B}\hat{A}$. Let ϕ_n be the eigenfunction of the operator $\hat{A}$ with eigenvalue a_n. Then we have

$$\hat{A}\hat{B}\phi_n = \hat{B}\hat{A}\phi_n = a_n(\hat{B}\phi_n), \tag{2.7.41}$$

where we made use of the eigenvalue equation for the operator $\hat{A}$: $\hat{A}\phi_n = a_n\phi_n$. Equation (2.7.41) says that $\hat{B}\phi_n$ is also an eigenfunction of $\hat{A}$ with the same eigenvalue a_n. Therefore, $\hat{B}\phi_n \propto \phi_n$, that is,

$$\hat{B}\phi_n = b_n\phi_n, \tag{2.7.42}$$

where b_n is the constant of proportionality. Equation (2.7.42) is nothing but the eigenvalue equation for the operator $\hat{B}$ with ϕ_n as the eigenfunction and b_n as the eigenvalue. Thus, ϕ_n is also an eigenfunction of $\hat{B}$ with eigenvalue b_n. The theorem is proved.

Theorem 2.7.1 can be generalized to the case of many mutually compatible observables $A, B, C, \ldots$. All the corresponding compatible operators, $\hat{A}, \hat{B}, \hat{C}, \ldots$, will possess a common set of eigenfunctions.

Theorem 2.7.2: If two observables A and B have a common set of eigenfunctions, they are compatible, that is, the corresponding operators $\hat{A}$ and $\hat{B}$ commute.

Proof: Let $\{\phi_n\}$ be the common set of eigenfunctions for the operators $\hat{A}$ and $\hat{B}$. Then we have $\hat{A}\phi_n = a_n\phi_n$ and $\hat{B}\phi_n = b_n\phi_n$, where a_n and b_n are the eigenvalues of the operators $\hat{A}$ and $\hat{B}$, respectively. We then have

$$\hat{A}\hat{B}\phi_n = \hat{A}(b_n\phi_n) = b_n(\hat{A}\phi_n) = b_n a_n\phi_n, \tag{2.7.43}$$

$$\hat{B}\hat{A}\phi_n = \hat{B}(a_n\phi_n) = a_n(\hat{B}\phi_n) = a_n b_n\phi_n. \tag{2.7.44}$$

Subtracting (2.7.44) from (2.7.43), we obtain

$$(\hat{A}\hat{B} - \hat{B}\hat{A})\phi_n = 0. \tag{2.7.45}$$

For the equation (2.7.45) to be valid for any ϕ_n, we must have

$$(\hat{A}\hat{B} - \hat{B}\hat{A}) = 0. \tag{2.7.46}$$

Thus, the operators $\hat{A}$ and $\hat{B}$ commute. The theorem is proved.

Definition: The set of hermitian operators $\hat{A}$, $\hat{B}$, $\hat{C}$,... is called a complete set of commuting operators (CSCO) if the operators mutually commute and the set of their common eigenfunctions is complete and unique.

A complete set of operators, may sometimes consist of only one operator. For instance, the position operator $\hat{x}$ of a spinless particle, moving in one spatial dimension, provides a complete set. Its momentum operator $\hat{p}$ also constitutes a complete set. However, if we combine them together, $\hat{x}$ and $\hat{p}$ do not form a complete set because they do not commute.

We shall now discuss the problem of measuring more than one observable on a system in a given state. Consider two observables represented by the operators $\hat{A}$ and $\hat{B}$. Suppose we want to measure these on a system which is in a state ψ. Since, in general, operators do not commute, the result obtained by measuring A first and then B will differ from the one obtained by measuring B first and then A. Let us find out the reason behind it.

Suppose $\hat{A}$ and $\hat{B}$ do not commute and ψ is the nth eigenstate of the operator $\hat{A}$ with eigenfunction ϕ_n and the corresponding eigenvalue a_n. If we measure A first, the measurement will, with certainty, yield the value a_n and the state of the system will change from ψ to ϕ_n (the nth eigenstate of $\hat{A}$). Since $\hat{A}$ and $\hat{B}$ do not commute, ϕ_n is not an eigenstate of $\hat{B}$. If we now measure B, the result of measurement cannot be predicted in advance. Any of the eigenvalues, b_m, of $\hat{B}$ can occur. The probability of obtaining b_m will be given by $|c_{nm}|^2$ where c_{nm} is the coefficient in the expansion of ϕ_n into a series with respect to the complete set of eigenfunctions, $\{\chi_m\}$, of $\hat{B}$

$$\phi_n = \sum_m c_{nm}\chi_m. \tag{2.7.47}$$

Now, we reverse the sequence of measurement and measure B first and then A. In this case, the measurement of B will yield one of the eigenvalues of $\hat{B}$, say b_j, and the system will collapse into the eigenstate χ_j. Since χ_j is not an eigenstate of $\hat{A}$, the measurement of A can result in any of the eigenvalues of $\hat{A}$. The probability of getting the value a_m for A will be given by $|c_{jm}|^2$, where c_{jm} are the coefficients in the expansion of χ_j into a series with respect to the complete set of eigenfunctions, $\{\phi_m\}$, of $\hat{A}$

$$\chi_j = \sum_n c_{jm}\phi_m. \tag{2.7.48}$$

Obviously, the results of the first set of measurements will, in general, be different from the results of the second set of measurements. Hence, the result of measurement of more than one observables on a system depends on the order in which they are measured.

Now assume that the operators $\hat{A}$ and $\hat{B}$ commute. Then according to Theorem 2.7.1, they will have a common set of eigenfunctions, say, $\{\phi_j\}$. Now if ψ is, say, the nth eigenstate of $\hat{A}$ (described by the eigenfunction ϕ_n) and we measure A, the result we shall get will be nothing but the eigenvalue a_n of $\hat{A}$. Clearly, the state of the system will not be altered by the measurement, i.e., the state remains to be the the nth eigenstate of $\hat{A}$. Since ϕ_n is also an eigenfunction of $\hat{B}$ with eigenvalue b_n, the measurement of B will now with certainty yield b_n. Conversely, if we measure B first, the result will be b_n and again the state of the system will not be altered. The subsequent measurement of A will with certainty yield a_n. In both the sequences of measurements, the result will be the same.

Corollary: Compatible observables can be measured simultaneously with any desired accuracy, while non-compatible observables cannot be measured simultaneously accurately.

Note that if an observable, A, represented by a hermitian operator $\hat{A}$ is measured on a system in a state ψ, the uncertainty, ΔA, in its measurement is defined as

$$\Delta A = \sqrt{\langle \hat{A}^2 \rangle - \langle \hat{A} \rangle^2}. \tag{2.7.49}$$

We shall prove in Chapter 4 that, if two hermitian operators, $\hat{A}$ and $\hat{B}$, do not commute the uncertainties in their measurements in a given state satisfy the following generalized Heisenberg uncertainty relation

$$\Delta A\, \Delta B \geq \frac{1}{2} |\langle [\hat{A}, \hat{B}] \rangle|. \tag{2.7.50}$$

Thus, if two observables A and B are compatible, the corresponding operators $\hat{A}$ and $\hat{B}$ will commute and the relation (2.7.50) tells us that both the observables can be measured simultaneously with high degrees of accuracy.

Example 2.7.4: A system is initially in the state

$$\psi_0 = \frac{1}{\sqrt{3}}\phi_1(x) + \frac{1}{\sqrt{2}}\phi_2(x) + \frac{1}{\sqrt{6}}\phi_3(x), \tag{2.7.51}$$

where ϕ_n are the normalized eigenstates of the system's Hamiltonian such that $\hat{H}\phi_n = n^2 \varepsilon_0 \phi_n$. Here, ε_0 is a constant with dimensions of energy and n is a positive energy.

(a) If energy is measured, what values will be obtained and with what probabilities?

(b) Consider an observable A, which is represented by operator $\hat{A}$, whose action on ϕ_n is defined by $\hat{A}\phi_n = (n+1)a_0\phi_n$. If A is measured, what values will be obtained and with what probabilities?

(c) Suppose that a measurement of the energy yields $4\varepsilon_0$. If we measure A immediately afterward, what value will be obtained?

Solution:

(a) A measurement of the energy in the state ϕ_n yields the expectation value of the Hamiltonian, given by

$$E_n = \langle \hat{H} \rangle = \int_{-\infty}^{+\infty} \phi_n^*(x) \hat{H} \phi_n(x)\, dx = n^2 \varepsilon_0, \tag{2.7.52}$$

where we have taken into account that the wave function ψ_0 is normalized to unity. Thus, the values of energy that can result, when measured in the state ψ_0, are, $E_1 = \varepsilon_0, E_2 = 4\varepsilon_0, E_3 = 9\varepsilon_0$.

The probability for obtaining the value E_n for energy is

$$P(E_n) = \frac{\left|\int_{-\infty}^{+\infty} \phi_n^*(x)\,\psi_0(x)\,dx\right|^2}{\int_{-\infty}^{+\infty} |\psi_0(x)|^2\,dx} = \left|\int_{-\infty}^{+\infty} \phi_n^*(x)\,\psi_0(x)\,dx\right|^2. \tag{2.7.53}$$

Taking into account that

$$\int_{-\infty}^{+\infty} \phi_m^*(x)\phi_n(x)\,dx = \delta_{mn}, \tag{2.7.54}$$

we obtain

$$P(E_1) = \left|\frac{1}{\sqrt{3}}\int_{-\infty}^{+\infty} \phi_1^*(x)\,\psi_0(x)\,dx\right|^2 = \frac{1}{3},$$
$$P(E_2) = \left|\frac{1}{\sqrt{2}}\int_{-\infty}^{+\infty} \phi_2^*(x)\,\psi_0(x)\,dx\right|^2 = \frac{1}{2},$$
$$P(E_3) = \left|\frac{1}{\sqrt{6}}\int_{-\infty}^{+\infty} \phi_3^*(x)\,\psi_0(x)\,dx\right|^2 = \frac{1}{6}. \tag{2.7.55}$$

(b) A measurement of A in the state ϕ_n yields the expectation value of $\hat{A}$, given by

$$a_n = \langle \hat{A} \rangle = \int_{-\infty}^{+\infty} \phi_n^*(x)\hat{A}\phi_n(x)\,dx = (n+1)a_0, \tag{2.7.56}$$

that is, $a_1 = 2a_0,\ a_2 = 3a_0,\ a_3 = 4a_0$. The probabilities corresponding to these values of A will be

$$P(a_1) = \frac{1}{3}, \quad P(a_2) = \frac{1}{2}, \quad P(a_3) = \frac{1}{6}. \tag{2.7.57}$$

(c) Since the average value of the Hamiltonian in the normalized state ϕ_2 is given by

$$\int_{-\infty}^{+\infty} \phi_2^*(x)\hat{H}\phi_2(x)\,dx = 4\varepsilon_0, \tag{2.7.58}$$

an energy measurement giving $4\varepsilon_0$ implies that the system, after measurement, is left in the state $\phi_2(x)$. Therefore a measurement of the observable A immediately afterward yields a value equal to

$$\langle \hat{A} \rangle_2 = \int_{-\infty}^{+\infty} \phi_2^*(x)\hat{A}\phi_2(x)\,dx = 3a_0. \tag{2.7.59}$$

Example 2.7.5: Two observables, $\hat{A}$, and $\hat{B}$ of a system are given by the following matrices, respectively:

$$A = \varepsilon_0 \begin{pmatrix} 3 & (\sqrt{2}-\sqrt{5}i) \\ (\sqrt{2}+\sqrt{5}i) & -3 \end{pmatrix}, \quad B = \hbar \begin{pmatrix} 1 & 1 \\ 1 & 1 \end{pmatrix},$$

where ε_0 has the dimensions of energy. (a) If we measure the observable A in some state ψ of a quantum system, what could be the possible results? (b) Find the eigenvectors of A. (c) If we measure the observable B in the state ψ, what could be the possible results? (d) Suppose a measurement of $\hat{A}$ in the state ψ yields a value equal to $4\varepsilon_0$. Immediately after that we measure $\hat{B}$. What is the probability of getting a value $2\hbar$?

Solution:

(a) According to the measurement postulate of quantum mechanics, the possible results of the measurement of $\hat{A}$ will be the eigenvalues of the operator $\hat{A}$. Therefore, let us find these eigenvalues. The characteristic equation for the determination of the eigenvalues, λ, is given by

$$|A - \lambda \hat{I}| = \begin{vmatrix} (3\varepsilon_0 - \lambda) & \varepsilon_0(\sqrt{2}-\sqrt{5}i) \\ \varepsilon_0(\sqrt{2}+\sqrt{5}i) & -(3\varepsilon_0+\lambda) \end{vmatrix} = 0. \quad \Rightarrow \lambda^2 - 16\varepsilon_0^2 = 0. \tag{2.7.60}$$

This leads to $\lambda_1 = 4\varepsilon_0$ and $\lambda_2 = -4\varepsilon_0$. Hence, the possible results of measurements are $A = 4\varepsilon_0$ and $A = -4\varepsilon_0$.

(b) Let us find the eigenvectors of A corresponding to these eigenvalues. For $\lambda = 4\varepsilon_0$, we have

$$\begin{pmatrix} -\varepsilon_0 & \varepsilon_0(\sqrt{2}-\sqrt{5}i) \\ \varepsilon_0(\sqrt{2}+\sqrt{5}i) & -7\varepsilon_0 \end{pmatrix} \begin{pmatrix} \alpha \\ \beta \end{pmatrix} = 0. \tag{2.7.61}$$

$$\Rightarrow \quad \varepsilon_0 \begin{pmatrix} -\alpha + (\sqrt{2}-\sqrt{5}i)\beta) \\ (\sqrt{2}+\sqrt{5}i)\alpha - 7\beta \end{pmatrix} = 0. \tag{2.7.62}$$

We get from here that β can be an arbitrary constant. Let $\beta = 1$. Then, $\alpha = (\sqrt{2} - \sqrt{5}i)$. Therefore, the normalized eigenvector of A, corresponding to the eigenvalue $\lambda_1 = 4\varepsilon_0$, is

$$\psi_1 = \frac{1}{2\sqrt{2}} \begin{pmatrix} \sqrt{2}-\sqrt{5}i \\ 1 \end{pmatrix}. \tag{2.7.63}$$

Similarly, we can calculate the normalized eigenvector of A with $\lambda_2 = -4\varepsilon_0$ to be

$$\psi_2 = \frac{1}{2\sqrt{2}} \begin{pmatrix} -1 \\ (\sqrt{2}+\sqrt{5}i) \end{pmatrix}. \tag{2.7.64}$$

It is easy to check that the eigenvectors ψ_1 and ψ_2 are orthonormal.

(c) The characteristic equation for the operator $\hat{B}$ reads:

$$|B-\lambda\hat{I}| = \begin{vmatrix} (\hbar-\lambda) & \hbar \\ \hbar & (\hbar-\lambda) \end{vmatrix} = 0. \quad \Rightarrow \quad (\hbar-\lambda) = \pm\hbar. \tag{2.7.65}$$

This leads to $\lambda_1 = 2\hbar$ and $\lambda_2 = 0$. Hence, the possible results of measurement of B are $B = 2\hbar$ and $B = 0$.

(d) The eigenvectors of $\hat{B}$ corresponding to the eigenvalues $2\hbar$ and 0 are

$$\phi_1 = \frac{1}{\sqrt{2}}\begin{pmatrix} 1 \\ 1 \end{pmatrix}, \quad \phi_2 = \frac{1}{\sqrt{2}}\begin{pmatrix} 1 \\ -1 \end{pmatrix}, \tag{2.7.66}$$

respectively.

Now, if the measurement of $\hat{A}$ gives the value $4\varepsilon_0$, then after the measurement, the state of the system will be ψ_1. Then according to the postulates of quantum mechanics, the probability of getting the value $2\hbar$ of $\hat{B}$ will be

$$P(2\hbar) = \frac{|(\phi_1,\psi_1)|^2}{(\psi_1,\psi_1)} = \left|\frac{1}{4}\left[\sqrt{2}-\sqrt{5}i+1\right]\right|^2 = \frac{8+2\sqrt{2}}{16} \approx 0.677. \tag{2.7.67}$$

Similarly, the probability of getting the value 0 for $\hat{B}$ is

$$P(0) = \frac{|(\phi_2,\psi_1)|^2}{(\psi_1,\psi_1)} = \left|\frac{1}{4}\left[\sqrt{2}-\sqrt{5}i-1\right]\right|^2 = \frac{8-2\sqrt{2}}{16} \approx 0.32322. \tag{2.7.68}$$

Example 2.7.6: A particle of mass m is confined in a one-dimensional infinite potential well of width a $(0 \le x \le a)$. It is prepared to be in the second excited state $(n = 3)$. Now the width of the potential well is suddenly changed to 4 times the initial width without affecting the state of the particle. If a measurement of the energy is carried out on the particle, what are the probabilities that the particle is found in the first excited state and the ground state of the new well?

Solution: Initially, the particle is in the second excited state of the original well. Hence, the initial wave function of the particle is (see Eq.(2.7.32))

$$\psi(x) = \sqrt{\frac{2}{a}}\,\sin\left(\frac{3\pi}{a}x\right). \tag{2.7.69}$$

The bound state wave functions, $\psi'_n(x)$, and the corresponding energy eigenvalues, E'_n, of the new well are obtained by replacing a with $4a$ in the formulae (2.7.32):

$$\psi'_n(x) = \sqrt{\frac{2}{a}}\,\sin\left(\frac{n\pi}{4a}x\right), \tag{2.7.70}$$

and

$$E'_n = \frac{n^2\pi^2\hbar^2}{32ma^2}, \quad n = 1,2,3,\ldots \tag{2.7.71}$$

Since the wave functions of the original as well as new wells are normalized to unity, the probability of finding the particle in the first excited state of the new well is given by

$$P_2 = \left|\int_0^a \psi_2'^*(x)\psi(x)\,dx\right|^2 = \left|\frac{2}{a}\int_0^a \sin\left(\frac{2\pi}{4a}x\right)\,\sin\left(\frac{3\pi}{a}x\right)\,dx\right|^2. \tag{2.7.72}$$

Using the well-known formula, $2\sin(A)\,\sin(B) = \cos(A-B) - \cos(A+B)$, the integrand can be written as

$$2\sin\left(\frac{3\pi}{a}x\right)\,\sin\left(\frac{\pi}{2a}x\right) = \cos\left(\frac{5\pi}{2a}x\right) - \cos\left(\frac{7\pi}{2a}x\right). \tag{2.7.73}$$

The integrals are then given by

$$\int_0^a \cos\left(\frac{5\pi}{2a}x\right)\,dx = \frac{2a}{5\pi}\,\sin\left(\frac{5\pi}{2a}x\right)\Bigg|_0^a = \frac{2a}{5\pi}\,\sin\left(\frac{5\pi}{2}\right) = \frac{2a}{5\pi}, \tag{2.7.74}$$

$$\int_0^a \cos\left(\frac{7\pi}{2a}x\right)\,dx = \frac{2a}{7\pi}\,\sin\left(\frac{7\pi}{2a}x\right)\Bigg|_0^a = \frac{2a}{7\pi}\,\sin\left(\frac{7\pi}{2}\right) = -\frac{2a}{7\pi}. \tag{2.7.75}$$

Therefore, the probability of finding the particle in the first excited state is

$$P_2 = \left|\frac{2}{5\pi} + \frac{2}{7\pi}\right|^2 = \left|\frac{24}{35\pi}\right|^2 = 0.04764 \approx 4.8\,\%. \tag{2.7.76}$$

The probability of finding the particle in the ground state of the new well is given by

$$P_1 = \left| \int_0^a \psi_1'^*(x)\psi(x)\,dx \right|^2 = \left| \frac{2}{a}\int_0^a \sin\left(\frac{\pi}{4a}x\right)\sin\left(\frac{3\pi}{a}x\right)dx \right|^2. \tag{2.7.77}$$

The integrand can again be written as

$$2\sin\left(\frac{3\pi}{a}x\right)\sin\left(\frac{\pi}{4a}x\right) = \cos\left(\frac{11\pi}{4a}x\right) - \cos\left(\frac{13\pi}{4a}x\right), \tag{2.7.78}$$

leading to the following values of the integrals

$$\begin{aligned}\int_0^a \cos\left(\frac{11\pi}{4a}x\right)dx &= \frac{4a}{11\pi}\sin\left(\frac{11\pi}{4a}x\right)\Bigg|_0^a \\ &= \frac{4a}{11\pi}\sin\left(\frac{11\pi}{4}\right) = \frac{2\sqrt{2}a}{11\pi},\end{aligned} \tag{2.7.79}$$

$$\begin{aligned}\int_0^a \cos\left(\frac{7\pi}{2a}x\right)dx &= \frac{4a}{13\pi}\sin\left(\frac{13\pi}{4a}x\right)\Bigg|_0^a \\ &= \frac{4a}{13\pi}\sin\left(\frac{13\pi}{4}\right) = -\frac{2\sqrt{2}a}{13\pi}.\end{aligned} \tag{2.7.80}$$

Therefore, the probability of finding the particle in the ground state is

$$P_1 = \left| \frac{2\sqrt{2}}{13\pi} + \frac{2\sqrt{2}}{11\pi} \right|^2 = \left| \frac{48\sqrt{2}}{143\pi} \right|^2 \approx 0.023. \tag{2.7.81}$$

Homework Problems

1. Which of the functions

   $$(i)\ \cos(4x), \quad (ii)\ \cosh(4x), \quad (iii)\ e^{-4x}, \quad (iv)\ e^{-4x^2}$$

   is (are) a genuine quantum mechanical eigenfunction (eigenfunctions) of the operator $\hat{A} = -(d^2/dx^2)$ and why?

2. Let the wave function of a particle be

   $$\psi(x,0) = A\,e^{-|x|/2b},$$

   where A and b are real constants. Normalize the wave function to find A.

3. A particle is confined to move in the region $0 < x < L$, where L is a positive constant. If the wave function of the particle is

$$\psi(x,0) = A\,(x - x^3),$$

where A is a real constant, find A. Calculate the probability of finding the particle in the region $L/3 \le x \le 2L/3$.

4. An electron is described by the wave function

$$\psi(x) = \begin{cases} 0 & x < 0 \\ Ce^{-x}\,(1 - e^{-x}) & x > 0 \end{cases}$$

where C is a constant. (a) Find the value of C that normalizes $\psi(x)$. (b) Where is the electron most likely to be found; that is, for what value of x is the probability of finding the electron largest?

5. The wave function of a particle

$$\psi(x,0) = \sqrt{\frac{\sqrt{\alpha}}{2\sqrt{\pi}}}\,x e^{-\alpha x^2/2},$$

satisfies the TISE for $E = \beta\hbar\omega$, where α, β and ω are real constants. Find the potential V as a function of x.

6. A particle, confined to move in the region $0 < x < L$, where L is a positive constant, is in a state described by the wave function

$$\psi(x,0) = \sqrt{\frac{1}{L}}\left[\sin\left(\frac{\pi x}{L}\right) + \sin\left(\frac{2\pi x}{L}\right)\right].$$

Calculate the probability of finding the particle in the interval from $x = 0$ to $x = L/2$.

7. Find the hermitian conjugate of the following operators and discuss their hermiticity: (a) $\frac{d}{dx}$, (b) $i\frac{d}{dx}$, and (c) $\hat{x}\frac{d}{dx} + 5$.

8. $\hat{A}$ and $\hat{B}$ are hermitian. Under what conditions is the linear combination $\alpha\hat{A} + \beta\hat{B}$, where α and β are arbitrary complex numbers, hermitian?

9. Show that the operator $\hat{A} = i(\hat{x}^2 + 1)\frac{d}{dx} + i\hat{x}$ is hermitian. Find its eigenfunction corresponding to the eigenvalue $\lambda = 0$. If a particle is in a state described by this eigenfunction, calculate the probability of finding the particle in the interval $-1 \le x \le +1$.

10. Check whether the operator,

$$\hat{A} = \begin{pmatrix} 7i & -1 \\ 2 & -6i \end{pmatrix},$$

is hermitian. Find its eigenvalues.

11. Find the operator hermitian conjugate to the operator

$$\hat{B} = \begin{pmatrix} 2 & 0 & 1 \\ 0 & 3 & 1 \\ 1 & 0 & 4 \end{pmatrix}.$$

Find the eigenvalues and the normalized eigenvectors of $\hat{B}$.

12. Find the eigenvalues and the normalized eigenvectors of the matrix operators

$$(a)\ \hat{A} = \begin{pmatrix} 0 & 1 \\ 1 & 0 \end{pmatrix}, \quad (b)\ \hat{B} = \begin{pmatrix} 0 & i \\ -i & 0 \end{pmatrix}.$$

13. Consider an operator $\hat{A} = \hat{x}\frac{d}{dx} + \alpha$, where α is a constant. Calculate

$$(a)\ [\hat{A}, \hat{x}], \quad (b)\ [\hat{A}, \hat{p}_x], \quad (c)\ \left[\frac{d}{dx}, \left[\hat{A}, \frac{d}{dx}\right]\right].$$

14. Using the fundamental commutator, $[\hat{x}, \hat{p}_x] = i\hbar$, show that

$$[\hat{x}^n, \hat{p}_x] = i\hbar n \hat{x}^{n-1}.$$

15. Show that

$$[f(\hat{x}), \hat{p}_x] = i\hbar \frac{\partial f(\hat{x})}{\partial \hat{x}},$$

where $f(\hat{x})$ is an arbitrary differentiable function of $\hat{x}$.

16. Find the inverse of the following operators

$$(a)\ \hat{A} = \begin{pmatrix} 1 & i \\ -i & 2 \end{pmatrix}, \quad (b)\ \hat{B} = \begin{pmatrix} 2 & 0 & 0 \\ 0 & -3 & i \\ 0 & -i & 1 \end{pmatrix}.$$

17. Show that the operators

$$\hat{A} = \begin{pmatrix} \cos\theta & \sin\theta \\ -\sin\theta & \cos\theta \end{pmatrix}, \quad (b)\ \hat{B} = \begin{pmatrix} \frac{1}{\sqrt{2}} & \frac{i}{\sqrt{2}} & 0 \\ -\frac{i}{\sqrt{2}} & \frac{i}{\sqrt{2}} & 0 \\ 0 & 0 & i \end{pmatrix}.$$

are unitary. Compute their eigenvalues and the corresponding normalized eigenvectors.

18. The Hamiltonian for a free particle in one dimension is given by

$$\hat{H} = -\frac{\hbar^2}{2m}\frac{\partial^2}{\partial x^2}.$$

Check whether the following functions are eigenstates of the Hamiltonian and if they are, write down the corresponding energy eigenvalues:

(a) $e^{ipx/\hbar}$, (b) $\sin\left(\frac{5\pi x}{L}\right)$, (c) $\sin\left(\frac{5\pi x}{L}\right) + 2\cos\left(\frac{5\pi x}{L}\right)$,

(d) $\sin\left(\frac{5\pi x}{L}\right) + 2\cos\left(\frac{7\pi x}{L}\right)$, (e) $\sin\left(\frac{5\pi x}{L}\right)\cos\left(\frac{5\pi x}{L}\right)$.

Here, L is a real constant and p is a constant with dimensions of momentum.

19. A particle is restricted to move in the region $0 < x < a$, where a is a positive constant. Initially ($t = 0$), it is in a state with the wave function

$$\psi(x) = \sqrt{\frac{1}{a}}\left[\sin\left(\frac{\pi x}{a}\right) + \sin\left(\frac{3\pi x}{a}\right)\right].$$

Calculate the probability of finding the particle in the interval $\frac{a}{3} < x < \frac{2a}{3}$.

20. Find the wave function for any $t > 0$ in Problem 19. Assume that $\psi(x,t)$ and calculate the probability density $\rho(x,t) = \psi^*(x,t)\psi(x,t)$.

21. The orthonormal states of a system corresponding to the energy eigenvalues, E_n, $n = 1,2,3,...$, are described by the wave functions $\phi_n(x)$, $n = 1,2,3,....$ When energy is measured on the system in the state $\psi(x,0)$ at $t = 0$, the values obtained are E_1 with probability $1/6$, E_2 with probability $1/2$, E_3 with probability $1/9$, and E_4 with probability $2/9$. (a) Write down the expression for $\psi(x,0)$ on the basis of the given information. (b) What will be the expression for ψ at a later time $t > 0$? (c) Show that the expectation value of the Hamiltonian does not depend on time.

22. The state of this system is given in terms of three functions $\phi_i(x)$, $i = 1,2,3$ as

$$\psi = \sqrt{\frac{1}{10}}\phi_1 + \sqrt{\frac{7}{10}}\phi_2 + \sqrt{\frac{2}{10}}\phi_3,$$

where ϕ_i constitute a complete set of orthonormal eigenstates of some observable $\hat{A}$.

(a) Verify that ψ is normalized. Then, calculate the probability of finding the system in any one of the states ϕ_1, ϕ_2, and ϕ_3. Verify that the total probability is equal to one.

(b) Consider an ensemble of 500 identical particles. The measurement is done on all of them to determine their states. Find the number of particles to be found in each of the states ϕ_1, ϕ_2 and ϕ_3.

23. An observable $\hat{A}$ has two eigenfunctions ψ_1 and ψ_2 with eigenvalues a_1 and a_2 respectively. Another observable $\hat{B}$ has two eigenfunctions ϕ_1 and ϕ_2 with eigenvalues b_1 and b_2 respectively. Eigenfunctions of both these observables are normalized and are related to each other through the following equations:

$$\psi_1 = \frac{3\phi_1 + 4\phi_2}{5}, \qquad \psi_2 = \frac{4\phi_1 - 3\phi_2}{5}.$$

(a) The observable $\hat{A}$ is measured on a system in the state ψ and the result is a_1. What is the state of the system immediately after the measurement?

The observable $\hat{B}$ is now measured on the system. What are the possible results and what are their respective probabilities?

24. The wave function of a particle of mass m moving in a one-dimensional infinite potential well of width a is

$$\psi(x,0) = i\sqrt{\frac{1}{2a}}\sin\left(\frac{\pi x}{a}\right) + \sqrt{\frac{1}{a}}\sin\left(\frac{3\pi x}{a}\right) - \sqrt{\frac{1}{2a}}\sin\left(\frac{\pi x}{a}\right).$$

Calculate the average values of the operators $\hat{x}, \hat{x}^2, \hat{p}_x$ and $\hat{p}_x^2$.

25. Consider the particle and its wave function given in the Problem 24. If we measure energy, what values would be obtained and with what probabilities? What will be the average value of the energy of the particle in the given state?

Chapter 3

One-dimensional Problems

In this chapter, we shall first discuss the important properties of stationary state solutions of the time-independent Schrödinger equation (TISE) in one spatial dimension and then take up some typical problems.

The TISE in one spatial dimension takes the form:

$$-\frac{\hbar^2}{2m}\frac{\partial^2\phi(x)}{\partial x^2}+V(x)\phi(x)=E\phi(x), \tag{3.0.1}$$

where $x \in (-\infty, +\infty)$ is the independent variable. The nature and the properties of the solutions to this equation depend on the interrelationship between the total energy, E, of the particle and the potential $V(x)$. Let us discuss some of the important concepts related to it.

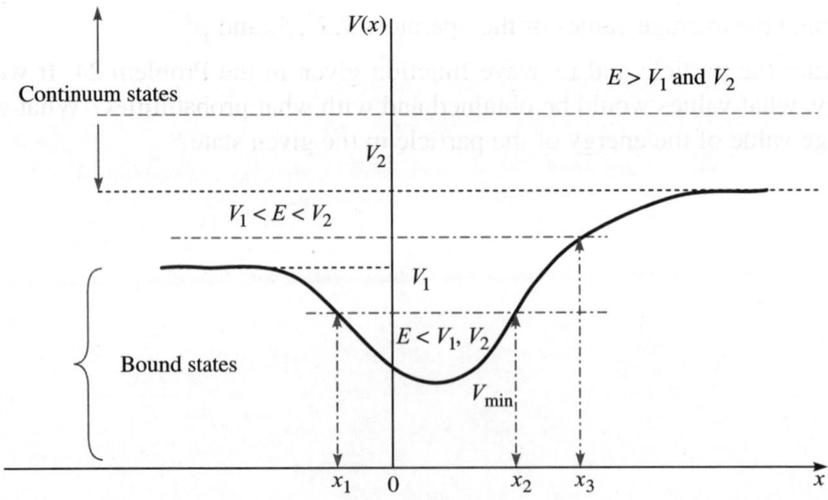

Figure 3.1 Various possibilities for the bound and scattering states of a particle, with total energy E, moving in an arbitrary one-dimensional potential $V(x)$.

3.1 Bound and Scattering States

Consider an arbitrary form of the potential $V(x)$, shown in Figure 3.1, which is general enough to allow for the illustration of all the desired features[1]. Without any loss of generality, the potential has been assumed to remain finite at spatial infinities: $\lim_{x\to-\infty} V(x) = V_1$ and $\lim_{x\to+\infty} V(x) = V_2$ and it has a minimum V_{min} at some point. The character of the energy states of the particle is completely determined by the energy E of the particle in comparison with the asymptotic values of the potential.

In general, the stationary state solutions are categorized as *bound state solutions* and *scattering state solutions*.

Bound states: Bound states occur whenever the particle is confined (or bound) at all energies to move within a finite and limited region of space. In the case of the potential shown in Figure 3.1, if the total energy E of the particle is greater than V_{min} but less than both the asymptotic values V_1 and V_2 of the potential, the particle's motion is restricted between the two classical turning points x_1 and x_2. The states corresponding to this energy range are called bound states.

Scattering states: If the total energy of the particle is either greater than V_1 and less than V_2 or greater than both V_1 and V_2, the particle's motion is not confined to a finite region of space and the states of the particle, corresponding to these ranges of the total energy, are called scattering states. Note that for the bound states to exist, the potential $V(x)$ must have at least one minimum that is lower than V_1.

Important properties of bound state energy levels and the wave functions in one dimension:

1. *The bound state energy levels of a system in one spatial dimension are discrete and nondegenerate.*

Proof: The solutions of the TISE must satisfy the boundary conditions at the classical turning points x_1 and x_2. The result is that acceptable solutions exist only for a discrete set of energy eigenvalues.

The proof of non-degeneracy goes as follows. Suppose there are two solutions $\phi_1(x)$ and $\phi_2(x)$ for the same energy eigenvalue E. They both must satisfy the TISE and we get

$$\phi_1'' = -\frac{2m}{\hbar^2}(E - V(x))\,\phi_1, \tag{3.1.1}$$

$$\phi_2'' = -\frac{2m}{\hbar^2}(E - V(x))\,\phi_2. \tag{3.1.2}$$

Equations (3.1.1) and (3.1.2) lead to

$$\frac{\phi_1''}{\phi_1} = \frac{\phi_2''}{\phi_2}, \qquad \Rightarrow \qquad \frac{d}{dx}(\phi_1'\phi_2 - \phi_2'\phi_1) = 0. \tag{3.1.3}$$

[1]Landau L.D. and Lifshitz E.M., Quantum Mechanics, Ch. III, p.61, Pergamon Press, 1977

Integrating 3.1.3 once over x and taking into account the fact that the wave functions and their first derivatives must vanish at infinity, we obtain

$$\frac{\phi_1'}{\phi_1} = \frac{\phi_2'}{\phi_2}. \tag{3.1.4}$$

Integrating once more over x and taking into account the boundary conditions, we arrive at

$$\phi_1 = C\phi_2, \tag{3.1.5}$$

where C is the integration constant. Since C can be absorbed in the normalization of the wave function, we conclude that $\phi_1 \equiv \phi_2$. The theorem is proved.

2. The ground state wave function has no nodes, that is, it does not become zero anywhere in the entire region $-\infty < x < +\infty$. The next higher energy bound state is called the first excited state and has one node, that is, it becomes zero only at one point in space. The second excited state has two nodes and so on. *In general, the nth bound state wave function, $\phi_n(x)$, in one spatial dimension has n nodes (that is, $\phi_n(x)$ vanishes n times), if $n = 0$ corresponds to the ground state and $(n-1)$ nodes if $n = 1$ corresponds to the ground state.*

The aforementioned property is proved by using the so-called variational principle. We shall not present it here. Instead, we refer the reader to the book, *Methods of Mathematical Physics, Vol. 1* by R. Courant and D. Hilbert.

Before moving on, let us try to solve the one-dimensional TISE and obtain the stationary state solutions in a couple of simple cases, which will illustrate the methodology and the peculiarities of quantum mechanics.

3.2 The Free Particle Solution

A free particle represents a typical example of a stationary state that corresponds to an unbounded motion (scattering state) both along the positive and the negative x directions. In this case, the external potential is absent, that is, $V(x) = 0$, and the TISE reads

$$-\frac{\hbar^2}{2m}\frac{d^2\phi(x)}{dx^2} = E\phi(x) \quad \Rightarrow \quad \frac{d^2\phi(x)}{dx^2} + k^2\phi(x) = 0, \tag{3.2.1}$$

where

$$k^2 = \frac{2mE}{\hbar^2}, E > 0. \tag{3.2.2}$$

Equation (3.2.1) has two linearly independent solutions:

$$\phi_{(+)}(x) = e^{ikx}, \quad \phi_{(-)}(x) = e^{-ikx}. \tag{3.2.3}$$

The general stationary state solution is the linear superposition given by

$$\psi(x,t) = A_{(+)}e^{i(kx-\omega t)} + A_{(-)}e^{-i(kx+\omega t)}, \tag{3.2.4}$$

where $A_{(+)}$ and $A_{(-)}$ are arbitrary, in general complex, constants and we have used the fact that $\omega = E/\hbar$. If we use the de-Broglie formula

$$p = \hbar k, \tag{3.2.5}$$

then the solution (3.2.4) can be written as

$$\psi(x,t) = A_{(+)}\, e^{\frac{i}{\hbar}(px-Et)} + A_{(-)}\, e^{-\frac{i}{\hbar}(px+Et)}. \tag{3.2.6}$$

The first term in the above equation represents a particle travelling to the right (positive x direction) and the second term represents a particle travelling to the left with well defined momenta $p_{\pm} = \pm\hbar k$ and energy $E_{\pm} = \hbar^2 k^2/2m$. The intensities of corresponding waves are $|A_+|^2$ and $|A_-|^2$, respectively. Since there are no boundary conditions, there are no restrictions on the values of k and E; all values of k and E give solutions to the TISE. *Thus, a free particle has a continuous energy spectrum.*

There is, however, some problems related to the free particle solution. Firstly, the probability densities corresponding to either solutions are constant

$$P_{\pm} = \left|A_{(\pm)}\right|^2, \tag{3.2.7}$$

that is, they depend neither on x nor on t. This is due to the fact that, for a state with definite values of momentum, $p_{\pm} = \pm\hbar k$, and energy $E_{\pm} = \hbar^2 k^2/2m$, there occurs a complete loss of information about the position of the particle and the instant of time at which it is located at that position. This is the consequence of Heisenberg's uncertainty principle, according to which, since the momentum and energy of a particle are known exactly ($\Delta p = 0$, and $\Delta E = 0$), there must be a total uncertainty about its position and time at which it is located at that position.

The second difficulty is in an apparent discrepancy between the speed of the wave and the speed of the particle it is supposed to represent. The speed of the right or the left moving plane wave is given by

$$v_p = \frac{\omega}{k} = \frac{E}{\hbar k} = \frac{\hbar k}{2m}. \tag{3.2.8}$$

The velocity of the particle, on the other hand, is given by

$$v = \frac{p}{m} = \frac{\hbar k}{m} = 2v_p. \tag{3.2.9}$$

This means that the particle travels twice as fast as the wave that represents it.

The third difficulty is that the free particle wave function cannot be normalized:

$$\int_{-\infty}^{+\infty} |\psi(x,t)|^2 \, dx = |A_\pm|^2 \int_{-\infty}^{+\infty} dx \;\rightarrow\; \infty. \tag{3.2.10}$$

Hence, as we have discussed earlier, these plane wave solutions of the free Schrödinger equation cannot be taken as quantum mechanical wave functions representing free particles. The natural question arises: Is there anyway out of this problem?

The answer is yes! What saves us is the fact that the Schrödinger equation is linear and superposition principle holds. Therefore, we can superpose a large number of plane wave solutions and the resulting function will be a solution of the Schrödinger equation. Such a solution turns out to be localized and is called a wave packet. Mathematically it is written as

$$\psi(x,t) = \frac{1}{\sqrt{2\pi}} \int_{-\infty}^{+\infty} \tilde{\psi}(k)\, e^{i(k(\omega)x-\omega t)}\, dk, \tag{3.2.11}$$

where the amplitude of the wave packet $\tilde{\psi}(k)$ is given by the Fourier transform of $\psi(x,0)$:

$$\tilde{\psi}(k) = \frac{1}{\sqrt{2\pi}} \int_{-\infty}^{+\infty} \psi(x,0)\, e^{-ik(\omega)x}\, dx. \tag{3.2.12}$$

The wave packet represented by the equation (3.2.11) is localized in space, namely at $x = 0$ (Figure 3.2). This is because of the fact that $\psi(x,t)$ is a superposition of an infinite number of plane waves that are, as we know, coherent and will interfere with each other. They add up constructively at $x = 0$, while their constructive interference diminishes as we move away from the point $x = 0$. The rapid oscillations of the exponential factor e^{ikx} ensures that the waves interfere destructively for $x \rightarrow \pm\infty$. Similarly, the function $\tilde{\psi}(k)$ represents a wave packet in k-space (momentum space). It is localized at $k = 0$ and vanishes at large values of k.

As a measure of the size of the packet in x-space, it is customary to define a half-width Δx corresponding to the half-maximum of $|\psi(x,t)|^2$. It is defined such that when x varies from 0 to $\pm\Delta x$, the function $|\psi(x)|^2$ drops down to $e^{-1/2}$ times its initial value:

$$\frac{|\psi(\pm\Delta x)|^2}{|\psi(0,0)|^2} = \frac{1}{e^{1/2}}. \tag{3.2.13}$$

Similarly, one defines a half-width Δk corresponding to the half-maximum of $|\tilde{\psi}(k)|^2$. In this case it is defined such that when k varies from k_0 to $k_0 \pm \Delta k$, the function $|\tilde{\psi}(k)|^2$ drops down to $e^{-1/2}$ times its initial value:

$$\frac{|\tilde{\psi}(\pm\Delta k)|^2}{|\tilde{\psi}(0)|^2} = \frac{1}{e^{1/2}} \tag{3.2.14}$$

The quantity Δk is the measure of the size of the packet in the momentum space.

The physical interpretation of the wave packet can be given as follows: $|\psi(x,t)|^2$ is the probability density for finding the particle at position x at a given t and $P(x)\,dx = |\psi(x,t)|^2\,dx$ gives the probability of finding the particle in the interval between x and $x + dx$. Similarly, $|\tilde{\psi}(k)|^2$ and $|\tilde{\psi}(k)|^2\,dk$ represent the probability density for measuring the wave vector k (or, equivalently momentum $p = k/\hbar$) of the particle and the probability of finding the particle's wave vector in the interval between k and $k + dk$. Note that it is not difficult to see from the equations (3.2.11) and (3.2.12) that if $\psi(x)$ is normalized to unity, so is $\tilde{\psi}(k)$.

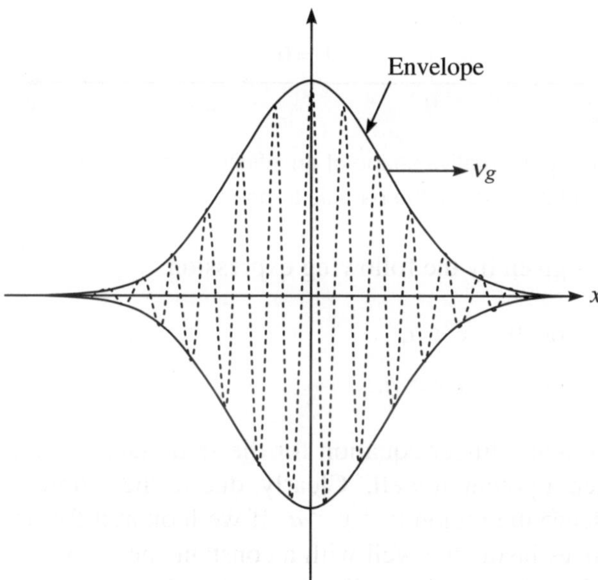

Figure 3.2 The snapshot of a localized wave packet.

The representation of a free particle by a wave packet overcomes the earlier mentioned difficulties related to the position, the instant of time at which the particle is located at that position and the normalization of the plane wave solutions. Since the position, and momentum of a particle, represented by a wave packet, are no longer known exactly (only probabilistic outcomes are possible), the difficulties related to position and time are automatically resolved. The difficulty, related to the speed of the particle being twice that of the speed of the de Broglie wave representing it, is also overcome because, now both the particle and the wave packet travel with the same speed equal to the group velocity, v_g, of the wave packet. Finally, the wave packet, given by (3.2.11), is normalizable.

3.3 Particle in an Infinite Potential Well

In the last section, we considered the free particle solution as our first example. Here, we consider a second example in which we make the situation a bit more involved by restricting the motion of a free particle to a small region of width a by putting walls of infinite potential at $x = 0$ and $x = a$ (see Figure 3.3). This is known as asymmetric infinite square well potential.

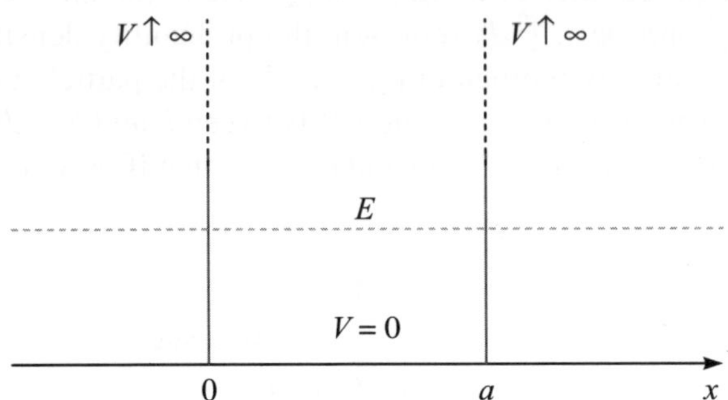

Figure 3.3 The representative shape of an infinite potential well $V(x)$ of width a. E is the total energy of the particle trapped in the potential.

Mathematically this is given by the following expression:

$$V(x) = \begin{cases} 0, & \text{for } 0 < x < a, \\ \infty, & \text{for } x \leq 0, x \geq a. \end{cases} \tag{3.3.1}$$

We want to solve the Schrödinger equation for the stationary states of a particle of mass m moving inside such a potential well. Clearly, due to the infinite walls, the particle is trapped and cannot leave the region $0 < x < a$. If we look at it from the classical point of view, the particle moves inside the well with a constant speed, $p/m = \pm\sqrt{2mE}/m$, back and forth getting reflected from the walls at $x = 0$ and $x = a$. Since the motion of the particle is confined inside the well, quantum mechanically, it corresponds to the case of a bound state problem. In order to find the bound state energies and wave functions, we must solve the TISE with appropriate boundary conditions. Since the particle cannot penetrate the regions $x < 0$ and $x > a$, the wave function of the particle must be zero in these regions: $\psi = 0$ for $x < 0$ and $x > a$.

The TISE

$$\frac{d^2\phi}{dx^2} + \frac{2m}{\hbar^2}(E - V)\phi = 0 \tag{3.3.2}$$

for the given case can be written as

$$\frac{\phi''}{\phi} = -\frac{2m}{\hbar^2}(E - V), \tag{3.3.3}$$

where the prime stands for ordinary derivative with respect to x. Inside the well, $V = 0$, and the solution is given by the linear combination

$$\phi(x) = A \sin(kx) + B \cos(kx), \tag{3.3.4}$$

where A and B are arbitrary constants and

$$k^2 = \frac{2mE}{\hbar^2}. \tag{3.3.5}$$

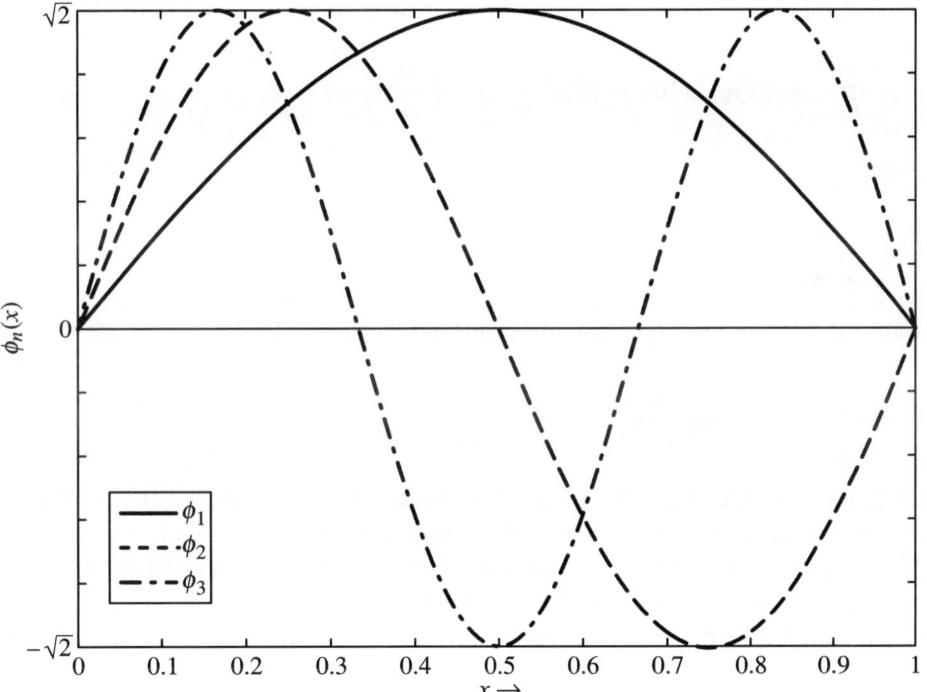

Figure 3.4 Spatial parts of the wave functions for the first three stationary states of a particle in the infinite square well potential with $a = 1$.

According to the standard conditions, the wave function has to be continuous across the boundaries and we must have $\phi \equiv 0$ for $x = 0$ and $x = a$. The first boundary condition $\phi(x = 0) = 0$ leads to $B = 0$. So, we are left with $\phi(x) = A \sin(kx)$. The second boundary condition yields

$$\sin(ka) = 0, \quad \Rightarrow \quad k_n = \frac{n\pi}{a}, \; n = 1, 2, 3, ... \tag{3.3.6}$$

Taking into account (3.3.6), we conclude that the boundary conditions can be satisfied only for the discrete values of energy

$$E_n = \frac{n^2\hbar^2\pi^2}{2ma^2}, \quad n = 1, 2, 3, \ldots, \tag{3.3.7}$$

where we have omitted $n = 0$ because it leads to an uninteresting result: $\phi_0(x) = 0$ and $E_0 = 0$. Thus, a particle, trapped inside an infinite potential well, can have only discrete set of energy eigenvalues given by (3.3.7). The corresponding eigenfunctions are

$$\phi_n(x) = B_n \sin\left(\frac{n\pi}{a}x\right). \tag{3.3.8}$$

The constant B_n is determined by the normalization condition

$$|B_n|^2 \int_{-\infty}^{+\infty} \phi_n^*(x)\phi_n(x)dx = |B_n|^2 \int_0^a \sin^2\left(\frac{\pi x}{a}n\right) dx = 1. \tag{3.3.9}$$

The result is

$$B_n = \sqrt{\frac{2}{a}}. \tag{3.3.10}$$

Therefore, the normalized eigenfunctions and the corresponding energies are

$$\psi_n(x,t) = \sqrt{\frac{2}{a}} \sin\left(\frac{\pi x}{a}n\right), \quad E_n = \frac{n^2\hbar^2\pi^2}{2ma^2}, \quad n = 1, 2, 3, \ldots . \tag{3.3.11}$$

We thus got an infinite sequence of discrete energy levels corresponding to the positive integer values of the quantum number n. The ground state corresponds to $n = 1$ with energy $E_1 = \hbar^2\pi^2/(2ma^2)$. The states with quantum numbers $n > 1$ are called the excited states. Their energies are equal to n^2 times the ground state energy.

The full stationary state solutions are

$$\psi_n(x,t) = \sqrt{\frac{2}{a}} \sin\left(\frac{\pi x}{a}n\right) e^{-i\frac{n^2\pi^2\hbar}{2ma^2}t}. \tag{3.3.12}$$

Note that, in view of the linearity of the Schrödinger equation, the most general stationary state solution for the given case can be written as

$$\psi(x,t) = \sum_{n=1}^{\infty} c_n \sqrt{\frac{2}{a}} \sin\left(\frac{\pi x}{a}n\right) e^{-i\frac{n^2\pi^2\hbar}{2ma^2}t}, \tag{3.3.13}$$

where c_n are arbitrary constants. The spatial parts of the wave functions, for the first three stationary states of a particle in the infinite square well potential with $a = 1$, are depicted in Fig.3.4.

Properties of the eigenfunctions: Let us enumerate the important properties of the obtained solutions. These properties are quite general and hold good for most of the potentials encountered in quantum mechanics.

1. The eigenfunction $\phi_n(x)$ has $(n-1)$ nodes (zero-crossing).
2. These functions are alternately symmetric and antisymmetric with respect to the centre of the well. For instance, as shown in Figure 3.4, the functions ϕ_1 and ϕ_3 are symmetric whereas the function ϕ_2 is antisymmetric. In general, the eigenfunctions ϕ_n with odd n are symmetric while those with even n are antisymmetric.
3. None of the energy levels is degenerate, that is, each energy level corresponds to a unique eigenfunction.
4. The eigenfunctions corresponding to different energy eigenvalues are orthogonal:

$$\int_{-\infty}^{+\infty} \phi_m^*(x)\phi_n(x)dx = \int_0^a \phi_m^*(x)\phi_n(x)dx = \delta_{mn}, \tag{3.3.14}$$

where δ_{mn} is the Kronecker delta:

$$\delta_{mn} = \begin{cases} 1 & \text{if } m = n \\ 0 & \text{if } m \neq n. \end{cases} \tag{3.3.15}$$

5. The eigenfunctions $\{\phi_n(x)\}, n = 1, 2, 3, \ldots$ constitute a complete set in the sense that an arbitrary function $f(x)$ can be expanded as a linear combination of these functions:

$$f(x) = \sum_{n=1}^{\infty} c_n \phi_n(x) = \sqrt{\frac{2}{a}} \sum_{n=1}^{\infty} c_n \sin\left(\frac{\pi x}{a} n\right), \tag{3.3.16}$$

where the coefficients c_n are calculated as

$$c_n = \int_0^a \phi_n^*(x) f(x) dx. \tag{3.3.17}$$

Note that, the ground state corresponds to $n = 1$ instead of $n = 0$. The reason behind it lies in *Heisenberg's uncertainty relation* between the position and momentum (see Eq. (3.10.12)) of the particle. If the particle has zero total energy, it will be at rest inside the well and we can, in principle, precisely determine its position and momentum simultaneously at a given instant of time. This is not permitted by the uncertainty relation.

Furthermore, since our particle is localized inside the well of width a, according to the uncertainty relation, the minimum uncertainty in the momentum of the particle is of the order of $\hbar/2a$, that is, $\Delta p = \hbar/2a$. This leads to a minimum possible value of the kinetic energy of the particle equal to $\hbar^2/8ma^2$, which is of the order of the ground state energy $E_1 = \pi^2\hbar^2/2ma^2$. This unavoidable minimum energy enforced by the uncertainty principle is known as the *zero-point energy*. The zero-point energy therefore reflects the necessity of minimum motion of the particle due to localization. It occurs for all bound state problems. In the case of binding potentials, the lowest energy state has energy higher than the minimum of the potential energy. This is in contrast to the situation in classical mechanics where the lowest possible energy is equal to the minimum of the potential energy with zero kinetic energy. In quantum mechanics, however, the lowest energy state does minimize the total energy $E = T + V$ but leads to a finite nonzero value of the kinetic energy.

Example 3.3.1: A particle in an infinite symmetrical potential well of width a $(-\frac{a}{2} \leq x \leq +\frac{a}{2})$ is initially ($t = 0$) in a state with the wave function

$$\psi(x,0) = A\left(1 - \frac{x^2}{a^2}\right), \tag{3.3.18}$$

where A is an arbitrary real constant. Find the wave function $\psi(x,t)$ at $t > 0$.

Solution: First, we normalize the wave function to find A. We have

$$\int_{-a}^{+a} |\psi(x,t)|^2\, dx = A^2 \int_{-a}^{+a} \left(1 - 2\frac{x^2}{a^2} + \frac{x^4}{a^4}\right) dx$$

$$= A^2\left(2a - \frac{4a}{3} + \frac{2a}{5}\right) = A^2 \frac{16a}{15} = 1. \tag{3.3.19}$$

This gives the constant A as

$$A = \frac{\sqrt{15}}{4\sqrt{a}}. \tag{3.3.20}$$

The general solution at $t > 0$ is given by the linear combination

$$\psi(x,t) = \sum_n c_n \phi_n(x)\, e^{-\frac{i}{\hbar}E_n t}, \tag{3.3.21}$$

where $\phi_n(x)$ are the normalized time independent solutions of the corresponding TISE. The coefficients, c_n, are to be calculated for a given $\psi(x,0)$. Since the potential is symmetric with respect to the centre of the well (at $x = 0$), the solutions are

$$\phi_n(x) = \begin{cases} \sqrt{\frac{2}{a}} \cos\left(\frac{n\pi x}{a}\right), & n = 1,3,5,... \\ \sqrt{\frac{2}{a}} \sin\left(\frac{n\pi x}{a}\right), & n = 2,4,6,... \end{cases} \tag{3.3.22}$$

For odd n, the coefficients c_n are

$$c_{1n} = A\sqrt{\frac{2}{a}}\int_{-\frac{a}{2}}^{+\frac{a}{2}} \cos\left(\frac{n\pi x}{a}\right) dx - \frac{A}{a^2}\sqrt{\frac{2}{a}}\int_{-\frac{a}{2}}^{+\frac{a}{2}} x^2 \cos\left(\frac{n\pi x}{a}\right) dx \equiv I_1 + I_2. \quad (3.3.23)$$

Taking the above integrals and using the expression for A, we obtain

$$I_1 = \frac{\sqrt{30}}{2n\pi} \sin\left(\frac{n\pi}{2}\right), \quad (3.3.24)$$

and

$$I_2 = \sqrt{30}\left(\frac{1}{8n\pi} - \frac{1}{n^3\pi^3}\right) \sin\left(\frac{n\pi}{2}\right). \quad (3.3.25)$$

For even n, the coefficients c_n are

$$c_{2n} = \left[A\sqrt{\frac{2}{a}}\int_{-\frac{a}{2}}^{+\frac{a}{2}} \sin\left(\frac{n\pi x}{2a}\right) dx - \frac{A}{a^2}\sqrt{\frac{2}{a}}\int_{-\frac{a}{2}}^{+\frac{a}{2}} x^2 \sin\left(\frac{n\pi x}{2a}\right) dx\right]. \quad (3.3.26)$$

In this case, both the integrals are zero because the integrands are odd functions of x. Therefore, the expansion coefficients are given by

$$c_n = \sqrt{30}\left(\frac{5}{8n\pi} - \frac{1}{n^3\pi^3}\right) \sin\left(\frac{n\pi}{2}\right), \qquad n = 1,3,5,\ldots \quad (3.3.27)$$

As a consequence, the wave function at $t > 0$ is given by the following linear combination

$$\psi(x,t) = \sum_n \sqrt{30}\left(\frac{5}{8n\pi} - \frac{1}{n^3\pi^3}\right) \sin\left(\frac{n\pi}{2}\right) \phi_n(x)\, e^{-i(n^2\pi^2\hbar/8ma^2)t}, \qquad n = 1,3,5,\ldots \quad (3.3.28)$$

3.4 Discontinuous Potentials and the Differentiability of the Wave Function

We have seen that any physically acceptable solution of the TISE must satisfy the standard conditions. Since the TISE involves an external potential, $V(x)$, the form of its solutions and the fulfillment of the standard conditions depend on the properties of the function $V(x)$. If $V(x)$ is finite and continuous everywhere, we can expect the solutions of the TISE to be finite, continuous and differentiable. However, if the potential has points of discontinuity, then we have to examine whether the wave function will be continuous and differentiable at these points or not. It is evident from the physical interpretation of the wave function that it has to be continuous everywhere irrespective of the fact whether or not the potential has discontinuity. However, the differentiability of the wave function is not guaranteed in advance and hence, must be examined. This is also important because of the fact that the general solution of TISE contains two integration constants to be determined by the boundary conditions and one of the boundary conditions involves the first-order derivative of the wave function.

(a) The potential has a finite jump (discontinuity), say, at $x = 0$:

$$V(x) = \begin{cases} 0 & \text{for} \quad x < 0 \\ V_0 > 0 & \text{for} \quad x \geq 0. \end{cases} \tag{3.4.1}$$

The wave function has to be continuous across $x = 0$. To check the continuity of the first derivative, we first replace the potential $V(x)$ by a smoothened potential $V_\varepsilon(x)$ in the interval $x \in [-\varepsilon, +\varepsilon]$ such that

$$\lim_{\varepsilon \to 0} V_\varepsilon(x) = V_0. \tag{3.4.2}$$

Here $\varepsilon \ll 1$ is an infinitesimal positive parameter. Integrating the time-independent Schrödinger equation in this interval over x, we obtain

$$\left(\frac{d\phi}{dx}\right)_{+\varepsilon} - \left(\frac{d\phi}{dx}\right)_{-\varepsilon} = -\frac{2mE}{\hbar^2}\int_{-\varepsilon}^{+\varepsilon} \phi(x)\,dx + \frac{2mE}{\hbar^2}\int_{-\varepsilon}^{+\varepsilon} V(x)\phi(x)\,dx. \tag{3.4.3}$$

If we take the limit $\varepsilon \to 0$ in (3.4.3), we get

$$\Delta\left(\frac{d\phi}{dx}\right) = -\frac{2mE}{\hbar^2}\lim_{\varepsilon \to 0}\int_{-\varepsilon}^{+\varepsilon} \phi(x)\,dx + \frac{2mE}{\hbar^2}\lim_{\varepsilon \to 0}\int_{-\varepsilon}^{+\varepsilon} V(x)\phi(x)\,dx. \tag{3.4.4}$$

The first term on the right-hand side of (3.4.4) is zero because $\phi(x)$ is continuous across $x = 0$ and hence, the integral goes to zero as ε becomes zero. The second term is also zero because

$$\lim_{\varepsilon \to 0}\int_{-\varepsilon}^{+\varepsilon} V(x)\phi(x)\,dx. = V_0 \lim_{\varepsilon \to 0}\int_{-\varepsilon}^{+\varepsilon} \phi(x)\,dx = 0. \tag{3.4.5}$$

As a result, we arrive at

$$\left(\frac{d\phi}{dx}\right)_{+\varepsilon} = \left(\frac{d\phi}{dx}\right)_{-\varepsilon}. \tag{3.4.6}$$

Thus, if the potential has a finite jump at a point, the wave function and its first derivative are continuous at the point of discontinuity. That is, the wave function is differentiable at the points of finite discontinuity of the potential.

(b) *The potential $V(x)$ is infinite in a region:* In this case, the particle cannot penetrate through the infinite barrier and the probability of finding the particle inside the barrier is zero. Therefore, the wave function must vanish everywhere in the region of infinite potential.

(c) *The potential becomes infinite at a point (that is, has a singularity at a point)*: We can model this situation by assuming $V(x) = -\alpha\,\delta(x-x_0)$, where α is a positive constant. The wave function will be continuous at $x = x_0$. In order to verify the continuity of the first derivative, we once again integrate the corresponding TISE in the vicinity of the point $x = x_0$. We get

$$\left(\frac{d\phi}{dx}\right)_{+\varepsilon} - \left(\frac{d\phi}{dx}\right)_{-\varepsilon} = -\frac{2m\alpha}{\hbar^2}\int_{-\varepsilon}^{+\varepsilon} \delta(x-x_0)\phi(x)\,dx = -\frac{2m\alpha}{\hbar^2}\phi(x_0). \qquad (3.4.7)$$

Thus, the first derivative of the wave function is not continuous across the point of singularity. Instead, it has a finite jump of $(-2m\alpha/\hbar^2)\phi(x_0)$ at $x = x_0$.

Example 3.4.1: A free particle of mass, m, and total energy, E, is incident from $x \to -\infty$ on a potential step given by

$$V(x) = \begin{cases} 0 & \text{for} \quad x < 0 \\ V_0 > 0 & \text{for} \quad x \geq 0, \end{cases} \qquad (3.4.8)$$

where $V_0 > E$ is a positive constant. Solve the corresponding TISE, apply the appropriate boundary conditions and determine the wave function.

Solution: The given potential divides the entire region $-\infty < x < +\infty$ into two halves: $x < 0$, where the potential is zero and $x > 0$, where the potential has a constant value V_0. We will call them Region 1 and Region 2, respectively. The corresponding stationary state wave functions in these regions are denoted as $\psi_1(x,t) = \phi_1(x)e^{-iEt/\hbar}$ and $\psi_2(x,t) = \phi_2(x)e^{-iEt/\hbar}$, respectively. In Region 1, the TISE

$$\frac{d^2\phi}{dx^2} + \frac{2mE}{\hbar^2}\phi = 0 \qquad (3.4.9)$$

has the following general solution

$$\phi(x) = Ae^{ik_1x} + Be^{-ik_1x}, \qquad (3.4.10)$$

where $k_1^2 = 2mE/\hbar^2$ and A and B are arbitrary constants. As a result,

$$\psi_1(x,t) = Ae^{i\left(kx - i\frac{E}{\hbar}t\right)} + Be^{-i\left(kx + i\frac{E}{\hbar}t\right)}. \qquad (3.4.11)$$

The first term of this solution represents the incident particle moving along the positive x-axis, while the second term represents the particle reflected by the potential barrier and moving along the negative x-axis.

In Region 2, the TISE reads

$$\frac{d^2\phi}{dx^2} - \frac{2m(V_0 - E)}{\hbar^2}\phi = 0. \tag{3.4.12}$$

Its general solution is

$$\phi(x) = Ce^{-k_2 x} + De^{k_2 x}, \tag{3.4.13}$$

where $k_2^2 = 2m(V_0 - E)/\hbar^2$ and C and D are arbitrary constants. Since the wave function must tend to zero at spatial infinities ($x \to \pm\infty$), we must put $D = 0$, otherwise the solution will diverge. Therefore, the stationary state solution in the second region can be written as

$$\psi_2(x,t) = Ce^{-k_2 x - i(E/\hbar)t}. \tag{3.4.14}$$

Since the potential has only a finite jump at $x = 0$, both the wave functions (ϕ_1 and ϕ_2) and their first-order derivatives must be continuous at $x = 0$. We thus have

$$A + B = C, \tag{3.4.15}$$

$$ik_1(A - B) = -k_2 C. \tag{3.4.16}$$

There is a small problem here because we have only two equations but three constants to be determined. Let us first determine the coefficients B and C in terms of the constant A and then see what we can do about A. From the equations (3.4.15) and (3.4.16), we have

$$1 + \frac{B}{A} = \frac{C}{A}, \tag{3.4.17}$$

$$1 - \frac{B}{A} = \frac{ik_2}{k_1}\frac{C}{A}. \tag{3.4.18}$$

Solving these equations for C/A, we get

$$C = \frac{2k_1}{k_1 + ik_2}A. \tag{3.4.19}$$

If we subtract (3.4.18) from (3.4.17) and use (3.4.19), we obtain

$$B = \frac{k_1 - ik_2}{k_1 + ik_2}A. \tag{3.4.20}$$

Now, without any loss of generality, we might assume that the incident particle's wave function (a wave packet) is normalized in such a way that $A = 1$. Then the required wave function is

$$\phi(x) = \begin{cases} e^{i(k_1x - i\omega t)} + \frac{k_1 - ik_2}{k_1 + ik_2} e^{-i(kx + i\omega t)} & x < 0, \\ \frac{2k_1}{k_1 + ik_2} e^{-(k_2 x + i\omega t)} & x > 0, \end{cases} \tag{3.4.21}$$

where $\omega = E/\hbar$.

Example 3.4.2: A particle of mass m and total energy, $-E$ ($E > 0$), is subject to the potential given by

$$V(x) = -\alpha\delta(x),$$

here α is a positive constant and $\delta(x)$ is the Dirac delta function. Solve the Schrödinger equation for the bound states and find the energy levels and the corresponding normalized wave functions. How many bound states can the particle have in such a potential?

Solution: Let us first solve the time-independent Schrödinger equation

$$-\frac{\hbar^2}{2m}\frac{d^2\phi}{dx^2} + V(x)\phi = E\phi \tag{3.4.22}$$

for the wave function $\phi(x)$. For $x < 0$ and $x > 0$, $V(x) = 0$ and we have

$$\frac{d^2\phi}{dx^2} - \frac{2m|E|}{\hbar^2}\phi = 0. \tag{3.4.23}$$

Since the standard conditions require the wave function to vanish for $x \to \pm\infty$, we have

$$\phi(x) = \begin{cases} Ae^{kx} & \text{for} \quad x < 0 \\ Be^{-kx} & \text{for} \quad x > 0, \end{cases} \tag{3.4.24}$$

where $k = \sqrt{2m|E|}/\hbar$ and A and B are real but arbitrary constants. The continuity of $\phi(x)$ at $x = 0$ yields

$$A = B. \tag{3.4.25}$$

The potential is infinite at $x = 0$. Therefore, as discussed earlier, the first derivative of the wave function will be discontinuous and we shall have

$$\left(\frac{d\phi}{dx}\right)_{+\varepsilon} - \left(\frac{d\phi}{dx}\right)_{-\varepsilon} = -\frac{2m\alpha}{\hbar^2}\int_{-\varepsilon}^{+\varepsilon} \delta(x)\phi(x)\,dx = -\frac{2m\alpha}{\hbar^2}\phi(0). \tag{3.4.26}$$

If we take the limit $\varepsilon \to 0$ and put $A = B$, we obtain

$$-2kA = -\frac{2m\alpha}{\hbar^2}\phi(0) = -\frac{2m\alpha}{\hbar^2}A \quad \Rightarrow \quad k = \frac{m\alpha}{\hbar^2}. \tag{3.4.27}$$

We thus see that there is only one bound state for the particle in this case whose energy is

$$E = -\frac{m\alpha}{2\hbar^2}. \tag{3.4.28}$$

The normalization of the wave function reads

$$\int_{-\infty}^{+\infty} |\psi(x)|^2 dx = A^2 \int_{-\infty}^{0} e^{2kx} dx + A^2 \int_{0}^{+\infty} e^{-2kx} dx = \frac{A^2}{k} = 1. \tag{3.4.29}$$

Hence, $A = \sqrt{k}$. The normalized wave function is thus given by

$$\phi(x) = \begin{cases} \sqrt{k}e^{kx} & \text{for} \quad x < 0, \\ \sqrt{k}e^{-kx} & \text{for} \quad x > 0. \end{cases} \qquad \text{or,} \quad \phi(x) = \sqrt{\frac{m\alpha}{\hbar^2}}\, e^{-\frac{m\alpha}{\hbar^2}|x|}. \tag{3.4.30}$$

3.5 Conservation of Probability and the Continuity Equation

Continuity equation in quantum mechanics

In Chapter 2, we talked about the statistical interpretation of the wave function in which the quantity $|\psi(x,t)|^2$ represents the probability density at a given instant of time. The argument that at any t, the particle is definitely somewhere in the universe led to the normalization condition for the wave function. Later, we also postulated the time-evolution of the wave function to be governed by the time-dependent Schrödinger equation. Therefore, it is natural to check whether the statistical interpretation of the wave function is consistent with its time-evolution. In other words, we want to answer the following question: If the wave function is normalized at $t = 0$, does it remain normalized at any $t > 0$? It turns out that it does. This is what we are going to show.

Consider, for simplicity, one-dimensional Schrödinger equations both for the wave function, ψ, and its complex conjugate function ψ^*. We have

$$i\hbar\frac{\partial \psi}{\partial t} = -\frac{\hbar^2}{2m}\frac{\partial^2 \psi}{\partial x^2} + V(x)\psi, \tag{3.5.1}$$

$$-i\hbar\frac{\partial \psi^*}{\partial t} = -\frac{\hbar^2}{2m}\frac{\partial^2 \psi^*}{\partial x^2} + V(x)\psi^*. \tag{3.5.2}$$

If we multiply the equation (3.5.1) by ψ^*, the equation (3.5.2) by ψ and subtract the second from the first, the result can be written in the form of a continuity equation:

$$\frac{\partial \rho(x,t)}{\partial t} + \frac{\partial j(x,t)}{\partial x} = 0, \tag{3.5.3}$$

where $\rho(x,t) = \psi^*(x,t)\psi(x,t) = |\psi(x,t)|^2$ is the probability density and $\vec{j}(x,t) = j(x,t)\,\hat{\imath}$ is the probability current density given by the following expression:

$$\vec{j}(x,t) = \hat{\imath}\frac{\hbar}{2im}\left(\psi^*(x,t)\frac{\partial \psi(x,t)}{\partial x} - \frac{\partial \psi^*(x,t)}{\partial x}\psi(x)\right). \tag{3.5.4}$$

Equation (3.5.3) represents the local conservation of quantum mechanical probability the same way as the continuity equation in electrodynamics represents the local charge conservation. If we integrate the continuity equation over x from $-\infty$ to $+\infty$, we get

$$\begin{aligned} \frac{d}{dt}\left(\int_{-\infty}^{+\infty} \rho\, dx\right) &= -\frac{\hbar}{2im}\int_{-\infty}^{+\infty}\frac{\partial}{\partial x}\left(\psi^*(x,t)\frac{\partial \psi(x,t)}{\partial x} - \frac{\partial \psi^*(x,t)}{\partial x}\psi(x)\right) dx, \\ &= -\frac{\hbar}{2im}\left[\psi^*(x,t)\frac{\partial \psi(x,t)}{\partial x} - \frac{\partial \psi^*(x,t)}{\partial x}\psi(x)\right]_{-\infty}^{+\infty}. \end{aligned} \tag{3.5.5}$$

Due to the standard conditions, the wave function and its first derivative must vanish at spatial infinities. Therefore, the right-hand side of (3.5.5) becomes zero, and we get

$$\frac{d}{dt}\left(\int_{-\infty}^{+\infty} \rho\, dx\right) = 0 \Rightarrow \int_{-\infty}^{+\infty} \psi^*\psi\, dx = C, \tag{3.5.6}$$

where C is the integration constant to be determined by the initial condition. If the wave function is normalized at $t = 0$, we have $C = 1$. In other words, the total probability of finding the particle at some point in space is independent of time and the total probability is conserved even if the wave function is changing in time according to the time-dependent Schrödinger equation. In addition, if we have normalized the wave function at $t = 0$, it remains normalized for all times to come.

In the three-dimensional case, the probability density, ρ, and the probability current density, $\vec{j}$, are given by

$$\rho(\vec{r},t) = \psi^*\psi \tag{3.5.7}$$

$$\vec{j}(\vec{r},t) = \frac{\hbar}{2im}\left[\psi^*\left(\vec{\nabla}\psi\right) - \left(\vec{\nabla}\psi^*\right)\psi\right], \tag{3.5.8}$$

where $\psi = \psi(\vec{r},t)$. Consequently, the continuity equation is written in the form

$$\frac{\partial \rho(\vec{r},t)}{\partial t} + \vec{\nabla}\cdot\vec{j}(\vec{r},t) = 0. \tag{3.5.9}$$

If we integrate the equation (3.5.9) over all space and use the divergence theorem, we obtain

$$\frac{d}{dt}\left(\int_{-\infty}^{+\infty}\psi^*\psi\, d^3r\right) = -\frac{\hbar}{2im}\oint_{S_\infty}\left[\psi^*(\vec{\nabla}\psi) - (\vec{\nabla}\psi^*)\psi\right]\cdot d\vec{s}. \tag{3.5.10}$$

The surface integral on the right-hand side vanishes because the wave function and its first-order derivatives must vanish at an infinitely remote surface. The result is once again the conservation of total probability:

$$\frac{d}{dt}\left(\int_{-\infty}^{+\infty}\psi^*\psi\, d^3r\right) = 0, \tag{3.5.11}$$

that is, the total probability of finding a particle somewhere in space is independent of time. Once again, if we have normalized the wave function at $t = 0$, it remains normalized for all times to come.

The physical interpretation of the continuity equation: Note that (3.5.9) can be interpreted in a more physically meaningful manner if we regard $\rho = |\psi|^2$ as the mean particle density and $\vec{j}$ as the mean particle flux density defined as the average number of particles per unit time passing through a unit area held perpendicular to the direction of motion of the particles. Now, if we multiply ρ by the mass m of a particle then $\rho_m = m|\psi|^2$ will be the average mass density and correspondingly $\vec{j}$ will represent the average current density of matter. Then the equation (3.5.9) can be thought of as a manifestation of the fact that the change in time of the average mass density in some infinitesimal volume, $d\tau$ (Figure 3.5(a)), is due to the inflow or outflow of this mass through the surface enclosing it. Equation (3.5.11) then says that the average number of particles inside $d\tau$ remains constant. In other words, the average number of particles per unit time entering the surface, enclosing $d\tau$, is equal to the average number of particles per unit time leaving this boundary surface.

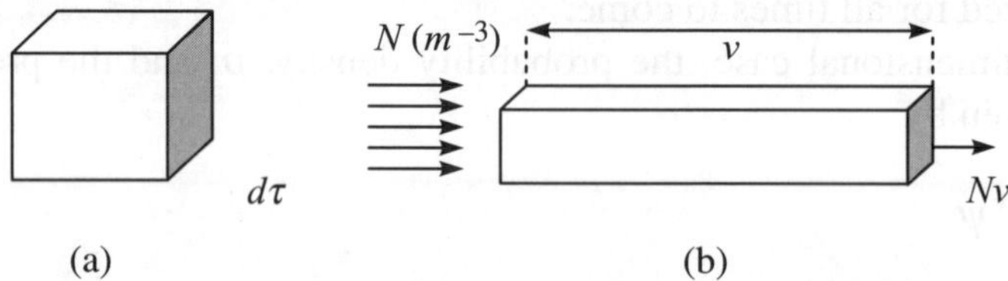

Figure 3.5 (a) Infinitesimal volume, (b) Cylinder of unit cross-section and length v.

For instance, suppose we have a stream of particles (all propagating along the x direction with a velocity $\vec{v}$) with density N. Then the particle current density will be given by $\vec{j} = Nv\hat{i}$. Consider now a cylindrical volume of unit cross-sectional area and length v lying along the direction of motion of the particles (Figure 3.5(b)). Since the length of the cylinder is v, the average number of particles entering the cylinder per second through the rear cross-sectional area will be equal to Nv and this will in turn be equal to the number of particles leaving the cylinder per second through the front cross-sectional area, so that the number of particles inside the cylinder at any instant of time is constant. The continuity equation (3.5.9) or (3.5.3) can thus be looked upon as expressing the local conservation of the number of particles.

Example 3.5.1: The wave function is given to be

$$\psi(x,t) = A\,e^{i(kx-\omega t)} + B\,e^{-i(kx+\omega t)},$$

where A and B are arbitrary complex constants. Calculate the probability density and the probability current density. Show the validity of the continuity equation. Give the physical interpretation of the obtained results.

Solution: The probability density $\rho(x,t)$ is given by

$$\begin{aligned}\rho(x,t) &= |\psi(x,t)|^2 = \left(A^*\,e^{-i(kx-\omega t)} + B^*\,e^{i(kx+\omega t)}\right)\left(A\,e^{i(kx-\omega t)} + B\,e^{-i(kx+\omega t)}\right) \\ &= |A|^2 + A^*\,B\,e^{-2ikx} + |B|^2 + B^*\,A\,e^{2ikx}. \end{aligned} \tag{3.5.12}$$

Also,

$$\begin{aligned}\psi^*\frac{\partial\psi}{\partial x} &= ik\left(A^*\,e^{-i(kx-\omega t)} + B^*\,e^{i(kx-\omega t)}\right)\left(A\,e^{i(kx-\omega t)} - B\,e^{-i(kx-\omega t)}\right) \\ &= ik\left(|A|^2 - |B|^2 - A^*\,B\,e^{-2i(kx-\omega t)} + B^*\,A\,e^{2i(kx-\omega t)}\right). \end{aligned} \tag{3.5.13}$$

Taking the complex conjugate of (3.5.13), we obtain

$$\begin{aligned}\psi\frac{\partial\psi^*}{\partial x} &= -ik\left(A\,e^{i(kx-\omega t)} + B\,e^{-i(kx-\omega t)}\right)\left(A^*\,e^{-i(kx-\omega t)} - B^*\,e^{i(kx-\omega t)}\right) \\ &= -ik\left(|A|^2 - |B|^2 + A^*\,B\,e^{-2i(kx-\omega t)} - B^*\,A\,e^{2i(kx-\omega t)}\right). \end{aligned} \tag{3.5.14}$$

From (3.5.13) and (3.5.14), we calculate the current density as

$$j_x = \frac{\hbar}{2mi}\left[\psi^*(x,t)\frac{\partial\psi}{\partial x} - \frac{\partial\psi^*}{\partial x}\psi(x,t)\right] = \frac{\hbar k}{m}\left(|A|^2 - |B|^2\right). \tag{3.5.15}$$

We see that the probability density is time independent, while the probability current density is time as well as space independent. Therefore, the continuity equation is satisfied identically. Physically, it shows that the number of particles passing through a unit area per second along the positive x direction is numerically equal to the number of particles crossing the same unit area per second along the negative x direction.

Example 3.5.2: At $t = 0$, a particle of mass m, free to move inside an infinite potential well with walls at $x = 0$ and $x = a$, is in a state that is a linear superposition of the ground state and the first excited state

$$\psi(x,0) = \frac{1}{\sqrt{2}}\left[\phi_1(x) + \phi_2(x)\right] = \frac{1}{\sqrt{a}}\left[\sin\left(\frac{\pi x}{a}\right) + \sin\left(\frac{2\pi x}{a}\right)\right],$$

Find the wave function at any $t > 0$. Check whether the continuity equation holds good for this state or not.

Solution: The wave function of the particle at $t > 0$ will be

$$\psi(x,t) = \frac{1}{\sqrt{a}}\left[\sin\left(\frac{\pi x}{a}\right)e^{-i\frac{E_1}{\hbar}t} + \sin\left(\frac{2\pi x}{a}\right)e^{-i\frac{E_2}{\hbar}t}\right] \tag{3.5.16}$$

The probability density is calculated to be

$$\rho(x,t) = |\psi(x,t)|^2 = \frac{1}{a}\left[\sin^2\left(\frac{\pi x}{a}\right) + \sin^2\left(\frac{2\pi x}{a}\right)\right].$$

$$+\frac{1}{a}\sin\left(\frac{\pi x}{a}\right)\sin\left(\frac{2\pi x}{a}\right)\left\{e^{i\frac{(E_1-E_2)}{\hbar}t} + e^{-i\frac{(E_1-E_2)}{\hbar}t}\right\}. \tag{3.5.17}$$

We simplify it further, by dividing the last term by 2 and using the Euler formula, to obtain

$$\rho(x,t) = \frac{1}{a}\left[\sin^2\left(\frac{\pi x}{a}\right) + \sin^2\left(\frac{2\pi x}{a}\right)\right]$$

$$+\frac{2}{a}\sin\left(\frac{\pi x}{a}\right)\sin\left(\frac{2\pi x}{a}\right)\cos\left[\frac{(E_1 - E_2)}{\hbar}t\right]. \tag{3.5.18}$$

Let us now calculate the probability current density. We have

$$\frac{\partial\psi}{\partial x} = \frac{1}{\sqrt{a}}\left[\frac{\pi}{a}\cos\left(\frac{\pi x}{a}\right)e^{-i\frac{E_1}{\hbar}t} + \frac{2\pi}{a}\cos\left(\frac{2\pi x}{a}\right)e^{-i\frac{E_2}{\hbar}t}\right], \tag{3.5.19}$$

$$\psi^*(x,t)\frac{\partial\psi}{\partial x} = \frac{\pi}{a^2}\sin\left(\frac{\pi x}{a}\right)\cos\left(\frac{\pi x}{a}\right) + \frac{2\pi}{a^2}\sin\left(\frac{\pi x}{a}\right)\cos\left(\frac{2\pi x}{a}\right)e^{i\frac{(E_1-E_2)}{\hbar}t}$$

$$+\frac{\pi}{a^2}\sin\left(\frac{2\pi x}{a}\right)\cos\left(\frac{\pi x}{a}\right)e^{-i\frac{(E_1-E_2)}{\hbar}t} + \frac{2\pi}{a^2}\sin\left(\frac{2\pi x}{a}\right)\cos\left(\frac{2\pi x}{a}\right). \tag{3.5.20}$$

Taking the complex conjugate of the last equation, we get

$$\frac{\partial \psi^*}{\partial x}\psi(x,t) = \frac{\pi}{a^2}\sin\left(\frac{\pi x}{a}\right)\cos\left(\frac{\pi x}{a}\right) + \frac{2\pi}{a^2}\sin\left(\frac{\pi x}{a}\right)\cos\left(\frac{2\pi x}{a}\right)e^{-i\frac{(E_1-E_2)}{\hbar}t}$$

$$+\frac{\pi}{a^2}\sin\left(\frac{2\pi x}{a}\right)\cos\left(\frac{\pi x}{a}\right)e^{i\frac{(E_1-E_2)}{\hbar}t} + \frac{2\pi}{a^2}\sin\left(\frac{2\pi x}{a}\right)\cos\left(\frac{2\pi x}{a}\right). \tag{3.5.21}$$

If we subtract (3.5.21) from (3.5.20), we obtain

$$\psi^*(x,t)\frac{\partial \psi}{\partial x} - \frac{\partial \psi^*}{\partial x}\psi(x,t) = \frac{2\pi}{a^2}\sin\left(\frac{\pi x}{a}\right)\cos\left(\frac{2\pi x}{a}\right)\left[e^{i\frac{(E_1-E_2)}{\hbar}t} - e^{-i\frac{(E_1-E_2)}{\hbar}t}\right]$$

$$-\frac{\pi}{a^2}\sin\left(\frac{2\pi x}{a}\right)\cos\left(\frac{\pi x}{a}\right)\left[e^{i\frac{(E_1-E_2)}{\hbar}t} - e^{-i\frac{(E_1-E_2)}{\hbar}t}\right]. \tag{3.5.22}$$

The probability current density j_x is therefore given by

$$\begin{aligned} j_x &= \frac{\hbar}{2mi}\left[\psi^*(x,t)\frac{\partial \psi}{\partial x} - \frac{\partial \psi^*}{\partial x}\psi(x,t)\right] \\ &= \frac{2\pi\hbar}{ma^2}\sin\left(\frac{\pi x}{a}\right)\cos\left(\frac{2\pi x}{a}\right)\sin\left[\frac{(E_1-E_2)}{\hbar}t\right] \\ &\quad - \frac{\pi\hbar}{ma^2}\sin\left(\frac{2\pi x}{a}\right)\cos\left(\frac{\pi x}{a}\right)\sin\left[\frac{(E_1-E_2)}{\hbar}t\right]. \end{aligned} \tag{3.5.23}$$

The time derivative of the probability density is

$$\frac{\partial \rho(x,t)}{\partial t} = -\frac{2}{a}\frac{(E_1-E_2)}{\hbar}\sin\left(\frac{\pi x}{a}\right)\sin\left(\frac{2\pi x}{a}\right)\sin\left[\frac{(E_1-E_2)}{\hbar}t\right]. \tag{3.5.24}$$

Since

$$\frac{(E_1-E_2)}{\hbar} = \frac{1}{\hbar}\frac{(\pi^2\hbar^2 - 4\pi^2\hbar^2)}{2ma^2} = -\frac{3\pi^2\hbar}{2ma^2}, \tag{3.5.25}$$

we get from (3.5.24)

$$\frac{\partial \rho(x,t)}{\partial t} = \frac{3\pi^2\hbar}{ma^3}\sin\left(\frac{\pi x}{a}\right)\sin\left(\frac{2\pi x}{a}\right)\sin\left[\frac{(E_1-E_2)}{\hbar}t\right]. \tag{3.5.26}$$

The x derivative of j_x can be calculated to be

$$\frac{\partial j_x}{\partial x} = \left[\frac{2\pi^2\hbar}{ma^3}\cos\left(\frac{\pi x}{a}\right)\cos\left(\frac{2\pi x}{a}\right) - \frac{4\pi^2\hbar}{ma^3}\sin\left(\frac{\pi x}{a}\right)\sin\left(\frac{2\pi x}{a}\right)\right]\sin\left[\frac{(E_1 - E_2)}{\hbar}t\right]$$

$$-\left[\frac{2\pi^2\hbar}{ma^3}\cos\left(\frac{2\pi x}{a}\right)\cos\left(\frac{\pi x}{a}\right) - \frac{\pi^2\hbar}{ma^3}\sin\left(\frac{2\pi x}{a}\right)\sin\left(\frac{\pi x}{a}\right)\right]\sin\left[\frac{(E_1 - E_2)}{\hbar}t\right],$$

$$\frac{\partial j_x}{\partial x} = -\frac{3\pi^2\hbar}{ma^3}\sin\left(\frac{\pi x}{a}\right)\sin\left(\frac{2\pi x}{a}\right)\sin\left[\frac{(E_1 - E_2)}{\hbar}t\right]. \tag{3.5.27}$$

From (3.5.26) and (3.5.27), we conclude that

$$\frac{\partial \rho(x,t)}{\partial t} + \frac{\partial j_x}{\partial x} = 0. \tag{3.5.28}$$

Hence, the continuity equation is indeed satisfied.

3.6 Symmetric Potential and Even and Odd Parity Solutions

In Chapter 2, we discussed about the parity operator and proved that it is hermitian as well as unitary. We also saw that its eigenfunctions had definite parity and they formed a complete set. In the following we shall study the properties of the solutions of the Schrödinger equation with symmetric potentials, i.e., with potentials that are invariant under parity transformation.

Consider the Schrödinger equation with a potential that is symmetric with respect to space inversion: $V(-x) = V(x)$. Clearly, when $V(x)$ is symmetric, the corresponding Hamiltonian,

$$\hat{H} = -\frac{\hbar^2}{2m}\frac{\partial^2 \psi}{\partial x^2} + V(x), \tag{3.6.1}$$

is also symmetric. In other words, $\hat{H}$ is an even operator. We have seen in Chapter 2 that even operators commute with the parity operator $\hat{\mathscr{P}}$. Therefore, for symmetric potentials $[\hat{\mathscr{P}}, \hat{H}] = 0$ and the Hamiltonian and the parity operator can have a common set of eigenfunctions.

Theorem 3.6.1: *The bound state wave functions of a particle moving in a one-dimensional symmetric potential have definite parity, that is, they are either even or odd.*

Proof: Although this theorem follows immediately from the fact that the parity operator, $\hat{\mathscr{P}}$, and the Hamiltonian, $\hat{H}$, are compatible and that the eigenfunctions of the parity operator has a definite parity, it is useful and instructive to prove the theorem in a straightforward way on the basis of the TISE and its solutions. It will, hopefully, make the point and the content of the theorem more lucid.

Consider now the TISE for the symmetric potential:

$$\left[-\frac{\hbar^2}{2m}\frac{d^2}{dx^2}+V(x)\right]\phi(x)=E\phi(x). \tag{3.6.2}$$

Let us now perform the spatial inversion by replacing x with $-x$. Then, $\hat{\mathscr{P}}\phi(x)\to\phi(-x)$ and $\hat{\mathscr{P}}V(x)\to V(-x)$. Since $V(-x)=V(x)$, the Hamiltonian commutes with the parity operator and we get

$$\left[-\frac{\hbar^2}{2m}\frac{d^2}{dx^2}+V(x)\right]\phi(-x)=E\phi(-x). \tag{3.6.3}$$

Thus, we see that the stationary Schrödinger equation (3.6.3) for the symmetric potential is satisfied by $\phi_1(-x)=\phi_1(x)$ as well as $\phi_2(-x)=-\phi_2(x)$. The former, denoted as $\phi^s(x)$, is called the *symmetric wave function* and has *even parity*, while the latter, denoted as $\phi^a(x)$, is called the *anti-symmetric wave function* and has *odd parity*.

Let us now recollect our earlier result that, in one spatial dimension, the bound state energy spectrum is discrete and non-degenerate. In view of this result, we conclude that the wave functions of a particle, moving in a one-dimensional symmetric potential, have a definite parity (either even or odd) . *The theorem is proved.*

Note that, if the spectrum of the Hamiltonian corresponding to a symmetric potential is degenerate, the energy eigenstates do not have definite parity.

Example 3.6.1 Solve the TISE for the potential

$$V(x)=\begin{cases}0, & \text{for } -a<x<a,\\ \infty, & \text{for } x\leq -a, x\geq a.\end{cases} \tag{3.6.4}$$

Find the energy eigenfunctions and the corresponding energy eigenvalues.

Solution: The given problem is once again the problem of a particle trapped inside an infinite square well potential. However, unlike the earlier one, the given well is symmetric with respect to the center at $x=0$: $V(-x)=V(x)$. Therefore, according to Theorem 3.6.1, the solutions of the corresponding TISE are either symmetric, $\phi(-x)=\phi(x)$, or anti-symmetric, $\phi(-x)=-\phi(x)$. In the former case the solutions are said to have *even parity*, while in the latter they are said to have *odd parity*.

As discussed earlier, the solutions in the regions on both sides of the well, that is, for $x<-a$ and $x>a$, must be identically equal to zero. Inside the well, the TISE has two linearly independent solutions $\phi^s(x)=A\cos(kx)$, which is symmetric, and $\phi^a(x)=B\sin(kx)$, which is anti-symmetric, where A and B are arbitrary constants and $k^2=2mE/\hbar^2$. In view of the above mentioned properties of the solutions, we treat the two cases separately.

Symmetric solutions: $\phi^s(x) = A\cos(kx)$. The continuity condition at the boundaries $x = \pm a$ leads to

$$A\cos(ka) = 0, \quad \Rightarrow \cos(ka) = 0, \quad \Rightarrow k_n a = \frac{n\pi}{2}, n = 1,3,5,... \tag{3.6.5}$$

The corresponding energies are: $E_n^s = n^2\pi^2\hbar^2/8ma^2$. The normalization of the wave function yields: $\phi_n^s(x) = \sqrt{1/a}\cos(n\pi x/2a)$.

Anti-symmetric solutions: $\phi^a(x) = B\sin(kx)$. The continuity condition at the boundaries $x = \pm a$ leads to

$$B\sin(ka) = 0, \quad \Rightarrow \sin(ka) = 0, \quad \Rightarrow k_n a = \frac{n\pi}{2}, n = 2,4,6,... \tag{3.6.6}$$

The corresponding energies are: $E_n^s = n^2\pi^2\hbar^2/8ma^2$. The normalization of the wave function yields: $\phi_n^a(x) = \sqrt{1/a}\sin(n\pi x/2a)$.

The two wave functions can be combined together and we have

$$\phi_n(x) = \sqrt{\frac{1}{a}}\sin\left[\frac{n\pi}{2a}(x+a)\right] = \begin{cases} (-1)^{\frac{n-1}{2}}\sqrt{\frac{1}{a}}\cos\left(\frac{n\pi x}{2a}\right) & \text{n=1,3,5,...} \\ (-1)^{\frac{n}{2}}\sqrt{\frac{1}{a}}\sin\left(\frac{n\pi x}{2a}\right) & \text{n=2,4,6,...} \end{cases} \tag{3.6.7}$$

The corresponding energies are

$$E_n = \frac{n^2\pi^2\hbar^2}{8ma^2}, n = 1,2,3,... \tag{3.6.8}$$

Note that, since the energy is inversely proportional to the square of the width of the potential, the bound state energies here differ from the corresponding energies of the asymmetric well by a factor of $(1/4)$.

If we take the total width of the symmetric potential to be a $(-a/2 < x < a/2)$, we shall get the corresponding wave functions and the energy levels as

$$\phi_n(x) = \begin{cases} \sqrt{\frac{2}{a}}\cos\left(\frac{n\pi x}{2a}\right) & \text{n=1,3,5,...} \\ \sqrt{\frac{2}{a}}\sin\left(\frac{n\pi x}{2a}\right) & \text{n=2,4,6,...,} \end{cases} \tag{3.6.9}$$

$$E_n = \frac{n^2\pi^2\hbar^2}{2ma^2}, n = 1,2,3,...,. \tag{3.6.10}$$

3.7 Particle in a Finite Square Well Potential

Consider the motion of a quantum particle in a finite potential well (Figure 3.6):

$$V(x) = \begin{cases} 0, & \text{if} \quad |x| \leq a \\ V_0, & \text{if} \quad |x| > a\,. \end{cases} \tag{3.7.1}$$

We are required to solve the TISE with this potential for the bound states, when the total energy, E, of the particle is less than V_0 and determine the eigenfunctions and the corresponding energy eigenvalues. This type of potential is considered as an approximate model for the solution of several problems in atomic and nuclear physics.

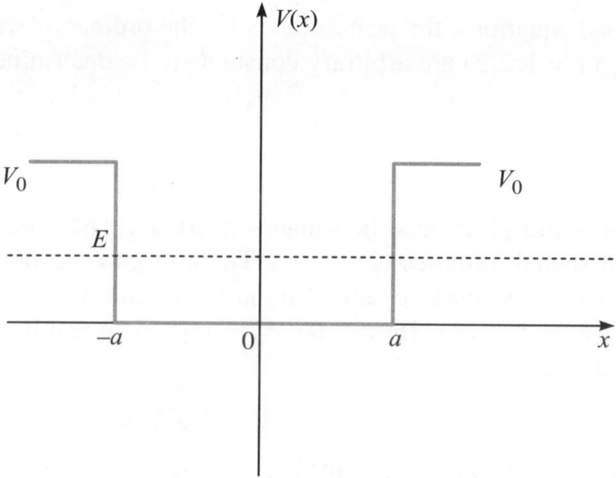

Figure 3.6 The representative shape of a finite potential well $V(x)$ of depth V_0.

Solution: The entire range of x from $-\infty$ to $+\infty$ can be divided into three regions: $-a \leq x \leq a$ (Region I), $x < -a$ (Region II) and $x > a$ (Region III). The general TISE reads

$$-\frac{\hbar^2}{2m}\frac{d^2\phi}{dx^2} + V(x)\phi = E\phi. \tag{3.7.2}$$

The TISE and the corresponding solutions in these regions can be written as:

Region I:

$$\phi_1'' + k_1^2\phi_1 = 0, \qquad k_1^2 = \frac{2mE}{\hbar^2},$$
$$\phi_1 = A_1 \cos(k_1 x) + B_1 \sin(k_1 x)\,. \tag{3.7.3}$$

Region II:

$$\phi_2'' - k_2^2\phi_2 = 0, \quad k_2^2 = \frac{2m(V_0 - E)}{\hbar^2},$$
$$\phi_2 = A_2\, e^{k_2 x} + B_2\, e^{-k_2 x}. \tag{3.7.4}$$

Region III:

$$\phi_3'' - k_2^2\phi_3 = 0,$$
$$\phi_3 = A_3\, e^{k_2 x} + B_3\, e^{-k_2 x}. \tag{3.7.5}$$

In the aforementioned equations, the prime stands for the ordinary derivative with respect to x, and A_j and B_j $(j = 1,2,3)$ are arbitrary constants to be determined by the boundary conditions.

Boundary conditions:

1. The full solution of the TISE must be square-integrable. That means that the solution must tend to zero at spatial infinities $(|x| \to \infty)$. Therefore, the second term in ϕ_2, which tends to infinity as $x \to -\infty$, must be zero. Similarly, the first term in ϕ_3, which tends to infinity as $x \to +\infty$, must be zero. Hence, $B_2 = A_3 = 0$. As a result, the total solution of the TISE can be written as

$$\phi(x) = \begin{cases} \phi_2 = A_2\, e^{k_2 x}, & x < -a \\ \phi_1 = A_1 \cos(k_1 x) + B_1 \sin(k_1 x), & -a \le x \le a \\ \phi_3 = B_3\, e^{-k_2 x}. & x > a \end{cases} \tag{3.7.6}$$

2. Since the TISE is second order in its spatial derivative with respect to x, the solutions belonging to different regions in x must be continuous and differentiable at the boundaries $x = \pm a$, that is, $\phi_1(-a) = \phi_2(-a), \phi_1'(-a) = \phi_2'(-a), \phi_1(a) = \phi_3(a)$ and $\phi_1'(a) = \phi_3'(a)$. These conditions lead to

$$A_2\, e^{-k_2 a} = A_1 \cos(k_1 a) - B_1 \sin(k_1 a), \tag{3.7.7}$$

$$k_2 A_2\, e^{-k_2 a} = k_1 A_1 \sin(k_1 a) + k_1 B_1 \cos(k_1 a), \tag{3.7.8}$$

$$B_3\, e^{-k_2 a} = A_1 \cos(k_1 a) + B_1 \sin(k_1 a), \tag{3.7.9}$$

$$-k_2 B_3\, e^{-k_2 a} = -k_1 A_1 \sin(k_1 a) + k_1 B_1 \cos(k_1 a). \tag{3.7.10}$$

If we add (3.7.7) and (3.7.9) and subtract (3.7.10) from (3.7.8), we get

$$(A_2 + B_3)\, e^{-k_2 a} = 2A_1 \cos(k_1 a), \tag{3.7.11}$$

$$k_2 \left(A_2 + B_3\right) e^{-k_2 a} = 2k_1 A_1 \sin\left(k_1 a\right). \tag{3.7.12}$$

Similarly if subtract (3.7.9) from (3.7.7) and add (3.7.8) and (3.7.10), we get

$$\left(A_2 - B_3\right) e^{-k_2 a} = -2B_1 \sin\left(k_1 a\right), \tag{3.7.13}$$

$$k_2 \left(A_2 - B_3\right) e^{-k_2 a} = 2k_1 B_1 \cos\left(k_1 a\right). \tag{3.7.14}$$

If $A_2 + B_3 \neq 0$ and $A_1 \neq 0$, then the equations (3.7.11) and (3.7.12) yield

$$k_2 = k_1 \tan(k_1 a). \tag{3.7.15}$$

Now, from (3.7.13) and (3.7.14), we have

$$B_1 \sin\left(k_1 a\right) = -\frac{k_1}{k_2} B_1 \cos\left(k_1 a\right) = -B_1 \frac{k_1^2}{k_2^2} \sin\left(k_1 a\right), \tag{3.7.16}$$

where we have made use of (3.7.15). We thus get

$$B_1 \left(1 + \frac{k_2^2}{k_1^2}\right) = 0, \quad \Rightarrow \quad B_1 = 0. \tag{3.7.17}$$

Equation (3.7.13) or (3.7.14) then yields $A_2 = B_3$. Taking all these results into account, we get that the full solution, corresponding to the case when $A_2 + B_3 \neq 0$ and $A_1 \neq 0$, is

$$\phi(x) = \begin{cases} A_2 e^{k_2 x} & \text{for} \quad x < -a \\ A_1 \cos\left(k_1 x\right) & \text{for} \quad -a \leq x \leq a \\ A_2 e^{-k_2 x} & \text{for} \quad x > a, \end{cases} \tag{3.7.18}$$

where A_1 and A_2 are arbitrary constants. It is not difficult to check that the given solution is a symmetric solution, that is, $\phi(-x) = \phi(x)$, and hence has positive parity. The boundary conditions, as shown earlier, lead to a transcendental equation, given by (3.7.15), for the determination of the energies of the bound states.

Since the potential is symmetric in x: $V(-x) = V(x)$, there is another solution to the TISE which is anti-symmetric. Let us determine that solution and the corresponding transcendental equation for the determination of the energy eigenvalues.

For this purpose, we make use of the equations (3.7.13) and (3.7.14). If $A_2 - B_3 \neq 0$ and $B_1 \neq 0$, we get

$$-k_1 \cot(k_1 a) = k_2. \tag{3.7.19}$$

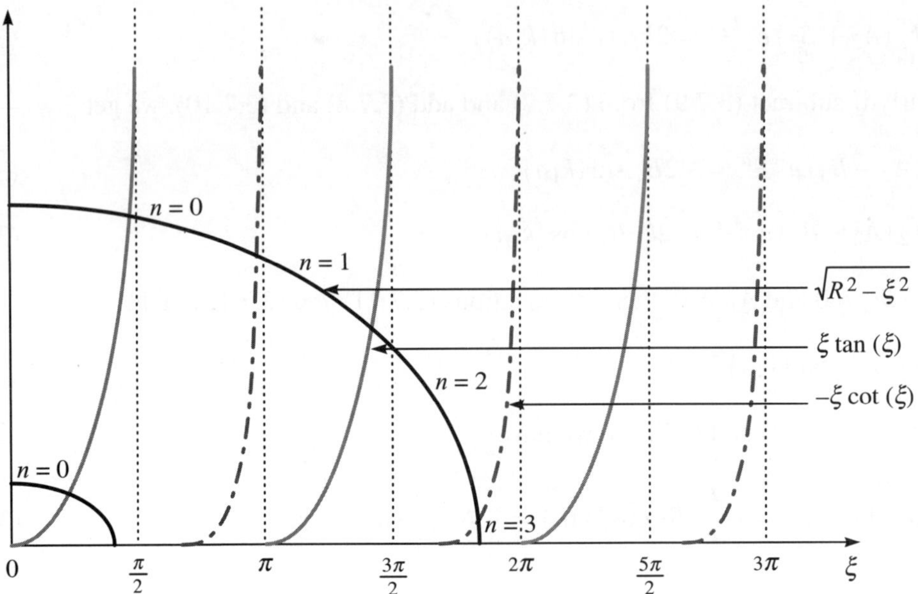

Figure 3.7 The graphical solutions for the finite square well potential: They are given by the points of intersection of the curves $\sqrt{R^2 - \alpha_n^2}$ with $\alpha_n \tan(\alpha_n)$ (solid lines) and $-\alpha_n \cot(\alpha_n)$ (dotted lines).

From (3.7.11) and (3.7.12), we derive

$$A_1 \cos(k_1 a) = \frac{k_1}{k_2} A_1 \sin(k_1 a) = -A_1 \frac{k_1^2}{k_2^2} \cos(k_1 a), \tag{3.7.20}$$

where we have made use of (3.7.19). Equations (3.7.20) leads to

$$A_1 \left(1 + \frac{k_2^2}{k_1^2}\right) = 0, \quad \Rightarrow \quad A_1 = 0. \tag{3.7.21}$$

Equation (3.7.11) or Equation (3.7.12) then yields $A_2 = -B_3$. Taking all these results into account, we get that the antisymmetric solution, corresponding to the case when $A_2 - B_3 \neq 0$ and $B_1 \neq 0$, is given by

$$\phi(x) = \begin{cases} A_2 e^{k_2 x} & \text{for} \quad x < -a \\ B_1 \sin(k_1 x) & \text{for} \quad -a \leq x \leq a \\ -A_2 e^{-k_2 x} & \text{for} \quad x > a, \end{cases} \tag{3.7.22}$$

where A_2 and B_1 are arbitrary constants. It is not difficult to check that the given solution is an anti-symmetric solution, that is, $\phi(-x) = -\phi(x)$, and hence has negative parity. The boundary conditions, as shown earlier, lead to a transcendental equation (3.7.19) for the determination of the energy eigenvalues for the corresponding bound states.

Clearly, for the given values of V_0 and a, (3.7.15) and (3.7.19), can be satisfied not for all values of E but for a selected set of values. This means that a particle, confined inside a potential well with finite height of the walls (which is the same as a potential well of finite depth), has a discrete energy spectrum.

Equations (3.7.15) and (3.7.19) are transcendental equations and cannot be solved analytically. However, they can be solved graphically as described here. Let us introduce new variables

$$\xi = k_1\, a = \sqrt{\frac{2mE}{\hbar^2}}\, a, \qquad \eta = k_2\, a = \sqrt{\frac{2m(V_0 - E)}{\hbar^2}}\, a. \tag{3.7.23}$$

Clearly, the following holds

$$\xi^2 + \eta^2 = R^2, \qquad R^2 = \frac{2ma^2 V_0}{\hbar^2}. \tag{3.7.24}$$

If we multiply (3.7.15) and (3.7.19) by a, they take the form

$$\xi\, \tan(\xi) = \eta, \tag{3.7.25}$$

$$-\xi\, \cot(\xi) = \eta. \tag{3.7.26}$$

Let ξ_n be the nth root of the transcendental equations (3.7.15) and (3.7.19). If we introduce the notation

$$\xi_n^2 = (k_1\, a)^2 = \frac{2ma^2 E_n}{\hbar^2}, \tag{3.7.27}$$

then $\eta = \sqrt{R^2 - \xi_n^2}$ and the equations (3.7.25) and (3.7.26) take the form

$$\xi_n \tan \xi_n = \sqrt{R^2 - \xi_n^2}. \qquad \text{(For even parity states)} \tag{3.7.28}$$

$$-\,\xi_n \cot \xi_n = \sqrt{R^2 - \xi_n^2}. \qquad \text{(For odd parity states)} \tag{3.7.29}$$

The left-hand sides of (3.7.28) and (3.7.29) contain trigonometric functions, while the right-hand sides represent a circle of radius R. The solutions are given by the points where the circle $\sqrt{R^2 - \xi_n^2}$ intersects the functions $\xi_n \tan \xi_n$ and $-\xi_n \cot \xi_n$. The solutions form a discrete set. Figure 3.7 contains the results of the solution of the equations (3.7.15) and (3.7.19) for two values of the radius, $R = 1$ and $R = 2$, which correspond to $V_0 a^2 = \hbar^2/2m$ and $V_0 a^2 = 2\hbar^2/m$, respectively. As depicted in Figure 3.7, the intersection of the small circle ($R = 1$) with the curve $\xi_n \tan \xi_n$ yields only one bound state, $n = 0$. The intersection of the larger circle ($R = 2$) with $\xi_n \tan \xi_n$ yields two bound states, $n = 0, 2$, and its intersection with $-\xi_n\, cot \xi_n$ yields two other bound states, $n = 1, 3$.

Hence, for $R = 2$, the system in all will have four bound states. *This analysis shows that the number of solutions depends on the value of R, which in turn depends on the depth of the well, V_0, and the width 2a of the well. Clearly, the deeper and wider the well, the greater the number of points of intersection of the curves and hence, greater will be the number of bound states of the particle inside the well.* Thus, there is always at least one bound state (that is, one intersection) no matter how small V_0 is. A closer look at Figure 3.7 shows that when

$$0 < R < \frac{\pi}{2}, \quad \text{that is,} \quad 0 < V_0 < \frac{\pi^2\hbar^2}{8ma^2}, \tag{3.7.30}$$

there is only one point of intersection of the circle with the function $\xi_n \tan \xi_n$ and there is only one bound state that we call $n = 0$ state. This is the ground state of the particle and happens to be an even parity state. When

$$\frac{\pi}{2} < R < \pi, \quad \text{that is,} \quad \frac{\pi^2\hbar^2}{8ma^2} < V_0 < \frac{\pi^2\hbar^2}{2ma^2}, \tag{3.7.31}$$

there are two bound states: an even state (the ground state) corresponding to $n = 0$ and the first odd parity state corresponding to $n = 1$. Now, if

$$\pi < R < \frac{3\pi}{2}, \quad \text{that is,} \quad \frac{\pi^2\hbar^2}{2ma^2} < V_0 < \frac{9\pi^2\hbar^2}{8ma^2}, \tag{3.7.32}$$

there exist three bound states: the ground state (even state), $n = 0$, the first excited state (odd state), corresponding to $n = 1$, and the second excited state (even state), which corresponds to $n = 2$. Similarly for

$$\frac{3\pi}{2} < R < 2\pi, \quad \text{that is,} \quad \frac{9\pi^2\hbar^2}{8ma^2} < V_0 < \frac{2\pi^2\hbar^2}{ma^2}, \tag{3.7.33}$$

there will be four bound states (two even and two odd) and so on and so forth. In general, for a given V_0, the width, $w_0 = 2a$, of the well that allows for n bound states is determined by

$$R = \frac{n\pi}{2}, \tag{3.7.34}$$

and equals

$$w_0 = \frac{\pi^2\hbar^2}{2mV_0}n^2. \tag{3.7.35}$$

In the limiting case of $ma^2V_0 \to \infty$ for a given a, the radius of the circle becomes infinite and the intersections occur at

$$\tan(k_1 a) = \infty \quad \Rightarrow \quad k_1 a = \frac{2n+1}{2}, n = 0, 1, 2, 3, ... \tag{3.7.36}$$

$$-\cot(k_2 a) = \infty \quad \Rightarrow \quad k_2 a = n\pi, n = 1, 2, 3, ... \tag{3.7.37}$$

If we combine the two, we obtain

$$k_1 a = \frac{n\pi}{2} \quad \Rightarrow \quad \frac{2mE_n}{\hbar^2} = \frac{n^2\pi^2}{4a^2}. \tag{3.7.38}$$

Finally, we arrive at

$$E_n = \frac{n^2\pi^2\hbar^2}{8ma^2}. \tag{3.7.39}$$

Thus, we recover the energy spectrum of the infinite potential well.

Before we wind up, let us talk a little about the so-called penetration depth in the classically forbidden region. When $E < V_0$, the regions $x < -a$ and $x > a$ are classically forbidden for the particle in the sense that it cannot penetrate into these regions. Consider $x > a$. The solution of the TISE in this region is $\phi(x) \sim e^{-k_2 x}$. Let us define

$$\phi(x) = \frac{\phi(0)}{e} = e^{-k_2\eta}, \tag{3.7.40}$$

where $x = \eta$ is the point where the wave function falls by a factor of $1/e$. Then, we have

$$\eta = \frac{1}{k_2} = \frac{\hbar}{\sqrt{2m(V_0 - E)}}. \tag{3.7.41}$$

η is called the penetration depth, that is, the distance to which the particle can penetrate into the classically forbidden region. Hence, the probability of finding the particle inside the forbidden regions on either side of the finite potential well is in principle non-zero.

Example 3.7.2 Find the number of bound states and the corresponding energies for the finite square well potential when $V_0 a^2 = \hbar^2/2m$.

Solution: In the given case $R = \left(2ma^2 V_0/\hbar^2\right)^{1/2} = 1$, which means $0 < R < \pi/2$ and there will be only one bound state corresponding to $n = 0$. The energy eigenvalue for this bound state is obtained by using the solution, ξ_0, of the equation

$$\xi_0 \tan(\xi_0) = \sqrt{1 - \xi_0^2}, \tag{3.7.42}$$

in the expression $k_2 a = \sqrt{1-\xi_0}$. We have

$$\xi_0^2\left(1+\tan^2\xi_0\right) = 1 \quad \Rightarrow \quad \cos^2(\xi_0) = \xi_0^2. \tag{3.7.43}$$

The numerical solution of the equation (3.7.43) yields $\xi_0 = 0.73909$. Therefore, $2mE_0a^2/\hbar^2 = (0.73909)^2 = 0.54625$, and the energy of the bound state $n = 0$ is given by

$$E_0 \approx \frac{0.54625}{2ma^2}\hbar^2 = \frac{0.273125}{ma^2}\hbar^2. \tag{3.7.44}$$

3.8 Potential Barrier and Tunneling

What we are going to discuss now is a very important phenomenon of *barrier penetration – tunneling*. Due to this effect, a micro-particle incident on one side of a potential barrier of height V_0 with a total energy $E < V_0$ can pass through the barrier and appear on the other side. This phenomenon does not have any classical analogue and represents a purely quantum mechanical effect and has been confirmed experimentally.

Consider an external potential field given by

$$V(x) = \begin{cases} V_0, & \text{for } 0 \le x \le a, \\ 0, & \text{otherwise.} \end{cases} \tag{3.8.1}$$

Assume that a particle of mass m, moving freely with a velocity $\vec{v} = v\hat{i}$, is incident on this barrier from the left, that is, from $x \to -\infty$. We are required to solve the corresponding time-independent Schrödinger equation and determine the reflection and transmission coefficients.

In general, both the cases with $E > V_0$ and $E < V_0$ are possible. However, as stated earlier, the case with the total energy $E < V_0$ corresponds to tunnelling and we take up this case.

For the solution of the problem, we divide the entire region $-\infty < x < +\infty$ into three parts: $-\infty < x < 0$ (Region 1), $0 < x < a$ (Region 2) and $a < x < +\infty$ (Region 3). The one-dimensional potential barrier of width a and height V_0 is shown in Figure 3.8. The TISE and the corresponding solutions in these regions can be written as:

Region 1:

$$\phi_1'' + k_1^2\phi_1 = 0, \quad k_1^2 = \frac{2mE}{\hbar^2},$$

$$\phi_1 = A\,e^{ik_1x} + B\,e^{-ik_1x}, \tag{3.8.2}$$

where A and B are arbitrary complex constants. Here the first term in the solution corresponds to the incident particle propagating along the positive x direction, while the second term describes the particle reflected from the potential and propagating along the negative x direction.

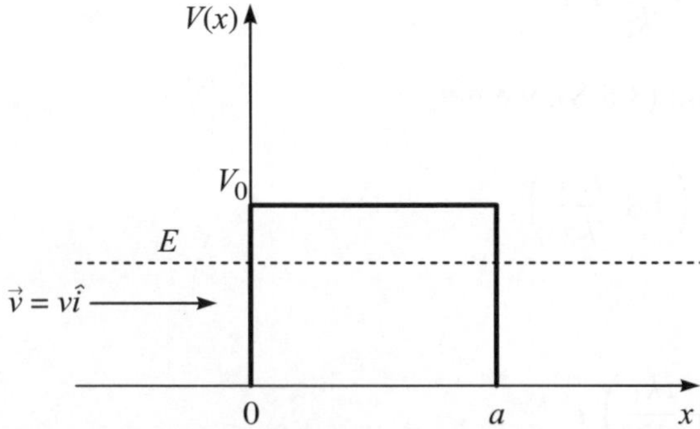

Figure 3.8 One-dimensional potential barrier of width a and height V_0.

Region 2:

$$\phi_2'' - k_2^2\phi_2 = 0, \quad k_2^2 = \frac{2m(V_0 - E)}{\hbar^2},$$

$$\phi_2 = C\,e^{k_2 x} + D\,e^{-k_2 x}, \tag{3.8.3}$$

where C and D are arbitrary complex constants.

Region 3:

$$\phi_3'' + k_1^2\phi_3 = 0,$$

$$\phi_3 = F\,e^{ik_1 x}. \tag{3.8.4}$$

The prime in the aforementioned equations stands for the ordinary derivative with respect to x. Here, F is an arbitrary complex constant and the solution represents the transmitted particle travelling along the positive x direction. Note that, because of the fact that the potential vanishes beyond $x = a$, there cannot be any reflected particle in this region and hence, we have taken only the forward propagating plane wave as solution.

Boundary conditions: The wave functions $\phi_1(x), \phi_2(x)$ and $\phi_3(x)$ have to be continuous in the entire region of x, as required by the standard conditions. Since the potential has a finite jump at $x = 0$ and $x = a$, the first derivatives of the wave functions with respect to x will also be continuous everywhere. These boundary conditions then yield

$$A + B = C + D, \tag{3.8.5}$$

$$(A - B) = -\frac{ik_2}{k_1}(C - D), \tag{3.8.6}$$

$$C e^{k_2 a} + D e^{-k_2 a} = F e^{ik_1 a}, \tag{3.8.7}$$

$$C e^{k_2 a} - D e^{-k_2 a} = \frac{i k_1}{k_2} F e^{ik_1 a}. \tag{3.8.8}$$

If we add up (3.8.7) and (3.8.8), we get

$$2C e^{k_2 a} = F e^{ik_1 a}\left(1 + \frac{i k_1}{k_2}\right). \tag{3.8.9}$$

Hence,

$$C = \frac{F}{2} e^{ik_1 a}\left(1 + \frac{i k_1}{k_2}\right) e^{-k_2 a}. \tag{3.8.10}$$

Now subtracting (3.8.8) from (3.8.7), we obtain

$$2D e^{-k_2 a} = F e^{ik_1 a}\left(1 - \frac{i k_1}{k_2}\right), \tag{3.8.11}$$

and therefore

$$D = \frac{F}{2} e^{ik_1 a}\left(1 - \frac{i k_1}{k_2}\right) e^{k_2 a}. \tag{3.8.12}$$

Substitution of C and D into the equation (3.8.5) yields

$$\begin{aligned} 1 + \frac{B}{A} &= \frac{F}{2A} e^{ik_1 a}\left[\left(1 + \frac{i k_1}{k_2}\right) e^{-k_2 a} + \left(1 - \frac{i k_1}{k_2}\right) e^{k_2 a}\right] \\ &= \frac{F}{A} e^{ik_1 a}\left[\frac{e^{k_2 a} + e^{-k_2 a}}{2} - \frac{i k_1}{k_2}\frac{\left(e^{k_2 a} - e^{-k_2 a}\right)}{2}\right] \\ &= \frac{F}{A} e^{ik_1 a}\left[\cosh(k_2 a) - \frac{i k_1}{k_2}\sinh(k_2 a)\right]. \end{aligned} \tag{3.8.13}$$

Similarly from (3.8.6), we get

$$1 - \frac{B}{A} = \frac{F}{2A} e^{ik_1 a}\left[\left(-\frac{i k_2}{k_1} + 1\right) e^{-k_2 a} + \left(\frac{i k_2}{k_1} + 1\right) e^{k_2 a}\right]$$

$$= \frac{F}{A}\, e^{ik_1 a} \left[\frac{e^{k_2 a} + e^{-k_2 a}}{2} + \frac{i k_2}{k_1} \frac{\left(e^{k_2 a} - e^{-k_2 a}\right)}{2} \right]$$

$$= \frac{F}{A}\, e^{ik_1 a} \left[\cosh(k_2 a) + \frac{i k_2}{k_1} \sinh(k_2 a) \right]. \tag{3.8.14}$$

Now, adding (3.8.13) and (3.8.14), we obtain

$$2 = \frac{F}{A}\, e^{ik_1 a} \left[2\cosh(k_2 a) + i \left(\frac{k_2}{k_1} - \frac{k_1}{k_2} \right) \sinh(k_2 a) \right]. \tag{3.8.15}$$

Similarly, subtraction of (3.8.14) from (3.8.13) leads to

$$2\frac{B}{A} = -i \frac{F}{A}\, e^{ik_1 a} \left(\frac{k_2}{k_1} + \frac{k_1}{k_2} \right) \sinh(k_2 a). \tag{3.8.16}$$

If we find the value of $\frac{F}{A}\, e^{ik_1 a}$ from (3.8.15) and put it into (3.8.16), we obtain

$$\frac{B}{A} = -\frac{i\left(\frac{k_2}{k_1} + \frac{k_1}{k_2}\right) \sinh(k_2 a)}{\left[2\cosh(k_2 a) + i\left(\frac{k_2}{k_1} - \frac{k_1}{k_2}\right) \sinh(k_2 a) \right]}. \tag{3.8.17}$$

The reflection coefficient is defined as

$$\mathscr{R} = \frac{\text{Reflected particle flux density}}{\text{Incident particle flux density}} = \frac{J_R}{J_I} = \frac{v_1 |B|^2}{v_1 |A|^2} = \frac{|B|^2}{|A|^2}. \tag{3.8.18}$$

It is given by

$$\mathscr{R} = \frac{\left(\frac{k_2^2 + k_1^2}{k_2 k_1}\right)^2 \sinh^2(k_2 a)}{\left[4\cosh^2(k_2 a) + \left(\frac{k_2^2 - k_1^2}{k_1 k_2}\right)^2 \sinh^2(k_2 a) \right]}. \tag{3.8.19}$$

The transmission coefficient, on the other hand, is defined as

$$\mathscr{T} = \frac{\text{Transmitted particle flux density}}{\text{Incident particle flux density}} = \frac{J_T}{J_I} = \frac{v_1 |B|^2}{v_1 |A|^2} = \frac{|F|^2}{|A|^2}. \tag{3.8.20}$$

Using (3.8.15) we arrive at

$$\mathscr{T} = \frac{4}{\left[4\cosh^2(k_2 a) + \left(\frac{k_2^2 - k_1^2}{k_1 k_2}\right)^2 \sinh^2(k_2 a) \right]}. \tag{3.8.21}$$

Further, making use of the well-known formula $\cosh^2 x - \sinh^2 x = 1$, we can rewrite the reflection and the transmission coefficients as

$$\mathscr{R} = \frac{\mathscr{T}}{4}\left(\frac{k_1^2+k_2^2}{k_2k_1}\right)^2 \sinh^2(k_2a), \tag{3.8.22}$$

$$\mathscr{T} = \frac{1}{\left[1+\frac{1}{4}\left(\frac{k_1^2+k_2^2}{k_2k_1}\right)^2 \sinh^2(k_2a)\right]} \tag{3.8.23}$$

Clearly, the transmission probability is finite. Therefore, we conclude that the probability that a quantum particle could penetrate a classically impenetrable barrier is non-zero. This is a purely quantum mechanical effect and is due to the wave aspect of microscopic objects. This barrier penetration effect is usually called *the tunneling effect* and has important physical implications. The radioactive decay and charge transport in electronic devices are typical examples of the quantum mechanical tunneling effect.

Using the expressions for k_1 and k_2 in terms of the physical parameters, we have

$$\left(\frac{k_1^2+k_2^2}{k_2k_1}\right)^2 = \left(\frac{V_0}{\sqrt{E(V_0-E)}}\right)^2 = \frac{V_0^2}{E(V_0-E)}. \tag{3.8.24}$$

Therefore, we can rewrite the expressions for the reflection and transmission coefficients as

$$\mathscr{R} = \mathscr{T}\frac{V_0^2}{4E(V_0-E)} \sinh^2\left(\frac{a}{\hbar}\sqrt{2m(V_0-E)}\right), \tag{3.8.25}$$

$$\mathscr{T} = \frac{1}{1+\frac{1}{4}\frac{V_0^2}{E(V_0-E)} \sinh^2\left(\frac{a}{\hbar}\sqrt{2m(V_0-E)}\right)}. \tag{3.8.26}$$

Let us consider the case when the energy of the incident particle is much smaller than the height of the barrier $E \ll V_0$. Then, we have

$$\frac{a}{\hbar}\sqrt{2m(V_0-E)} = \frac{a\sqrt{2mV_0}}{\hbar}\sqrt{1-\frac{E}{V_0}} \gg 1, \tag{3.8.27}$$

and we can write

$$\sinh\left(\frac{a}{\hbar}\sqrt{2m(V_0-E)}\right) \sim \frac{1}{2}e^{\frac{a\sqrt{2mV_0}}{\hbar}\sqrt{1-\frac{E}{V_0}}} = \frac{1}{2}\, e^{(a/\hbar)\sqrt{2m(V_0-E)}}. \tag{3.8.28}$$

Therefore, in the low energy limit, the transmission coefficient $\mathscr{T}$ is given by

$$\mathscr{T} = \frac{16E}{V_0}\left(1 - \frac{E}{V_0}\right) e^{-(2a/\hbar)\sqrt{2m(V_0-E)}}. \tag{3.8.29}$$

Also, when $E \sim V_0$, it is not difficult to deduce the following expressions for the reflection and transmission coefficients:

$$\mathscr{R} = \left(1 + \frac{2\hbar^2}{ma^2V_0}\right)^{-1}, \tag{3.8.30}$$

$$\mathscr{T} = \left(1 + \frac{ma^2V_0}{2\hbar^2}\right)^{-1}. \tag{3.8.31}$$

We, thus, see that even if the energy of the particle is much smaller than the barrier height, there is a finite probability that the particle can tunnel through the barrier and appear on the other side of it. Classically, such a phenomenon is not possible. The region $0 < x < a$ is forbidden for a particle with energy less than the barrier height V_0. Quantum mechanically, such tunneling effect is permissible and the apparent paradox arising out of it can be resolved with the help of Heisenberg's uncertainty principle (see Section 3.10).

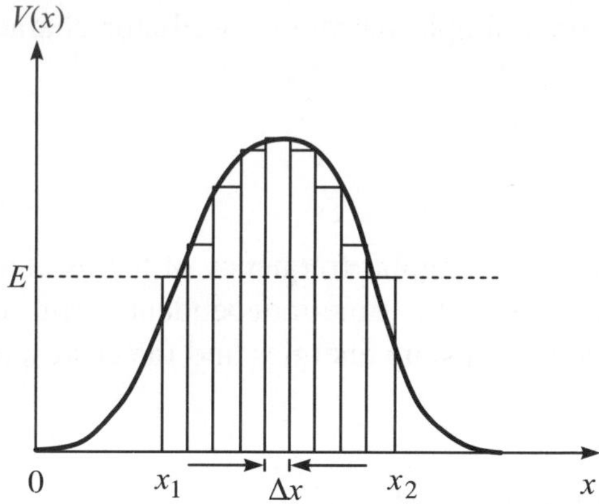

Figure 3.9 A general one-dimensional potential barrier $V = V(x)$.

Note that in the given example we considered the constant value for the potential barrier. In a more general case, the potential barrier is not a constant but can be a function of x: $V = V(x)$ (Figure 3.9). Unlike the constant potential barrier, in this case, the analytical solution is not possible for potentials with an arbitrary dependence on x. However, an approximate formula for the transmission coefficient can be derived by dividing the classically forbidden region between the turning points x_1 and x_2 into N (N large enough to approximate the curve $V(x)$) small rectangular sequence of barriers, each of width Δx. In each of these rectangular barriers, we can assume the potential to be constant. Then for each of them, the transmission coefficient can be written as:

$$\mathscr{T}_i \sim \exp\left[-\frac{2\Delta x_i}{\hbar}\sqrt{2m(V(x_i)-E)}\right], \tag{3.8.32}$$

where Δx_i is the width of the ith rectangular barrier with a constant height $V(x_i)$.

The transmission coefficient for the entire potential is then given by the following limit:

$$\mathscr{T} \approx \exp\left[-\frac{2}{\hbar}\lim_{\Delta x_i \to 0}\sum_{i=1} f(x_i)\,\Delta x_i\right], \tag{3.8.33}$$

where $f(x_i) = \sqrt{2m(V(x_i)-E)}$. As a result, we obtain

$$\mathscr{T} \approx \exp\left[-\frac{2}{\hbar}\int_{x_1}^{x_2} dx\sqrt{2m(V(x)-E)}\right]. \tag{3.8.34}$$

Note that the aforementioned approximate analysis is valid and gives satisfactory results only if the potential is a smooth and slowly varying function of x.

3.9 One-dimensional Harmonic Oscillator

Consider the one-dimensional simple harmonic oscillator characterized by the potential energy

$$V(x) = \frac{1}{2}m\omega^2 x^2, \tag{3.9.1}$$

where m is the mass and ω is the angular frequency of the oscillator, which is assumed to be constant. We want to solve the time-independent Schrödinger equation for this potential and determine the bound state energies and the corresponding eigenfunctions of the oscillator. We have

$$-\frac{\hbar^2}{2m}\frac{d^2\phi(x)}{dx^2} + \frac{1}{2}m\omega^2 x^2\phi(x) = E\phi(x), \tag{3.9.2}$$

which can be rewritten as

$$\phi''(x) + \frac{2m}{\hbar^2}\left[E - \frac{1}{2}m\omega^2 x^2\right]\phi(x) = 0, \tag{3.9.3}$$

where the prime stands for the ordinary derivative with respect to x. Let us introduce the following abbreviations

$$\lambda = \frac{2mE}{\hbar^2}, \quad \alpha = \frac{m\omega}{\hbar}. \tag{3.9.4}$$

Then (3.9.1) takes the form

$$\phi'' + [\lambda - \alpha^2 x^2]\phi = 0. \tag{3.9.5}$$

This is a second order ordinary differential equation with variable coefficients. Therefore, in order to have an idea about the behaviour of the solution at large values of x, let $\alpha x \gg 1$ so that we can neglect the term $\lambda\phi$ in comparison with the term $\alpha^2 x^2 \phi$. We then have

$$\phi'' - \alpha^2 x^2 \phi = 0. \tag{3.9.6}$$

For $\phi = \exp(-\gamma x^2)$, (3.9.6) yields

$$[-2\gamma + 4\gamma^2 x^2 - \alpha^2 x^2]\exp(-\gamma x^2) = 0. \tag{3.9.7}$$

Note that, for large x, we can neglect 2γ in comparison with the other two terms in (3.9.7). Consequently, we obtain

$$\gamma = \frac{\alpha}{2}. \tag{3.9.8}$$

Therefore, we look for the solution of the equation (3.9.5) in the form

$$\phi(x) = e^{-\alpha x^2/2} f(x), \tag{3.9.9}$$

where $f(x)$ is an arbitrary function of x to be determined. We have

$$\phi' = (-\alpha x f + f')e^{-\alpha x^2/2}, \tag{3.9.10}$$

$$\phi'' = [(-\alpha f - \alpha x f' + f'') + \alpha^2 x^2 f - \alpha x f']e^{-\alpha x^2/2}. \tag{3.9.11}$$

From (3.9.5) and (3.9.9)-(3.9.11), we arrive at the following differential equation for the function $f(x)$

$$f'' - 2\alpha x f' + (\lambda - \alpha) f = 0. \tag{3.9.12}$$

Introducing the dimensionless variable

$$\xi = \sqrt{\alpha} x, \tag{3.9.13}$$

we get

$$\frac{d}{dx} = \sqrt{\alpha}\frac{d}{d\xi}, \quad \frac{d^2}{dx^2} = \alpha\frac{d^2}{d\xi^2}. \tag{3.9.14}$$

As a result (3.9.12) can be rewritten as

$$f'' - 2\xi f' + \left(\frac{\lambda}{\alpha} - 1\right) f = 0, \tag{3.9.15}$$

where *prime* stands for ordinary derivative with respect to ξ. We look for the series solution of (3.9.15) in the following form

$$f(x) = \sum_{k=\nu}^{\infty} a_k \xi^k, \tag{3.9.16}$$

where the value of ν will be determined later. From (3.9.15) and (3.9.16), we get

$$\sum_{k=\nu}^{\infty} \left[k(k-1)a_k \xi^{k-2} - 2k a_k \xi^k + \left(\frac{\lambda}{\alpha} - 1\right) a_k \xi^k \right] = 0. \tag{3.9.17}$$

Writing the series on the left-hand side in the order of increasing powers of ξ, we obtain

$$\begin{aligned} &\nu(\nu-1)a_\nu \xi^{\nu-2} + \nu(\nu+1)a_{\nu+1}\xi^{\nu-1} + (\nu+1)(\nu+2)a_{\nu+2}\xi^\nu \\ &- 2\nu a_\nu \xi^\nu + \left(\frac{\lambda}{\alpha} - 1\right) a_\nu \xi^\nu + ... = 0. \end{aligned} \tag{3.9.18}$$

For this equation to hold good, the coefficient before each power of ξ must be equal to zero. We have

$$\nu(\nu-1) = 0 \quad \Rightarrow \quad \nu = 0, 1, \tag{3.9.19}$$

$$\nu(\nu+1) = 0 \quad \Rightarrow \quad \nu = 0, -1. \tag{3.9.20}$$

The value -1 of ν is not acceptable because, in that case, the series (3.9.16) will start with the term $\sim \xi^{-1}$ that blows up at $\xi = 0$. Hence, ν can take only two values 0 and 1.

Equating the coefficient of ξ^ν equal to zero, we arrive at the recursion relation for the coefficients of the series

$$a_{\nu+2} = \frac{2\nu - \left(\frac{\lambda}{\alpha} - 1\right)}{(\nu+1)(\nu+2)} a_\nu. \tag{3.9.21}$$

Consequently, we shall have two possible solutions for $f(\xi)$:

$$f_1(\xi) \sim a_0 + a_2\xi^2 + a_4\xi^4 + a_6\xi^6 + ..., \tag{3.9.22}$$

and

$$f_2(\xi) \sim a_1\xi + a_3\xi^3 + a_5\xi^5 + ..., \tag{3.9.23}$$

Let us take the first of the solutions that starts with $\nu = 0$ and see how it behaves for large values of ξ. For that, let us determine the behaviour of the ratio $a_{\nu+2}/a_\nu$ for $\nu \to \infty$. We have

$$\lim_{\nu\to\infty} \frac{a_{\nu+2}}{a_\nu} = \lim_{\nu\to\infty} \frac{\nu\left(2 - \frac{\left(\frac{\lambda}{\alpha}-1\right)}{\nu}\right)}{\nu^2(1+1/\nu)(1+2/\nu)} = \frac{2}{\nu}. \tag{3.9.24}$$

For comparison, consider the series

$$e^{\xi^2} = \sum_{\sigma=0}^{\infty} b_\sigma \xi^\sigma = 1 + \frac{\xi^2}{1!} + \frac{\xi^4}{2!} + \frac{\xi^6}{3!} + ... + \frac{\xi^\sigma}{\frac{\sigma}{2}!} + \frac{\xi^{\sigma+2}}{\left(\frac{\sigma}{2}+1\right)!} + ... \tag{3.9.25}$$

For this exponential series,

$$\lim_{\sigma\to\infty} \frac{b_{\sigma+2}}{b_\sigma} = \lim_{\sigma\to\infty} \frac{\frac{\sigma}{2}!}{\left(\frac{\sigma}{2}+1\right)!} = \lim_{\xi\to\infty} \frac{\frac{\sigma}{2}!}{\left(\frac{\sigma}{2}+1\right)\frac{\sigma}{2}!} \approx \frac{2}{\sigma}. \tag{3.9.26}$$

Therefore, for large values of ξ, the series (3.9.22) behaves as the exponential series given by (3.9.25). The same applies to the series (3.9.23). Consequently, for large values of ξ, the function $f(\xi)$ blows up because

$$f(\xi) \approx e^{-\frac{\xi^2}{2}} \cdot e^{\xi^2} \sim e^{\frac{\xi^2}{2}}. \tag{3.9.27}$$

Thus, the infinite series solution (3.9.16), whose coefficients are determined by the recursion relation (3.9.21), does not satisfy the boundary conditions and hence, cannot be the acceptable solution. However, the situation can be retrieved if the infinite series can be converted into a polynomial so that the exponential factor $e^{-\frac{\xi^2}{2}}$, standing before $f(\xi)$ (see Eq.(3.9.9)), could force the wave function $\phi(\xi)$ to tend to zero for $\xi \to \pm\infty$. For this to happen, the series has to be truncated at some term, say n^{th} term. In that case, the numerator in (3.9.21) would be zero for $\nu = n$. As a consequence, we get

$$2n - \frac{\lambda}{\alpha} - 1 = 0, \quad \Rightarrow \quad \frac{\lambda}{\alpha} = 2n+1. \tag{3.9.28}$$

Substituting the values of λ and α, we obtain

$$\frac{2mE_n}{\hbar^2} = \frac{m\omega}{\hbar}(2n+1). \tag{3.9.29}$$

Equation (3.9.29) leads to the quantization of energy of the harmonic oscillator:

$$E_n = \hbar\omega\left(n + \frac{1}{2}\right), \quad n = 0, 1, 2, 3, ... \tag{3.9.30}$$

Note that this formula for the quantized energy of the oscillator differs from the one obtained in the old quantum theory

$$E_n = n\hbar\omega, \quad n = 0, 1, 2, 3, ... \tag{3.9.31}$$

by the fact that it possesses a non-zero energy in the lowest quantum state with $n = 0$. This energy is called the zero-point energy, E_0, whose value is given by

$$E_0 = \frac{1}{2}\hbar\omega. \tag{3.9.32}$$

Let us go back to our problem of finding the solutions to the differential equation (3.9.15). Evidently, the solutions satisfying the standard conditions can now be written as

$$\phi_n(\xi) = N_n e^{-\xi^2/2} H_n(\xi), \tag{3.9.33}$$

where N_n is the normalization constant and $H_n(\xi)$ is the polynomial of degree n whose coefficients are given by (3.9.21) under the condition $\lambda/\alpha = 2n+1$. These polynomials for different n values are known as Hermite polynomials. The coefficient before the term in the polynomial containing ξ^n is obtained by taking $\nu = n-2$ in (3.9.21). It is given by

$$a_n = \frac{2(n-2)+1-(2n+1)}{n(n-1)} a_{n-2} = -\frac{4}{n(n-1)} a_{n-2}. \tag{3.9.34}$$

Therefore, we have

$$a_{n-2} = -\frac{n(n-1}{4} a_n \equiv -\frac{n(n-1)}{1\times 2^2} a_n. \tag{3.9.35}$$

Similarly, we can compute

$$a_{n-4} = -\frac{(n-2)(n-1)}{8} a_{n-2} = \frac{n(n-1)(n-2)(n-3)}{1\times 2\times 2^2} a_n, \tag{3.9.36}$$

and so on and so forth. As a result, the polynomial will be given by

$$H_n(\xi) = a_n \left[\xi^n - \frac{n(n-1)}{1\times 2^2}\xi^{n-2} + \frac{n(n-1)(n-2)(n-3)}{1\times 2\times 2^2}\xi^{n-4} - ... \right]. \tag{3.9.37}$$

If we put $a_n = 2^n, n = 0, 1, 2, 3, ...,$ we obtain the formulae for the polynomials of the corresponding degree. A few of these are given here for illustration:

$$\begin{aligned} &H_0(\xi) = 1, && H_1(\xi) = 2\xi, \\ &H_2(\xi) = 4\xi^2 - 2, && H_3(\xi) = 8\xi^3 - 12\xi, \\ &H_4(\xi) = 16\xi^4 - 48\xi^2 + 12, && H_5(\xi) = 32\xi^5 - 160\xi^3 + 120\xi. \end{aligned} \tag{3.9.38}$$

Rodriguez's formula for the Hermite polynomials: The following compact formula for computing the Hermite polynomials is known as Rodriguez's formula:

$$H_n(\xi) = (-1)^n e^{\xi^2} \frac{d^n \left(e^{-\xi^2}\right)}{d\xi^n}. \tag{3.9.39}$$

It allows us to generate the required polynomial of any degree by simply plugging in the value of n and simplifying the expression. For instance,

$$n = 1: \ H_1(\xi) = -e^{\xi^2} \frac{d\left(e^{-\xi^2}\right)}{d\xi} = (2\xi)\, e^{\xi^2}.e^{-\xi^2} = 2\xi, \tag{3.9.40}$$

$$n = 2: \ H_2(\xi) = (-1)^2 e^{\xi^2} \frac{d^2\left(e^{-\xi^2}\right)}{d\xi^2} = -2e^{\xi^2} \frac{d}{d\xi}\left(\xi e^{-\xi^2}\right) = 4\xi^2 - 2. \tag{3.9.41}$$

Recurrence formula for Hermite polynomials: Let us, for the convenience in calculations, derive a recurrence formula for the polynomials themselves. Using Rodriguez's formula, can write

$$H_{n+1}(\xi) = (-1)^{(n+1)}\, e^{\xi^2} \frac{d^{(n+1)}\left(e^{-\xi^2}\right)}{d\xi^{(n+1)}}. \tag{3.9.42}$$

Using the following formulae

$$\frac{d^{(n+1)}\left(e^{-\xi^2}\right)}{d\xi^{(n+1)}} = -2\frac{d^n}{d\xi^n}\left(\xi e^{-\xi^2}\right), \tag{3.9.43}$$

$$\frac{d^n}{d\xi^n}[f(\xi)g(\xi)] = f\frac{d^n g}{d\xi^n} + n\frac{df}{d\xi}\frac{d^{n-1}g}{d\xi^{n-1}} + \frac{n(n-1)}{2}\frac{d^2 f}{d\xi^2}\frac{d^{n-2}g}{d\xi^{n-2}} + ..., \tag{3.9.44}$$

we get

$$\frac{d^n}{d\xi^n}\left(\xi e^{-\xi^2}\right) = \xi\frac{d^n}{d\xi^n}\left(e^{-\xi^2}\right) + n\frac{d^{n-1}\left(e^{-\xi^2}\right)}{d\xi^{n-1}}. \tag{3.9.45}$$

As a result, we have

$$H_{n+1}(\xi) = (-1)^n 2\xi\, e^{\xi^2}\frac{d^n}{d\xi^n}\left(e^{-\xi^2}\right) + (-1)^n 2n\, e^{\xi^2}\frac{d^{n-1}}{d\xi^{n-1}}\left(e^{-\xi^2}\right)$$

$$= (-1)^n 2\xi \, e^{\xi^2} \frac{d^n}{d\xi^n}\left(e^{-\xi^2}\right) - (-1)^{n-1} 2n \, e^{\xi^2} \frac{d^{n-1}}{d\xi^{n-1}}\left(e^{-\xi^2}\right). \tag{3.9.46}$$

Using the Rodriguez's formula once again, we arrive at the desired recurrence relation

$$H_{n+1}(\xi) = 2\xi \, H_n(\xi) - 2n \, H_{n-1}(\xi). \tag{3.9.47}$$

Being the eigenfunctions of a hermitian operator, the eigenfunctions of the harmonic oscillator corresponding to different eigenvalues are orthogonal. Using this, we can calculate the normalization constant N_n as

$$\begin{aligned} \int_{-\infty}^{+\infty} |\phi_n(\xi)|^2 \, d\xi &= (-1)^n \frac{N_n^2}{\sqrt{\alpha}} \int_{-\infty}^{+\infty} e^{-\xi^2} e^{\xi^2} \frac{d^n e^{-\xi^2}}{d\xi^n} H_n(\xi) \, d\xi \\ &= (-1)^n \frac{N_n^2}{\sqrt{\alpha}} \int_{-\infty}^{+\infty} \frac{d^n e^{-\xi^2}}{d\xi^n} H_n(\xi) \, d\xi. \end{aligned} \tag{3.9.48}$$

Integrating by parts, we obtain

$$\begin{aligned} \int_{-\infty}^{+\infty} |\phi_n(\xi)|^2 \, d\xi &= (-1)^n \frac{N_n^2}{\sqrt{\alpha}} \int_{-\infty}^{+\infty} \frac{d^n e^{-\xi^2}}{d\xi^n} H_n(\xi) \, d\xi = (-1)^n \frac{N_n^2}{\sqrt{\alpha}} \frac{d^{n-1} e^{-\xi^2}}{d\xi^{n-1}} H_n(\xi) \Bigg|_{-\infty}^{+\infty} \\ &+ (-1)^{n-1} \frac{N_n^2}{\sqrt{\alpha}} \int_{-\infty}^{+\infty} \frac{d^{n-1} e^{-\xi^2}}{d\xi^{n-1}} \frac{dH_n(\xi)}{d\xi} \, d\xi. \end{aligned} \tag{3.9.49}$$

$H_n(\xi)$, according to Rodriguez's formula, contains $e^{-\xi^2}$ and its derivatives. Since the function $e^{-\xi^2}$ and all its derivatives tend to zero at $|\xi| = \pm\infty$, the first term on the right-hand side in (3.9.49) vanishes. As a result

$$\int_{-\infty}^{+\infty} |\phi_n(\xi)|^2 \, d\xi = (-1)^{n-1} \frac{N_n^2}{\sqrt{\alpha}} \int_{-\infty}^{+\infty} \frac{d^{n-1} e^{-\xi^2}}{d\xi^{n-1}} \frac{dH_n(\xi)}{d\xi} \, d\xi. \tag{3.9.50}$$

Therefore, if we integrate (3.9.50) by parts another $(n-1)$ number of times, we finally get that

$$\int_{-\infty}^{+\infty} |\phi_n(\xi)|^2 \, d\xi = \frac{N_n^2}{\sqrt{\alpha}} \int_{-\infty}^{+\infty} e^{-\xi^2} \frac{d^n H_n(\xi)}{d\xi^n} \, d\xi. \tag{3.9.51}$$

Since $H_n(\xi) = 2^n \xi^n + ...$ is a polynomial of degree n,

$$\frac{dH_n(\xi)}{d\xi^n} = 2^n n!, \tag{3.9.52}$$

and we obtain

$$\int_{-\infty}^{+\infty} |\phi_n(\xi)|^2\, d\xi = 2^n n! \frac{N_n^2}{\sqrt{\alpha}} \int_{-\infty}^{+\infty} e^{-\xi^2}\, d\xi = 2^n n! \frac{N_n^2}{\sqrt{\alpha}} \sqrt{\pi}. \tag{3.9.53}$$

The normalization condition then yields

$$N_n = \sqrt{\frac{\alpha^{1/2}}{2^n n! \pi^{1/2}}}. \tag{3.9.54}$$

Hence, the full stationary state solutions to the Schrödinger equation for the harmonic oscillator potential are

$$\psi(\xi,t) = \sqrt{\frac{\alpha^{1/2}}{2^n n! \pi^{1/2}}}\, e^{-\xi^2/2} H_n(\xi)\, e^{-\frac{i}{\hbar}E_n t}, \tag{3.9.55}$$

where

$$E_n = \hbar\omega \left(n + \frac{1}{2}\right), \quad n = 0,1,2,3,... \tag{3.9.56}$$

are the corresponding stationary state energies.

3.10 Heisenberg's Uncertainty Relation

We have proved earlier that two operators which have the same set of eigenfunctions commute. If we combine this with the fact that a dynamical variable can have a definite value in its eigenstates only, we come to the conclusion that *for two or more dynamical variables to have definite values simultaneously their corresponding operators must commute.* On the other hand, we have seen that many of the operators of interest in quantum mechanics do not commute. Therefore, it is quite natural to ask the following question: *What if we measure two non-compatible observables A and B, one after the other in a given state, how will the inaccuracy in their measurements be related?* The answer to this fundamental question is provided by Heisenberg's uncertainty principle which we are now going to derive.

In this regard, we must first decide the way we are going to characterize the accuracy of measurement. Assume that we have conducted a large number of measurements of some physical quantity a and obtained a series of its numerical values $a_1, a_2, a_3, ...$ whose average value we denote by $\langle a \rangle$. In probability theory, the deviation of a value a_k of a random variable from its average value, $\langle a \rangle$, is usually characterized by the root-mean-square deviation defined as

$$\Delta a = \sqrt{\langle (a_k - \langle a \rangle)^2 \rangle} = \sqrt{\langle a_k^2 \rangle - \langle a \rangle^2}. \tag{3.10.1}$$

Although the uncertainty relation can be derived for any pair of non-commuting hermitian operators $\hat{A}$ and $\hat{B}$ in general, here, for the sake of simplicity, we shall deduce it for $\hat{x}$ and $\hat{p}_x$ only.

We shall characterize the inaccuracy in the measurements of the x coordinate and the corresponding x-component of the momentum, p_x, by their root-mean-square deviations:

$$\Delta x = \sqrt{\langle \hat{x}^2 \rangle - \langle \hat{x} \rangle^2}, \tag{3.10.2}$$

$$\Delta p_x = \sqrt{\langle \hat{p}_x^2 \rangle - \langle \hat{p}_x \rangle^2}. \tag{3.10.3}$$

Let us choose the system of coordinates in which the origin lies at $\langle x \rangle$ and the system is moving with a speed equal to p/m along the x-direction. In such a system $\langle x \rangle = 0$, $\langle \hat{p}_x \rangle = 0$ and we shall have $\Delta x = \sqrt{\langle \hat{x}^2 \rangle}$ and $\Delta p = \sqrt{\langle \hat{p}_x^2 \rangle}$.

Consider the following inequality

$$\int \left| \alpha x \psi + \beta \frac{d\psi}{dx} \right|^2 dx \geq 0, \tag{3.10.4}$$

where α and β are two auxiliary real variables.

The integrand is given by

$$\begin{aligned} \left| \alpha x \psi + \beta \frac{d\psi}{dx} \right|^2 &= \alpha^2 x^2 \psi^* \psi + \alpha\beta x \left(\psi^* \frac{d\psi}{dx} + \psi \frac{d\psi^*}{dx} \right) + \beta^2 \frac{d\psi^*}{dx} \frac{d\psi}{dx} \\ &= \alpha^2 x^2 \psi^* \psi + \alpha\beta x \frac{d}{dx} (\psi^* \psi) + \beta^2 \frac{d\psi^*}{dx} \frac{d\psi}{dx}. \end{aligned} \tag{3.10.5}$$

From (3.10.4) and (3.10.5), we arrive at the following inequality

$$A\alpha^2 - B\alpha\beta + C\beta^2 \geq 0, \tag{3.10.6}$$

where

$$A = \int x^2 \psi^* \psi dx > 0, \quad B = -\int x \frac{d}{dx} (\psi^* \psi) dx, \quad C = \int \frac{d\psi^*}{dx} \frac{d\psi}{dx} dx. \tag{3.10.7}$$

For inequality (3.10.6) to hold (which is equivalent to the positivity of the quadratic form $(\alpha - \alpha_1)(\alpha - \alpha_2)$, $\alpha_i, i = 1,2$ being the roots of the quadratic equation $A\alpha^2 - B\alpha\beta + C\beta^2 = 0$), the roots $\alpha_i, i = 1,2$ must be complex. This, in turn, requires that

$$B^2 - 4AC \leq 0, \quad \text{or,} \quad 4AC \geq B^2. \tag{3.10.8}$$

Now, since $\langle x \rangle = 0$, the expectation value $\langle \hat{x}^2 \rangle = A$. Consider the integral B. If we integrate by parts and take into account that the wave function tends to zero at $x = \pm\infty$, we get

$$B = -\int x \frac{d}{dx}(\psi^* \psi) dx = \int \psi^* \psi dx = 1. \tag{3.10.9}$$

Similarly, we get for C:

$$C = -\int \psi^* \frac{d^2 \psi}{dx^2} dx = \frac{1}{\hbar^2} \langle \hat{p}_x^2 \rangle. \tag{3.10.10}$$

Substituting the above values of A, B and C into Eq.(3.10.8), we obtain

$$\sqrt{\langle \hat{x}^2 \rangle} \sqrt{\langle \hat{p}_x^2 \rangle} \geq \frac{\hbar}{2}. \tag{3.10.11}$$

Or, using (3.10.2) and (3.10.3), we have the relation

$$\Delta x \, \Delta p_x \geq \frac{\hbar}{2}. \tag{3.10.12}$$

The same relations result for the other two coordinates, y and z, and the corresponding components of the linear momentum, p_y and p_z:

$$\Delta y \, \Delta p_y \geq \frac{\hbar}{2}, \tag{3.10.13}$$

$$\Delta z \, \Delta p_z \geq \frac{\hbar}{2}. \tag{3.10.14}$$

The inequalities (3.10.12)–(3.10.14) are known as Heisenberg's uncertainty relations and represent the constraint on the accuracy in the simultaneous measurement of coordinate and momentum. They show that the product of uncertainties in the measurement of a coordinate and the corresponding component of momentum cannot be less than $\hbar/2$. Consequently, an exact knowledge of the position of a quantum system, at a given instant of time, makes its momentum indeterminate and vice versa.

Example 3.10.1: Consider a particle of mass m in an infinite potential well of width a whose wave function is given by

$$\psi(x) = \begin{cases} Ax(a^2 - x^2) & \text{for} \quad 0 < x < a, \\ 0 & \quad\quad \text{elsewhere,} \end{cases} \tag{3.10.15}$$

where A is a real constant. (a) Find A so that $\psi(x)$ is normalized. (b) Calculate the position and momentum uncertainties, Δx and Δp, and the product $\Delta x \Delta p$. (c) Calculate the probability of finding $5\pi^2\hbar^2/2ma^2$ for a measurement of the energy.

Solution: The normalization of the wave function reads

$$\int_0^a A^2 x^2 (a^2 - x^2)^2 \, dx = A^2 \int_0^a (a^4 x^2 - 2a^2 x^4 + x^6) \, dx$$

$$A^2 \left[a^4 \left. \frac{x^3}{3} \right|_0^a - 2a^2 \left. \frac{x^5}{5} \right|_0^a + \left. \frac{x^7}{7} \right|_0^a \right] = \frac{8A^2 a^7}{105} = 1. \tag{3.10.16}$$

Hence, the value of A is

$$A = \sqrt{\frac{105}{8a^7}}. \tag{3.10.17}$$

The average value of $\hat{x}$ and $\hat{x}^2$ will be

$$\begin{aligned}\langle \hat{x} \rangle &= \frac{105}{8a^7} \int_0^a (a^4 x^3 - 2a^2 x^5 + x^7) \, dx \\ &= \frac{105}{8a^7} \left[a^4 \left. \frac{x^4}{4} \right|_0^a - 2a^2 \left. \frac{x^6}{6} \right|_0^a + \left. \frac{x^8}{8} \right|_0^a \right] = \frac{35\,a}{64}.\end{aligned} \tag{3.10.18}$$

$$\begin{aligned}\langle \hat{x}^2 \rangle &= \frac{105}{8a^7} \int_0^a (a^4 x^4 - 2a^2 x^6 + x^8) \, dx \\ &= \frac{105}{8a^7} \left[a^4 \left. \frac{x^5}{5} \right|_0^a - 2a^2 \left. \frac{x^7}{7} \right|_0^a + \left. \frac{x^9}{9} \right|_0^a \right] = \frac{a^2}{3}.\end{aligned} \tag{3.10.19}$$

The uncertainty in x is given by

$$\Delta x = \sqrt{\langle \hat{x}^2 \rangle - \langle \hat{x} \rangle^2} = \sqrt{\frac{a^2}{3} - \left(\frac{35a}{64} \right)^2} \approx 0.185\,a. \tag{3.10.20}$$

The average value of $\hat{p}_x$ and $\hat{p}_x^2$ will be

$$\begin{aligned}\langle \hat{p}_x \rangle &= -iA^2 \hbar \int_0^a (a^4 x - 4a^2 x^3 + 3x^5) \, dx \\ &= -iA^2 \hbar \left[a^4 \left. \frac{x^2}{2} \right|_0^a - 4a^2 \left. \frac{x^4}{4} \right|_0^a + 3 \left. \frac{x^6}{6} \right|_0^a \right] = 0.\end{aligned} \tag{3.10.21}$$

$$\begin{aligned}\langle \hat{p}_x^2 \rangle &= 6\hbar^2 A^2 \int_0^a (a^2 x^2 - x^4) \, dx \\ &= 6\hbar^2 \frac{105}{8a^7} \left[a^2 \left. \frac{x^3}{3} \right|_0^a - \left. \frac{x^5}{5} \right|_0^a \right] = \frac{21\hbar^2}{2a^2}.\end{aligned} \tag{3.10.22}$$

The uncertainty in p_x is given by

$$\Delta p_x = \sqrt{\langle \hat{p}_x^2 \rangle - \langle \hat{p}_x \rangle^2} = \sqrt{\frac{21\hbar^2}{2a^2}} \approx 3.24\,\frac{\hbar}{a}. \tag{3.10.23}$$

The product $\Delta x\, \Delta p_x \approx 0.6\hbar > 0.5\hbar$.

Example 3.10.2: Consider a system whose Hamiltonian $\hat{H}$ and an operator $\hat{A}$, representing an observable A, are given by the matrices

$$H = \mathscr{E}_0 \begin{pmatrix} 1 & -1 & 0 \\ -1 & 1 & 0 \\ 0 & 0 & -1 \end{pmatrix}, \quad A = \alpha \begin{pmatrix} 0 & 4 & 0 \\ 4 & 0 & 1 \\ 0 & 1 & 0 \end{pmatrix},$$

where $\mathscr{E}_0$ and α are constants. $\mathscr{E}_0$ has the dimensions of energy.
(a) If we measure the energy, what values will obtain? (b) Suppose that when we measure energy, we obtain a value of $-\mathscr{E}_0$. Immediately afterward, we measure A. What values shall we get for A and what are the probabilities corresponding to each of these values? (c) Find the uncertainty in the measurement of A.

Solution: According to the postulate of quantum mechanics, the possible results of measurement of energy will be the eigenvalues of the Hamiltonian. Hence, let us first find these eigenvalues. The characteristic equation reads

$$\mathscr{E}_0 \begin{vmatrix} (1-\lambda) & -1 & 0 \\ -1 & (1-\lambda) & 0 \\ 0 & 0 & -(1+\lambda) \end{vmatrix} = \mathscr{E}_0 (1-\lambda) \begin{vmatrix} 1-\lambda & 0 \\ 0 & -(1+\lambda) \end{vmatrix} + \mathscr{E}_0 \begin{vmatrix} -1 & 0 \\ 0 & -(1+\lambda) \end{vmatrix} = 0 \tag{3.10.24}$$

The simplification leads to

$$-\lambda^3 + \lambda^2 + 2\lambda = 0, \;\Rightarrow\; \lambda_1 = 0,\; \lambda_2 = -\mathscr{E}_0,\; \lambda_3 = 2\mathscr{E}_0. \tag{3.10.25}$$

Hence, the values of energy that can result are $E_1 = 0, E_2 = -\mathscr{E}_0, E_3 = 2\mathscr{E}_0$.

Let us now calculate the eigenvectors of the Hamiltonian. Consider first the eigenvalue $\lambda_1 = 0$. We have

$$\mathscr{E}_0 \begin{pmatrix} 1 & -1 & 0 \\ -1 & 1 & 0 \\ 0 & 0 & -1 \end{pmatrix} \begin{pmatrix} a \\ b \\ c \end{pmatrix} = \begin{pmatrix} 0 \\ 0 \\ 0 \end{pmatrix}. \tag{3.10.26}$$

$$\Rightarrow \quad \begin{cases} a-b=0, \\ -a+b=0, \\ -c=0. \end{cases} \quad \Rightarrow \quad a=b \text{ and } c=0. \tag{3.10.27}$$

If we take $a=b=1$ and normalized the eigenvector, we get

$$|\phi_1\rangle = \frac{1}{\sqrt{2}}\begin{pmatrix} 1 \\ 1 \\ 0 \end{pmatrix}. \tag{3.10.28}$$

Similarly, the other two eigenvectors of H are

$$|\phi_2\rangle = \begin{pmatrix} 0 \\ 0 \\ 1 \end{pmatrix} \quad \text{and} \quad |\phi_3\rangle = \frac{1}{\sqrt{2}}\begin{pmatrix} -1 \\ 1 \\ 0 \end{pmatrix}. \tag{3.10.29}$$

It is easy to check that these eigenvectors of H are orthonormal: $\langle \phi_j|\phi_j\rangle = \delta_{ij}$.

(b) If a measurement of the energy yields $-\mathscr{E}_0$, this means that the system is left in the state $|\phi_2\rangle$. When we next measure the observable, A, the system is in the state $|\phi_2\rangle$. The result we obtain for A is given by any of the eigenvalues of $\hat{A}$. A diagonalization of of the matrix A yields three non-degenerate values: $\alpha_1 = -\sqrt{17}\,\alpha, \alpha_2 = 0$, and $\alpha_3 = \sqrt{17}\,\alpha$; their respective eigenvectors are given by

$$|\alpha_1\rangle = \frac{1}{\sqrt{34}}\begin{pmatrix} 4 \\ -\sqrt{17} \\ 1 \end{pmatrix}, \quad |\alpha_2\rangle = \frac{1}{\sqrt{17}}\begin{pmatrix} 1 \\ 0 \\ -4 \end{pmatrix}, \tag{3.10.30}$$

$$|\alpha_3\rangle = \frac{1}{\sqrt{2}}\begin{pmatrix} 4 \\ \sqrt{17} \\ 1 \end{pmatrix}. \tag{3.10.31}$$

Thus, when measuring A on a system, which is in the state $|\phi_2\rangle$, the probability of finding $-\sqrt{17}a$ is given by

$$P(\alpha_1) = |\langle|\alpha_1|\phi_2\rangle|^2 = \left|\frac{1}{\sqrt{34}}\begin{pmatrix} 1 & -\sqrt{17} & 1 \end{pmatrix}\begin{pmatrix} 0 \\ 0 \\ 1 \end{pmatrix}\right|^2 = \frac{1}{34}. \tag{3.10.32}$$

Similarly, the probabilities of measuring 0 and $\sqrt{17}a$ are

$$P(\alpha_2) = |\langle|\alpha_1|\phi_2\rangle|^2 = \left|\frac{1}{\sqrt{17}}\begin{pmatrix} 1 & 0 & -4 \end{pmatrix}\begin{pmatrix} 0 \\ 0 \\ 1 \end{pmatrix}\right|^2 = \frac{16}{17}, \tag{3.10.33}$$

and

$$P(\alpha_3) = |\langle|\alpha_1|\phi_2\rangle|^2 = \left|\frac{1}{\sqrt{34}}\,(4 \quad \sqrt{17} \quad 1)\begin{pmatrix}0\\0\\1\end{pmatrix}\right|^2 = \frac{1}{34}. \tag{3.10.34}$$

(c) Since the system, when measuring A is in the state $|\phi_2\rangle$, the uncertainty ΔA is given by

$$\Delta A = \sqrt{\langle\phi_2|\hat{A}^2|\phi_2\rangle - (\langle\phi_2|\hat{A}|\phi_2\rangle)^2}. \tag{3.10.35}$$

We have

$$\langle\phi_2|\hat{A}|\phi_2\rangle = \alpha \,(0 \quad 0 \quad 1)\begin{pmatrix}0&4&0\\4&0&1\\0&1&0\end{pmatrix}\begin{pmatrix}0\\0\\1\end{pmatrix} = 0, \tag{3.10.36}$$

$$\langle\phi_2|A^2|\phi_2\rangle = \alpha^2\,(0 \quad 0 \quad 1)\begin{pmatrix}16&0&4\\0&17&0\\4&0&1\end{pmatrix}\begin{pmatrix}0\\0\\1\end{pmatrix} = \alpha^2. \tag{3.10.37}$$

The uncertainty in the measurement of A in the state ϕ_2 is

$$\Delta A = \sqrt{\langle\phi_2|A^2|\phi_2\rangle - \langle\phi_2|A|\phi_2\rangle^2} = \alpha. \tag{3.10.38}$$

Example 3.10.3: Consider a particle of mass m that is moving in a one-dimensional infinite square well potential with walls at $x = 0$ and $x = a$. At $t = 0$, it is in the state characterized by the wave function

$$\psi(x,0) = \frac{1}{\sqrt{2}}\left[\phi_1(x) + \phi_3(x)\right],$$

where $\phi_1(x)$ and $\phi_3(x)$ are the normalized wave functions for the ground and the second excited states, respectively. (a) Find the average value of $\hat{x}$, $\hat{x}^2$, $\hat{P}_x$ and $\hat{p}_x^2$ in the state $\psi(x,0)$. (b) Check whether the uncertainties in the position and the momentum of the particle satisfy the uncertainty relation or not.

Solution: Since $\phi_1(x)$ and $\phi_3(x)$ are normalized, the wave function ψ is also normalized:

$$\int_0^a |\psi(x,0)|^2\,dx = \frac{1}{2}\left[\int_0^a |\phi_1(x)|^2\,dx + \int_0^a |\phi_3(x)|^2\,dx\right] = \frac{(1+1)}{2} = 1. \tag{3.10.39}$$

Therefore, the average values of $\hat{x}$ is given by

$$\langle \hat{x} \rangle = \frac{1}{a}\left[\int_0^a x\sin^2\left(\frac{\pi x}{a}\right) dx + \int_0^a x\sin^2\left(\frac{3\pi x}{a}\right)\right] dx$$

$$= \frac{1}{2a}\int_0^a \left[2x - x\cos\left(\frac{2\pi x}{a}\right) - x\cos\left(\frac{6\pi x}{a}\right)\right] dx = \frac{a}{2}. \tag{3.10.40}$$

The average value of $\hat{x}^2$ is

$$\langle \hat{x}^2 \rangle = \frac{1}{a}\left[\int_0^a x^2\sin^2\left(\frac{\pi x}{a}\right) dx + \int_0^a x^2\sin^2\left(\frac{3\pi x}{a}\right)\right] dx$$

$$= \frac{1}{2a}\int_0^a \left[2x^2 - x^2\cos\left(\frac{2\pi x}{a}\right) - x^2\cos\left(\frac{6\pi x}{a}\right)\right] dx$$

$$= \frac{1}{2a}\left\{\frac{2a^3}{3} - \frac{a}{2\pi}\left[x^2\sin\left(\frac{2\pi x}{a}\right)\right]_0^a + \frac{a}{\pi}\int_0^a x\sin\left(\frac{2\pi x}{a}\right) dx\right\}$$

$$-\frac{1}{2a}\left\{\frac{a}{6\pi}\left[x^2\sin\left(\frac{6\pi x}{a}\right)\right]_0^a - \frac{a}{3\pi}\int_0^a x\sin\left(\frac{6\pi x}{a}\right) dx\right\}$$

$$= \frac{1}{2a}\left\{\frac{2a^3}{3} + \frac{a}{\pi}\int_0^a x\sin\left(\frac{2\pi x}{a}\right) dx + \frac{a}{3\pi}\int_0^a x\sin\left(\frac{6\pi x}{a}\right) dx\right\}. \tag{3.10.41}$$

Taking into account that

$$\int_0^a x\,\sin\left(\frac{2\pi}{a}x\right) dx = -\frac{a^2}{2\pi}, \quad \int_0^a x\,\sin\left(\frac{6\pi}{a}x\right) dx = -\frac{a^2}{6\pi}, \tag{3.10.42}$$

we obtain

$$\langle \hat{x}^2 \rangle = \frac{a^2}{3} - \frac{a^2}{4\pi^2} - \frac{a^2}{36\pi^2} = \frac{a^2}{3} - \frac{5a^2}{18\pi^2}. \tag{3.10.43}$$

The average value of $\hat{p}_x$ is

$$\langle \hat{p}_x \rangle = -\frac{i\hbar}{a}\int_0^a \left[\sin\left(\frac{\pi x}{a}\right) + \sin\left(\frac{3\pi x}{a}\right)\right]\left[\frac{\pi}{a}\cos\left(\frac{\pi x}{a}\right) + \frac{3\pi}{a}\cos\left(\frac{3\pi x}{a}\right)\right] dx$$

$$= -\frac{i\pi\hbar}{2a^2}\int_0^a \left[3\sin\left(\frac{6\pi x}{a}\right) + 4\sin\left(\frac{4\pi x}{a}\right) - \sin\left(\frac{2\pi x}{a}\right)\right] dx = 0. \tag{3.10.44}$$

Now we calculate the average value of $\hat{p}_x^2$:

$$\langle \hat{p}_x^2 \rangle = -\frac{\hbar^2}{a}\int_0^a \left[\sin\left(\frac{\pi x}{a}\right)+\sin\left(\frac{3\pi x}{a}\right)\right]\left[-\frac{\pi^2}{a^2}\sin\left(\frac{\pi x}{a}\right)-\frac{9\pi^2}{a^2}\sin\left(\frac{3\pi x}{a}\right)\right]dx$$
$$= \frac{\hbar^2\pi^2}{a^3}\int_0^a \sin^2\left(\frac{\pi x}{a}\right)dx + \frac{9\hbar^2\pi^2}{a^3}\int_0^a \sin\left(\frac{\pi x}{a}\right)\sin\left(\frac{3\pi x}{a}\right)dx$$
$$+ \int_0^a \frac{9\hbar^2\pi^2}{a^3}\sin^2\left(\frac{3\pi x}{a}\right)dx + \frac{\hbar^2\pi^2}{a^3}\int_0^a \sin\left(\frac{\pi x}{a}\right)\sin\left(\frac{3\pi x}{a}\right)dx. \quad (3.10.45)$$

The second and the fourth integral on the right-hand side of (3.10.48) equal zero because of the orthonormality of the eigenfunctions of the infinite potential well. For the rest of the integrals, using the trigonometrical formula $\sin^2 x = (1-\cos(2x))/2$, we obtain

$$\langle \hat{p}_x^2 \rangle = \frac{\pi^2\hbar^2}{a^3}\left(\frac{a}{2}+\frac{9a}{2}\right) = \frac{5\pi^2\hbar^2}{a^2}. \quad (3.10.46)$$

The uncertainty in the measurement of the position is

$$\Delta x = \sqrt{\langle \hat{x}^2\rangle - \langle \hat{x}\rangle^2} = \sqrt{\frac{a^2}{3} - \frac{5a^2}{18\pi^2} - \frac{a^2}{4}} = \frac{a}{6\pi}\sqrt{3\pi^2-10} \approx 0.74\frac{a}{\pi}. \quad (3.10.47)$$

Similarly, the uncertainty in the measurement of momentum is

$$\Delta p_x = \sqrt{\langle \hat{p}_x^2\rangle - \langle \hat{p}_x\rangle^2} = \sqrt{\frac{5\pi^2\hbar^2}{a^2}} = \frac{\pi\hbar}{a}\sqrt{5}. \quad (3.10.48)$$

As a consequence,

$$\Delta x\,\Delta p_x = 0.74 \times \frac{a}{\pi} \times \frac{\pi\hbar}{a}\sqrt{5} \approx 1.655\,\hbar > \frac{\hbar}{2}. \quad (3.10.49)$$

Hence, the product $\Delta x\,\Delta p_x$ satisfies the uncertainty relation.

3.11 Quantum–Classical Correspondence and Ehrenfest's Theorem

In this section, we shall discuss the possibility of establishing a connection between classical and quantum mechanics. Intuitively, we expect the average value of an observable to play the key role in this regard. Therefore, we should try to deduce the time-evolution of the expectation value of an operator from the given formalism of quantum mechanics and see whether it compares with the time-evolution of the corresponding classical dynamical variable or not. Since Ehrenfest's theorem holds good in general, while deducing the theorem, we shall not confine ourselves to one spatial dimension.

Time derivative of an operator: In general, an operator, representing an observable, depends on time as a parameter. We wish to determine its time derivative. Since an observable cannot have a definite value at a given instant of time (the measurement can yield any one of its eigenvalues with some corresponding probability). Therefore, it is not possible to define the time derivative of an operator in the usual way of mathematical analysis:

$$\frac{d\hat{A}(t)}{dt} = \lim_{\Delta t \to 0} \frac{\hat{A}(t+\Delta t) - \hat{A}(t)}{\Delta t}. \tag{3.11.1}$$

However, the expectation (average) value of the observable A, given by $\langle \hat{A} \rangle$, can have a definite value at a given instant t. Therefore, for defining the time derivative of an operator, we must use its expectation value rather than the operator itself. Hence, we adopt the following proposal:

The time derivative of the expectation value, $\langle \hat{A} \rangle$, of the observable, is equal to the expectation value of the time derivative of the operator $\hat{A}$ itself. That means:

$$\frac{d\langle \hat{A} \rangle}{dt} = \left\langle \frac{d\hat{A}}{dt} \right\rangle. \tag{3.11.2}$$

In the context of quantum mechanics, this proposal should be viewed as the definition of the dynamical variable dA/dt whose operator in quantum mechanics is given by $d\hat{A}/dt$.

According to the formalism of quantum mechanics, we have

$$\langle \hat{A} \rangle = \int_{-\infty}^{+\infty} \psi^*(\vec{r},t)\hat{A}\psi(\vec{r},t)\, d\tau, \tag{3.11.3}$$

where $d\tau = dxdydz$. Therefore,

$$\frac{d\langle \hat{A} \rangle}{dt} = \int_{-\infty}^{+\infty} \left(\frac{\partial \psi^*}{\partial t}\hat{A}\psi + \psi^* \frac{\partial \hat{A}}{\partial t}\psi + \psi^* \hat{A} \frac{\partial \psi}{\partial t} \right) d\tau. \tag{3.11.4}$$

Using the time-dependent Schrödinger equations, we have

$$\frac{\partial \psi}{\partial t} = \frac{1}{i\hbar}\hat{H}\psi, \quad \frac{\partial \psi^*}{\partial t} = -\frac{1}{i\hbar}\psi^*\hat{H}^\dagger = -\frac{1}{i\hbar}\psi^*\hat{H}, \tag{3.11.5}$$

where $\hat{H}$ is the Hamiltonian operator which is hermitian and we have used the general formula $(\hat{F}\hat{G})^\dagger = \hat{G}^\dagger \hat{F}^\dagger$ for any two operators (matrices) $\hat{F}$ and $\hat{G}$. From (3.11.4) and (3.11.5), we get

$$\frac{d\langle\hat{A}\rangle}{dt} = \int_{-\infty}^{+\infty} \psi^* \left[\frac{\partial\hat{A}}{\partial t} + \frac{1}{i\hbar}\left(-\hat{H}\hat{A}+\hat{A}\hat{H}\right)\right] \psi\, d\tau. \tag{3.11.6}$$

Recollecting that

$$\frac{\partial\langle\hat{A}\rangle}{\partial t} = \left\langle\frac{\partial\hat{A}}{\partial t}\right\rangle \quad \text{where} \quad \left\langle\frac{\partial\hat{A}}{\partial t}\right\rangle = \int_{-\infty}^{+\infty} \psi^*(\vec{r},t)\frac{\partial\hat{A}}{\partial t}\psi(\vec{r},t)\, d\tau, \tag{3.11.7}$$

we arrive at

$$\frac{d\langle\hat{A}\rangle}{dt} = \frac{\partial\langle\hat{A}\rangle}{\partial t} + \frac{1}{i\hbar}\int_{-\infty}^{+\infty} \psi^*(\vec{r},t)\left(-\hat{H}\hat{A}+\hat{A}\hat{H}\right)\psi(\vec{r},t)\, d\tau. \tag{3.11.8}$$

Equation (3.11.8) can be written as

$$\frac{d\langle\hat{A}\rangle}{dt} = \frac{\partial\langle A\rangle}{\partial t} + \frac{1}{i\hbar}\left\langle[\hat{A},\hat{H}]\right\rangle, \tag{3.11.9}$$

where $[\hat{A},\hat{H}] = \hat{A}\hat{H} - \hat{H}\hat{A}$ is the commutator of the operator $\hat{A}$ with the Hamiltonian $\hat{H}$ and

$$\left\langle[\hat{A},\hat{H}]\right\rangle = \int_{-\infty}^{+\infty} \psi^*(\vec{r},t)\left(\hat{A}\hat{H} - \hat{H}\hat{A}\right)\psi(\vec{r},t)\, d\tau, \tag{3.11.10}$$

the average value of the commutator in the state $\psi(\vec{r},t)$ at a given instant t. In the case when there is no explicit dependence of the operator $\hat{A}$ on time, we have

$$\frac{d\langle\hat{A}\rangle}{dt} = \frac{1}{i\hbar}\left\langle[\hat{A},\hat{H}]\right\rangle. \tag{3.11.11}$$

A comparison of (3.11.9) and (3.11.11) with the Poisson bracket formulation of classical mechanics leads to an important conclusion which is known as Ehrenfest's theorem in quantum mechanics.

Ehrenfest's theorem: The average values of observables in quantum mechanics obey the classical equations of motion.

This theorem, to some extent, establishes a '*bridge*' between classical mechanics and quantum mechanics, which is impossible to have otherwise.

Quantum mechanical version of Newton's equations of motion: It turns out that it is possible to write the equations of motion for the expectation values of the position momentum operators in a manner completely analogous to the equations of motion in classical mechanics.

Since the position operator, $\hat{x}$, and the momentum operator, $\hat{p}_x$, do not depend explicitly on time, it follows from (3.11.11) that

$$\frac{d\langle \hat{x} \rangle}{dt} = \frac{1}{i\hbar}\langle [\hat{x}, \hat{H}] \rangle, \tag{3.11.12}$$

$$\frac{d\langle \hat{p}_x \rangle}{dt} = \frac{1}{i\hbar}\langle [\hat{p}_x, \hat{H}] \rangle. \tag{3.11.13}$$

Here,

$$\hat{H} = \frac{\hat{p}_x^2}{2m} + V(x) \tag{3.11.14}$$

is the Hamiltonian operator and $\hat{V}(\hat{x}) = V(x)$ is the potential energy operator for the particle. Simplifying the commutator on the right hand-side of (3.11.12), we have

$$[\hat{x}, \hat{H}] = \left[\hat{x}, \frac{\hat{p}_x^2}{2m} + V(x)\right] = \frac{1}{2m}[\hat{x}, \hat{p}_x^2] + [\hat{x}, V(x)]. \tag{3.11.15}$$

Using the distributive property $[\hat{A}, \hat{B}^2] = \hat{B}[\hat{A}, \hat{B}] + [\hat{A}, \hat{B}]\hat{B}$ and the fact that $[\hat{x}, V(x)] = 0$, we get

$$[\hat{x}, \hat{H}] = \frac{1}{2m}\hat{p}_x[\hat{x}, \hat{p}_x] + [\hat{x}, \hat{p}_x]\hat{p}_x = \frac{1}{2m}(2i\hbar\, \hat{p}_x) = i\hbar\frac{\hat{p}_x}{m}. \tag{3.11.16}$$

As a result, the time evolution equation for $\langle \hat{x} \rangle$ reads

$$\frac{d\langle \hat{x} \rangle}{dt} = \frac{\langle \hat{p}_x \rangle}{m}. \tag{3.11.17}$$

We see that the relation between the time derivative of $\langle \hat{x} \rangle$ and the expectation value of the momentum operator in quantum mechanics is exactly the same as that between momentum p_x and velocity $v_x = \dot{x}$ in classical mechanics.

Let us now compute the commutator $[\hat{p}_x, \hat{H}]$. Since

$$[\hat{p}_x, \hat{p}_x^2] = \hat{p}_x[\hat{p}_x, \hat{p}_x] + [\hat{p}_x, \hat{p}_x]\hat{p}_x = 0, \tag{3.11.18}$$

and

$$[\hat{p}_x, f(x)] = -i\hbar\frac{\partial f(x)}{\partial x}, \tag{3.11.19}$$

for any function $f(x)$ of x, we obtain

$$[\hat{p}_x, \hat{H}] = -i\hbar\frac{\partial V(x)}{\partial x}. \tag{3.11.20}$$

Making use of Ehrenfest's theorem (3.11.11), we obtain

$$\frac{d\langle\hat{p}_x\rangle}{dt} = -\left\langle\frac{\partial V(x)}{\partial x}\right\rangle. \tag{3.11.21}$$

Differentiating (3.11.17) once with respect to time and making use of (3.11.21), we arrive at

$$\frac{d^2\langle\hat{x}\rangle}{dt^2} = -\left\langle\frac{\partial V(x)}{\partial x}\right\rangle. \tag{3.11.22}$$

Equation (3.11.22), written for the expectation values of the position operator and the force as the gradient of the potential, is the quantum mechanical version of Newton's equations of motion.

3.12 Periodic Potentials, Bloch's Theorem and Energy Bands

In this sub-section we shall discuss the solutions of the TISE for the case in which the potential is a periodic function of x. It has some very useful applications in solid state physics.

A typical periodic potential is shown in Fig.3.10. As shown, the potential is zero over a distance a, peaks at $V(x) = V_0$ over a distance b and then repeats itself. It is evident that

$$V(x+c) = V(x). \tag{3.12.1}$$

where $c = a+b$ is the period. Since the potential is a periodic function of x with a period c, the Schrödinger equation is invariant under space translations

$$x \to x+nc, \quad n = 0, \pm 1, \pm 2, \pm 3, ... \tag{3.12.2}$$

This invariance imposes certain restriction on the form of the allowable solution of the Schrödinger equation. To determine this restriction, let us introduce an operator $\hat{D}$, called the space translation operator, which while acting on a function $f(x)$ shifts it horizontally along the x direction over a distance c:

$$\hat{D}f(x) = f(x+c). \tag{3.12.3}$$

For instance, acting on the potential function $V(x)$, it shifts the entire potential over a distance c: $\hat{D}V(x) = V(x+c)$. Repeated applications this operator leads to

$$\hat{D}f(x) = f(x+c), \hat{D}^2 f(x) = f(x+2c), \hat{D}^3 f(x) = f(x+3c), \ldots, \hat{D}^n f(x) = f(x+nc). \tag{3.12.4}$$

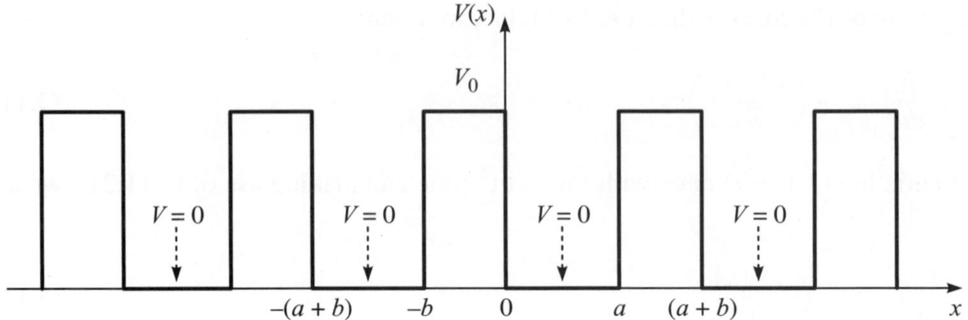

Figure 3.10 This is the sketch of a representative periodic potential with a separation of a between the peaks of width b.

Consider now the following

$$
\begin{aligned}
(\hat{D}\hat{H})\psi(x) = \hat{D}(\hat{H}\psi) &= \hat{D}\left(-\frac{\hbar^2}{2m}\frac{\partial^2}{\partial x^2}+V(x)\right)\psi(x) \\
&= \left(-\frac{\hbar^2}{2m}\frac{\partial^2}{\partial x^2}+V(x+c)\right)\psi(x+c) \\
&= \left(-\frac{\hbar^2}{2m}\frac{\partial^2}{\partial x^2}+V(x)\right)\psi(x+c) \\
&= \hat{H}\left(\hat{D}\psi(x)\right) = \left(\hat{H}\hat{D}\right)\psi(x).
\end{aligned} \tag{3.12.5}
$$

In obtaining the above result we have used the fact that

$$
\frac{\partial}{\partial(x+c)} = \frac{\partial}{\partial x}\frac{\partial x}{\partial(x+c)} = \frac{\partial}{\partial x}. \tag{3.12.6}
$$

Thus, the Hamiltonian and the translation operator commute: $[\hat{D},\hat{H}] = 0$. It means that, if $\psi(x)$ is an eigenfunction of the Hamiltonian with energy E (i.e., $\hat{H}\psi(x) = E\psi(x)$), then $\hat{D}\psi(x)$ is also an eigenfunction of the Hamiltonian with the same energy E:

$$
\hat{H}\left(\hat{D}\psi(x)\right) = \left(\hat{H}\hat{D}\right)\psi(x) = \left(\hat{D}\hat{H}\right)\psi(x) = E\left(\hat{D}\psi(x)\right), \tag{3.12.7}
$$

This, in turn means that, if the energy spectrum is non-degenerate, $\psi(x+c)$ and $\psi(x)$ must represent the same state of the system. Therefore, $\psi(x+c)$ can differ from $\psi(x)$ only by a constant factor:

$$
\psi(x+c) = \alpha\,\psi(x), \tag{3.12.8}
$$

where α is a constant of magnitude unity.

$$\alpha = \exp\left(\frac{2\pi i\ell}{n}\right), \qquad \ell = 0, 1, 2, 3, \ldots \tag{3.12.9}$$

Defining now

$$\kappa = \frac{2\pi\ell}{nc}, \tag{3.12.10}$$

we arrive at

$$\psi(x+nc) = e^{i\kappa c}\psi(x). \tag{3.12.11}$$

Now, any function $\psi(x)$, satisfying the above condition, can be written as

$$\psi(x) = e^{i\kappa x}u_\kappa(x), \tag{3.12.12}$$

where $u_\kappa(x)$ is a periodic function of x of period c: $u_\kappa(x+c) = u_\kappa(x)$. To ensure that it is really so, we write

$$\psi(x+c) = e^{i\kappa\,(x+c)}u_\kappa(x+c) = e^{i\kappa c}\,e^{i\kappa x}\,u_\kappa(x+c). \tag{3.12.13}$$

Therefore, if $u_\kappa(x+c) = u_\kappa(x)$,

$$\psi(x+c) = e^{i\kappa(x+c)}u_\kappa(x+c) = e^{i\kappa c}\,e^{i\kappa x}\,u_\kappa(x) = e^{i\kappa c}\,\psi(x). \tag{3.12.14}$$

The above result is a fundamental result for condensed matter physics and it is known as *Bloch's theorem.* It states that any solution to the Schrödinger equation, with a periodic potential of period c, must have the form given by equation (3.12.14).

Consider now the case of a particle (mass m and total energy $E < V_0$) subject to the above periodic potential. If we introduce

$$k_1^2 = \frac{2mE}{\hbar^2}, \tag{3.12.15}$$

$$k_2^2 = \frac{2m(V_0 - E)}{\hbar^2}, \tag{3.12.16}$$

the solutions of the time-independent Schrödinger equations in the relevant regions can be written as

$$\psi(x) = A\,\cos(k_1 x) + B\,\sin(k_1 x), \qquad (0 < x < a), \tag{3.12.17}$$

$$\psi(x) = C\,\cosh(k_2 x) + D\,\sinh(k_2 x), \qquad (-b < x < 0), \tag{3.12.18}$$

where A, B, C and D are arbitrary constants. They must be chosen such that both $\psi(x)$ and $\psi'(x)$ are continuous at the boundaries, where the potential has a finite jump, and abide by Bloch's theorem.

At $x = 0$, we have

$$A = C, \tag{3.12.19}$$

$$k_1 B = k_2 D. \tag{3.12.20}$$

Furthermore, using the Bloch theorem (with $n = 1$), we get

$$\psi(a) = e^{iKc}\,\psi(-b), \tag{3.12.21}$$

$$\psi'(a) = e^{iKc}\,\psi'(-b), \tag{3.12.22}$$

where

$$K = \frac{2\pi\ell}{(a+b)}. \tag{3.12.23}$$

The boundary conditions (3.12.21) and (3.12.22) lead to

$$A\cos(k_1 a) + B\sin(k_1 a) = e^{iKc}\left[C\cosh(k_2 b) - D\sinh(k_2 b)\right], \tag{3.12.24}$$

$$-k_1 A\sin(k_1 a) + k_1 B\cos(k_1 a) = e^{iKc}\left[-k_2 C\sinh(k_2 b) + k_2 D\cosh(k_2 b)\right]. \tag{3.12.25}$$

The algebraic equations (3.12.19), (3.12.20), (3.12.24), and (3.12.25), can be written as a matrix equation: $\mathscr{M}X = 0$, where $X = (A\ B\ C\ D)^T$ is a column matrix and

$$\mathscr{M} = \begin{pmatrix} 1 & 0 & -1 & 0 \\ 0 & k_1 & 0 & -k_2 \\ \cos(k_1 a) & \sin(k_1 a) & -e^{iKc}\cosh(k_2 b) & e^{iKc}\sinh(k_2 b) \\ -k_1\sin(k_1 a) & k_1\cos(k_1 a) & k_2 e^{iKc}\sinh(k_2 b) & -k_2 e^{iKc}\cosh(k_2 b) \end{pmatrix}. \tag{3.12.26}$$

For the non-trivial solutions the determinant of the matrix, $|\mathscr{M}|$, must be zero:

$$|\mathscr{M}| = \begin{vmatrix} 1 & 0 & -1 & 0 \\ 0 & k_1 & 0 & -k_2 \\ \cos(k_1 a) & \sin(k_1 a) & -e^{iKc}\cosh(k_2 b) & e^{iKc}\sinh(k_2 b) \\ -k_1\sin(k_1 a) & k_1\cos(k_1 a) & k_2 e^{iKc}\sinh(k_2 b) & -k_2 e^{iKc}\cosh(k_2 b) \end{vmatrix} = 0. \tag{3.12.27}$$

Using the Gaussian decomposition rule for the determinants, we have

$$|\mathscr{M}| = \begin{vmatrix} k_1 & 0 & -k_2 \\ \sin(k_1 a) & -e^{iKc}\cosh(k_2 b) & e^{iKc}\sinh(k_2 b) \\ k_1\cos(k_1 a) & k_2\, e^{iKc}\sinh(k_2 b) & -k_2\, e^{iKc}\cosh(k_2 b) \end{vmatrix}$$

$$- \begin{vmatrix} 0 & k_1 & -k_2 \\ \cos(k_1 a) & \sin(k_1 a) & e^{iKc}\sinh(k_2 b) \\ -k_1\sin(k_1 a) & k_1\cos(k_1 a) & -k_2\, e^{iKc}\cosh(k_2 b) \end{vmatrix} = 0. \qquad (3.12.28)$$

Further simplification of the determinants leads to

$$\begin{aligned} & k_1\left[k_2\, e^{2iKc}\cosh^2(k_2 b) - k_2\, e^{2iKc}\sinh^2(k_2 b)\right] - k_2\left[k_2\, e^{iKc}\sinh(k_2 b)\sin(k_1 a)\right. \\ & \left. + k_1\, e^{iKc}\cosh(k_2 b)\cos(k_1 a)\right] + k_1\left[-k_2\, e^{iKc}\cosh(k_2 b)\cos(k_1 a)\right. \\ & \left. + k_1\, e^{iKc}\sinh(k_2 b)\sin(k_1 a)\right] + k_2\left[k_1\cos^2(k_1 a) + k_1\sin^2(k_1 a)\right] = 0. \qquad (3.12.29) \end{aligned}$$

Opening up the brackets, we get

$$\begin{aligned} & k_1 k_2\, e^{2iKc} - k_2^2\, e^{iKc}\sinh(k_2 b)\sin(k_1 a) - k_1 k_2\, e^{iKc}\cosh(k_2 b)\cos(k_1 a) \\ & - k_1 k_2\, e^{iKc}\cosh(k_2 b)\cos(k_1 a) + k_1^2\, e^{iKc}\sinh(k_2 b)\sin(k_1 a) + k_1 k_2 = 0 \qquad (3.12.30) \end{aligned}$$

From here, collecting similar terms, we obtain

$$(k_1^2 - k_2^2)\sinh(k_2 b)\sin(k_1 a) - 2k_1 k_2\cosh(k_2 b)\cos(k_1 a) + k_1 k_2\left[e^{iKc} + e^{-iKc}\right] = 0. \qquad (3.12.31)$$

Equation (3.12.31) yields the following transcendental equation for the determination of the energy eigenvalues

$$\frac{(k_2^2 - k_1^2)}{2k_1 k_2}\sinh(k_2 b)\sin(k_1 a) + \cosh(k_2 b)\cos(k_1 a) = \cos[K(a+b)]. \qquad (3.12.32)$$

In general, the equation (3.12.32) cannot be solved analytically. For given values of the model parameters a, b and V_0, it can be solved numerically. It is usually done by using one of the standard root finding algorithms for a given value of K. As a result of the numerical solution, one gets the values of k_1 using which one can calculate the energy eigenvalues as

$$E = \frac{\hbar^2 k_1^2}{2m}. \qquad (3.12.33)$$

***Energy Bands*:** Note that, for practical purposes, the above transcendental equation can be simplified by imposing some reasonable restrictions on the model parameters.

Assume that the width of the potential tends to zero while the height tends to infinity such that $V_0 b$ remains constant. In such a limit

$$\lim_{b\to 0} \sinh(k_2 b) = k_2 b, \qquad \lim_{b\to 0} \cos(k_2 b) = 1.$$

Here, we have gone to the leading order in the Taylor expansions of the hyperbolic trigonometric functions on the left-hand side, and simply let $b = 0$ on the right-hand side. We obtain

$$\frac{(k_2^2 - k_1^2)}{2k_1} b \sin(k_1 a) + \cos(k_1 a) = \cos[Ka]. \tag{3.12.34}$$

We then find it convenient to define the dimensionless quantity, $P = \frac{mV_0 ba}{2}$, which determines the effective strength of the potential. Then we have

$$F(k_1 a) = \cos[Ka], \tag{3.12.35}$$

where

$$F(k_1 a) = P\frac{\sin(k_1 a)}{k_1} + \cos(k_1 a). \tag{3.12.36}$$

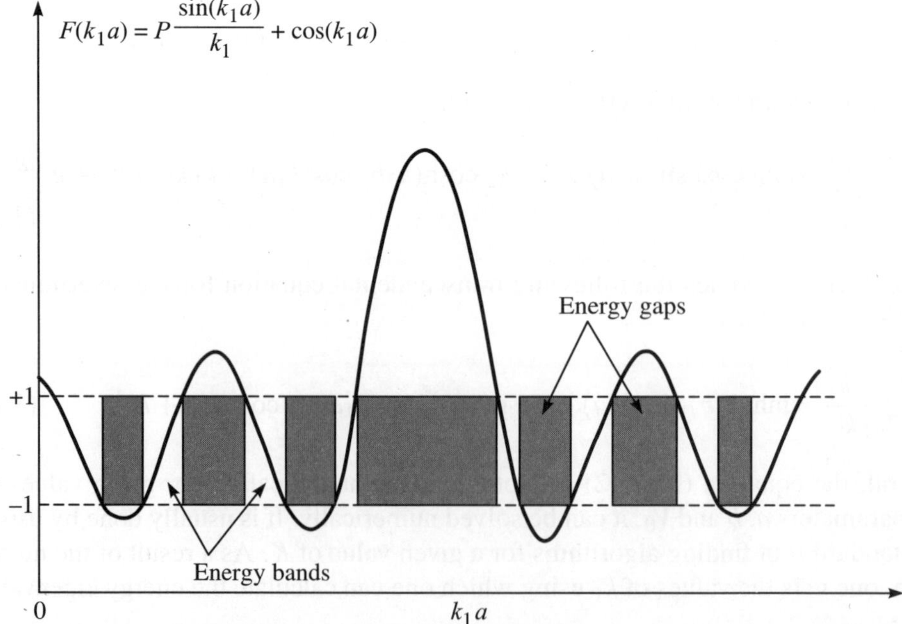

Figure 3.11 A schematic representation of the allowed and forbidden energy bands.

In this form, for a fixed value of P, it is rather simple to plot the left-hand side as a function of k_1a.When the value of the left hand side of the equation (3.12.35) (that is the value of $F(k_1a)$) is between -1 and $+1$ (which is the range of $\cos(Ka)$ for real arguments) there will be a value of K for which a solution exists. When the value of $F(k_1a)$ is outside of this interval, there will be no real K for which (3.12.35) is satisfied. In other words, for the interval of k_1 in which the values of $F(k_1a)$ lower than -1 or greater than $+1$, there will not be any acceptable energy eigenfunction. The resultant spectrum of solutions will then have gaps in the admissible energies. It means that, for certain energies, there will be no proper solutions to the the Schrödinger equation with periodic potential. Thus the energy spectrum will consist of bands of continuous energies separated by energy gaps. The situation has been depicted schematically in Fig.3.11.

Homework Problems

1. Show that the group velocity for a non-relativistic free electron is also given by $v_g = p/m_e = v_0$, where v_0 is the electron's velocity.
2. The dispersion relation for free relativistic electron waves is

$$\omega(k) = \sqrt{k^2c^2 + (m_ec^2/\hbar)^2},$$

   Obtain expressions for the phase velocity v_p and group velocity v_g of these waves and show that their product is a constant, independent of k. From your result, what can you conclude about v_g if $v_p > c$?
3. If the wave function of the particle at $t = 0$ is given by

$$\psi(x,0) = \left(\frac{2}{\pi a^2}\right)^{1/4} e^{-x^2/a^2},$$

   use (3.2.12) to calculate $\tilde{\psi}(k)$. Calculate Δx and Δk at $t = 0$. Now, use $\tilde{\psi}(k)$ in (3.2.11) to compute $\psi(x,t)$ for any $t > 0$ by expanding the dispersion relation $k(\omega)$ into a Taylor series and taking the required integral. On the basis of the obtained results, comment on the half-width Δx of the packet for $t > 0$.
4. If the ground-state energy of an electron in a one-dimensional infinite square well potential were of the same magnitude as that of hydrogen in the ground state, how would the width of the box compare to the Bohr radius?
5. Consider a potential well having an infinite wall at $x = 0$ and a wall of height V_0 at $x = L$. For the case $E < V_0$, obtain solutions to the Schrödinger equation inside the well ($0 \le x \le L$) and in the region beyond ($x > L$) that satisfy the appropriate boundary conditions at $x = 0$ and $x = \infty$. Enforce the proper matching conditions at $x = L$ to find an equation for the allowed energies of this system. Are there conditions for which no solution is possible? Explain.

6. An electron, trapped at a defect in a crystal, can be modeled as one moving freely in a one-dimensional infinite potential well of width, say, $L = 1$ nm with infinite walls at $x = 0$ and $x = L$. (a) Sketch the wavefunctions and probability densities for the $n = 1$ and $n = 2$ states. (b) For the $n = 1$ and $n = 2$ states, calculate the probability of finding the electron between $x_1 = 0.15$ nm and $x_2 = 0.35$ nm. (c) Calculate the energies in electron volts of the $n = 1$ and $n = 2$ states.
7. A laser emits light of wavelength $\lambda = 800$ nm. If this light is due to transitions from the $n = 2$ state to the $n = 1$ state of an electron in an infinite potential well, find the width, L, of the well.
8. Consider an electron trapped in an infinite potential well, with length $L = 20$ nm (1 nm $=10^{-9}$ m, the electron mass is $m_e = 9.11 \times 10^{-31}$ kg). What is the energy E_5 of the fifth level? And whats the energy of the first (lowest) level? If electron drops down from the fourth to the first level, the corresponding energy difference has to be released, for example in the form of a photon. What would be the wavelength of this photon?
9. A free particle of mass, m, and total energy, E, is incident from $x \to -\infty$ on a potential step given by

$$V(x) = \begin{cases} 0 & \text{for} \quad x < 0 \\ V_0 > 0 & \text{for} \quad x \geq 0. \end{cases} \tag{3.12.37}$$

where $V_0 > 0$ is a positive constant with the dimensions of energy. Solve the corresponding TISE, apply the appropriate boundary conditions and determine the wave function for the case when $E > V_0$.

10. A particle of mass m and total energy, $-E$ with $E > 0$, is subject to the potential given by

$$V(x) = -\alpha\delta(x-a),$$

where α and a are positive constants. Solve the Schrödinger equation for the bound states and find the normalized wave functions and the corresponding energies. How many bound states the particle can have in such a potential?

11. A particle of mass m and total energy E is subject to a potential given by

$$V(x) = \begin{cases} \infty & \text{for} \quad x < 0 \\ 0 & \text{for} \quad 0 < x < a, \\ V_0 > 0 & \text{for} \quad x \geq a. \end{cases} \tag{3.12.38}$$

where $E < V_0$. Find the energy levels and the corresponding wave functions.

12. A particle of mass m and total energy, $-E$ ($E > 0$), is subject to the potential given by

$$V(x) = -\alpha\,[\delta(x-a)+\delta(x+a)]$$

where α and a are positive constants. Sketch the potential. Determine the number of bound states supported by this potential. Find the allowed values of energy for $\alpha = \hbar^2/ma$ and $\alpha = \hbar^2/4ma$.

13. A particle, moving freely in an infinite potential well of width a ($0 < x < a$), is in a state described by the wave function

$$\psi(x,t) = \frac{1}{\sqrt{a}}\sin\left(\frac{3\pi x}{a}\right)e^{-\frac{i}{\hbar}E_3 t} + \frac{1}{\sqrt{a}}\sin\left(\frac{5\pi x}{a}\right)e^{-\frac{i}{\hbar}E_5 t}.$$

Calculate the probability current density.

14. Using the continuity equation find the expression for dP_{ab}/dt in terms of the current density, where P_{ab} is the probability of locating the particle in the interval $a \le x \le b$.

15. A particle is free to move in a confined region of space $-a \le x \le a$. At $t = 0$ it is in a state

$$\psi(x,0) = \frac{1}{\sqrt{5a}}\cos\left(\frac{\pi x}{2a}\right) + \frac{2}{\sqrt{5a}}\sin\left(\frac{\pi x}{a}\right).$$

(a) Find $\psi(x,t)$ at a later time t. (b) Calculate the probability density $\rho(x,t)$ and the probability current density $j_x(x,t)$. (c) Verify that the probability is conserved:

$$\frac{\partial\rho}{\partial t} + \frac{\partial j_x(x,t)}{\partial x} = 0.$$

16. Consider an electron of energy 5.1 eV approaching an energy barrier of height 6.8 eV and thickness $L = 750$ pm. What is the transmission coefficient $\mathscr{T}$?

17. A 1500 kg car moving at 20 m/s approaches a hill 24 m high and 30 m long. What is the probability that the car will tunnel quantum mechanically through the hill?

18. Find the reflection and the transmission coefficients for a particle of mass m incident from the left, i.e., from $x \to -\infty$ on the step-potential

$$V(x) = \begin{cases} 0 & \text{for } x < 0, \\ V_0 & \text{for } x > 0, \end{cases}$$

where V_0 is a real positive constant. Consider both the cases when (i) the energy of the incident particle is less than the height of the barrier, i.e., $E < V_0$ and (ii) the energy of the particle is greater than the height of the barrier, i.e., $E > V_0$.

19. Find the reflection and the transmission coefficients for a particle of mass m incident from the left, i.e., from $x \to -\infty$ on the attractive delta potential

$$V(x) = -\alpha\delta(x),$$

where α is a real positive constant. The energy of the incident particle is negative i.e., $E < 0$.

20. The wave function of a particle is given by

$$\psi(x) = \sqrt{\frac{30}{a^5}}\,(ax - x^2), \quad (0 < x < a).$$

Calculate the average values of $\hat{x}$ and $\hat{x}^2$. Compute the uncertainty in measurement of the position of the particle in this state.

21. A particle is in the ground state of the harmonic oscillator with the classical frequency ω, when suddenly the spring constant quadruples without changing the state of the particle, so that $\omega' = 2\omega$. If the energy of the particle is now measured, what is the probability of getting the value $\hbar\omega/2$? What is the probability of getting the value $\hbar\omega$?

22. A particle is trapped inside a harmonic oscillator potential. At $t = 0$, the particle's wave function is

$$\psi(x,0) = A\,[\phi_0(x) + \phi_1(x)],$$

where ϕ_0 and ϕ_1 are the ground state and the first excited state wave functions of the particle.

(a) Find $\psi(x,t)$ and the probability density $\rho(x,t) = |\psi(x,t)|^2$.

(b) Calculate the expectation value of $\hat{x}$ at $t > 0$. Find the period of its time variation.

(c) Recall Ehrenfest's theorem and use the result of (b) to obtain the average value of the momentum in the state $\psi(x,t)$.

23. Consider the one dimensional harmonic oscillator potential, with energy levels: $E_n = \left(n + \frac{1}{2}\right), n = 0, 1, 2, 3, ...,$ which are non-degenerate. How do the energy levels and the degeneracy change, if the potential is modified as

$$V(x) = \begin{cases} \frac{1}{2}m\omega^2 x^2, & \text{for } x < 0, \\ \infty, & \text{for } x \geq 0? \end{cases}$$

24. The Hamiltonian of a two dimensional harmonic oscillator is given by

$$\hat{H} = \frac{1}{2}(\hat{p}_x^2 + \hat{p}_y^2) + \frac{1}{2}m\omega^2(x^2 + y^2).$$

Use the separation of variables to reduce the solution of the corresponding TISE to the solution of the TISE in one spatial dimension and determine the wave functions and the corresponding energy levels.

25. Find the average value of $\hat{x}$, $\hat{x}^2$, $\hat{p}_x$, and $\hat{p}_x^2$ in the second excited state of the harmonic oscillator. Check the validity of the Heisenberg's uncertainty relation.
26. A particle of mass m is subjected to a symmetric infinite potential well

$$V(x) = \begin{cases} 0 & -L/2 < x < L/2 \\ \infty & x < -L/2, x > L/2, \end{cases}$$

where L is a real positive constant. Calculate the product $\Delta x \Delta p_x$ for the lowest even and the lowest odd states of the particle in this well.
27. Consider a particle in a one-dimensional harmonic oscillator potential. At time t = 0 the particle is in the state

$$\psi(x,0) = \frac{1}{\sqrt{2}} \left[\phi_0(x) + \phi_1(x)\right],$$

where ϕ_0 and ϕ_1 are the ground state and the first excited state wave functions of a particle in the harmonic oscillator potential. Evaluate the expectation values $\langle \hat{x} \rangle$ and $\langle \hat{p}_x \rangle$ as functions of time.
28. Electrons move in a series of equidistant δ-barriers, with distance between two neighboring ones equal to ℓ. Each potential has the form of $V(x) = \alpha\delta(x)$, so that the total potential can be written as

$$V(x) = \sum_{N=-\infty}^{+\infty} \alpha\,\delta(x - N\ell), \quad \alpha > 0,$$

where N is an integer. (a) Find the energy eigenstates. (b) Derive the quantization condition for the energy eigenstates. (c) Show graphically that the allowed energy states form bands.

Chapter 4

Algebraic Formulation of Quantum Mechanics

As we know quantum mechanics was initially developed in two seemingly different but physically equivalent forms: matrix mechanics (Heisenberg, 1925) and wave mechanics (Schrödinger, 1926). Later, Dirac invented a more general formalism and showed that both these formulations of quantum mechanics (i.e., wave mechanics and matrix mechanics) could be cast into a single algebraic framework with the help of the theory of linear vector spaces and the operators acting in them. It turned out that Dirac's formalism was a unified one in the sense that, if it was constructed using a continuous basis, it led to Schrödinger's wave mechanics, while its representation in a discrete basis resulted into Heisenberg's matrix mechanics.

The linear vector space that one has to use for such a formulation has to be a complex one (the elements are complex numbers/functions) because of the complex nature of the wave function discussed earlier. Apart from that it turned out that the complex linear vector space in quantum mechanics has to have some specific properties: it has to be an infinite dimensional vector space equipped with an inner (scalar) product. It is called a Hilbert space and we shall denote it by $\mathscr{H}$. A rigorous definition of the Hilbert space will be given later.

4.1 Linear Vector Spaces

Let us start by defining a linear vector space and listing out its properties. In general, a linear vector space consists of a set of elements $\psi, \phi, \chi, \ldots$ (called vectors) and a set of numbers $a, b, c, \ldots$ (called scalars), a set of rules each for the addition and multiplication of vectors. The meaning of a vector is assumed to be independent of the coordinate system chosen to represent its components.

Definition: A linear vector space V is a set of elements $\psi, \phi, \chi, \ldots$, called vectors, for which the following properties hold:

1. V is closed under addition. This means that if two vectors ψ and ϕ belong to V then their sum, written as $\psi + \phi$, also belongs to V.

2. A vector ψ can be multiplied by a scalar a to yield a new, well-defined vector $a\psi$ that belongs to V,
3. The addition of vectors is commutative, that is, $\psi + \phi = \phi + \psi$.
4. The addition of vectors is associative, that is, $\psi + (\phi + \chi) = (\psi + \phi) + \chi$.
5. There exists a unique element called 0 that satisfies $\psi + 0 = \psi$ for every element $\psi \in V$.
6. There exists an identity element, E, in V such that $E\psi = \psi$ for every element $\psi \in V$.
7. The multiplication of a vector by scalars is associative, that is, $(ab)\psi = a(b\psi)$.
8. The multiplication of a vector by a scalar is linear, that is, $a(\psi + \phi) = a\psi + a\phi$, $\psi(a+b) = a\psi + b\psi$.
9. For each ψ in V , there exists a unique additive inverse $(-\psi)$ such that $\psi + (-\psi) = 0$.

If the vectors and the scalars associated with a given vector space are real, we say that we are working with a *real vector space*. On the other hand, if the vectors and the scalars are complex, then we say that we are working with a *complex vector space*. The vector spaces used in quantum mechanics are complex.

Basis and dimension of a linear vector space: *Basis in linear vector space* In order to define what we mean by a basis and the dimensionality of a linear vector space, we must introduce the concept of linear independence of a set of vectors.

Linear independence of vectors: Consider a set of n vectors, $\{\phi_1, \phi_2, \phi_1, \phi_3, ..., \phi_n\}$, and their linear combination $a_1\phi_1 + a_2\phi_2 + a_3\phi_3 + ... + a_n\phi_n$, where $a_j, j = 1,2,3,\ldots,n$ are all constants. The vectors of this set are said to be linearly independent if the equation

$$a_1\phi_1 + a_2\phi_2 + a_3\phi_3 + ... + a_n\phi_n = 0 \tag{4.1.1}$$

hold only if $a_1 = a_2 = ... = a_n = 0$. If this condition is not met, we say that the set is linearly dependent.

Note that if a set of vectors is linearly dependent, one of the vectors can be expressed as a linear combination of the others. For instance, assume that

$$a\psi + b\phi + \ldots + c\chi = 0, \tag{4.1.2}$$

where not all of the scalars are zero. Then one of the vectors can be expressed in terms of the other vectors as follows. Let a be non-zero. Then, we have

$$\psi = p\phi + \ldots + q\chi, \tag{4.1.3}$$

where

$$p = -\frac{b}{a}, \ldots, q = -\frac{c}{a}. \tag{4.1.4}$$

Definition: A linear vector space, V, is said to have dimension n, if the maximum number of linearly independent vectors in V equals n.

If this number n is finite, the linear vector space is called finite. On the other hand, if it is possible to find any number (as large as possible) of linearly independent vectors in it, then it is called infinite.

Basis: Any set of n linearly independent vectors, $\{\phi_i\}, i = 1,2,3,\ldots,n$, belonging to the n-dimensional linear vector space, V, is called its basis. The elements, $\phi_1, \phi_2, \phi_3, \ldots$, of this set are called the basis vectors.

Moreover, a basis is said to be complete if it spans the entire space; that is, there is no need to introduce any additional basis vector. It also means that every vector ψ of a linear vector space V, with a complete basis, can be written as a unique linear combination of the basis vectors:

$$\psi = c_1\phi_1 + c_2\phi_1 + c_3\phi_3 + \ldots + c_n\phi_n, \tag{4.1.5}$$

where the expansion coefficients $c_i, i = 1,2,3,\ldots,n$ are called the components of the vector ψ in the basis $\{\phi_i\}$.

Note that the basis may be discrete, i.e., consisting of discrete vectors $\{\phi_i\}$, or it may consist of vectors which are functions of one or more continuous parameters. In the latter case we have an infinite set of continuous basis vectors and, correspondingly, the vector space is an infinite dimensional one. For instance, the space of all continuous functions $\phi_i(x)$, where x takes continuous values in a finite or an infinite interval, constitutes an example of an infinite dimensional space with continuous basis.

Some examples of linear vector spaces

1. The set of vectors, $\{\vec{a}, \vec{b}, \vec{c}, \ldots\}$, in the familiar three dimensional space. The addition of vectors and the multiplication of a vector by a real scalar are defined in accordance with the rules of vector algebra. The additive unit vector is the null vector $\vec{0}$: $\vec{a} + \vec{0} = \vec{a}$ for any $\vec{a} \in V$. For an arbitrary $\vec{a} \in V$ there is an additive inverse given by $-\vec{a}$: $\vec{a} + (-\vec{a}) = \vec{0}$. All other aforementioned requirements are also satisfied.
2. The set of all continuous function defined on the interval $a \leq x \leq b$, where a and b are constants. The addition of vectors and the multiplication by a constant are defined according to the rules of calculus.
3. The set of n complex numbers placed in a well-defined order: $\psi = (\psi_1, \psi_2, \psi_3, \ldots, \psi_n)$. The addition of vectors and the multiplication of a vector by a scalar are given by

$$\psi + \phi = (\psi_1 + \phi_1, \psi_2 + \phi_2, \psi_3 + \phi_3, \dots \psi_n + \phi_n), \tag{4.1.6}$$

and

$$a\psi = (a\psi_1, a\psi_2, a\psi_3, \dots, a\psi_n). \tag{4.1.7}$$

The unit element is the vector

$$0 = (0, 0, 0, \dots, 0), \tag{4.1.8}$$

while the role of the element inverse to a given element ψ is played by the vector

$$-\psi = (-\psi_1, -\psi_2, -\psi_3, \dots, -\psi_n). \tag{4.1.9}$$

Example 4.1.1: Show that the vectors

$$\psi_1 = \begin{pmatrix} 5 \\ 3 \\ 4 \end{pmatrix}, \quad \psi_2 = \begin{pmatrix} 1 \\ 2 \\ 3 \end{pmatrix}, \quad \psi_3 = \begin{pmatrix} 7 \\ 7 \\ 10 \end{pmatrix}$$

are linearly dependent.

Solution: For these vectors to be linearly independent, their linear combination $a\psi_1 + b\psi_2 + c\psi_3$ must be zero only if $a = b = c = 0$. Let us check whether it is really the case. For this purpose, we put $a\psi_1 + b\psi_2 + c\psi_3 = 0$ and determine the values of a, b and c for which this equation can be satisfied. Hence, we have

$$a\begin{pmatrix} 5 \\ 3 \\ 4 \end{pmatrix} + b\begin{pmatrix} 1 \\ 2 \\ 3 \end{pmatrix} + c\begin{pmatrix} 7 \\ 7 \\ 10 \end{pmatrix} = \begin{pmatrix} 5a + b + 7c \\ 3a + 2b + 7c \\ 4a + 3b + 10c \end{pmatrix} = \begin{pmatrix} 0 \\ 0 \\ 0 \end{pmatrix}. \tag{4.1.10}$$

Equation (4.1.10) leads to a system of three algebraic equations for a, b and c. The solution of these equations yields: $a = -c, b = -2c$. Clearly, (4.1.10) holds good for $a = b = c = 0$. However, (4.1.10) is also satisfied for non-zero values of the constants a, b and c. For instance, it is satisfied for $a = 1, b = 2$ and $c = -1$, which shows that ψ_3 is a linear combination of the other two vectors: $\psi_3 = \psi_1 + 2\psi_2$. Hence, the given system of vectors is linearly dependent.

Example 4.1.2: Check whether the system of vectors

$$\psi_1 = \begin{pmatrix} 5 \\ 0 \\ 0 \end{pmatrix}, \quad \psi_2 = \begin{pmatrix} 0 \\ -2 \\ 0 \end{pmatrix}, \quad \psi_3 = \begin{pmatrix} 0 \\ 0 \\ 10 \end{pmatrix}$$

is linearly independent or not.

Solution: Just as in Example 4.1.1, we put $a\psi_1 + b\psi_2 + c\psi_3 = 0$ and determine the constants. We get

$$a\begin{pmatrix} 5 \\ 0 \\ 0 \end{pmatrix} + b\begin{pmatrix} 0 \\ \text{-2} \\ 3 \end{pmatrix} + c\begin{pmatrix} 0 \\ 0 \\ 10 \end{pmatrix} = \begin{pmatrix} 5a \\ -2b \\ 10c \end{pmatrix} = \begin{pmatrix} 0 \\ 0 \\ 0 \end{pmatrix}. \tag{4.1.11}$$

The only set of solutions for the constants is $a = b = c = 0$. Hence, the given set of vectors is linearly independent.

Note that, so far, the only condition that has been imposed on the basis vectors is their linear independence. However, it is desirable to have an orthonormal basis consisting of basis vectors that are not only pairwise orthogonal but also have unit length. In order to incorporate these two concepts, we shall have to introduce and define a mathematical operation called *inner product* (scalar product) of vectors. We shall discuss that in the next section.

4.2 Dirac Notation

We have already stated that vector spaces in quantum mechanics are complex. Therefore, we assume the elements of our n-dimensional linear vector space to be complex. We also assume the vector space to have a fixed basis $\{\phi_i\}, i = 1,2,3,\ldots,n$. For example, in this basis, a vector ψ belonging to our n-dimensional linear vector space is given by its n components $\psi_i, i = 1,2,3,\ldots,n$, which are complex numbers.

Dirac notation: We introduce the notation $|\psi\rangle$ for a vector ψ belonging to an n-dimensional linear vector space V, and we call it a *ket vector* or simply *a ket*. As mentioned above, in a fixed basis $\{\phi_i\}, i = 1,2,3,\ldots,n$, a ket will be characterized by its complex components $\psi_i, i = 1,2,3,\ldots,n$. It is convenient to arrange these components in to a column vector and write a ket as a column vector:

$$|\psi\rangle = \begin{pmatrix} \psi_1 \\ \psi_2 \\ \psi_3 \\ \cdot \\ \cdot \\ \cdot \\ \psi_n \end{pmatrix}. \tag{4.2.1}$$

Dual vector: The familiar notion of a "scalar product" is incorporated by introducing a *dual vector*, written as $\langle\psi|$, for each of the vectors, $|\psi\rangle$, of V. In Dirac's language, it is called a *bra vector*. The bra $\langle\psi|$ dual to a ket $|\psi\rangle$ is constructed by transposing the ket (that is, we write it as a row vector) followed by complex conjugation. In other words:

$$\text{If} \quad |\psi\rangle = \begin{pmatrix} \psi_1 \\ \psi_2 \\ \psi_3 \\ \cdot \\ \cdot \\ \cdot \\ \psi_n \end{pmatrix}, \quad \text{then} \quad \langle\psi| = \begin{pmatrix} \psi_1^* & \psi_2^* & \psi_3^* & \cdot & \cdot & \cdot & \psi_n^* \end{pmatrix}. \tag{4.2.2}$$

This method of complex conjugation is known as *hermitian conjugation* or *dagger operation*: $\langle\psi| = (|\psi\rangle)^\dagger$. There is a one-to-one correspondence between bras (constructed in this manner) and kets, that is, for a given ket $|\psi\rangle$, there is a unique bra $\langle\psi|$. In addition, the following relations hold good:

(a) If $|\lambda\rangle = \alpha|\mu\rangle$, then $\langle\lambda| = \alpha^*\langle\mu|$.

(b) If $|\lambda\rangle = |\alpha\mu\rangle + \beta|\nu\rangle$, then $\langle\lambda| = \alpha^*\langle\mu| + \beta^*\langle\nu|$.

The set of bras, dual to the kets of V, also forms a linear vector space, which is called the dual (to V) vector space. It is denoted as V^*.

The inner (or, scalar) product: The inner product (also called the scalar product) of two vectors $|\psi\rangle$ and $|\phi\rangle$ (written as $\langle\phi|\psi\rangle$) is defined by the following expression:

$$\langle\phi|\psi\rangle = \begin{pmatrix} \phi_1^* & \phi_2^* & \phi_3^* & \cdot & \cdot & \cdot & \phi_n^* \end{pmatrix} \begin{pmatrix} \psi_1 \\ \psi_2 \\ \psi_3 \\ \cdot \\ \cdot \\ \cdot \\ \psi_n \end{pmatrix}$$

$$= (\phi_1^*\psi_1 + \phi_2^*\psi_2 + \phi_3^*\psi_3 + ... + \phi_n^*\psi_n) = \sum_{i=1}^{n} \phi_i^*\psi_i. \tag{4.2.3}$$

We call $\langle\phi|\psi\rangle$ a '*bracket*'. Evidently, the procedure is to take the bra, $\langle\phi| \in V^*$, corresponding to $|\phi\rangle \in V$ and multiply it with the ket $|\psi\rangle \in V$ according to the rules of matrix multiplication. Therefore, if $|\psi\rangle$ and $|\phi\rangle$ belong to the same vector space, the products of the type $|\psi\rangle|\phi\rangle$ and $\langle\psi|\langle\phi|$ are not defined; they are in fact forbidden because there are no rules in matrix algebra for the multiplication of two column matrices or two row matrices. However, if $|\psi\rangle$ and $|\phi\rangle$ belong to two different vector spaces, a product of the type $|\psi\rangle|\phi\rangle$ ($\langle\psi|\langle\phi|$) is written as $|\psi\rangle \otimes |\phi\rangle$ ($\langle\psi| \otimes \langle\phi|$) and understood as the tensor product of two vectors.

A vector space that also has an inner product is referred to as an *inner product space* or *Euclidean space*.

Properties of the inner product: Let $(|\psi\rangle, |\psi\rangle, |\omega\rangle, \ldots$ be the vectors belonging to a complex vector space V and let α and β be complex numbers. Then the inner product satisfies the following properties:

$$\langle\phi|\psi\rangle = \langle\psi|\phi\rangle^*. \tag{4.2.4}$$

If $|\lambda\rangle = \alpha|\mu\rangle$, then for any ket $|\psi\rangle$,

$$\langle\lambda|\psi\rangle = \langle\psi|\lambda\rangle^* = \langle\psi|\alpha\mu\rangle^* = (\alpha\langle\psi|\mu\rangle)^* = \alpha^*\langle\psi|\mu\rangle^* = \alpha^*\langle\mu|\psi\rangle. \tag{4.2.5}$$

$$\langle\psi|\left(\alpha\,|\phi\rangle + \beta\,|\omega\rangle\right) = \alpha\,\langle\psi|\phi\rangle + \beta\,\langle\phi|\,\omega\rangle, \tag{4.2.6}$$

$$\left(\langle\alpha\psi| + \langle\beta\omega|\right)|\phi\rangle = \alpha^*\,\langle\psi|\phi\rangle + \beta^*\,\langle\omega|\phi\rangle, \tag{4.2.7}$$

$$\langle\psi|\psi\rangle \geq 0. \tag{4.2.8}$$

In (4.2.8), the equality holds, if and only if $|\psi\rangle = 0$. *If the inner product between two vectors is zero, $\langle\phi|\psi\rangle = 0$, we say that the vectors are orthogonal.*

Norm of a vector: The square root of the inner product of a vector with itself is called the norm, and is written as:

$$\|\psi\| = \sqrt{\langle\psi|\psi\rangle}. \tag{4.2.9}$$

A vector $|\psi\rangle$ is said to be normalized if its norm is equal to 1:

$$\|\psi\| = \sqrt{\langle\psi|\psi\rangle} = 1. \tag{4.2.10}$$

Orthonormal and complete basis: An orthonormal basis consists of the basis vectors $\{|\phi_i\rangle\}, i = 1, 2, 3, \ldots, n$, which have a unit norm and are pairwise orthogonal:

$$\langle\phi_i|\phi_j\rangle = \delta_{ij}, \quad \|\phi_i\| = \sqrt{\langle\phi_i|\phi_i\rangle} = 1, \tag{4.2.11}$$

where δ_{ij} is the Kronecker delta. The completeness of an orthonormal basis is mathematically expressed in terms of a closure relation, which is what we are going to derive.

Let us first assume the basis to be discrete. An arbitrary vector, $|\psi\rangle$, belonging to the linear vector space can be expanded in this basis as

$$|\psi\rangle = \sum_i c_i\,|\phi_i\rangle, \tag{4.2.12}$$

where the expansion coefficients $c_i = \langle\phi_i|\psi\rangle$ are called the components of the vector ψ in the basis $\{|\phi_i\rangle\}$. Note that if $|\psi\rangle$ is normalized to unity, i.e., $\langle\psi|\psi\rangle = 1$ then

$$\begin{aligned}\langle\psi|\psi\rangle &= \sum_i\sum_j\langle\phi_i|c_i^* c_j|\phi_j\rangle = \sum_i\sum_j c_i^* c_j\langle\phi_i|\phi_j\rangle \\ &= \sum_i\sum_j c_i^* c_j\,\delta_{ij} = \sum_i |c_i|^2 = 1. \end{aligned} \tag{4.2.13}$$

Since $\langle\phi_i|\psi\rangle^\dagger = \langle\psi|\phi_i\rangle$, (4.2.13) can be written as

$$\sum_i |c_i|^2 = \sum_i c^* c_i = \sum_i \langle\psi|\phi_i\rangle\langle\phi_i|\psi\rangle = \langle\psi\left(\sum_i |\phi_i\rangle\langle\phi_i|\right)\psi\rangle = 1. \tag{4.2.14}$$

Since $|\psi\rangle$ is normalized to unity,

$$\sum_i |\phi_i\rangle\langle\phi_i| = \hat{I}. \tag{4.2.15}$$

The relation (4.2.15) is known as *completeness condition* or *closure relation* for the basis vectors.

In the case of a continuous basis in which the vector functions depend on a continuous parameter α, the closure relation reads:

$$\int d\alpha\,|\phi(\alpha)\rangle\langle\phi(\alpha)| = \hat{I}. \tag{4.2.16}$$

We shall always assume that we have an orthonormal and complete basis in our linear vector space unless stated otherwise.

Finally, let us note that in an orthonormal basis $\{\phi_i\}, i = 1,2,3,\ldots,n$, an arbitrary ket, $|\psi\rangle$ (belonging to the vector space) is represented by a column matrix whose elements are the expansion coefficients $c_i, i = 1,2,3,\ldots,n$:

$$|\psi\rangle = \begin{pmatrix} \langle\phi_1|\psi\rangle \\ \langle\phi_2|\psi\rangle \\ \langle\phi_3|\psi\rangle \\ . \\ . \\ . \\ \langle\phi_n|\psi\rangle \end{pmatrix} = \begin{pmatrix} c_1 \\ c_2 \\ c_3 \\ . \\ . \\ . \\ c_n \end{pmatrix}. \tag{4.2.17}$$

Example 4.2.1: Suppose that $|\phi_1\rangle, |\phi_2\rangle, |\phi_3\rangle$ constitute an orthonormal basis, that is, $\langle\phi_i|\phi_j\rangle = \delta_{ij}$. Consider the following kets given in this basis:

$$|\psi\rangle = 3i|\phi_1\rangle + 2|\phi_2\rangle + i|\phi_3\rangle, \tag{4.2.18}$$

$$|\phi\rangle = 2|\phi_1\rangle - 3|\phi_2\rangle + 5|\phi_3\rangle. \tag{4.2.19}$$

(a) Find $\langle\psi|$ and $\langle\phi|$.
(b) Compute the inner product $\langle\phi|\psi\rangle$ and show that $\langle\phi|\psi\rangle = \langle\psi|\phi\rangle^*$.
(c) Let $a = 2+3i$ and compute $|a\psi\rangle$.
(d) Find $|\psi+\phi\rangle$ and $|\psi-\phi\rangle$.

Solution: (a) Using the properties discussed earlier,

$$\langle\psi| = (3i)^*\langle\phi_1| + 2\langle\phi_2| + (i)^*|\phi_3\rangle = -3i\langle\phi_1| + 2\langle\phi_2| - i\langle\phi_3| \tag{4.2.20}$$

$$\langle\phi| = 2\langle\phi_1| - 3\langle\phi_2| + 4\langle\phi_3|. \tag{4.2.21}$$

(b) To compute the inner product, we rely on the fact that the basis is orthonormal,

$$\langle\phi_i|\phi_j\rangle = \delta_{ij}. \tag{4.2.22}$$

And so we obtain

$$\begin{aligned}\langle\phi|\psi\rangle &= (2\langle\phi_1| - 3\langle\phi_2| + 4\langle\phi_3|)(3i|\phi_1\rangle + 2|\phi_2\rangle + i|\phi_3\rangle)\\ &= 6i\langle\phi_1|\phi_1\rangle + 4\langle\phi_1|\phi_2\rangle + 2i\langle\phi_1|\phi_3\rangle - 9i\langle\phi_2|\phi_1\rangle - 6\langle\phi_2|\phi_2\rangle - 3i\langle\phi_2|\phi_3\rangle\\ &\quad + 12i\langle\phi_3|\phi_1\rangle + 8\langle\phi_3|\phi_2\rangle + 4i\langle\phi_3|\phi_3\rangle\\ &= 6i - 6 + 4i = -6 + 10i.\end{aligned} \tag{4.2.23}$$

Now the inner product $\langle\psi|\phi\rangle$ is

$$\begin{aligned}\langle\psi|\phi\rangle &= (-3i|\phi_1\rangle + 2|\phi_2\rangle - i|\phi_3\rangle)(2\langle\phi_1| - 3\langle\phi_2| + 4\langle\phi_3|)\\ &= -6i\langle\phi_1|\phi_1\rangle + 9i\langle\phi_1|\phi_2\rangle - 12i\langle\phi_1|\phi_3\rangle + 4\langle\phi_2|\phi_1\rangle - 6\langle\phi_2|\phi_2\rangle + 8\langle\phi_2|\phi_3\rangle\\ &\quad - 2i\langle\phi_3|\phi_1\rangle + 3i\langle\phi_3|\phi_2\rangle - 4i\langle\phi_3|\phi_3\rangle\\ &= -6i - 6 - 4i = -6 - 10i = \langle\phi|\psi\rangle^*.\end{aligned} \tag{4.2.24}$$

(c) To compute $|a\psi\rangle$, we multiply each coefficient in the expansion by a:

$$\begin{aligned}|a\,\psi\rangle &= (2+3i)(3i|\phi_1\rangle + 2|\phi_2\rangle + i|\phi_3\rangle) = 6i|\phi_1\rangle + 4|\phi_2\rangle + 2i|\phi_3\rangle\\ &\quad - 9|\phi_1\rangle + 6i|\phi_2\rangle - 3|\phi_3\rangle = (-9+6i)|\phi_1\rangle + (4+6i)|\phi_2\rangle - (3-2i)|\phi_3\rangle.\end{aligned} \tag{4.2.25}$$

(d) To compute $|\psi \pm \phi\rangle$, we add (subtract) each term of $|\psi\rangle$ to (from) the respective term of $|\phi\rangle$ to get

$$|\psi + \phi\rangle = (2+3i)\,|\phi_1\rangle - |\phi_2\rangle + (5+i)\,|\phi_3\rangle, \tag{4.2.26}$$

$$|\psi - \phi\rangle = (-2+3i)\,|\phi_1\rangle + 5\,|\phi_2\rangle + (-5+i)\,|\phi_3\rangle. \tag{4.2.27}$$

Example 4.2.2: Consider the ket vectors and the basis given in Example 4.2.1. Compute the matrices corresponding to the kets $|\psi\rangle$ and $|\phi\rangle$ in that basis.

Solution: Since the basis is orthonormal, $\langle\phi_i|\phi_j\rangle = \delta_{ij}$ and we have for $|\psi\rangle$:

$$\langle\phi_1|\psi\rangle = 3i\langle\phi_1|\phi_1\rangle + 2\langle\phi_1|\phi_2\rangle + i\langle\phi_1|\phi_3\rangle = 3i, \tag{4.2.28}$$

$$\langle\phi_2|\psi\rangle = 3i\langle\phi_2|\phi_1\rangle + 2\langle\phi_2|\phi_2\rangle + i\langle\phi_2|\phi_3\rangle = 2, \tag{4.2.29}$$

$$\langle\phi_3|\psi\rangle = 3i\langle\phi_3|\phi_1\rangle + 2\langle\phi_3|\phi_2\rangle + i\langle\phi_3|\phi_3\rangle = i. \tag{4.2.30}$$

Similarly for $|\phi\rangle$:

$$\langle\phi_1|\phi\rangle = 2\langle\phi_1|\phi_1\rangle - 3\langle\phi_1|\phi_2\rangle + 5\langle\phi_1|\phi_3\rangle = 2, \tag{4.2.31}$$

$$\langle\phi_2|\phi\rangle = 2\langle\phi_2|\phi_1\rangle - 3\langle\phi_2|\phi_2\rangle + 5\langle\phi_2|\phi_3\rangle = -3, \tag{4.2.32}$$

$$\langle\phi_3|\phi\rangle = 2\langle\phi_3|\phi_1\rangle - 3\langle\phi_3|\phi_2\rangle + 5\langle\phi_3|\phi_3\rangle = 5. \tag{4.2.33}$$

The column matrices representing the kets $|\psi\rangle$ and $|\phi\rangle$ in the given basis are, respectively

$$|\psi\rangle = \begin{pmatrix} 3i \\ 2 \\ i \end{pmatrix}, \qquad |\phi\rangle = \begin{pmatrix} 2 \\ -3 \\ 5 \end{pmatrix}. \tag{4.2.34}$$

More properties of the inner product of vectors: Let us discuss some more important properties of the inner product of vectors. They will be very useful later while discussing some important physical consequences of the fundamental postulates of quantum mechanics.

Theorem 4.2.1: The scalar product of two vectors $|\psi\rangle$ and $|\phi\rangle$ satisfies

$$2\mathrm{Re}(\langle\psi|\phi\rangle) \le \langle\psi|\psi\rangle + \langle\phi|\phi\rangle. \tag{4.2.35}$$

Proof: For any ket, $|\alpha\rangle$, we have $\langle\alpha|\alpha\rangle \ge 0$. So if $|\alpha\rangle = |\psi - \phi\rangle$, we get

$$0 \le \langle\alpha|\alpha\rangle = \langle\psi|\psi\rangle - \langle\psi|\phi\rangle - \langle\phi|\psi\rangle^* + \langle\phi|\phi\rangle. \tag{4.2.36}$$

Since $\langle\psi|\phi\rangle$ is a complex number and $2\mathrm{Re}(z) = z + z^*$ for any complex number z, we get

$$0 \leq \langle\psi|\psi\rangle - 2\mathrm{Re}(\langle\psi|\phi\rangle) + \langle\phi|\phi\rangle. \tag{4.2.37}$$

The last equation leads to the desired result:

$$2\mathrm{Re}(\langle\psi|\phi\rangle) \leq \langle\psi|\psi\rangle + \langle\phi|\phi\rangle. \tag{4.2.38}$$

Theorem 4.2.2: The absolute value of the scalar product of two vectors $|\psi\rangle$ and $|\phi\rangle$ is less than or equal to the product of the norms of the vectors:

$$|\langle\psi|\phi\rangle| \leq \sqrt{\langle\psi|\psi\rangle}\sqrt{\langle\phi|\phi\rangle}. \tag{4.2.39}$$

Proof: Once again we use the fact that for any ket, $|\alpha\rangle$, we have $\langle\alpha|\alpha\rangle \geq 0$. Let

$$|\alpha\rangle = |\phi\rangle - \frac{\langle\psi|\phi\rangle}{\langle\psi|\psi\rangle}|\psi\rangle. \tag{4.2.40}$$

We have

$$\langle\alpha|\alpha\rangle = \langle\phi|\phi\rangle - \frac{\langle\psi|\phi\rangle}{\langle\psi|\psi\rangle}\langle\phi|\psi\rangle - \frac{\langle\phi|\psi\rangle}{\langle\psi|\psi\rangle}\langle\psi|\phi\rangle + \frac{\langle\psi|\phi\rangle\langle\phi|\psi\rangle}{\langle\psi|\psi\rangle\langle\psi|\psi\rangle}\langle\psi|\psi\rangle. \tag{4.2.41}$$

We know from the first property of the scalar product that

$$\langle\phi|\psi\rangle\langle\psi|\phi\rangle = |\langle\phi|\psi\rangle|^2. \tag{4.2.42}$$

Therefore, we can rewrite (4.2.41) as

$$\langle\alpha|\alpha\rangle = \langle\phi|\phi\rangle - 2\frac{|\langle\phi|\psi\rangle|^2}{\langle\psi|\psi\rangle} + \frac{|\langle\phi|\psi\rangle|^2}{\langle\psi|\psi\rangle} = \langle\phi|\phi\rangle - \frac{|\langle\phi|\psi\rangle|^2}{\langle\psi|\psi\rangle} \geq 0. \tag{4.2.43}$$

From (4.2.43), we arrive at

$$|\langle\psi|\phi\rangle|^2 \leq \langle\psi|\psi\rangle\langle\phi|\phi\rangle. \tag{4.2.44}$$

Taking the square root in (4.2.44), we obtain the required inequality

$$|\langle\psi|\phi\rangle| \leq \sqrt{\langle\psi|\psi\rangle}\sqrt{\langle\phi|\phi\rangle}. \tag{4.2.45}$$

The inequality (4.2.45) is known as *Schwartz inequality*.

Theorem 4.2.3: For any two vectors $|\psi\rangle$ and $|\phi\rangle$ belonging to V, the norm of the sum of the vectors is less than or equal to the sum of the norms of the individual vectors.

$$\sqrt{\langle\psi+\phi|\psi+\phi\rangle} \leq \sqrt{\langle\psi|\psi\rangle} + \sqrt{\langle\phi|\phi\rangle}. \tag{4.2.46}$$

Proof: Let z be a complex number. Then $|\mathrm{Re}(z)| \leq |z|$. Since the inner product is a complex number, we have $|\mathrm{Re}(\langle\psi|\phi\rangle)| \leq |\langle\psi|\phi\rangle|$. To derive the result, we use this fact together with the Schwartz inequality. First, we expand the inner product $\langle\psi+\phi|\psi+\phi\rangle$:

$$\langle\psi+\phi|\psi+\phi\rangle = \langle\psi|\psi\rangle + \langle\psi|\phi\rangle + \langle\phi|\psi\rangle + \langle\phi|\phi\rangle. \tag{4.2.47}$$

We note that $\langle\psi|\phi\rangle + \langle\phi|\psi\rangle = \langle\psi|\phi\rangle + \langle\psi|\phi\rangle^* = 2\mathrm{Re}(\langle\psi|\phi\rangle)$. Therefore, we have

$$\langle\psi+\phi|\psi+\phi\rangle = \langle\psi|\psi\rangle + \langle\phi|\phi\rangle + 2\mathrm{Re}(\langle\psi|\phi\rangle). \tag{4.2.48}$$

Using $|\mathrm{Re}(z)| \leq |z|$ now, we obtain

$$\langle\psi|\psi\rangle + \langle\phi|\phi\rangle + 2\mathrm{Re}(\langle\psi|\phi\rangle) \leq \langle\psi|\psi\rangle + \langle\phi|\phi\rangle + 2|\langle\psi|\phi\rangle|. \tag{4.2.49}$$

If we apply the Schwartz inequality for the third term on the right-hand side of (4.2.49), we have

$$\langle\psi|\psi\rangle + \langle\phi|\phi\rangle + 2|\langle\psi|\phi\rangle| \leq \langle\psi|\psi\rangle + \langle\phi|\phi\rangle + 2\sqrt{\langle\psi|\psi\rangle}\sqrt{\langle\phi|\phi\rangle}. \tag{4.2.50}$$

Thus, we get that

$$\langle\psi+\phi|\psi+\phi\rangle \leq \langle\psi|\psi\rangle + \langle\phi|\phi\rangle + 2\sqrt{\langle\psi|\psi\rangle}\sqrt{\langle\phi|\phi\rangle} = \left(\sqrt{\langle\psi|\psi\rangle} + \sqrt{\langle\phi|\phi\rangle}\right)^2. \tag{4.2.51}$$

From (4.2.51) we have the required result

$$\sqrt{\langle\psi+\phi|\psi+\phi\rangle} \leq \sqrt{\langle\psi|\psi\rangle} + \sqrt{\langle\phi|\phi\rangle}. \tag{4.2.52}$$

The inequality (4.2.52) is known as *the triangular inequality.*

In quantum mechanics, the linear vector spaces are, as a rule, infinite-dimensional. The so-called Hilbert space plays an exceptional role among all the infinite-dimensional linear vector spaces. A Hilbert space is equipped with an inner product that is essentially positive and allows to introduce metric relationship among various quantities. In this sense, a Hilbert space is a natural generalization of Euclidean spaces to infinite-dimensional spaces. In the following, we shall present the mathematical definition of a Hilbert space.

4.3 Hilbert Space

A Hilbert space $\mathscr{H}$ is a collection of vectors, $\psi, \phi, \chi, \ldots$ and scalars, $a, b, c, \ldots$ that satisfies the following properties.

1. $\mathscr{H}$ is an infinite-dimensional linear vector space, that is, it has infinite dimensions and possesses all the properties of a linear vector space discussed earlier.
2. There exists in $\mathscr{H}$ a real inner product which is finite and satisfies all the aforementioned properties.

   Besides these, a Hilbert space satisfies the following specific properties:

3. $\mathscr{H}$ is separable. It means that there exists a Cauchy sequence $\{\psi_n\} \in \mathscr{H}$, $n = 1,2,3,...,$ such that for every $\psi \in \mathscr{H}$ and $\varepsilon > 0$, there is at least one ψ_n of the sequence for which

$$\|\psi - \psi_n\| < \varepsilon. \tag{4.3.1}$$

4. $\mathscr{H}$ is complete. It means that every Cauchy sequence of elements $\{\psi_n\} \in \mathscr{H}$ converges to an element of $\mathscr{H}$. In other words, the relation

$$\lim_{n,m\to\infty} \|\psi_n - \psi_m\| = 0, \tag{4.3.2}$$

   implies a unique limit $\psi \in \mathscr{H}$ for every Cauchy sequence $\{\psi_n\}$ belonging to $\mathscr{H}$, that is,

$$\lim_{n\to\infty} \|\psi - \psi_n\| = 0. \tag{4.3.3}$$

Examples of Hilbert space:

(i) A set of complex vectors with infinite components:

$$|\psi\rangle = \begin{pmatrix} \psi_1 \\ \psi_2 \\ \psi_3 \\ \cdot \\ \cdot \\ \cdot \\ \psi_n \\ \cdot \\ \cdot \\ \cdot \end{pmatrix}$$

satisfying the condition

$$\langle\psi|\psi\rangle = \sum_{j=1}^{\infty} |\psi_j|^2 < \infty. \tag{4.3.4}$$

The addition of vectors and the multiplication of a vector by a scalar are defined as

$$|\psi\rangle + |\phi\rangle = \begin{pmatrix} \psi_1 \\ \psi_2 \\ \psi_3 \\ \cdot \\ \cdot \\ \cdot \\ \psi_n \\ \cdot \\ \cdot \\ \cdot \end{pmatrix} + \begin{pmatrix} \phi_1 \\ \phi_2 \\ \phi_3 \\ \cdot \\ \cdot \\ \cdot \\ \phi_n \\ \cdot \\ \cdot \\ \cdot \end{pmatrix} = \begin{pmatrix} \psi_1 + \phi_1 \\ \psi_2 + \phi_2 \\ \psi_3 + \phi_3 \\ \cdot \\ \cdot \\ \cdot \\ \psi_n + \phi_n \\ \cdot \\ \cdot \\ \cdot \end{pmatrix}, \ a|\psi\rangle = \begin{pmatrix} a\psi_1 \\ a\psi_2 \\ a\psi_3 \\ \cdot \\ \cdot \\ \cdot \\ a\psi_n \\ \cdot \\ \cdot \\ \cdot \end{pmatrix}. \tag{4.3.5}$$

The scalar product of two vectors $|\psi\rangle$ and $|\phi\rangle$ is defined as

$$\langle\phi|\psi\rangle = \sum_{j=1}^{\infty} \phi_j^* \, \psi_j. \tag{4.3.6}$$

(ii) The set of complex functions, $\{\psi(x), \phi(x), \ldots, \chi(x), \ldots\}$, that are square-integrable on the entire x-axis:

$$\int_{-\infty}^{+\infty} |\psi(x)|^2 \, dx < \infty, \qquad \int_{-\infty}^{+\infty} |\phi(x)|^2 \, dx < \infty, \ldots \tag{4.3.7}$$

The addition of functions and the multiplication of a function by a number are determined according to the general rules of calculus. The inner product is defined as

$$\langle\phi(x)|\psi(x)\rangle = \int_{-\infty}^{+\infty} \phi^*(x)\,\psi(x)\,dx. \tag{4.3.8}$$

Example 4.3.1: Show that for a system of orthonormal vectors $\{|\phi_1\rangle, |\phi_2\rangle, |\phi_3\rangle, \ldots, |\phi_m\rangle, \ldots\}$ to be complete, it is necessary and sufficient that the relation

$$\langle\psi|\phi\rangle = \sum_{m=1}^{\infty} \langle\psi|\phi_m\rangle\langle\phi_m|\phi\rangle \tag{4.3.9}$$

holds good for arbitrary kets $|\psi\rangle$ and $|\phi\rangle$.

Proof: Let us first show that (4.3.9) is a sufficient condition. For this, we will have to prove the following. Given that the condition holds, we must show that there does not exist a non-zero vector χ that is orthogonal to all the vectors of the given system $\{|\phi_i\rangle\}, i = 1,2,3,\ldots,m,\ldots$. Let us assume that there exists such a vector; then, we can take $|\psi\rangle = |\phi\rangle = |\chi\rangle$ and obtain

$$\langle\chi|\chi\rangle = \sum_m \langle\chi|\phi_m\rangle\langle\phi_m|\chi\rangle = 0, \tag{4.3.10}$$

whence $|\chi\rangle = 0$. This contradicts our assumption. Hence, there does not exist any non-zero vector χ that is orthogonal to all the orthonormal vectors $|\phi_i\rangle$. Therefore, the given system is complete.

To show that the given relation is a necessary condition, we shall have to use the completeness of the given system of orthonormal vectors. Since $\{|\phi_i\rangle\}, i = 1,2,3,\ldots, m,\ldots$ is complete, we can expand

$$|\psi\rangle = \sum_{m=1}^{\infty} a_m|\phi_m\rangle, \qquad a_m = \langle\phi_m|\psi\rangle, \tag{4.3.11}$$

$$|\phi\rangle = \sum_{m=1}^{\infty} b_m|\phi_m\rangle, \qquad b_m = \langle\phi_m|\phi\rangle. \tag{4.3.12}$$

Then, we have

$$\begin{aligned}\langle\psi|\phi\rangle &= \sum_{m,n=1}^{\infty} \langle\phi_m|\psi\rangle^\dagger\,(|\phi_m\rangle)^\dagger\langle\phi_n|\phi\rangle\,|\phi_n\rangle = \sum_{m,n=1}^{\infty} \langle\psi|\phi_m\rangle\langle\phi_n|\phi\rangle\,\underbrace{\langle\phi_m|\phi_n\rangle}_{\delta_{mn}} \\ &= \sum_{m,n=1}^{\infty} \langle\psi|\phi_m\rangle\langle\phi_n|\phi\rangle\,\delta_{mn} = \sum_{m=1}^{\infty} \langle\psi|\phi_m\rangle\langle\phi_m|\phi\rangle. \end{aligned} \tag{4.3.13}$$

Thus, from the completeness of the system of orthonormal vectors, there follows the validity of the relation in (4.3.9).

4.4 Observables and Operators

The measurable physical characteristics of a system, such as position, momentum, energy etc, are called observables and are represented by operators. Mathematically, an operator, $\hat{O}$, can be defined as a map $\hat{O} : V \to V$ that takes a vector, belonging to a vector space V, to another vector also belonging to V, where V is a linear vector space over the complex field $\mathscr{C}$. In general, an operator is characterized by its action on the basis vectors of V and hence, in a chosen basis, it is represented by a matrix. The eigenvalues of this matrix give the possible results of measurements of the observable represented by the operator, while the eigenvectors of this matrix give us a basis that we can use to represent a general state of the quantum system.

An operator, representing an observable A, is usually denote by capital letter A with a *hat* over it, i.e., by $\hat{A}$. The action of an arbitrary operator $\hat{A}$ on a ket $|\psi\rangle \in V$ is written as:

$$\hat{A}\,|\psi\rangle = |\phi\rangle\,. \tag{4.4.1}$$

where $|\phi\rangle$ also belongs to V.

The product of an operator $\hat{A}$ and a number (complex) a is an operator $a\hat{A}$, which takes a vector $|\psi\rangle \in V$ into the vector $a(\hat{A}\psi\rangle) \in V$:

$$(a\hat{A})|\psi\rangle = a(\hat{A}|\psi\rangle). \tag{4.4.2}$$

The sum, $\hat{C}$, of two operators $\hat{A}$ and $\hat{B}$ is defined as

$$\hat{C}\,|\psi\rangle = (\hat{A}+\hat{B})\,|\psi\rangle = \hat{A}\,|\psi\rangle + \hat{B}\,|\psi\rangle\,. \tag{4.4.3}$$

It means that we act on $|\psi\rangle$ with $\hat{A}$ and $\hat{B}$ one-by-one and then add up the results. The new vector, thus obtained, is the vector that would result, if we act on $|\psi\rangle$ directly with $\hat{C}$.

The operators in quantum mechanics are linear. An operator $\hat{A}$ is said to be linear on V if for given complex scalars α and β in $\mathscr{C}$ and vectors $|\psi\rangle$ and $|\phi\rangle$ in V, the following holds

$$\hat{A}(\alpha\,|\psi\rangle + \beta\,|\phi\rangle) = \alpha\hat{A}\,|\psi\rangle + \beta\hat{A}\,|\phi\rangle\,. \tag{4.4.4}$$

The product of two linear operators $\hat{A}$ and $\hat{B}$ acts on a vector in the following manner:

$$(\hat{A}\hat{B})\,|\psi\rangle = \hat{A}(\hat{B}\,|\psi\rangle). \tag{4.4.5}$$

It means that the product operator $\hat{A}\hat{B}$ acting on the ket $|\psi\rangle$ transforms it into another ket $|\phi\rangle$, which is obtained by first applying $\hat{B}$ to $|\psi\rangle$ and then acting on the resulting ket by $\hat{A}$. In other words, if $\hat{B}\,|\psi\rangle = |\chi\rangle$, we have

$$(\hat{A}\hat{B})\,|\psi\rangle = \hat{A}(\hat{B}\,|\psi\rangle) = \hat{A}\,|\chi\rangle = |\phi\rangle\,. \tag{4.4.6}$$

In general, the product of two operators is not commutative, i.e.,

$$\hat{A}\hat{B} \neq \hat{B}\hat{A}. \tag{4.4.7}$$

The difference $(\hat{A}\hat{B} - \hat{B}\hat{A})$ is an operator, called the commutator of the operators $\hat{A}$ and $\hat{B}$, which is written as

$$[\hat{A},\hat{B}] = \hat{A}\hat{B} - \hat{B}\hat{A}. \tag{4.4.8}$$

The operator $(\hat{A}\hat{B}+\hat{B}\hat{A})$ is called the anticommutator of the operators $\hat{A}$ and $\hat{B}$. It is written as

$$[\hat{A},\hat{B}]_+ = \hat{A}\hat{B}+\hat{B}\hat{A}. \tag{4.4.9}$$

Up till now we have talked about the operators acting in the linear vector space V of *ket* vectors. In an exactly analogous way, one can introduce operators acting in the linear vector space V^* of *bra* vectors. Consider an operator $\hat{A}$ acting in V

$$|\phi\rangle = \hat{A}|\psi\rangle. \tag{4.4.10}$$

Let us introduce an operator $\hat{A}^\dagger$ which acts in dual space V^* by taking the bra $\langle\psi|$, corresponding to the ket $|\psi\rangle$, into the bra $\langle\phi|$, corresponding to $|\phi\rangle$:

$$\langle\phi| = \langle\psi|\hat{A}^\dagger. \tag{4.4.11}$$

The operator $\hat{A}^\dagger$ is called the operator hermitian conjugate (adjoint) to the operator $\hat{A}$. By definition, a conjugate operator acts on the bra vectors from right. In other words, an operator $\hat{A}$ stands to the left of a ket, while the hermitian conjugate operator $\hat{A}^\dagger$ stands to the right of the corresponding bra. Therefore, by definition, $\hat{A}|\psi\rangle$ and $\langle\psi|\hat{A}^\dagger$ are valid expressions, but $\hat{A}\langle\psi|$ and $|\psi\rangle\hat{A}^\dagger$ are not.

Multiplying (4.4.10) from left by $\langle\chi|$ and (4.4.11) from right by $|\chi\rangle$ and comparing the left hand-sides of the resulting equations, we arrive at an important result:

$$\langle\psi|\hat{A}^\dagger|\chi\rangle^* = \langle\chi|\hat{A}|\psi\rangle. \tag{4.4.12}$$

Equation (4.4.11) can be taken to be the defining equation for the operator $\hat{A}^\dagger$, conjugate (adjoint) to the operator $\hat{A}$. Note that (4.4.11) is completely equivalent to the definition of a hermitian conjugate operator given in Chapter 2 (see (2.3.1)).

The identity operator: The simplest operator of all is the identity operator, $\hat{I}$, which does nothing to a ket:

$$\hat{I}|\psi\rangle = |\psi\rangle. \tag{4.4.13}$$

Outer product: The outer product between a ket and a bra is written as:

$$|\psi\rangle\langle\phi|. \tag{4.4.14}$$

This expression is an operator. If we apply it to a ket $|\chi\rangle$, it produces a new ket that is proportional to $|\psi\rangle$

$$(|\psi\rangle\langle\phi|)|\chi\rangle = |\psi\rangle\langle\phi|\chi\rangle. \tag{4.4.15}$$

The product $\langle\phi|\chi\rangle$ is simply a complex number, say, α and hence,

$$(|\psi\rangle\langle\phi|)|\chi\rangle = \alpha|\psi\rangle. \tag{4.4.16}$$

The closure relation: We have seen that, given an orthonormal basis $\{|\phi_i\rangle\}, i = 1,2,$ $3,...n$, we can expand a ket $|\psi\rangle$ as

$$|\psi\rangle = \sum_{i=1}^{n} c_i|\phi_i\rangle = \sum_{i=1}^{n} \langle\phi_i|\psi\rangle|\phi_i\rangle, \tag{4.4.17}$$

which can also be written as

$$|\psi\rangle = \sum_{i=1}^{n} (|\phi_i\rangle\langle\phi_i|)|\psi\rangle. \tag{4.4.18}$$

Equation (4.4.18) implies that

$$\sum_{i=1}^{n} |\phi_i\rangle\langle\phi_i| = \hat{I}. \tag{4.4.19}$$

This is the required closure relation; the same as obtained earlier.

4.5 Matrix Representation of Operators

In a linear vector space equipped with an orthonormal and complete basis $\{|\phi_i\rangle\}, i = 1,2,3,\ldots$, an operator $\hat{F}$ is represented by a square matrix F. The matrix elements F_{ij} of the matrix F are given by the following inner products:

$$F_{ij} = \langle\phi_i|\hat{F}|\phi_j\rangle, \quad i,j = 1,2,3,\ldots \tag{4.5.1}$$

The above expression for the matrix elements follows from the fact that, using the completeness of the basis, we can write

$$\hat{F} = \sum_{i,j} |\phi_i\rangle\langle\phi_i|\hat{F}|\phi_j\rangle\langle\phi_j| = \sum_{i,j} F_{ij}|\phi_i\rangle\langle\phi_j|. \tag{4.5.2}$$

Obviously, the square matrix F has countably infinite numbers of columns and rows and it is written as

$$F = (F_{ij}) = \begin{pmatrix} F_{11} & F_{12} & F_{13} & . & . & . \\ F_{21} & F_{22} & F_{23} & . & . & . \\ . & . & . & . & . & . \\ . & . & . & . & . & . \\ . & . & . & . & . & . \\ F_{n1} & F_{n2} & F_{n3} & . & . & . \end{pmatrix}. \tag{4.5.3}$$

When the vector space, in which the operator $\hat{F}$ acts, is n dimensional, the components of the operator can be arranged into an $n \times n$ matrix, where F_{ij} is the element at the intersection of the ith row and j^{th} column:

$$\hat{F} = (F_{ij}) = \begin{pmatrix} F_{11} & F_{12} & F_{13} & . & . & F_{1n} \\ F_{21} & F_{22} & F_{23} & . & . & F_{2n} \\ . & . & . & . & . & . \\ . & . & . & . & . & . \\ . & . & . & . & . & . \\ F_{n1} & F_{n2} & F_{n3} & . & . & F_{nn} \end{pmatrix}. \tag{4.5.4}$$

Or,

$$(F_{ij}) = \begin{pmatrix} \langle\phi_1|\hat{F}|\phi_1\rangle & \langle\phi_1|\hat{F}|\phi_2\rangle & . & . & \langle\phi_1|\hat{F}|\phi_n\rangle \\ \langle\phi_2|\hat{F}|\phi_1\rangle & \langle\phi_2|\hat{F}|\phi_2\rangle & . & . & \langle\phi_2|\hat{F}|\phi_n\rangle \\ . & . & . & . & . \\ . & . & . & . & . \\ . & . & . & . & . \\ \langle\phi_n|\hat{F}|\phi_1\rangle & \langle\phi_n|\hat{F}|\phi_2\rangle & . & . & \langle\phi_n|\hat{F}|\phi_n\rangle \end{pmatrix}. \tag{4.5.5}$$

Representation in a continuous basis: Up till now, we have considered the representation of kets, bras and operators in a discrete basis and saw that they are represented by *discrete matrices*. It is not difficult to generalize the formalism to include representations of these quantities in the *continuous basis*.

Consider a continuous basis given by the kets $\{|\phi_k\rangle\}$, where the subscript k takes continuous values. The orthonormality condition for these basis vectors is given by

$$\langle\phi_k|\phi_{k'}\rangle = \delta(k-k'), \tag{4.5.6}$$

where $\delta(k-k')$ is the Dirac delta function, defined by

$$\delta(x) = \frac{1}{2\pi}\int_{-\infty}^{+\infty} e^{ikx}dk. \tag{4.5.7}$$

The completeness condition reads

$$\int_{-\infty}^{+\infty} |\phi_k\rangle\langle\phi_k|dk = \hat{I}, \tag{4.5.8}$$

where $\hat{I}$ is the unit operator. The expansion of a state vector $|\psi\rangle$ in terms of the complete set of basis vectors $\{|\phi_k\rangle\}$ can be accomplished as follows

$$|\psi\rangle = \left(\int_{-\infty}^{+\infty} dk|\phi_k\rangle\langle\phi_k|\right)|\psi\rangle = \int_{-\infty}^{+\infty} dk\langle\phi_k|\psi\rangle|\phi_k\rangle \equiv \int_{-\infty}^{+\infty} a_k|\phi_k\rangle\, dk, \tag{4.5.9}$$

where $a_k = \langle \phi_k | \psi \rangle$ gives the projection of $|\psi\rangle$ on $|\phi_k\rangle$. Note that the norm of the discrete basis vectors is finite, however the norm of the kets $|\phi_k\rangle$ is infinite:

$$\langle \phi_k | \phi_{k'} \rangle \propto \int_{-\infty}^{+\infty} dk \quad \rightarrow \quad \infty. \tag{4.5.10}$$

This implies that the basis kets $|\phi_k\rangle$ do not belong to a Hilbert space because they are not square-integrable. However, these kets do constitute a legitimate basis because the norm of any state vector $|\psi\rangle$ in this basis is finite. We have

$$\langle \psi | \psi \rangle = \int_{-\infty}^{+\infty} dk'' \int_{-\infty}^{+\infty} dk' \;\; a_{k'}^* \, a_{k''} \, \delta(k-k')\delta(k'-k'') = |a_k|^2 = |\langle \phi_k | \psi \rangle|^2 < \infty, \tag{4.5.11}$$

in view of the finiteness of the scalar product $\langle \phi_k | \psi \rangle$. Thus, in a continuous basis, $\{|\phi_k\rangle\}$, an arbitrary ket, $|\psi\rangle$, can be represented by the following column matrix

$$|\psi\rangle = \begin{pmatrix} \cdot \\ \cdot \\ \cdot \\ \langle \phi_k | \psi \rangle \\ \cdot \\ \cdot \\ \cdot \end{pmatrix} \tag{4.5.12}$$

which has infinite number of continuous (non-countable) components $\langle \phi_k | \psi \rangle$. Similarly, an arbitrary bra $\langle \psi |$ is represented by the following row matrix

$$\langle \psi | = \begin{pmatrix} \cdot & \cdot & \cdot & \langle \phi_k | \psi \rangle & \cdot & \cdot & \cdot \end{pmatrix}. \tag{4.5.13}$$

The operators are represented by square continuous matrices whose rows and columns have continuous and infinite number of components:

$$\hat{F} = \begin{pmatrix} \cdot & \cdot & \cdot & \cdot & \cdot \\ \cdot & \cdot & \cdot & \cdot & \cdot \\ \cdot & \cdot & F(k,k') & \cdot & \cdot \\ \cdot & \cdot & \cdot & \cdot & \cdot \\ \cdot & \cdot & \cdot & \cdot & \cdot \end{pmatrix}, \tag{4.5.14}$$

The trace of an operator: The trace of an operator $\hat{F}$ is defined as the sum of the diagonal elements of its matrix in a given basis. It is denoted by Tr(F):

$$\begin{aligned}\text{Tr}(F) &= F_{11} + F_{22} + F_{33} + ... + F_{nn} = \sum_{i=1}^{n} T_{ii} \\ &= \langle \phi_1 | \hat{F} | \phi_1 \rangle + \langle \phi_2 | \hat{F} | \phi_2 \rangle + \langle \phi_3 | \hat{F} | \phi_3 \rangle + ... + \langle \phi_n | \hat{F} | \phi_n \rangle .\end{aligned} \tag{4.5.15}$$

The trace is cyclic, that is,

$$\text{Tr}(FGH) = \text{Tr}(GHF) = \text{Tr}(HFG). \tag{4.5.16}$$

Let us prove that $\text{Tr}(FGH) = \text{Tr}(GHF)$. Using the closure relation of the basis, we have

$$\begin{aligned}\text{Tr}(FGH) &= \sum_{i=1}^{n} \langle \phi_i | FGH | \phi_i \rangle = \sum_{i=1}^{n} \langle \phi_i | F\ (\hat{I})\ G\ (\hat{I})\ H | \phi_i \rangle \\ &= \sum_{i=1}^{n} \langle \phi_i | F \sum_{j=1}^{n} | \phi_j \rangle \langle \phi_j | G \sum_{k=1}^{n} | \phi_k \rangle \langle \phi_k | H | \phi_i \rangle \\ &= \sum_{i=1}^{n} \sum_{j=1}^{n} \sum_{k=1}^{n} \langle \phi_i | F | \phi_j \rangle \langle \phi_j | G | \phi_k \rangle \langle \phi_k | H | \phi_i \rangle \\ &= \sum_{i=1}^{n} \sum_{j=1}^{n} \sum_{k=1}^{n} \langle \phi_j | G | \phi_k \rangle \langle \phi_k | H | \phi_i \rangle \langle \phi_i | F | \phi_j \rangle .\end{aligned} \tag{4.5.17}$$

This can be rewritten as

$$\begin{aligned}\text{Tr}(FGH) &= \sum_{j=1}^{n} \langle \phi_j | G \sum_{k=1}^{n} | \phi_k \rangle \langle \phi_k | H \sum_{i=1}^{n} | \phi_i \rangle \langle \phi_i | F | \phi_j \rangle \\ &= \sum_{j=1}^{n} \langle \phi_j | GHF | \phi_j \rangle \\ &= \text{Tr}(GHF).\end{aligned} \tag{4.5.18}$$

Similarly, one can prove that $\text{Tr}(GHF) = \text{Tr}(HFG)$.

Example 4.5.1: In a linear vector space equipped with an orthonormal basis, $\{|\phi_1\rangle, |\phi_2\rangle, |\phi_3\rangle\}$, an operator $\hat{A}$ acts on the basis vectors and the results are as follows:

$$\hat{A} | \phi_1 \rangle = 2 | \phi_1 \rangle , \tag{4.5.19}$$

$$\hat{A}\,|\phi_2\rangle = 3\,|\phi_1\rangle - i\,|\phi_3\rangle, \tag{4.5.20}$$

$$\hat{A}\,|\phi_3\rangle = -\,|\phi_2\rangle. \tag{4.5.21}$$

Compute the matrix representation of the operator in the given basis.

Solution: The matrix representation of the operator is given by:

$$A = \begin{pmatrix} \langle\phi_1|\hat{F}\,|\phi_1\rangle & \langle\phi_1|\hat{A}\,|\phi_2\rangle & \langle\phi_1|\hat{A}\,|\phi_3\rangle \\ \langle\phi_2|\hat{A}\,|\phi_1\rangle & \langle\phi_2|\hat{A}\,|\phi_2\rangle & \langle\phi_2|\hat{A}\,|\phi_3\rangle \\ \langle\phi_3|\hat{A}\,|\phi_1\rangle & \langle\phi_3|\hat{A}\,|\phi_2\rangle & \langle\phi_3|\hat{A}\,|\phi_3\rangle \end{pmatrix}$$

$$= \begin{pmatrix} 2\,\langle\phi_1|\phi_1\rangle & 3\,\langle\phi_1|\phi_1\rangle - i\,\langle\phi_1|\phi_3\rangle & -\,\langle\phi_1|\phi_2\rangle \\ 2\,\langle\phi_2|\phi_1\rangle & 3\,\langle\phi_2|\phi_1\rangle - i\,\langle\phi_2|\phi_3\rangle & -\,\langle\phi_2|\phi_2\rangle \\ 2\,\langle\phi_3|\phi_1\rangle & 3\,\langle\phi_3|\phi_1\rangle - i\,\langle\phi_3|\phi_3\rangle & -\,\langle\phi_3|\phi_2\rangle \end{pmatrix}. \tag{4.5.22}$$

Since the basis is orthonormal, we have $\langle\phi_i|\phi_j\rangle = \delta_{ij}$ and so the matrix representation of the operator $\hat{A}$ in this basis is:

$$A = \begin{pmatrix} 2 & 3 & 0 \\ 0 & 0 & -1 \\ 0 & -i & 0 \end{pmatrix}. \tag{4.5.23}$$

Example 4.5.2: The outer product $|\psi\rangle\langle\phi|$ is an operator, and therefore can be represented by a matrix. Show this for:

$$|\psi\rangle = \begin{pmatrix} 2 \\ 3i \\ 4 \end{pmatrix}, \quad |\phi\rangle = \begin{pmatrix} -1 \\ 0 \\ i \end{pmatrix}. \tag{4.5.24}$$

Solution: We have

$$\langle\phi| = \begin{pmatrix} -1 & 0 & -i \end{pmatrix}. \tag{4.5.25}$$

Therefore, we have

$$|\psi\rangle\langle\phi| = \begin{pmatrix} 2 \\ 3i \\ 4 \end{pmatrix} \times \begin{pmatrix} -1 & 0 & -i \end{pmatrix} = \begin{pmatrix} -2 & 0 & -2i \\ -3i & 0 & 3 \\ -4 & 0 & -4i \end{pmatrix}. \tag{4.5.26}$$

4.6 Hermitian and Unitary Operators

An operator $\hat{A}$ is called hermitian, if it is equal to its hermitian conjugate operator. That is if

$$\hat{A}^\dagger = \hat{A}, \tag{4.6.1}$$

the operator $\hat{A}$ is hermitian. Recalling the definition of the hermitian conjugate operator (4.4.11)), we conclude that for hermitian operator, we must have

$$\langle\psi|\hat{A}|\phi\rangle^\dagger = \langle\phi|\hat{A}|\psi\rangle. \tag{4.6.2}$$

Let us see how do we get the hermitian conjugate of an operator in practice. Let the matrix $A = (A_{ij})$, where i stands for the number of rows and j for the number of columns, represent an operator $\hat{A}$ in a linear vector space with an orthonormal basis $\{|\phi_i\rangle\}$. The first step is to find the matrix A^T which is transposed of the matrix A. It is obtained by interchanging the rows and columns of A, i.e.,

$$A^T = \begin{pmatrix} A_{11} & A_{12} & A_{13} \\ A_{21} & A_{22} & A_{23} \\ A_{31} & A_{32} & A_{33} \end{pmatrix}^T = \begin{pmatrix} A_{11} & A_{21} & A_{31} \\ A_{12} & A_{22} & A_{32} \\ A_{13} & A_{23} & A_{33} \end{pmatrix}. \tag{4.6.3}$$

The second and the final step is to find the matrix complex conjugate to the matrix A^T. Given a matrix, its complex conjugate is obtained by replacing each of the elements of the matrix by its complex conjugate. Hence, we have

$$\left(A^T\right)^* = \begin{pmatrix} A_{11} & A_{12} & A_{13} \\ A_{21} & A_{22} & A_{23} \\ A_{31} & A_{32} & A_{33} \end{pmatrix}^* = \begin{pmatrix} A_{11}^* & A_{12}^* & A_{13}^* \\ A_{21}^* & A_{22}^* & A_{23}^* \\ A_{31}^* & A_{32}^* & A_{33}^* \end{pmatrix}. \tag{4.6.4}$$

Thus, for any operator $\hat{F}$, the corresponding hermitian conjugate operator, $\hat{F}^\dagger$, is given by the matrix $F^\dagger$ which is hermitian conjugate to the matrix F:

$$F^\dagger = \begin{pmatrix} F_{11}^* & F_{21}^* & F_{31}^* \\ F_{12}^* & F_{22}^* & F_{32}^* \\ F_{13}^* & F_{23}^* & F_{33}^* \end{pmatrix}. \tag{4.6.5}$$

For a hermitian operator, $\hat{F}$, the matrix F is equal to its hermitian conjugate matrix $F^\dagger$.

The hermitian conjugate of a product of operators $\hat{A}, \hat{B}, \hat{C}...$ is given by reversing their order, and then forming the hermitian conjugation of each operator:

$$(\hat{A},\hat{B},\hat{C}...)^\dagger = (\hat{C}^\dagger\hat{B}^\dagger\hat{A}^\dagger...). \tag{4.6.6}$$

The following properties hold good for hermitian conjugation:

$$(\hat{A}+\hat{B})^\dagger = (\hat{A}^\dagger+\hat{B}^\dagger), \tag{4.6.7}$$

$$(a\hat{A})^\dagger = a^*\hat{A}^\dagger, \tag{4.6.8}$$

$$(\hat{A}\hat{B}\,|\psi\rangle)^\dagger = \langle\psi|\,\hat{B}^\dagger\hat{A}^\dagger), \tag{4.6.9}$$

$$\langle a\psi\hat{F}| = a^*\,\langle\psi|\,\hat{F}^\dagger, \tag{4.6.10}$$

where we have used that, for any two matrices F and G, $(F+G)^T = F^T+G^T$, and $(FG)^T = G^T F^T$.

It is clear from the last property that, if an operator is inside the bra, it is replaced by its hermitian conjugate when taken out of it. A scalar, however, is simply replaced by its complex conjugate when taken out of a bra.

Using the aforementioned properties, we can write the hermitian conjugate of any combination of operators and scalars by following the rule:

1. *Replace any constants by their complex conjugates.*
2. *Replace kets by their associated bras, and bras by their associated kets.*
3. *Replace each operator by its hermitian conjugate.*
4. *Reverse the order of all factors in the expression.*

Eigenvalues and eigenvectors

We have discussed earlier that to each physical observable (energy, momentum, angular momentum etc.), there corresponds an operator, which can be represented by a matrix and that the eigenvalues of this matrix are the possible results of measurement for that observable. For example, if the Hamiltonian, H, for a system is given, we can form a matrix representing the Hamiltonian operator $\hat{H}$. If the measurement of energy of the system is carried out, the result will be one of the eigenvalues of this matrix. On the other hand, the eigenvectors (corresponding to these eigenvalues) can be taken to be a basis in the Hilbert space of the possible quantum states of the system. An arbitrary state of the system can then be expanded in terms of these basis states. Such a representation of an arbitrary state in terms of the eigenstates of an observable is frequently used for solving various problems in quantum mechanics. Therefore, in what follows, we shall briefly review the concepts and the techniques that are required to determine the eigenvalues and eigenvectors of a matrix.

Let $\hat{F}$ be a linear operator on a complex linear vector space $\mathscr{V}$, and let λ be a complex number. We say that λ is an eigenvalue of $\hat{F}$ if:

$$\hat{F}\,|u\rangle = \lambda\,|u\rangle \tag{4.6.11}$$

for a vector $|u\rangle \in \mathscr{V}$. The vector $|u\rangle$ is called the eigenvector of $\hat{F}$. If for every eigenvalue there is a single eigenvector, the spectrum of the operator is called *non-degenerate*. If for a given eigenvalue, λ, there are more than one eigenfunctions, the eigenvalue λ is said to be *degenerate*. If, for instance, for a given eigenvalue, λ_k, there exist m linearly independent eigenfunctions, then the eigenvalue λ_k is said to be *m-fold degenerate*.

To find the eigenvalues and eigenvectors of the matrix F, representing the operator $\hat{F}$ in $\mathscr{V}$, we set up the characteristic equation for the determination of λ by equating the determinant of the matrix $(F - \lambda I)$, I being the unit matrix, to zero:

$$\det(F - \lambda I) = 0. \tag{4.6.12}$$

Equation (4.6.12) is a polynomial equation of degree n, where n is the rank of the matrix F. The solution of this equation gives us the eigenvalues λ. The eigenvectors are found according to the general methods in the theory of matrices. Let us consider some examples.

Example 4.6.1: Find the operator, $\hat{F}^\dagger$, hermitian conjugate to the operator $\hat{F}$, which given by the matrix

$$F = \begin{pmatrix} 5 & 0 & 0 \\ 0 & 1 & -2i \\ 0 & 2i & -1 \end{pmatrix}. \tag{4.6.13}$$

Using the result, verify whether $\hat{F}$ is a hermitian operator or not.

Solution: According to our prescription, we have to first find the transposed matrix. It is given by

$$F^T = \begin{pmatrix} 5 & 0 & 0 \\ 0 & 1 & 2i \\ 0 & -2i & -1 \end{pmatrix}. \tag{4.6.14}$$

The next step is to take the complex conjugate of the matrix F^T. We have

$$F^\dagger = \begin{pmatrix} 5 & 0 & 0 \\ 0 & 1 & -2i \\ 0 & 2i & -1 \end{pmatrix}, \tag{4.6.15}$$

The operator $\hat{F}^\dagger$ is given by the matrix $F^\dagger$. Since $F^\dagger = F$, the operator, $\hat{F}$, is hermitian.

Example 4.6.2: Let the Hamiltonian for a system be given by:

$$\hat{H} = \begin{pmatrix} \varepsilon_1 & \varepsilon_2 \\ \varepsilon_2 & \varepsilon_1 \end{pmatrix}, \tag{4.6.16}$$

where ε_1 and ε_2 are constants of the dimensions of energy. Find the eigenvalues and the corresponding eigenvectors of $\hat{H}$ and, thus, set up the basis in the state space of the system.

Solution: To find the eigenvalues of the Hamiltonian, we set

$$\det(H-\lambda I)=\det\begin{pmatrix}\varepsilon_1-\lambda & \varepsilon_2\\ \varepsilon_2 & \varepsilon_1-\lambda\end{pmatrix}=0. \tag{4.6.17}$$

Simplifying the above equation, we have

$$\lambda^2-2\varepsilon_1\lambda+(\varepsilon_1^2-\varepsilon_2^2)=0. \tag{4.6.18}$$

The solutions to this equation yield the eigenvalues of H:

$$\lambda_1=\varepsilon_1+\varepsilon_2,\ \ \lambda_1=\varepsilon_1-\varepsilon_2. \tag{4.6.19}$$

Let $|\alpha_1\rangle=\begin{pmatrix}a\\ b\end{pmatrix}$ be the eigenvector corresponding to λ_1. We have

$$\begin{pmatrix}\varepsilon_1 & \varepsilon_2\\ \varepsilon_2 & \varepsilon_1\end{pmatrix}\begin{pmatrix}a\\ b\end{pmatrix}=(\varepsilon_1+\varepsilon_2)\begin{pmatrix}a\\ b\end{pmatrix}. \tag{4.6.20}$$

Equation (4.6.20) leads to

$$\varepsilon_1 a+\varepsilon_2 b=(\varepsilon_1+\varepsilon_2)a, \tag{4.6.21}$$

$$\varepsilon_2 a+\varepsilon_1 b=(\varepsilon_1+\varepsilon_2)b, \tag{4.6.22}$$

which in turn yields $a=b$. As a result,

$$|\alpha_1\rangle=\begin{pmatrix}a\\ a\end{pmatrix}=a\begin{pmatrix}1\\ 1\end{pmatrix}. \tag{4.6.23}$$

Consequently, the normalized eigenvector corresponding to λ_1 is

$$|\alpha_1\rangle=\frac{1}{\sqrt{2}}\begin{pmatrix}1\\ 1\end{pmatrix}. \tag{4.6.24}$$

Similarly, the normalized eigenvector corresponding to the eigenvalue λ_2 is given by

$$|\alpha_2\rangle=\frac{1}{\sqrt{2}}\begin{pmatrix}1\\ -1\end{pmatrix}. \tag{4.6.25}$$

The orthonormal vectors $|\alpha_1\rangle$ and $|\alpha_2\rangle$ constitute the required basis.

Before we proceed further, let us state and prove some important theorems about the properties of the eigenvalues and the eigenfunctions of a hermitian operator. These are essentially the same as we discussed earlier; however, here we shall put them into Dirac notation.

Theorem 4.6.1: *The eigenvalues of a hermitian operator are real.*

Proof: Let $\hat{F}$ be a hermitian operator, and suppose that $|\psi\rangle$ is an eigenvector of $\hat{F}$ with eigenvalue λ. Then,

$$\langle\psi|\hat{F}|\psi\rangle = \lambda\langle\psi|\psi\rangle. \tag{4.6.26}$$

Using the properties of hermitian conjugation, discussed earlier, we get from (4.6.26),

$$\langle\psi|\hat{F}|\psi\rangle^* = \langle\psi|\hat{F}^\dagger|\psi\rangle = \lambda^*\langle\psi|\psi\rangle. \tag{4.6.27}$$

Given that $\hat{F}$ is hermitian ($\hat{F} = \hat{F}^\dagger$), we have

$$\lambda\langle\psi|\psi\rangle = \lambda^*\langle\psi|\psi\rangle \quad \Rightarrow \quad (\lambda - \lambda^*)\langle\psi|\psi\rangle = 0. \tag{4.6.28}$$

Since $\langle\psi|\psi\rangle \neq 0$, the equation (4.6.28) yields: $\lambda = \lambda^*$. That is, λ is real and the theorem is proved.

Theorem 4.6.2: *The eigenvectors of a hermitian operator, corresponding to distinct eigenvalues are orthogonal.*

Proof: Let $\hat{F}$ be a hermitian operator and let $|\psi_m\rangle$ and $|\psi_n\rangle$ be two of its eigenvectors, corresponding to two distinct eigenvalues λ_m and λ_n, respectively. That is,

$$\hat{F}|\psi_m\rangle = \lambda_m|\psi_m\rangle, \quad \hat{F}|\psi_n\rangle = \lambda_n|\psi_n\rangle, \; (\lambda_m \neq \lambda_n). \tag{4.6.29}$$

Case 1. The eigenvalues of $\hat{F}$ are non-degenerate.

Using (4.6.29), we obtain

$$\langle\psi_m|\hat{F}|\psi_n\rangle^\dagger = \lambda_n^*\langle\psi_m|\psi_n\rangle^\dagger = \lambda_n\langle\psi_n|\psi_m\rangle, \tag{4.6.30}$$

$$\langle\psi_n|\hat{F}|\psi_m\rangle = \lambda_m\langle\psi_n|\psi_m\rangle, \tag{4.6.31}$$

where we have used the fact that the eigenvalues of a hermitian operator are real. Since $\hat{F}$ is hermitian

$$\langle\psi_m|\hat{F}|\psi_n\rangle^\dagger = \langle\psi_n|\hat{F}|\psi_m\rangle. \tag{4.6.32}$$

Therefore, from (4.3.30) and (4.3.31) it follows that

$$(\lambda_m - \lambda_n)\langle\psi_n|\psi_m\rangle = 0. \tag{4.6.33}$$

Since $\lambda_m \neq \lambda_n$, we have $\langle\psi_n|\psi_m\rangle = 0$. That is, the eigenvectors $|\psi_m\rangle$ and $|\psi_n\rangle$ are orthogonal.

Case 2. The eigenvalues of $\hat{F}$ are degenerate.

It is clear from (4.6.33) that, if $\lambda_m = \lambda_n$, the scalar product $\langle\psi_n|\psi_m\rangle$ can be non-zero. Hence, the eigenvectors corresponding to the same eigenvalue will not, in general, be orthogonal. What to do in such a situation? It turns out that it is possible to construct a set of orthonormal eigenvectors using the linear combinations of the non-orthogonal eigenvectors. Let us show this in a particular case of two fold degeneracy.

Let $|\psi_1\rangle$ and $|\psi_2\rangle$ be two linearly independent eigenvectors of $\hat{F}$ with the same eigenvalue λ, that is,

$$\hat{F}|\psi_1\rangle = \lambda|\psi_1\rangle, \quad \hat{F}|\psi_2\rangle = \lambda|\psi_2\rangle. \tag{4.6.34}$$

Consider the linear combination

$$|\psi\rangle = c_1|\psi_1\rangle + c_2|\psi_2\rangle. \tag{4.6.35}$$

It is easy to check that ψ is an eigenvector of $\hat{F}$ with the eigenvalue λ:

$$\hat{F}|\psi\rangle = \hat{F}(c_1|\psi_1\rangle + c_2|\psi_2\rangle) = c_1\hat{F}|\psi_1\rangle + c_2\hat{F}|\psi_2\rangle = \lambda(c_1|\psi_1\rangle + c_2|\psi_2\rangle). \tag{4.6.36}$$

We now want to construct those linear combinations of $|\psi_1\rangle$ and $|\psi_2\rangle$ that will be orthogonal and normalized. Let us start with the normalization of the first eigenvector ψ_1 and introduce a normalized vector $|\phi_1\rangle$ by

$$|\phi_1\rangle = |\psi_1\rangle/\sqrt{\langle\psi_1|\psi_1\rangle}. \tag{4.6.37}$$

Now, consider the following linear combination

$$|\chi\rangle = a|\phi_1\rangle + |\psi_2\rangle, \tag{4.6.38}$$

where a is, in general, a complex coefficient, and demand it to be orthogonal to $|\phi_1\rangle$. This determines the coefficient a:

$$a = -\langle\phi_1|\psi_2\rangle. \tag{4.6.39}$$

Thus, the function given by (4.3.38), with the coefficient a given by (4.3.39), is orthogonal to $|\phi_1\rangle$. We now normalize $|\chi\rangle$ to get

$$|\phi_2\rangle = |\chi\rangle/\sqrt{\langle\chi|\chi\rangle}. \tag{4.6.40}$$

Thus, from the linearly independent eigenfunctions $|\psi_1\rangle$ and $|\psi_2\rangle$ of the two-fold degenerate eigenvalue λ of the operator $\hat{F}$, we have been able to construct two orthonormal eigenfunctions ϕ_1 and ϕ_2 corresponding to the same eigenvalue λ. Clearly, this procedure can easily be generalized to the case of k-fold degeneracy ($k > 2$).

In view of the two theorems proved here, we shall always assume that the eigenfunctions of a hermitian operator satisfy the orthonormality condition.

Corollary: *The eigenfunctions of a hermitian operator define a complete set of mutually orthonormal functions. This set is unique if the operator has no degenerate eigenvalues and is not unique if there is even one degenerate eigenvalue.*

Theorem 4.6.3: If two hermitian operators $\hat{F}$ and $\hat{G}$ commute (that is, $\hat{F}\hat{G} = \hat{G}\hat{F}$) and if $\hat{F}$ has no degenerate eigenvalues, then each eigenvector of $\hat{F}$ is an eigenvector of $\hat{G}$.

Proof: Since $\hat{F}$ has no degenerate eigenvalues, to each eigenvalue of $\hat{F}$ there corresponds only one eigenvector. Consider the equation

$$\hat{F}|\psi_n\rangle = \lambda_n|\psi_n\rangle, \tag{4.6.41}$$

where $|\psi_n\rangle$ is the eigenvector of $\hat{F}$ with eigenvalue λ_n. Since $\hat{F}$ commutes with $\hat{G}$, we have

$$\hat{G}\hat{F}|\psi_n\rangle = \hat{F}\hat{G}|\psi_n\rangle, \quad \text{or,} \quad \hat{F}\left(\hat{G}|\psi_n\rangle\right) = \lambda_n\left(\hat{G}|\psi_n\rangle\right). \tag{4.6.42}$$

That is, $\hat{G}|\psi_n\rangle$ is also an eigenvector of $\hat{F}$. Since the operator $\hat{F}$ has unique eigenvectors, $\hat{G}|\psi_n\rangle$ must be proportional to $|\psi_n\rangle$. Therefore, we must have

$$\hat{G}|\psi_n\rangle = \kappa_n|\psi_n\rangle, \tag{4.6.43}$$

that is, $|\psi_n\rangle$ is also an eigenvector of $\hat{G}$ with eigenvalue κ_n. The theorem is proved.

Note that if λ_n happens to be a degenerate eigenvalue, then $|\psi_n\rangle$ is not necessarily an eigenvector of $\hat{G}$. In this case, the only thing we can be sure of is that $\hat{G}|\psi_n\rangle$ is an eigenvector of $\hat{F}$ with eigenvalue λ_n. It does not mean that the operators $\hat{F}$ and $\hat{G}$ do not have any common set of eigenvectors. In the theory of linear vector spaces, it is proved that, in such cases, there exist an infinite number of sets of orthonormal eigenvectors that are common to both these operators.

Skew-hermitian Operator: An operator $\hat{F}$ is skew-hermitian or anti-hermitian if:

$$\hat{F}^\dagger = -\hat{F}. \tag{4.6.44}$$

Theorem 4.6.4: The eigenvalues of a skew-hermitian operator are either zero or purely imaginary.

Proof: Let $\hat{F}$ be a skew-hermitian operator, and suppose that $|\psi\rangle$ is an eigenvector of $\hat{F}$ with eigenvalue λ. Then,

$$\langle\psi|\hat{F}|\psi\rangle = \lambda\langle\psi|\psi\rangle. \tag{4.6.45}$$

On the other hand,

$$\langle\psi|\hat{F}^\dagger|\psi\rangle = \lambda^*\langle\psi|\psi\rangle. \tag{4.6.46}$$

Since the operator is skew-hermitian, $\hat{F} = -\hat{F}^\dagger$, we have

$$\lambda\langle\psi|\psi\rangle = -\lambda^*\langle\psi|\psi\rangle \quad \Rightarrow \quad (\lambda+\lambda^*)\langle\psi|\psi\rangle = 0. \tag{4.6.47}$$

The last equation yields: $\lambda = -\lambda^*$. This equation can be satisfied when either λ is identically equal to zero, or when it is purely imaginary. That means that the eigenvalues of a skew-hermitian operator are either zero or purely imaginary.

Unitary Operator: An operator $\hat{U}$ is called a unitary operator, if the following holds

$$\hat{U}^\dagger\hat{U} = \hat{U}\hat{U}^\dagger = \hat{I}, \tag{4.6.48}$$

where $\hat{I}$ is the unit operator. Clearly, the hermitian conjugate of a unitary operator is equal to the inverse of the operator: $\hat{U}^\dagger = \hat{U}^{-1}$. Another important characteristic of unitary matrices, corresponding to unitary operators, is that the rows or columns of a given matrix form an orthonormal set.

Theorem 4.6.5: The eigenvalues of a unitary operator are complex numbers with moduli equal to 1; the eigenvectors of a unitary operator that has no degenerate eigenvalues are mutually orthogonal.

Proof: Let $|\phi_m\rangle$ and $|\phi_n\rangle$ be the eigenvectors of a unitary operator corresponding to the eigenvalues λ_m and λ_n, respectively. We then have

$$\left(\langle\phi_m|\hat{U}^\dagger\right)\left(\hat{U}|\phi_n\rangle\right) = \lambda_m^*\lambda_n\langle\phi_m|\phi_n\rangle. \tag{4.6.49}$$

Since $\hat{U}^\dagger\hat{U} = \hat{I}$, this equation can be written as

$$(\lambda_m^*\lambda_n - 1)\langle\phi_m|\phi_n\rangle = 0. \tag{4.6.50}$$

If $m = n$, $\langle\phi_m|\phi_n\rangle > 0$, and we have $|\lambda_m|^2 = 1$ and hence, $|\lambda_m| = 1$. However, if $m \neq n$, the only possibility is that $\langle\phi_m|\phi_n\rangle = 0$. That is, the eigenvectors corresponding to different eigenvalues are orthogonal. The theorem is proved.

In view of the aforementioned proved theorems, it is tempting to expect that the eigenvectors of a hermitian operator form an orthonormal basis in the Hilbert space in which the given operator acts. In order to ascertain this, we must check the following:

1. Does the set of eigenvectors satisfy the completeness relation?
2. Are they orthonormal?

Let us demonstrate it by taking an example. Consider an operator represented by the following hermitian matrix

$$F = \begin{pmatrix} 0 & 1 \\ 1 & 0 \end{pmatrix}. \tag{4.6.51}$$

The eigenvalues of the matrix are found in the usual way:

$$\det(F-\lambda I)=\det\begin{pmatrix}0-\lambda & 1\\ 1 & 0-\lambda\end{pmatrix}=\lambda^2-1=0, \;\Rightarrow\; \lambda_1=+1,\ \lambda_2=-1. \tag{4.6.52}$$

Let $|\psi_1\rangle=\begin{pmatrix}a\\ b\end{pmatrix}$ be the eigenvector of F corresponding to $\lambda=+1$. We then have

$$F=\begin{pmatrix}0 & 1\\ 1 & 0\end{pmatrix}\begin{pmatrix}a\\ b\end{pmatrix}=\begin{pmatrix}a\\ b\end{pmatrix} \;\Rightarrow\; a=b \;\Rightarrow\; |\psi_1\rangle=a\begin{pmatrix}1\\ 1\end{pmatrix}. \tag{4.6.53}$$

The normalization condition yields $a=\frac{1}{\sqrt{2}}$. Therefore, the first normalized eigenvector of F is

$$|\psi_1\rangle=\frac{1}{\sqrt{2}}\begin{pmatrix}1\\ 1\end{pmatrix}. \tag{4.6.54}$$

Similarly, the second normalized eigenvector of F, corresponding to $\lambda_2=-1$, is given by

$$|\psi_2\rangle=\frac{1}{\sqrt{2}}\begin{pmatrix}1\\ -1\end{pmatrix}. \tag{4.6.55}$$

It is easily determined that the eigenvectors form an orthonormal set:

$$\langle\psi_1|\psi_1\rangle=\frac{1}{\sqrt{2}}\begin{pmatrix}1 & 1\end{pmatrix}\begin{pmatrix}1\\ -1\end{pmatrix}=1, \tag{4.6.56}$$

$$\langle\psi_1|\psi_2\rangle=-\frac{1}{2}\begin{pmatrix}1 & 1\end{pmatrix}\begin{pmatrix}1\\ -1\end{pmatrix}=0, \tag{4.6.57}$$

$$\langle\psi_2|\psi_2\rangle=\frac{1}{\sqrt{2}}\begin{pmatrix}1 & -1\end{pmatrix}\begin{pmatrix}1\\ -1\end{pmatrix}=1. \tag{4.6.58}$$

Now we need to find out if these vectors satisfy the completeness relation:

$$|\psi_1\rangle\langle\psi_1|+|\psi_2\rangle\langle\psi_2|=I, \tag{4.6.59}$$

where I is the unit matrix. We have

$$\begin{aligned}|\psi_1\rangle\langle\psi_1|+|\psi_2\rangle\langle\psi_2|&=\frac{1}{2}\begin{pmatrix}1\\ 1\end{pmatrix}\begin{pmatrix}1 & 1\end{pmatrix}+\frac{1}{2}\begin{pmatrix}1\\ -1\end{pmatrix}\begin{pmatrix}1 & -1\end{pmatrix}\\ &=\frac{1}{2}\left[\begin{pmatrix}1 & 1\\ 1 & 1\end{pmatrix}+\begin{pmatrix}1 & -1\\ -1 & 1\end{pmatrix}\right]=\frac{1}{2}\begin{pmatrix}2 & 0\\ 0 & 2\end{pmatrix}=I.\end{aligned} \tag{4.6.60}$$

We see that the completeness relation is satisfied, and that these eigenvectors are orthonormal. Therefore, we conclude that they form a basis. Any 2×1 vector can be written as a linear combination of these basis vectors.

4.7 Change of Basis and Unitary Transformations

One can choose one or the other set of basis vectors in the Hilbert space $\mathscr{H}$ of states of a quantum mechanical system to represent the state vectors and the operators belonging to $\mathscr{H}$. Obviously, when we change from one basis to the other, the components of a state vector change. Therefore, it is important to ascertain that the change in basis is done in such a way that the basic physical consequences remain unchanged. Evidently, for this to be the case, the norm of the state vector in the new basis must not change. The latter requirement can be fulfilled if the transformation matrix (i.e., the matrix that executes the change of basis) must be unitary. Let us check whether this holds good or not.

Let $\{|\phi_n\rangle\}$ and $\{|\chi_n\rangle\}$ be two bases in $\mathscr{H}$. Assume that we change from the so-called original (old) basis $\{|\phi_n\rangle\}$ to the new basis $\{|\chi_n\rangle\}$. The main task is to determine the components of the state vector $|\psi_m\rangle$ in the new basis, $\{|\chi_n\rangle\}$, if they are known in the old basis, $\{|\phi_n\rangle\}$.

Using the general procedure, we can expand each ket $|\psi_m\rangle$ (defined in the old basis set $\{|\phi_n\rangle\}$) in terms of the new basis set $\{|\chi_n\rangle\}$ as

$$|\psi_m\rangle = \left(\sum_n |\chi_n\rangle\langle\chi_n|\right)|\psi_m\rangle = \sum_n U_{nm}|\chi_n\rangle, \tag{4.7.1}$$

where

$$U_{nm} = \langle\chi_n|\psi_m\rangle. \tag{4.7.2}$$

The matrix U giving the transformation from the old basis to the new basis is given by

$$U = \begin{pmatrix} \langle\chi_1|\psi_1\rangle & \langle\chi_1|\psi_2\rangle & \langle\chi_1|\psi_3\rangle & . & . & . & \langle\chi_1|\psi_n\rangle \\ \langle\chi_2|\psi_1\rangle & \langle\chi_2|\psi_2\rangle & \langle\chi_2|\psi_3\rangle & . & . & . & \langle\chi_2|\psi_n\rangle \\ \langle\chi_3|\psi_1\rangle & \langle\chi_3|\psi_2\rangle & \langle\chi_3|\psi_3\rangle & . & . & . & \langle\chi_3|\psi_n\rangle \\ . & . & . & . & . & . & . \\ . & . & . & . & . & . & . \\ \langle\chi_n|\psi_1\rangle & \langle\chi_n|\psi_2\rangle & \langle\chi_n|\psi_3\rangle & . & . & . & \langle\chi_n|\psi_n\rangle \end{pmatrix}. \tag{4.7.3}$$

Let us prove that the matrix U is indeed a unitary matrix. We have

$$\begin{aligned} \langle\psi_i|U^\dagger U|\psi_j\rangle &= \langle\psi_i|U^\dagger\left(\sum_n |\chi_n\rangle\langle\chi_n|\right)U|\psi_j\rangle = \sum_n \langle\psi_i|U^\dagger|\chi_n\rangle\langle\chi_n|U|\psi_j\rangle \\ &= \sum_n \langle\chi_n|U|\psi_i\rangle^\dagger\langle\chi_n|U|\psi_j\rangle = \sum_n U_{ni}^\dagger U_{nj}. \end{aligned} \tag{4.7.4}$$

On the other hand, according to (4.7.2),

$$\sum_n U_{ni}^\dagger U_{nj} = \sum_n \langle \chi_n|\psi_i\rangle^\dagger \langle \chi_n|\psi_j\rangle = \sum_n \langle \psi_i|\chi_n\rangle\langle \chi_n|\psi_j\rangle = \langle \psi_i|\psi_j\rangle. \tag{4.7.5}$$

Equations (4.7.4) and (4.7.5) lead to

$$\langle \psi_i|U^\dagger U|\psi_j\rangle = \langle \psi_i|\psi_j\rangle, \tag{4.7.6}$$

that is,

$$U^\dagger U = \hat{I}. \tag{4.7.7}$$

Therefore, the transformation matrix U is unitary. Thus, we see that a change in basis can be done with the help of a unitary transformation.

Transformation properties of vectors and operators under a unitary transformation: The ket $|\psi\rangle$ and the bra $\langle\psi|$ transform as

$$|\psi'\rangle = \hat{U}|\psi\rangle, \quad \langle\psi'| = \langle\psi|\hat{U}^\dagger. \tag{4.7.8}$$

Consider the transformation of the equation

$$|\phi\rangle = \hat{F}|\psi\rangle, \tag{4.7.9}$$

where $\hat{F}$ is an operator. We have

$$|\phi'\rangle = \hat{F}'|\psi'\rangle. \tag{4.7.10}$$

On the other hand, under the unitary transformation, the kets $|\psi\rangle$ and $|\phi\rangle$ themselves go into

$$|\psi'\rangle = \hat{U}|\psi\rangle, \quad |\phi'\rangle = \hat{U}|\phi\rangle. \tag{4.7.11}$$

Therefore, we get

$$\hat{U}|\phi\rangle = \hat{F}'\hat{U}|\psi\rangle. \tag{4.7.12}$$

Multiplying both sides of this equation by $\hat{U}^\dagger = \hat{U}^{-1}$ and taking into account that $\hat{U}^\dagger\hat{U} = \hat{U}\hat{U}^\dagger = \hat{I}$, we arrive at

$$|\phi\rangle = \hat{U}^\dagger\hat{F}'\hat{U}|\psi\rangle. \tag{4.7.13}$$

Subtracting (4.7.13) from (4.7.9) and taking into account that $|\psi\rangle$ is an arbitrary vector, we get

$$\hat{F} = \hat{U}^\dagger \hat{F}' \hat{U} = \hat{U}^{-1} \hat{F}' \hat{U}, \quad \text{or,} \quad \hat{F}' = \hat{U} \hat{F} \hat{U}^\dagger = \hat{U} \hat{F} \hat{U}^{-1}. \tag{4.7.14}$$

This transformation of an operator under a unitary transformation is called a *similarity transformation*. It is frequently used in linear algebra to diagonalize a given matrix.

Properties of unitary transformation

1. If $\hat{F}$ is a hermitian operator, its transformed operator $\hat{F}'$ is also hermitian:

$$(\hat{F}')^\dagger = (\hat{U}\hat{F}\hat{U}^\dagger)^\dagger = (\hat{U}^\dagger)^\dagger \hat{F}^\dagger \hat{U}^\dagger = \hat{U}\hat{F}\hat{U}^\dagger = \hat{F}'. \tag{4.7.15}$$

2. The eigenvalues of $\hat{F}$ and that of $\hat{F}'$ are the same: If $\hat{F}|\psi\rangle = \lambda|\psi\rangle$, then

$$\hat{F}'|\psi'\rangle = \hat{U}\hat{F}\hat{U}^\dagger\hat{U}|\psi\rangle = \hat{U}\hat{F}|\psi\rangle = \lambda\hat{U}|\psi\rangle = \lambda|\psi'\rangle. \tag{4.7.16}$$

3. The commutator of two operators $\hat{F}$ and $\hat{G}$ remains unchanged under a unitary transformation: If $[\hat{F},\hat{G}] = \alpha$, where α is a complex number, then

$$\begin{aligned}[\hat{F}',\hat{G}'] &= [\hat{U}\hat{F}\hat{U}^\dagger,\hat{U}\hat{G}\hat{U}^\dagger] = \hat{U}\hat{F}\hat{U}^\dagger\hat{U}\hat{G}\hat{U}^\dagger - \hat{U}\hat{G}\hat{U}^\dagger\hat{U}\hat{F}\hat{U}^\dagger \\ &= \hat{U}[\hat{F}',\hat{G}']\hat{U}^\dagger = \hat{U}\alpha\hat{U}^\dagger = \alpha = [\hat{F},\hat{G}].\end{aligned} \tag{4.7.17}$$

4. The following relations hold good for any number of operators

$$\hat{F} = \alpha\hat{G} + \beta\hat{H} \quad \Rightarrow \quad \hat{F}' = \alpha\hat{G}' + \beta\hat{H}', \tag{4.7.18}$$

$$\hat{F} = \alpha\hat{X}\hat{Y}\hat{Z} \quad \Rightarrow \quad \hat{F}' = \alpha\hat{X}'\hat{Y}'\hat{Z}', \tag{4.7.19}$$

   where the prime over an operator stands for the transform of the operator under the unitary transformation.

5. Since a scalar remains invariant under a unitary transformation, complex numbers, such as $\langle\psi|\hat{F}|\phi\rangle$, remain unchanged under a unitary transformation:

$$\langle\psi'|\hat{F}'|\phi'\rangle = (\langle\psi|\hat{U}^\dagger)(\hat{U}\hat{F}\hat{U}^\dagger)(\hat{U}|\phi\rangle) = \langle\psi|(\hat{U}^\dagger\hat{U})\hat{F}(\hat{U}^\dagger\hat{U})|\phi\rangle = \langle\psi|\hat{F}|\phi\rangle. \tag{4.7.20}$$

   Taking $\hat{F} = \hat{I}$, we conclude that the inner products of the type $\langle\psi|\phi\rangle$ remain invariant. This tells us that the norm of a vector, $\||\psi\rangle\| = \sqrt{\langle\psi|\psi\rangle}$ remains invariant under a unitary transformation.

6. Using these results, it is easy to prove that

$$(\hat{U}\hat{F}\hat{U}^\dagger)^n = \hat{U}\hat{F}^n\hat{U}^\dagger, \tag{4.7.21}$$

$$\hat{U}f(\hat{F},\hat{G},\hat{H},...)\hat{U}^\dagger = f(\hat{U}\hat{F}\hat{U}^\dagger,\hat{U}\hat{G}\hat{U}^\dagger,\hat{U}\hat{H}\hat{U}^\dagger,...) = f(\hat{F}',\hat{G}',\hat{H}'), \tag{4.7.22}$$

for any number of operators $\hat{F},\hat{G},\hat{H}$ and so on.

Therefore, we conclude that a unitary transformation does not change the physics of the system. It only transforms one description into another physically equivalent description.

4.8 The Projection Operator

An operator $\hat{P}$ is said to be a projection operator if it is hermitian and equal to its own square

$$\hat{P} = \hat{P}^\dagger, \quad \hat{P}^2 = \hat{P}. \tag{4.8.1}$$

Clearly, the unit operator $\hat{I}$ satisfies these properties and is an example of a projection operator.

Consider an operator, $\hat{A}$, equal to the outer product of a ket and its corresponding bra:

$$\hat{A} = |\phi\rangle\langle\phi|. \tag{4.8.2}$$

By definition it acts on a ket $|\psi\rangle$ through the rule

$$\hat{A}|\psi\rangle = (|\phi\rangle\langle\phi|)\,|\psi\rangle = |\phi\rangle\,\langle\phi|\psi\rangle. \tag{4.8.3}$$

The claim is that if $|\phi\rangle$ is normalized to unity, the operator $\hat{A}$ is a projection operator. Let us check it. We have

$$\hat{A}^\dagger = \{|\phi\rangle\langle\phi|\}^\dagger = |\phi\rangle\langle\phi| = \hat{A}, \tag{4.8.4}$$

and

$$\hat{A}^2 = \{|\phi\rangle\langle\phi|\}\{|\phi\rangle\langle\phi|\} = |\phi\rangle\{\langle\phi|\phi\rangle\}\langle\phi|. \tag{4.8.5}$$

So, if $\langle\phi|\phi\rangle = 1$,

$$\hat{A}^2 = |\phi\rangle\langle\phi| = \hat{A}. \tag{4.8.6}$$

Since both the required properties are satisfied, $\hat{A} = |\phi\rangle\langle\phi|$ is a projection operator. That is $\hat{A} = \hat{P}$.

Let us mention here that using the closure relation and the definition of the projection operator, we can rewrite the expansion of a general ket in terms of the basis kets. We have

$$|\psi\rangle = \left(\sum_i |\phi_i\rangle\langle\phi_i|\right)|\psi\rangle = \sum_i |\phi_i\rangle\langle\phi_i|\psi\rangle = \sum_i c_i\,|\phi_i\rangle, \tag{4.8.7}$$

where $c_i = \langle\phi_i|\psi\rangle$, as earlier, are the components of the ket in the basis $\{|\phi_i\rangle\}$. As mentioned in the section 4.2, these components are arranged in to a column and the ket is written as a column vector. Since a Hilbert space is an infinite dimensional vector space, a ket will have an infinite number of components which can again be arranged into a column and the ket can be represented as a column vector with infinite number of rows. This is what we did in section 4.3 while discussing the concept of a Hilbert space.

Consider the sum of two projection operators $\hat{P}_1$ and $\hat{P}_2$: $\hat{P} = \hat{P}_1 + \hat{P}_2$. Let us check whether this can be a projection operator or not. Since $\hat{P}_1$ and $\hat{P}_2$ are projection operators, $\hat{P}^\dagger = (\hat{P}_1 + \hat{P}_2)^\dagger = \hat{P}_1^\dagger + \hat{P}_2^\dagger = \hat{P}_1 + \hat{P}_2 = \hat{P}$. So, $\hat{P}$ is hermitian. Let us check the second property. We have

$$\hat{P}^2 = \hat{P}_1^2 + \hat{P}_2^2 + \hat{P}_1\hat{P}_2 + \hat{P}_2\hat{P}_1. \tag{4.8.8}$$

We see that only if $\hat{P}_1\hat{P}_2 = \hat{P}_2\hat{P}_1 = 0$, we have $\hat{P}^2 = \hat{P}$. Therefore, we conclude that the sum of two projection operators $\hat{P}_1$ and $\hat{P}_2$ is a projection operator if and only if their product is zero.

Two projection operators are said to be orthogonal if their product is zero. Thus, the sum of two projection operators $\hat{P}_1$ and $\hat{P}_2$ is a projection operator if and only $\hat{P}_1$ and $\hat{P}_2$ are orthogonal.

It is easy to check that for a sum of projection operators $\hat{P}_1 + \hat{P}_2 + \hat{P}_2 + \ldots$ to be a projection operator, it is necessary and sufficient that these projection operators be mutually orthogonal.

Consider the product of two projection operators $\hat{P} = \hat{P}_1\hat{P}_2$. We want to find out the condition under which this product is a projection operator. Since $\hat{P}_1$ and $\hat{P}_2$ are projection operators, we have

$$\hat{P}^\dagger = (\hat{P}_1\hat{P}_2)^\dagger = \hat{P}_2^\dagger\hat{P}_1^\dagger = \hat{P}_2\hat{P}_1, \tag{4.8.9}$$

$$\hat{P}^2 = (\hat{P}_1\hat{P}_2)^2 = \hat{P}_1\hat{P}_2\hat{P}_1\hat{P}_2 = \hat{P}_1(\hat{P}_2\hat{P}_1)\hat{P}_2. \tag{4.8.10}$$

It is quite clear from the aforementioned equations that $\hat{P}$ will satisfy the required properties for being a projection operator only if $\hat{P}_1$ and $\hat{P}_2$ commute. It is also clear that, if $\hat{P}_1$ and $\hat{P}_2$ commute, $\hat{P}$ does satisfy the required properties for being a projection operator. Thus, for the product of two projection operators to be a projection operator, it is necessary and sufficient that the two projection operators commute.

Example 4.8.1: Consider the two quantum mechanical states given by the state vectors $|\psi\rangle = 5i|\phi_1\rangle + 2|\phi_2\rangle$ and $|\chi\rangle = -\frac{i}{\sqrt{3}}|\phi_1\rangle + \sqrt{\frac{2}{3}}|\phi_2\rangle$, where the two vectors $|\phi_1\rangle$ and $|\phi_2\rangle$ form a complete and orthonormal basis. Check whether $|\psi\rangle\langle\psi|$ and $|\chi\rangle\langle\chi|$ are projection operators or not.

Solution: We get

$$
\begin{aligned}
|\psi\rangle\langle\psi| &= (-5i\langle\phi_1| + 2\langle\phi_2|)(5i|\phi_1\rangle + 2|\phi_2\rangle) \\
&= 25\langle\phi_1|\phi_1\rangle - 10i\langle\phi_1|\phi_2\rangle + 10i\langle\phi_2|\phi_1\rangle + 4\langle\phi_2|\phi_2\rangle,
\end{aligned} \tag{4.8.11}
$$

$$
\begin{aligned}
|\chi\rangle\langle\chi| &= \left(\frac{i}{\sqrt{3}}\langle\phi_1| + \sqrt{\frac{2}{3}}\langle\phi_2|\right)\left(-\frac{i}{\sqrt{3}}|\phi_1\rangle + \sqrt{\frac{2}{3}}|\phi_2\rangle\right) \\
&= \frac{1}{3}\langle\phi_1|\phi_1\rangle + \frac{\sqrt{2}i}{3}\langle\phi_1|\phi_2\rangle - \frac{\sqrt{2}i}{3}\langle\phi_2|\phi_1\rangle + \frac{2}{3}\langle\phi_2|\phi_2\rangle.
\end{aligned} \tag{4.8.12}
$$

Let us check the two required properties for an operator to be a projection operator. We have

$$
(|\psi\rangle\langle\psi|)^\dagger = 25\langle\phi_1|\phi_1\rangle - 10i\langle\phi_1|\phi_2\rangle + 10i\langle\phi_2|\phi_1\rangle + 4\langle\phi_2|\phi_2\rangle = |\psi\rangle\langle\psi|, \tag{4.8.13}
$$

$$
(|\chi\rangle\langle\chi|)^\dagger = \frac{1}{3}\langle\phi_1|\phi_1\rangle + \frac{\sqrt{2}i}{3}\langle\phi_1|\phi_2\rangle - \frac{\sqrt{2}i}{3}\langle\phi_2|\phi_1\rangle + \frac{2}{3}\langle\phi_2|\phi_2\rangle = |\chi\rangle\langle\chi|. \tag{4.8.14}
$$

However, since

$$
(|\psi\rangle\langle\psi|)^2 = |\psi\rangle\langle\psi|\psi\rangle\langle\psi| = 25\,|\psi\rangle\langle\psi|, \tag{4.8.15}
$$

$$
(|\chi\rangle\langle\chi|)^2 = |\chi\rangle\langle\chi|\chi\rangle\langle\chi| = |\chi\rangle\langle\chi|. \tag{4.8.16}
$$

$(|\psi\rangle\langle\psi|)^2 \neq |\psi\rangle\langle\psi|$. Hence, only $|\chi\rangle\langle\chi|$ is the projection operator.

The expectation value of an operator: The expectation value of an operator $\hat{F}$, in a state described by the wave function $|\psi\rangle$, is defined as

$$
\langle\hat{F}\rangle = \frac{\langle\psi|\hat{F}|\psi\rangle}{\langle\psi|\psi\rangle} = \frac{\int_{-\infty}^{+\infty} \psi^*(\vec{r})\,[\hat{A}\psi(\vec{r})]\,d^3x}{\int_{-\infty}^{+\infty} \psi^*(\vec{r})\psi(\vec{r})\,d^3x}. \tag{4.8.17}
$$

If the wave function is normalized to unity, then

$$
\langle\hat{F}\rangle = \langle\psi|\hat{F}|\psi\rangle = \int_{-\infty}^{+\infty} \psi^*(\vec{r})\,[\hat{A}\psi(\vec{r})]\,d^3x. \tag{4.8.18}
$$

Example 4.8.2: A particle is in a state $|\psi\rangle = 2i|\phi_1\rangle - |\phi_2\rangle + 4i|\phi_3\rangle$, where $\{|\phi_1\rangle, |\phi_2\rangle, |\phi_3\rangle\}$ constitute an orthonormal basis. An operator, $\hat{F}$, is given by $\hat{F} = |\phi_1\rangle\langle\phi_1| - 2i|\phi_1\rangle\langle\phi_2| + |\phi_3\rangle\langle\phi_3|$. Find $\langle\hat{F}\rangle$ in this state.

Solution: We have

$$\langle\psi| = -2i\langle\phi_1| - \langle\phi_2| - 4i\langle\phi_3|. \tag{4.8.19}$$

$$\begin{aligned}\hat{F}|\psi\rangle &= (|\phi_1\rangle\langle\phi_1| - 2i|\phi_1\rangle\langle\phi_2| + |\phi_3\rangle\langle\phi_3|)(2i|\phi_1\rangle - |\phi_2\rangle + 4i|\phi_2\rangle) \\ &= 2i|\phi_1\rangle\langle\phi_1|\phi_1\rangle + 2i|\phi_1\rangle\langle\phi_2|\phi_2\rangle + 4i|\phi_3\rangle\langle\phi_3|\phi_3\rangle \\ &= 4i|\phi_1\rangle + 4i|\phi_3\rangle,\end{aligned} \tag{4.8.20}$$

where we have used the orthonormality of the given basis. Therefore, we get

$$\langle\hat{F}\rangle = \frac{\langle\psi|\hat{F}|\psi\rangle}{\langle\psi|\psi\rangle} = \frac{8\langle\phi_1|\phi_1\rangle + 16\langle\phi_3|\phi_3\rangle}{21} = \frac{24}{21}. \tag{4.8.21}$$

Finally, let us mention here that the bra–ket formalism and the associated linear algebra developed earlier can be generalized to the case of a continuous basis in the Hilbert space using the representations of the vectors and the operators in a continuous basis discussed earlier. Commonly used continuous bases are (i) the complete set of eigenvectors $|\vec{r}\rangle$ of the position operator $\hat{\vec{r}}$ and (ii) the complete set of eigenvectors $|\vec{p}\rangle$ of the momentum operator $\hat{\vec{p}}$. Later, when we derive the Schrödinger equation in the coordinate as well as in the momentum basis, we shall talk about it in more detail.

4.9 Coordinate and Momentum Representations of the State Vector and the Schrödinger Equation

In this section we shall derive the coordinate and the momentum representations of the state vector, $|\psi\rangle$, and the Schrödinger equation

$$i\hbar\frac{\partial|\psi\rangle}{\partial t} = \hat{H}|\psi\rangle, \tag{4.9.1}$$

where $|\psi\rangle$ belongs to the Hilbert space of states $\mathscr{H}$ of the system and $\hat{H}$ is the Hamiltonian operator. The state vector, $|\psi\rangle$, can be represented by its components in a given basis. The choice of a particular basis determines a particular representation for the state vector and the operators acting in the Hilbert space including the Hamiltonian operator $\hat{H}$. As a consequence, we get a particular representation of the Schrödinger equation.

In general, since the eigenvectors of a hermitian operator constitute an orthonormal and complete set of vectors, the easiest way to construct a particular basis in $\mathscr{H}$ is to choose a suitable observable, solve the eigenvalue problem for it and take the set of its normalized eigenvectors as a basis. However, as mentioned earlier, we are interested in describing the system in such a way that we have the maximum possible information about the system at any given instant of time. This maximal information depends upon the set of commuting observables of the system: greater the number of commuting observables in the set, greater will be the number of dynamical variables of the system that can be measured simultaneously accurately and hence greater will be the information about the system. Therefore, we first try to determine the maximal set of commuting observables for a quantum system.

Let us consider a collection of particles in one spatial dimension. Since, according to the uncertainty principle, classically conjugate dynamical variables can not be measured simultaneously accurately, the set of measuring apparatus has to be divided into groups which can detect particles in terms of a given particular set of mechanical characteristics. For instance, for our ensemble of particles, we can easily divide the set of measuring apparatus into at least two groups: One which sorts out particles in terms of their co-ordinates, i.e., in terms of any function of coordinates, say, the potential energy $V(x,y,z)$, and the other which sorts them out in terms of their momenta, i.e., kinetic energy. Clearly, in accordance with the uncertainty relation, the first group of apparatus excludes the selection in terms of momenta while the second excludes the selection of particles in terms of their coordinates. This fact, as we shall see below, leads to very stringent limitation on the choice of the required maximal set of commuting observables.

In what follows, while deriving the coordinate and momentum representations for the Schödinger equation, we shall use the generalized coordinates $q = \{q_i\}, i = 1,2,3,\ldots$ and the generalized momenta $p = \{p_i\}, i = 1,2,3,\ldots$.

We know that the observables are functions of the fundamental operators: the position operator $\hat{q}$ and the momentum operator $\hat{p}$. They can, in general, be represented by arbitrary functions of the type $F(\hat{q},\hat{p})$. Therefore, the algebra of observables is determined by the algebra of $\hat{q}$ and $\hat{p}$:

$$[\hat{q},\hat{p}] = i\hbar\hat{I}, \quad , \tag{4.9.2}$$

which must hold in any representation we wish to construct. Using the definition of a function of operators and the aforementioned commutation relation it is easy to show that

$$[\hat{q},F(\hat{q},\hat{p})] = i\hbar\frac{\partial F}{\partial \hat{p}}, \quad \textit{and} \quad [\hat{p},F(\hat{q},\hat{p})] = -i\hbar\frac{\partial F}{\partial \hat{q}}, \tag{4.9.3}$$

where $F(\hat{q},\hat{p})$ is an arbitrary function of $\hat{q}$ and $\hat{p}$. Consequently, $\hat{q}$ commutes with $F(\hat{q},\hat{p})$ iff F depends on $\hat{q}$ only and $\hat{p}$ commutes with F iff F depends on $\hat{p}$ alone. Thus for a given system the maximal set of commuting observables consists either of $\hat{q}$ (and the observables which are functions of $\hat{q}$), or of $\hat{p}$ (and the observables that are functions of $\hat{p}$). As a result, we have only two choices for a basis in the Hilbert space: (i) the basis consisting of the eigenvectors of the position operator $\hat{q}$ and (ii) the basis consisting of the eigenvectors of the momentum operator $\hat{p}$. The corresponding representations are, respectively, called the co-ordinate representation and the momentum representation.

The co-ordinate Representation: The eigenfunctions of the operator $\hat{q}$ satisfy

$$\hat{q}|q'\rangle = q'|q'\rangle, \tag{4.9.4}$$

where, following Dirac, we have used the same letter to represent the eigenkets and the eigenvalues of $\hat{q}$. The eigenvalues $(q', q'', q''', ...) \in R^1$ are continuous. The eigenfunctions, $|q'\rangle$, are normalized as:

$$\langle q'|q''\rangle = \delta(q' - q''), \tag{4.9.5}$$

where $\delta(q' - q'')$ is the Dirac delta function. The completeness condition for the eigenkets, $|q'\rangle$, of $\hat{q}$ reads as

$$\int |q'\rangle dq' \langle q'| = \hat{I}. \tag{4.9.6}$$

We take this complete set of eigenvectors of the position operator as our basis set in the Hilbert space. In this basis the state vector is characterized by its component $\langle q'|\psi\rangle = \psi(q', t)$, defined by the following relation

$$|\psi(t)\rangle = \int dq' |q'\rangle\langle q'|\psi(t)\rangle = \int dq' \psi(q', t)|q'\rangle. \tag{4.9.7}$$

$\psi(q', t)$, which is a function of position and time, is called the wave function of the system.

Also, the matrix elements of $\hat{q}$ in this basis are given by

$$\langle q''|\hat{q}|q'\rangle = q'\langle q''|q'\rangle = q'\delta(q'' - q'). \tag{4.9.8}$$

The time evolution of the system is governed by the Schrödinger equation (4.9.1), where the Hamiltonian is given by

$$\hat{H} = \frac{\hat{p}^2}{2m} + V(\hat{q}) \tag{4.9.9}$$

The Hamiltonian operator will be defined in this basis only if we determine how $\hat{q}$ and $\hat{p}$ act on the state vector in this basis.

The action of $\hat{q}$ on a state vector is determined by taking the inner product of $\hat{q}|\psi\rangle$ with the basis vectors:

$$\langle q'|\hat{q}|\psi\rangle = \int dq'' \delta(q' - q'')\langle q''|\hat{q}|\psi\rangle = \int dq'' \langle q'|q''\rangle \hat{q}\langle q''|\psi\rangle$$

$$= \int dq'' \langle q'|q''\rangle q'\langle q''|\psi\rangle = q' \int dq'' \langle q'|q''\rangle\langle q''|\psi\rangle = q'\langle q'|\psi\rangle = q'\psi(q'). \tag{4.9.10}$$

Hence, the action of the operator $\hat{q}$ on the wave function results in the multiplication of the wave function by the corresponding eigenvalue: $\hat{q}|\psi(q',t) = q'\psi(q',t)$.

The next step is to determine the result of action of the momentum operator, $\hat{p}$, on the state vector. For this purpose we need the matrix element, $\langle q'|\hat{p}|q''\rangle$, of the momentum operator in this basis. To have this matrix element we make use of the fundamental algebra

$$[\hat{q},\hat{p}] = i\hbar\hat{I}, \tag{4.9.11}$$

and calculate its matrix element as shown below:

$$\langle q'|\hat{q}\hat{p} - \hat{q}\hat{p}|q''\rangle = \langle q'|i\hbar\hat{I}|q''\rangle = i\hbar\langle q'|q''\rangle = i\hbar\delta(q'-q''). \tag{4.9.12}$$

The left hand-side of (4.9.12) is given by

$$\langle q'|\hat{q}\hat{p}|q''\rangle - \langle q'|\hat{p}\hat{q}|q''\rangle = (q'-q'')\langle q'|\hat{p}|q''\rangle \tag{4.9.13}$$

Therefore, using the identity $x\delta'(x) = -\delta(x)$ and combining (4.9.12) and (4.9.13), we arrive at

$$(q'-q'')\langle q'|\hat{p}|q''\rangle = i\hbar\delta(q'-q'') = -i\hbar(q'-q'')\frac{\partial}{\partial q'}\delta(q'-q'') \tag{4.9.14}$$

As a result, we have

$$\langle q'|\hat{p}|q''\rangle = -i\hbar\frac{\partial}{\partial q'}\delta(q'-q''). \tag{4.9.15}$$

As a consequence, the action of $\hat{p}$ on the state vector is determined as

$$\begin{aligned}\hat{p}\psi(q') &= \langle q'|\hat{p}|\psi\rangle = \int dq''(-i\hbar\frac{\partial}{\partial q'}\delta(q'-q''))\langle q''|\psi\rangle \\ &= \int dq''(-i\hbar\frac{\partial}{\partial q'}\delta(q'-q''))\psi(q'') = \left(-i\hbar\frac{\partial}{\partial q'}\right)\psi(q'). \end{aligned}\tag{4.9.16}$$

Hence, the action of the operator $\hat{p}$ results into the partial differentiation of the wave function (standing on the right of it), with respect the co-ordinate multiplied by the factor $-i\hbar$.

Schrödinger equation in coordinate representation: Having determined the action of the fundamental operators $\hat{q}$ and $\hat{p}$ on the state vector, let us go back to the Schrödinger equation (4.9.1) and find its projection on the basis vectors. This will give us the required form of the Schrödinger equation in this basis. We have

$$\begin{aligned}
\langle q'|\hat{H}|\psi\rangle &= \left\langle q'|\frac{\hat{p}^2}{2m}+V(\hat{q})|\psi\right\rangle = \frac{1}{2m}\int dq''\langle q'|\hat{p}^2|q''\rangle\langle q''|\psi\rangle \\
&+ \int dq''\langle q'|V(\hat{q})|q''\rangle\langle q''|\psi\rangle \\
&= \frac{1}{2m}\int dq''(-i\hbar)^2\frac{\partial^2}{\partial q'^2}\delta(q'-q'')\psi(q'') + \int dq''\langle q'|V(\hat{q})|q''\rangle\psi(q'') \\
&= -\frac{\hbar^2}{2m}\frac{\partial^2}{\partial q'^2}\psi(q') + \int dq'' V(q'')\langle q'|q''\rangle\psi(q'') \\
&= \left[-\frac{\hbar^2}{2m}\frac{\partial^2}{\partial q'^2}+V(q')\right]\psi(q')
\end{aligned} \tag{4.9.17}$$

The left-hand side of (4.9.1) gives

$$i\hbar\frac{\partial\langle q'|\psi\rangle}{\partial t} = i\hbar\frac{\partial\psi(q')}{\partial t}. \tag{4.9.18}$$

Therefore, the Schrödinger equation in the coordinate representation takes the form

$$i\hbar\frac{\partial\psi(q)}{\partial t} = \left[-\frac{\hbar^2}{2m}\frac{\partial^2}{\partial q^2}+V(q)\right]\psi(q), \tag{4.9.19}$$

where we have dropped the prime, since q' is an arbitrary generalized coordinate. In the usual case of our three-dimensional space covered with the Cartesian system of coordinates (x,y,z), the Schrödinger equation will read

$$i\hbar\frac{\partial\psi(\vec{r},t)}{\partial t} = -\frac{\hbar^2}{2m}\vec{\nabla}^2\psi(\vec{r},t)+V(\vec{r})\,\psi(\vec{r},t), \tag{4.9.20}$$

where $\vec{\nabla}$ is the gradient operator:

$$\vec{\nabla} = \hat{i}\frac{\partial}{\partial x}+\hat{j}\frac{\partial}{\partial y}+\hat{k}\frac{\partial}{\partial z}. \tag{4.9.21}$$

The form of the Schrödinger equation, given by the equation (4.9.20), leads to the following expression for the Hamiltonian operator in co-ordinate representation:

$$\hat{H} = -\frac{\hbar^2}{2m}\vec{\nabla}^2+V(\vec{r}). \tag{4.9.22}$$

Thus the Hamiltonian operator is completely defined, if we know the external potential field in which the particle is moving. Once the Hamiltonian is given, the solution of a quantum mechanical problem reduces to the solution of the partial differential equation of the parabolic type. The complexity arising in the solution of the problem depends entirely on the form of the potential.

Schrödinger equation in momentum representation: In this case we choose the complete set of eigenvectors of the momentum operator, $\hat{p}$, as the basis in $\mathscr{H}$. As in the previous case, we have

$$\hat{p}|p'\rangle = p'|p'\rangle, \tag{4.9.23}$$

$$\langle p'|p''\rangle = \delta(p' - p'') \tag{4.9.24}$$

$$\int |p'\rangle dp' \langle p'| = \hat{I}, \tag{4.9.25}$$

where $(p', p'', p''', ... \in R^1$ are the momentum eigenvalues. The expansion of the state vector $|\psi\rangle$ reads

$$|\psi(t)\rangle = \int dp'|p'\rangle\langle p'|\psi(t)\rangle = \int dp'\psi(p',t)|p'\rangle. \tag{4.9.26}$$

The function $\psi(p',t)$ is a function of momentum and time and is called the wave function of the system in momentum representation. Following exactly the same steps as in the case of coordinate representation, we get that the action of the operator $\hat{p}$ on the wave function results in the multiplication of the wave function by the corresponding eigenvalue: $\hat{p}\psi(p',t) = p'\psi(p',t)$ and the action of the operator $\hat{q}$ results into the partial differentiation of the wave function (standing on the right of it), with respect the momentum multiplied by the factor $-i\hbar$. Therefore, projecting the Schrödinger equation for the state vector on the basis $\{|p'\rangle\}$, we obtain

$$i\hbar\frac{\partial\psi(p,t)}{\partial t} = \left[\frac{p^2}{2m} + V\left(-i\hbar\frac{\partial}{\partial q}\right)\right]\psi(p,t). \tag{4.9.27}$$

This is the Schrödinger equation in momentum representation. In the usual case of our three dimensional space covered with the Cartesian system of co-ordinates (x,y,z), the Schrödinger equation will read

$$i\hbar\frac{\partial\psi(\vec{p},t)}{\partial t} = \frac{\vec{p}^2}{2m}\psi(\vec{p},t) + V\left(-i\hbar\vec{\nabla}_p\right)\psi(\vec{p},t), \tag{4.9.28}$$

where $\vec{\nabla}_p$ is the gradient operator with respect to the momentum variable:

$$\vec{\nabla}_p = \hat{i}\frac{\partial}{\partial p_x} + \hat{j}\frac{\partial}{\partial p_y} + \hat{k}\frac{\partial}{\partial p_z}. \tag{4.9.29}$$

Connection between the coordinate and momentum representations of the state vector

It turns out, that the coordinate and the momentum representations of the state vector are related through a unitary transformation and hence are equivalent. For instance, we can go from the momentum representation (p-representation) to the coordinate representation (q-representation) by the following transformation

$$\langle q|\psi\rangle = \int \langle q|p\rangle\langle p|\psi\rangle\, dp, \quad or, \quad \psi(q) = \int \langle q|p\rangle \psi(p)\, dp, \tag{4.9.30}$$

where $\langle q|p\rangle$ is the transfer function that effects transition from the momentum representation to the coordinate representation. Similarly, we can write

$$\langle p|\psi\rangle = \int \langle p|q\rangle\langle q|\psi\rangle\, dq, \quad or, \quad \psi(p) = \int \langle p|q\rangle \psi(q)\, dq, \tag{4.9.31}$$

where $\langle p|q\rangle$ is the transfer function that realizes the transition from the coordinate representation to the momentum representation.

We can determine the transfer functions. For instance, from (4.9.23), we have

$$\begin{aligned} p'\langle q'|p'\rangle &= \langle q'|\hat{p}|p'\rangle = \int \langle q'|\hat{p}|q''\rangle\langle q''|p'\rangle\, dq'', \\ &= -i\hbar \int \frac{\partial}{\partial q'}\delta(q'-q'')\langle q''|p'\rangle\, dq'' = -i\hbar\frac{\partial}{\partial q'}\left(\langle q'|p'\rangle\right). \end{aligned} \tag{4.9.32}$$

Equation (4.9.32) is a differential equation for $\langle q'|p'\rangle$ as a function of the variable q'. Its general solution can be written as

$$\langle q'|p'\rangle = A\, e^{\frac{i}{\hbar}q'\,p'}, \tag{4.9.33}$$

where A is an arbitrary constant. In quantum mechanics A is taken to be $(2\pi\hbar)^{-\frac{1}{2}}$. Thus,

$$\langle q|p\rangle = \frac{1}{\sqrt{2\pi\hbar}}\, e^{\frac{i}{\hbar}q\,p}. \tag{4.9.34}$$

Analogously, we can get the transfer function $\langle q'|p'\rangle$ corresponding to the inverse transformation as

$$\langle p|q\rangle = \frac{1}{\sqrt{2\pi\hbar}}\, e^{\frac{-i}{\hbar}q\,p}. \tag{4.9.35}$$

Note that

$$\langle q|p\rangle = \langle p|q\rangle^{\dagger}. \tag{4.9.36}$$

Let us rewrite the transformation equations (4.9.30) and (4.9.31) in the following symbolic forms

$$\langle q|\psi\rangle = \hat{S}(q,p)\,\langle p|\psi\rangle, \tag{4.9.37}$$

where $\hat{S}(q,p)$ is an integral operator whose kernel is given by the transfer function $\langle q|p\rangle$. Then the inverse transformation can be written as

$$\langle p|\psi\rangle = \hat{S}^{-1}(q,p)\langle q|\psi\rangle. \tag{4.9.38}$$

On the other hand, we have

$$\langle p|\psi\rangle = \int dq\langle p|q\rangle\langle q|\psi\rangle = \int dq\langle q|p\rangle^{\dagger}\langle q|\psi\rangle = \hat{S}^{\dagger}(q,p)\langle q|\psi\rangle. \tag{4.9.39}$$

Equations (4.9.38) and (4.9.39) lead to

$$\hat{S}^{-1}(q,p) = \hat{S}^{\dagger}(q,p) \;\; or, \;\; \hat{S}^{\dagger}\hat{S} = \hat{S}\hat{S}^{\dagger} = \hat{I} \tag{4.9.40}$$

Hence, the integral operator $\hat{S}$ is a unitary operator.

Note that such a transformation of the wave function is equivalent to changing the basis in the Hilbert space by a matrix $S = (S_{ij})$. Consequently an observable $\hat{A}$ changes to $\hat{A}' = \hat{S}\hat{A}\hat{S}^{\dagger}$. It is worth mentioning here that, under such a unitary transformation, the forms of the wave function and that of the observables do change, however the physical state of the system remains unaltered because the operator $\hat{S}$ is time-independent.

4.10 Basic Postulates of Quantum Mechanics

We now have the basic mathematical tools to formulate the general framework of quantum mechanics. The formulation given here is based on the fundamental concepts of quantum mechanics, discussed in the preceding chapters, and simply reformulates these concepts in algebraic language.

Postulate 1: *The state of a quantum mechanical system, at a given instant of time, is described by a vector, $|\psi(t)\rangle$, in the abstract Hilbert space $\mathscr{H}$ of the system.*

$|\psi(t)\rangle$ is called the state vector and is assumed to contain all the information about the system, that is, it gives a complete description of the system at a given instant of time. Any superposition of the state vectors, $\sum_j c_j|\psi_j\rangle, c_j = const$, is also a state vector belonging to $\mathscr{H}$.

As discussed earlier, the state vector is not determined uniquely: it can be multiplied by an arbitrary complex number without changing the physical state of the system. Because of that, we would assume that $|\psi(t)\rangle$ is normalized to unity.

So far as the solution of a concrete problem is concerned, it is convenient to work with a set of numbers rather than with the abstract vector $|\psi(t)\rangle$. For this, it is sufficient to choose a suitable orthonormal basis $\{|\phi_j\rangle\}, \langle\phi_i|\phi_j\rangle = \delta_{ij}$ in $\mathscr{H}$ and, using the completeness condition for the basis vectors, expand $|\psi(t)\rangle$ in terms of this basis:

$$|\psi(t)\rangle = \sum_j (|\phi_j\rangle\langle\phi_j|)\,|\psi\rangle = \sum_j \psi_j|\phi_j\rangle. \tag{4.10.1}$$

In this case, the state vector will be given by the set of complex numbers $\{\psi_j\} = \langle\phi_j|\psi\rangle$: the components of $|\psi(t)\rangle$ in the chosen basis.

Postulate 2. Observables: *A measurable physical quantity, A (called an observable or dynamical variable), is represented by a linear hermitian operator $\hat{A}$ acting in the Hilbert space of state vectors.*

In a given basis, the operator $\hat{A}$ is determined by a matrix A whose elements (complex numbers) are given by

$$A_{jk} = \langle\phi_j|\hat{A}|\phi_k\rangle. \tag{4.10.2}$$

Depending on the chosen basis, the matrix A can have a discrete or a continuous representation, discussed earlier.

Postulate 3. Measurement of an observable: *The measurement of an observable A in a given state may be represented formally by the action of the operator $\hat{A}$ on the state vector $|\psi(t)\rangle$. The only possible outcome of such a measurement is one of the eigenvalues, $\{a_j\}, j = 1,2,3,\ldots,$ of $\hat{A}$.*

If the result of the measurement of A is a_n, the state of the system, immediately after the measurement is given by

$$|\psi\rangle_{\text{after}} = \langle\phi_n|\psi\rangle\ |\phi_n\rangle, \tag{4.10.3}$$

where $|\phi_n\rangle$ is the eigenvector of the operator $\hat{A}$ corresponding to the eigenvalue $a_n$and $\langle\psi_n|\psi\rangle$ stands for the inner product of $|\phi_n\rangle$ and $|\psi\rangle$ in the basis consisting of the eigenvectors of $\hat{A}$.

What is characteristic of quantum mechanics is the fact that we cannot a priori predict the result of a measurement. We can only talk about the probability of a given result.

Postulate 4. Probabilistic outcome of a measurement: *If a measurement of an observable A is made in a state $|\psi(t)\rangle$ of the quantum mechanical system, the probability of obtaining one of the non-degenerate discrete eigenvalues a_j of the corresponding operator $\hat{A}$ is given by*

$$P(a_j) = \frac{|\langle\phi_j|\psi\rangle|^2}{\langle\psi|\psi\rangle}, \tag{4.10.4}$$

where $|\phi_j\rangle$ is the eigenfunction of $\hat{A}$ with eigenvalue a_j. If the state vector is normalized to unity, $P(a_j) = |\langle\phi_j|\psi\rangle|^2$.

If the eigenvalue a_j is m-fold degenerate, this probability is given by

$$P(a_j) = \frac{\sum_{i=1}^{m} |\langle \phi_j^i | \psi \rangle|^2}{\langle \psi | \psi \rangle}. \tag{4.10.5}$$

If the operator $\hat{A}$ possesses a continuous eigenspectrum $\{a\}$, the probability that the result of measurement will yield a value between a and $a + da$ is given by

$$dP(a) = \frac{|\langle \phi(a) | \psi \rangle|^2}{\langle \psi | \psi \rangle}\, da = \frac{|\langle \phi(a) | \psi \rangle|^2}{\int_{-\infty}^{+\infty} |\psi(a')|^2\, da'}\, da. \tag{4.10.6}$$

Postulate 5. Time-evolution of the state vector: *The time evolution of the state vector is governed by the time-dependent Schrödinger equation:*

$$i\hbar \frac{\partial |\psi\rangle}{\partial t} = \hat{H} |\psi\rangle, \tag{4.10.7}$$

where $\hat{H}$ is the Hamiltonian operator corresponding to the total energy of the system.

The principle of superposition: Since the time-evolution of the system is described by the Schrödinger equation, which is a linear differential equation in $|\psi\rangle$, any linear combination of the solutions of the Schrödinger equation (4.10.7) also satisfies this equation and can be used to describe a state of the system.

In other words, if $|\psi_1\rangle, |\psi_2\rangle, |\psi_3\rangle, ..., |\psi_n\rangle, ...$ describe various possible states of the system, then

$$|\psi\rangle = \sum_i c_i |\psi_i\rangle, \tag{4.10.8}$$

where c_i are complex numbers, also represents a possible physical state of the system. The probability density for this superposition state is given by

$$P = \left| \sum_i c_i |\psi_i\rangle \right|^2. \tag{4.10.9}$$

If the states $|\psi_i\rangle$ are mutually orthogonal, then P is equal to the sum of the probability densities, P_i, for the individual states

$$P = \left| \sum_i c_i |\psi_i\rangle \right|^2 = \sum_i |c_i|^2 = \sum_i |P_i|^2. \tag{4.10.10}$$

Measurement and the average value of dynamical variables: As we have discussed earlier, the measurement of an observable A, in a in a state $|\psi\rangle$, is represented by the action of the operator $\hat{A}$ on the state vector $|\psi\rangle$. As a result, we obtain one one of the eigenvalues $\hat{A}$ and the system makes a transition to the corresponding eigenstate of the operator $\hat{A}$. However, a priori, we cannot say which of the eigenvalues of $\hat{A}$ will result. The only thing which we can say is the probability of obtaining an eigenvalue of $\hat{A}$.

Since, a general state $|\psi\rangle$ can be represented by a linear superposition of eigenstates ϕ_n of $\hat{A}$ and each measurement can produce a result a_j with a probability P_j (see Postulate 4), a sensible thing will be to make a large number of identical copies of the given system (all in the same state $|\psi\rangle$), perform the measurement on each of them and then take the average value of the results as the final outcome of the measurement.

Assume that we have a large number of identical copies of a given system, say, of non-interacting hydrogen atoms. Besides that, assume that all these systems are in the same quantum state $|\psi\rangle$. We measure a dynamical variable A on all these systems separately (one-by-one). Now, we ask the question: What will be the average value of A in the state $|\psi\rangle$? The answer to this question is given by the following theorem.

The average value of the dynamical variable A in the state $|\psi\rangle$ is given by

$$\langle\hat{A}\rangle = \frac{\langle\psi|\hat{A}|\psi\rangle}{\langle\psi|\psi\rangle}. \tag{4.10.11}$$

Using the complete set of eigenvectors $\{|\phi_n\rangle\}$ of $\hat{A}$ we can write $\langle\hat{A}\rangle$ as

$$\langle\hat{A}\rangle = \frac{1}{\langle\psi|\psi\rangle}\sum_n\sum_m\langle\psi|\phi_m\rangle\langle\phi_m|\hat{A}|\phi_n\rangle\langle\phi_n|\psi\rangle = \frac{1}{\langle\psi|\psi\rangle}\sum_n a_n|\langle\phi_n|\psi\rangle|^2 = \sum_n a_n P_n^2, \tag{4.10.12}$$

where we have used that $\langle\phi_m|\hat{A}|\phi_n\rangle = a_n\langle\phi_m|\phi_n\rangle = a_n\delta_{mn}$.

In the case of continuous spectrum of $\hat{A}$, the average value is given by

$$\langle\hat{A}\rangle = \frac{\int_{-\infty}^{+\infty} a|\psi(a)|^2 da}{\int_{-\infty}^{+\infty}|\psi(a')|^2 da'} = \int_{-\infty}^{+\infty} a dp(a). \tag{4.10.13}$$

4.11 Generalized Heisenberg Uncertainty Relation

We have earlier discussed that for two observables $\hat{A}$ and $\hat{B}$ to be measured simultaneously accurately, it is necessary and sufficient that they commute. If they do not commute, there is a correlation between the uncertainties in their measurement: if one of them is measured with a greater degree of accuracy, there is a corresponding larger degree of inaccuracy in the measurement of the other. In what follows, we shall give a general proof of the uncertainty relation between the results of measurements of two observables represented by two non-commuting hermitian operators.

Let $\hat{A}$ and $\hat{B}$ be two hermitian operators representing two observables A and B. Suppose we measure them on a large number of identical systems; all prepared in the same state described by the state vector $|\psi\rangle$. As usual, we assume $|\psi\rangle$ to be normalized to unity. Let $\langle\hat{A}\rangle = \langle\psi|\hat{A}|\psi\rangle$ and $\langle\hat{B}\rangle = \langle\psi|\hat{B}|\psi\rangle$ be the average values $\hat{A}$ and $\hat{B}$ with respect to the state $|\psi\rangle$. Then, the uncertainties in their measurements are given by the root mean square deviations:

$$\Delta A = \sqrt{\langle\hat{A}^2\rangle - \langle\hat{A}\rangle^2}, \tag{4.11.1}$$

$$\Delta B = \sqrt{\langle\hat{B}^2\rangle - \langle\hat{B}\rangle^2}. \tag{4.11.2}$$

Let us introduce two operators

$$\Delta\hat{A} = \hat{A} - \langle\hat{A}\rangle, \quad \Delta\hat{B} = \hat{B} - \langle\hat{B}\rangle, \tag{4.11.3}$$

whose action on an arbitrary state vector $|\psi\rangle$ is given by

$$\Delta\hat{A}\,|\psi\rangle = |\phi\rangle, \quad \Delta\hat{B}\,|\psi\rangle = |\chi\rangle. \tag{4.11.4}$$

Clearly the operators $\Delta\hat{A}$ and $\Delta\hat{B}$ are hermitian. Therefore, using the Schwartz inequality for the state vectors $|\phi\rangle$ and $|\chi\rangle$, given by

$$\langle\phi|\phi\rangle\langle\chi|\chi\rangle \geq |\langle\phi|\chi\rangle|^2, \tag{4.11.5}$$

we get

$$\langle(\Delta\hat{A})^2\rangle\langle(\Delta\hat{B})^2\rangle \geq |\langle\Delta\hat{A}\;\Delta\hat{B}\rangle|^2, \tag{4.11.6}$$

where we have used the fact that

$$\langle\phi|\phi\rangle = \langle\psi|(\Delta\hat{A})^2|\psi\rangle, \quad \langle\chi|\chi\rangle = \langle\psi|(\Delta\hat{B})^2|\psi\rangle, \quad \langle\phi|\chi\rangle = \langle\Delta\hat{A}\,\Delta\hat{B}\rangle. \tag{4.11.7}$$

The product $\Delta\hat{A}\Delta\hat{B}$ can be written as

$$\Delta\hat{A}\Delta\hat{B} = \frac{1}{2}[\Delta\hat{A},\Delta\hat{B}] + \frac{1}{2}[\Delta\hat{A},\Delta\hat{B}]_+ = \frac{1}{2}[\hat{A},\hat{B}] + \frac{1}{2}[\Delta\hat{A},\Delta\hat{B}]_+. \tag{4.11.8}$$

Note that the commutator $[\hat{A},\hat{B}]$ is anti-hermitian, whereas the anti-commutator $[\Delta\hat{A},\Delta\hat{B}]_+$ is hermitian. Since the expectation value of a hermitian operator is real, while that of an anti-hermitian operator is imaginary, we have

$$\langle\Delta\hat{A}\Delta\hat{B}\rangle = \frac{1}{2}\langle[\hat{A},\hat{B}]\rangle + \frac{1}{2}i|\langle[\Delta\hat{A},\Delta\hat{B}]_+\rangle|. \tag{4.11.9}$$

Therefore, we get

$$|\langle \Delta\hat{A}\,\Delta\hat{B}\rangle|^2 = \frac{1}{4}|\langle[\hat{A},\hat{B}]\rangle|^2 + \frac{1}{4}|\langle[\Delta\hat{A},\Delta\hat{B}]_+\rangle|^2. \tag{4.11.10}$$

$$|\langle \Delta\hat{A}\,\Delta\hat{B}\rangle|^2 \geq \frac{1}{4}|\langle[\hat{A},\hat{B}]\rangle|^2. \tag{4.11.11}$$

Taking into account (4.11.11), we get from (4.11.6) that

$$\langle(\Delta\hat{A})^2\rangle\langle(\Delta\hat{B})^2\rangle \geq \frac{1}{4}|\langle[\hat{A},\hat{B}]\rangle|^2. \tag{4.11.12}$$

Now $\langle(\Delta\hat{A})^2\rangle = \langle\hat{A}^2\rangle + \langle\hat{A}\rangle^2 - 2\langle\hat{A}\rangle^2 = \langle\hat{A}^2\rangle - \langle\hat{A}\rangle^2 = (\Delta A)^2$. Similarly, $\langle(\Delta\hat{B})^2\rangle = (\Delta B)^2$. As a consequence, using (4.11.12), we arrive at the inequality

$$\Delta A \Delta B \geq \frac{1}{2}|\langle[\hat{A},\hat{B}]\rangle|, \tag{4.11.13}$$

which represents the generalized form of Heisenberg's uncertainty relation for any two non-commuting hermitian operators $\hat{A}$ and $\hat{B}$.

Time-Energy uncertainty relation: The generalized uncertainty relation (4.11.13) can be used to derive the time-energy uncertainty relation. Let the operator $\hat{B}$ represent the time-independent Hamiltonian operator, $\hat{H}$, of a quantum system and let $\hat{A}$ be another time-independent observable of the same system. Let $|\psi\rangle$ be the state vector of the system at some given instant of time. If we apply Schwartz inequality to the vectors $(\hat{A} - \langle\hat{A}\rangle)|\psi\rangle$ and $(\hat{H} - \langle\hat{H}\rangle)|\psi\rangle$ and repeat the same steps of calculations which led to (4.11.13), we shall arrive at the following inequality

$$\Delta A\,\Delta E \geq \frac{1}{2}|\langle[\hat{A},\hat{H}]\rangle|, \tag{4.11.14}$$

where ΔA and ΔE are the uncertainties in the measurements of the observable A and energy E , respectively.

Let us recollect that the time evolution of $\langle\hat{A}\rangle$ is governed by Ehrenfest's equation

$$\frac{d\langle\hat{A}\rangle}{dt} = \frac{1}{i\hbar}\langle[\hat{A},\hat{H}]\rangle. \tag{4.11.15}$$

Therefore, the inequality (4.11.14) can now be written as

$$\Delta A\,\Delta E \geq \frac{1}{2}\left|i\hbar\frac{d\langle\hat{A}\rangle}{dt}\right|, \tag{4.11.16}$$

Or,

$$\frac{\Delta A}{\left|\frac{d\langle\hat{A}\rangle}{dt}\right|}\Delta E \geq \frac{\hbar}{2}. \tag{4.11.17}$$

Let us introduce a time interval $\Delta\tau$

$$\Delta\tau = \frac{\Delta A}{\left|\frac{d\langle\hat{A}\rangle}{dt}\right|}, \tag{4.11.18}$$

which represents the characteristic interval of time for any noticeable change in the statistical distribution of the results of the measurement of the observable A to occur. In general, $\Delta\tau$ is the characteristic time of evolution of the physical properties of a quantum system.

Evidently, we can introduce a characteristic time for each of the dynamical variables of the system. Let Δt be the smallest of all such characteristic time intervals of a quantum system. Then, if $|t - t'| < \Delta t$, the statistical distribution of the results of any measurements done on the system at the instant t', will practically be indistinguishable from the statistical distribution of the results of the same measurements carried out at the instant t. Therefore, Δt is taken to be the characteristic time of evolution of the quantum system itself[1]. With this definition of the characteristic time of evolution of the system, the inequality (4.11.17) takes the form

$$\Delta t\,\Delta E \geq \frac{\hbar}{2}, \tag{4.11.19}$$

and represents the so-called time-energy uncertainty relation.

Example 4.11.1: Find the uncertainty relations between the components of the position vector $\vec{r}$ and momentum $\vec{p}$ of a particle.

Solution: We know that the components of $\hat{\vec{r}}$ and $\hat{\vec{p}}$ satisfy the following commutation relations:

$$[r_j, \hat{p}_k] = i\hbar\delta_{jk}, \tag{4.11.20}$$

where $j,k = 1,2,3$. Note that, $r_j = (r_1, r_2, r_3) = (x, y, z)$, $\hat{\vec{p}} = (\hat{p}_1, \hat{p}_2, \hat{p}_3) = (\hat{p}_x, \hat{p}_y, \hat{p}_z)$ and δ_{ij} is the Kronecker delta ($\delta_{ij} = 0$, if $i \neq j$, $\delta_{ij} = 1$, if $i = j$). Therefore, using the uncertainty relation (4.11.13), we get

$$\Delta r_j \Delta p_k \geq \frac{\hbar}{2}\delta_{jk}, \tag{4.11.21}$$

which is the well known Heisenberg's uncertainty relation.

[1] Albert Messiah, Quantum Mechanics, Volume 1, North-Holland Publishing Company, Amsterdam, 1967.

4.12 Time-evolution Operator and Pictures of Quantum Mechanics

We have seen that we can have various representations of the state vector and the operators depending on the basis chosen in $\mathscr{H}$. As it turns out, they are all equivalent and related to one another by a unitary transformation. We also saw that under such a unitary transformation, the forms of the wave function and that of the observables change, but the physical state of the system remains unaltered because the unitary operator $\hat{S}$ is time-independent.

We now want to know whether it is possible to do the same for the time-evolution of the quantum system. In what follows, we shall show that it is possible to describe the time-evolution of the state vector by a time-dependent unitary operator, $\hat{U}(t)$. $\hat{U}(t)$ is called the time-evolution operator or, simply, the evolution operator. It turns out that there are more than one ways to do it. Each of such descriptions is called *a picture of quantum mechanics*.

The Schrödinger picture: In this picture, the state vector, $|\psi(t)\rangle$, of a quantum system depends explicitly on time, while the observables (operators of physical characteristics) of the system are time-independent. The time evolution of the state vector is controlled by the Schrödinger equation

$$i\hbar\frac{\partial|\psi(t)\rangle}{\partial t} = \hat{H}|\psi(t)\rangle, \tag{4.12.1}$$

and can be represented in terms of a time *evolution operator* (propagator), $\hat{U}(t,t_0)$, as

$$|\psi(t)\rangle = \hat{U}(t,t_0)|\psi(t_0)\rangle. \tag{4.12.2}$$

The condition of conservation of the norm of the wave function under this representation reads

$$\begin{aligned}\langle\psi(t)|\psi(t)\rangle &= \langle\hat{U}(t,t_0)\psi(t_0)|\hat{U}(t,t_0)\psi(t_0)\rangle \\ &= \langle\psi(t_0)|\hat{U}^\dagger(t,t_0)\hat{U}(t,t_0)|\psi(t_0)\rangle = \langle\psi(t_0)|\psi(t_0)\rangle.\end{aligned} \tag{4.12.3}$$

This requires the evolution operator, $\hat{U}(t,t_0)$, to be unitary:

$$\hat{U}^\dagger(t,t_0)\hat{U}(t,t_0) = \hat{U}(t,t_0)\hat{U}^\dagger(t,t_0) = \hat{I}. \tag{4.12.4}$$

In addition, the evolution operator also satisfies the following properties

$$\hat{U}(t,t) = \hat{I}, \tag{4.12.5}$$

$$\hat{U}^\dagger(t,t_0) = \hat{U}^{-1}(t,t_0) = \hat{U}(t_0,t), \tag{4.12.6}$$

$$\hat{U}(t_k,t_j)\hat{U}(t_j,t_i) = \hat{U}(t_k,t_i), \quad t_k > t_j > t_i. \tag{4.12.7}$$

The last of the above properties is due to the time translation invariance of the Schrödinger equation.

The propagator can be determined as follows. Substitution of (4.12.2) in the Schrödinger equation (4.12.1) yields

$$i\hbar\frac{\partial \hat{U}(t,t_0)}{\partial t} = \hat{H}\hat{U}(t,t_0). \tag{4.12.8}$$

If the Hamiltonian, $\hat{H}$, is time independent, the solution of (4.12.8) satisfying the initial condition, $\hat{U}(t_0,t_0) = \hat{I}$, can be written as

$$\hat{U}(t,t_0) = e^{-\frac{i}{\hbar}(t-t_0)\hat{H}}. \tag{4.12.9}$$

Using (4.12.9), equation(4.12.2) can be rewritten as

$$|\psi(t)\rangle = e^{-\frac{i}{\hbar}(t-t_0)\hat{H}}|\psi(t_0)\rangle. \tag{4.12.10}$$

The meaning of equation (4.12.10) is the following. We have to expand the wave function $\psi(q,0)$ into a series with respect to the eigenfunctions, $\phi_m(q), m = 1,2,3,\ldots$, of the Hamiltonian

$$\psi(q,t_0) = \sum_m c_m\phi_m(q), \tag{4.12.11}$$

use the definition of the exponential operator in the form of Mclaurent series

$$e^{-\frac{i}{\hbar}\hat{H}(t-t_0)} = \sum_{n=0}^{\infty}\frac{1}{n!}\left(\frac{-i}{\hbar}\hat{H}(t-t_0)\right)^n \tag{4.12.12}$$

and act on the wave function. If we do that and take into account that ϕ_m are eigenfunctions of the Hamiltonian ($\hat{H}\phi_m = E_m^0\phi_m$) and sum up the resulting series, we get the wave function at time t:

$$\begin{aligned}\psi(q,t) &= \sum_{n=0}^{\infty}\frac{1}{n!}\left(\frac{-i\hat{H}}{\hbar}(t-t_0)\right)^n\sum_m c_m\phi_m \\ &= \sum_m c_m\phi_m\sum_{n=0}^{\infty}\frac{1}{n!}\left(\frac{-iE_m^0}{\hbar}(t-t_0)\right)^n = \sum_m c_m\phi_m e^{-\frac{i}{\hbar}E_m^0(t-t_0)}.\end{aligned} \tag{4.12.13}$$

The Heisenberg picture: In this picture, the state vector, $|\psi\rangle$, is time-independent, while the observables are time-dependent. This is accomplished by defining the Heisenberg state vector, $|\psi_H\rangle$, as

$$|\psi_H\rangle = \hat{U}^\dagger(t,t_0)|\psi(t)\rangle, \tag{4.12.14}$$

where $|\psi(t)\rangle$ is the state vector in the Schrödinger picture. With such a definition $|\psi_H\rangle$ turns out to be time-independent

$$|\psi_H\rangle = \hat{U}^\dagger(t,t_0)|\psi(t)\rangle. = \hat{U}^{-1}(t,t_0)|\psi(t)\rangle = e^{\frac{i}{\hbar}(t-t_0)\hat{H}}|\psi(t)\rangle = |\psi(t_0)\rangle, \quad (4.12.15)$$

If we compare (4.12.10) and (4.12.14), we conclude that the definition (4.12.14) is equivalent to going over to a basis (in the Hilbert space) which is translating in time in the sense opposite to that in the Schrödinger picture. As a consequence, the state vector $|\psi_H\rangle$ gets frozen in time. This leads to

$$\frac{d|\psi_H\rangle}{dt} = 0. \quad (4.12.16)$$

Since (4.12.14) represents a unitary transformation of the state vector, physical properties of a quantum system in both the the Schrödinger and the Heisenberg pictures should be the same. For instance, consider the average value of time-independent observable, $\hat{A}_S$, in the Schrödinger picture

$$\begin{aligned} \langle\hat{A}_S\rangle &= \left\langle\psi(t)\left|\hat{A}_S\right|\psi(t)\right\rangle = \left\langle\hat{U}(t,t_0)\psi_H\left|\hat{A}_S\right|\hat{U}(t,t_0)\psi_H\right\rangle \\ &= \left\langle\psi_H\left|\left(\hat{U}^\dagger(t,t_0)\hat{A}_S\hat{U}(t,t_0)\right)\right|\psi_H\right\rangle \end{aligned} \quad (4.12.17)$$

The requirement of the unchanged average value of $\hat{A}$ in both the pictures gives

$$\hat{A}_H(t) = \hat{U}^\dagger(t,t_0)\hat{A}_S(t_0)\hat{U}(t,t_0) = e^{\frac{i}{\hbar}(t-t_0)\hat{H}}\hat{A}_S(t_0)e^{-\frac{i}{\hbar}(t-t_0)\hat{H}}. \quad (4.12.18)$$

Or,

$$\hat{A}_S(t_0) = \hat{U}\hat{A}_H(t)\hat{U}^\dagger(t,t_0)(t,t_0) = e^{-\frac{i}{\hbar}(t-t_0)\hat{H}}\hat{A}_H(t)e^{\frac{i}{\hbar}(t-t_0)\hat{H}}. \quad (4.12.19)$$

Equations (4.12.18) and (4.12.19) show that the observables in the Heisenberg and the Schrödinger pictures are related through a *similarity transformation.*

The Heisenberg's equation of motion for an observable is obtained from (4.12.18) by simply differentiating it with respect to time

$$\begin{aligned} \frac{d\hat{A}_H}{dt} &= \frac{i}{\hbar}e^{\frac{i}{\hbar}(t-t_0)\hat{H}}\hat{H}\hat{A}_Se^{-\frac{i}{\hbar}(t-t_0)\hat{H}} - \frac{i}{\hbar}e^{\frac{i}{\hbar}(t-t_0)\hat{H}}\hat{A}_S\hat{H}e^{-\frac{i}{\hbar}(t-t_0)\hat{H}} \\ &= \frac{i}{\hbar}\left(\left\{e^{\frac{i}{\hbar}(t-t_0)\hat{H}}\hat{H}e^{-\frac{i}{\hbar}(t-t_0)\hat{H}}\right\}\left\{e^{\frac{i}{\hbar}(t-t_0)\hat{H}}\hat{A}_Se^{-\frac{i}{\hbar}(t-t_0)\hat{H}}\right\}\right. \\ &\quad \left. -\left\{e^{\frac{i}{\hbar}(t-t_0)\hat{H}}\hat{A}_Se^{-\frac{i}{\hbar}(t-t_0)\hat{H}}\right\}\left\{e^{\frac{i}{\hbar}(t-t_0)\hat{H}}\hat{H}e^{-\frac{i}{\hbar}(t-t_0)\hat{H}}\right\}\right) \\ &= \frac{i}{\hbar}\left(\hat{H}_H\hat{A}_H - \hat{A}_H\hat{H}_H\right). \end{aligned} \quad (4.12.20)$$

Since the evolution operator, $e^{-\frac{i}{\hbar}(t-t_0)\hat{H}}$, commutes with the Hamiltonian, we have $\hat{H}_H = \hat{H}$. Therefore, the Heisenberg's equation of motion can be written as

$$\frac{d\hat{A}_H}{dt} = \frac{1}{i\hbar}\left[\hat{A}_H, \hat{H}\right]. \tag{4.12.21}$$

If, in addition, $\hat{A}_H$ depends explicitly on time, the equations of motion takes the form

$$\frac{d\hat{A}_H}{dt} = \frac{\partial \hat{A}_H}{\partial t} + \frac{1}{i\hbar}\left[\hat{A}_H, \hat{H}\right]. \tag{4.12.22}$$

It remind us of the equations of motion of a dynamical variable, A, in the Poisson bracket formalism

$$\frac{dA}{dt} = \frac{\partial A}{\partial t} + \{A, H\}, \tag{4.12.23}$$

in which the Poisson bracket, $\{A,H\}$ has been replaced by the commutator of the corresponding operators divided by $i\hbar$.

Interaction picture: The interaction picture, the same way as the Heisenberg's picture, is useful for the solution of the problems involving time-dependent Hamiltonians. In this picture, both the state vector, $|\psi_I(t)\rangle$, and the observables depend explicitly on time. In the cases when the total Hamiltonian, $\hat{H}$, can be separated into a time-independent part, $\hat{H}_0$, and a time-dependent part, $\hat{W}(t)$ (interaction Hamiltonian), the state vector, $|\psi_I(t)\rangle$, is defined through

$$|\psi_I\rangle = \hat{U}_0^\dagger(t,t_0)|\psi(t)\rangle = \hat{U}_0^{-1}(t,t_0)|\psi(t)\rangle \equiv e^{\frac{i}{\hbar}(t-t_0)\hat{H}_0}|\psi(t)\rangle, \tag{4.12.24}$$

where $|\psi(t)\rangle$ is the state vector in the Schrödinger picture. The equation of motion for the state vector is obtained as follows. Differentiating $|\psi_I\rangle$ with respect to time, we obtain

$$\frac{\partial|\psi_I\rangle}{\partial t} = \frac{i}{\hbar}e^{\frac{i}{\hbar}(t-t_0)\hat{H}_0}\hat{H}_0|\psi(t)\rangle + e^{\frac{i}{\hbar}(t-t_0)\hat{H}_0}\frac{\partial|\psi(t)\rangle}{\partial t}. \tag{4.12.25}$$

Using the equation of motion (4.12.1), for $|\psi(t)\rangle$ in the Schrödinger's picture, and a bit of algebra we obtain

$$i\hbar\frac{\partial|\psi_I(t)\rangle}{\partial t} = \hat{W}_I(t)|\psi_I(t)\rangle, \tag{4.12.26}$$

where $\hat{W}_I(t) = e^{\frac{i}{\hbar}(t-t_0)\hat{H}_0}\hat{W}(t)e^{-\frac{i}{\hbar}(t-t_0)\hat{H}_0}$ is the time-dependent part of the total Hamiltonian in the interaction picture.

Defining an observable, $\hat{A}_I(t)$, in the interaction picture by

$$\hat{A}_I(t) = e^{\frac{i}{\hbar}(t-t_0)\hat{H}_0}\hat{A}e^{-\frac{i}{\hbar}(t-t_0)\hat{H}_0}, \tag{4.12.27}$$

where $\hat{A}$ is the corresponding observable in the Schrödinger's picture, and following the same calculations as in the case of Heisenberg's picture, we arrive at the following equation of motion for an observable in the interaction picture

$$i\hbar\frac{d\hat{A}_I}{dt} = \left[\hat{A}_I, \hat{H}_0\right]. \tag{4.12.28}$$

We see that, in this picture, the time evolution of the state vector is governed by the time-dependent interaction Hamiltonian $\hat{W}_I(t)$ only, while the time variation of an observable is controlled only by the time-independent part, $\hat{H}_0$, of the total Hamiltonian, $\hat{H}$.

We would like to note here that all the three pictures of quantum mechanics, discussed above, are equivalent because they are related trough unitary transformations. Depending on the problem at hand, one can choose to work with any one of them for relatively easier and faster solution of the problem.

4.13 Algebraic Treatment of One-dimensional Harmonic Oscillator

The harmonic oscillator: We are now going to discuss the one-dimensional harmonic oscillator that serves as one of the most important models (if not the most important model) in quantum theory and can be solved analytically.

The Hamiltonian for the one-dimensional harmonic oscillator (a particle of mass m attached to a spring) is given by

$$\hat{H} = \frac{\hat{p}^2}{2m} + \frac{1}{2}m\omega^2\hat{x}^2 = -\frac{\hbar^2}{2m}\frac{d^2}{dx^2} + \frac{1}{2}m\omega^2\hat{x}^2, \tag{4.13.1}$$

where x represents the displacement of the oscillator from the point of equilibrium (which is taken to be at the origin of the coordinate system) and ω is its angular frequency. The corresponding time-independent Schrödinger equation reads

$$-\frac{\hbar^2}{2m}\frac{d^2\phi(x)}{dx^2} + \frac{1}{2}m\omega^2\hat{x}^2\phi(x) = E\phi(x). \tag{4.13.2}$$

Our main aim, in this section, is to use the algebraic method for obtaining the energy eigenvalues and the corresponding bound state wave functions.

Let us introduce the following operators

$$\hat{a} = \frac{1}{\sqrt{2m\hbar\omega}}\left(i\hat{p} + m\omega\hat{x}\right), \tag{4.13.3}$$

$$\hat{a}^\dagger = \frac{1}{\sqrt{2m\hbar\omega}}\left(-i\hat{p}+m\omega\hat{x}\right). \tag{4.13.4}$$

Consider the product

$$\hat{a}\hat{a}^\dagger = \frac{1}{2m\hbar\omega}\left(i\hat{p}+m\omega\hat{x}\right)\left(-i\hat{p}+m\omega\hat{x}\right) = \frac{1}{2m\hbar\omega}\left(\hat{p}^2+m^2\omega^2\hat{x}^2+im\omega\left[\hat{x},\hat{p}\right]\right),$$

$$= \frac{1}{2m\hbar\omega}\left(\hat{p}^2+m^2\omega^2+m\hbar\omega\right) = \frac{1}{\hbar\omega}\left(\frac{\hat{p}^2}{2m}+\frac{1}{2}m\omega^2\hat{x}^2\right)+\frac{1}{2} = \frac{\hat{H}}{\hbar\omega}+\frac{1}{2}. \tag{4.13.5}$$

From (4.13.5), we get that

$$\hat{H} = \hbar\omega\left(\hat{a}\hat{a}^\dagger-\frac{1}{2}\right). \tag{4.13.6}$$

Similarly, we have

$$\hat{a}^\dagger\hat{a} = \frac{1}{2m\hbar\omega}\left(-i\hat{p}+m\omega\hat{x}\right)\left(+i\hat{p}+m\omega\hat{x}\right) = \frac{1}{2m\hbar\omega}\left(\hat{p}^2+m^2\omega^2\hat{x}^2-im\omega\left[\hat{x},\hat{p}\right]\right),$$

$$= \frac{1}{2m\hbar\omega}\left(\hat{p}^2+m^2\omega^2-m\hbar\omega\right) = \frac{1}{\hbar\omega}\left(\frac{\hat{p}^2}{2m}-\frac{1}{2}m\omega^2\hat{x}^2\right)+\frac{1}{2} = \frac{\hat{H}}{\hbar\omega}-\frac{1}{2} \tag{4.13.7}$$

and

$$\hat{H} = \hbar\omega\left(\hat{a}^\dagger\hat{a}+\frac{1}{2}\right) \tag{4.13.8}$$

Two important points are to be mentioned here. Firstly, equations (4.13.5) and (4.13.7) lead to the following commutation relation between $\hat{a}$ and $\hat{a}^\dagger$:

$$[\hat{a},\hat{a}^\dagger] = 1. \tag{4.13.9}$$

Secondly, the Schrödinger equation (4.13.2) is completely equivalent to any of the following equations

$$\hbar\omega\left(\hat{a}^\dagger\hat{a}+\frac{1}{2}\right)\phi = E\phi, \quad \text{or} \quad \hbar\omega\left(\hat{a}\hat{a}^\dagger-\frac{1}{2}\right)\phi = E\phi. \tag{4.13.10}$$

Theorem 4.13.1: Assume that ϕ_n is an eigenfunction of the Hamiltonian $\hat{H}$ with energy E_n. Then, $\hat{a}^\dagger\phi_n$ is an eigenfunction of the Hamiltonian with energy $(E_n+\hbar\omega)$.

Proof: We have

$$\hat{H}\hat{a}^{\dagger}\phi_n = \hbar\omega\left(\hat{a}^{\dagger}\hat{a}+\frac{1}{2}\right)\hat{a}^{\dagger}\phi_n = \hbar\omega\left(\hat{a}^{\dagger}\hat{a}\hat{a}^{\dagger}+\frac{1}{2}\hat{a}^{\dagger}\right)\phi_n$$
$$= \hbar\omega\left\{\hat{a}^{\dagger}\left(1+\hat{a}^{\dagger}\hat{a}\right)+\frac{1}{2}\hat{a}^{\dagger}\right\} = \hat{a}^{\dagger}\left[\hbar\omega\left(\hat{a}^{\dagger}\hat{a}+\frac{1}{2}\right)\phi_n+\hbar\omega\phi_n\right]$$
$$= \hat{a}^{\dagger}\left[\hat{H}\phi_n+\hbar\omega\phi_n\right] = \hat{a}^{\dagger}\left[E_n\phi_n+\hbar\omega\phi_n\right] = \left(E_n+\hbar\omega\right)\hat{a}^{\dagger}\phi_n. \quad (4.13.11)$$

Since $\hat{H}\hat{a}^{\dagger}\phi_n = (E_n+\hbar\omega)\,\hat{a}^{\dagger}\phi_n$, the theorem is proved.

Theorem 4.13.2: Assume that ϕ_n is an eigenfunction of the Hamiltonian $\hat{H}$ with energy E_n. Then, $\hat{a}\phi_n$ is an eigenfunction of the Hamiltonian with energy $(E_n-\hbar\omega)$.

Proof: We have

$$\hat{H}\hat{a}\phi_n = \hbar\omega\left(\hat{a}\hat{a}^{\dagger}-\frac{1}{2}\right)\hat{a}\phi_n = \hbar\omega\left(\hat{a}\hat{a}^{\dagger}\hat{a}-\frac{1}{2}\hat{a}\right)\phi_n$$
$$= \hbar\omega\left\{\hat{a}\left(\hat{a}\hat{a}^{\dagger}-1\right)-\frac{1}{2}\hat{a}\right\}\phi_n = \hat{a}\left[\hbar\omega\left(\hat{a}\hat{a}^{\dagger}-\frac{1}{2}\right)\phi_n-\hbar\omega\phi_n\right]$$
$$= \hat{a}\left[\hat{H}\phi_n-\hbar\omega\phi_n\right] = \hat{a}\left[E_n\phi_n-\hbar\omega\phi_n\right] = \left(E_n-\hbar\omega\right)\hat{a}^{\dagger}\phi_n. \quad (4.13.12)$$

Since $\hat{H}\hat{a}^{\dagger}\phi_n = (E_n-\hbar\omega)\,\hat{a}^{\dagger}\phi_n$, the theorem is proved.

Thus, while acting on the eigenfunction ϕ_n of $\hat{H}$ with energy E_n, the operator $\hat{a}$ lowers the energy by one unit of $\hbar\omega$, the operator $\hat{a}^{\dagger}$ increases the energy by one unit of $\hbar\omega$. Hence, if we set out with $\phi_n(x)$, describing the nth energy state of the oscillator (with energy E_n), we can generate all possible states of the oscillator, with energies higher than E_n as well as lower than E_n, by repeatedly acting on $\phi_n(x)$ with $\hat{a}^{\dagger}$ and $\hat{a}$, respectively. The operators $\hat{a}^{\dagger}$ and $\hat{a}$ are called ladder operators because they permit us to ascend or descend in energy. The operator $\hat{a}^{\dagger}$ is also known as creation operator, while the operator $\hat{a}$ is also called annihilation operator.

However, a paradoxical situation arises if we continue to act with the annihilation operator infinitely. If we do so, eventually we shall reach a state with energy less than zero, which for the harmonic oscillator does not exist. Thus, we have the situation where $\hat{a}\phi_n$ is a solution of the Schrödinger equation but the corresponding state does not exist. It means that the given procedure fails at some point or the other. What is the way out? All what we said earlier suggests that there must exist the lowest energy state (lowest rung in the ladder) whose wave function $\phi_0(x)$ must satisfy the equation

$$\hat{a}\phi_0(x) = 0. \quad (4.13.13)$$

We can use this to determine $\phi_0(x)$. We have

$$\frac{1}{\sqrt{2m\hbar\omega}}\left(i\hat{p}+m\omega\hat{x}\right)\phi_0(x)=0. \tag{4.13.14}$$

Or,

$$\frac{d\phi_0(x)}{dx}=-\frac{m\omega}{\hbar}x\phi_0(x). \tag{4.13.15}$$

Integrating, we get

$$\int\frac{d\phi_0(x)}{\phi_0(x)}=-\frac{m\omega}{\hbar}\int xdx. \Rightarrow \phi_0(x)=A_0\,e^{-\frac{m\omega}{2\hbar}x^2}\equiv A_0\,e^{-\frac{x^2}{2x_0^2}}, \tag{4.13.16}$$

where A_0 is a constant to be determined and $x_0=\sqrt{\hbar/m\omega}$ with the dimensions of length. To find the energy of this state, let us put this solution into the first of the Schrödinger equations (4.13.10). We have

$$\hbar\omega\left(\hat{a}^\dagger\hat{a}+\frac{1}{2}\right)\phi_0(x)=\frac{\hbar\omega}{2}\phi_0(x), \tag{4.13.17}$$

where we have taken into account the fact that $\hat{a}\phi_0(x)=0$. Hence, the energy of this state, called the ground state, is $\hbar\omega/2$.

Once we have determined the ground state eigenfunction and energy, we can find the eigenfunction and the corresponding energy of any excited state of the oscillator by successively applying the creation operator to the ground state wave function. For instance, the wave function of the first excited state is obtained as

$$\phi_1(x)=\hat{a}^\dagger\phi_0(x)=\frac{1}{\sqrt{2m\omega\hbar}}\left(-\hbar\frac{d}{dx}+m\omega x\right)A_0\,e^{-\frac{x^2}{2x_0^2}}=\sqrt{\frac{2}{\sqrt{\pi x_0^3}}}\,x\,e^{-\frac{x^2}{2x_0^2}}. \tag{4.13.18}$$

Note that the wave function $\phi_1(x)$ is normalized to unity

$$\int_{-\infty}^{+\infty}\phi_1^2(x)\,dx=\frac{2}{\sqrt{\pi x_0^3}}\int_{-\infty}^{+\infty}x^2\,e^{-\frac{x^2}{x_0^2}}\,dx=\frac{2}{\sqrt{\pi x_0^3}}\frac{\sqrt{\pi x_0^3}}{2}=1. \tag{4.13.19}$$

Since, by acting on an eigenstate of the Hamiltonian, the creation operator increases its energy by one unit of $\hbar\omega$, the energy of the first excited state is $\hbar\omega/2+\hbar\omega=3\hbar\omega/2$. Similarly, we can apply $\hat{a}^\dagger$ to $\phi_1(x)$ to get the wave function of the second excited state $\phi_2(x)$, and so on and so forth.

The number operator and the energy eigenfunctions: Let us introduce an operator $\hat{N} = \hat{a}^\dagger \hat{a}$. It is called the *occupation number operator* or, simply, the *number operator*.

First, we notice that the number operator commutes with the Hamiltonian:

$$[\hat{N},\hat{H}] = \left[\hat{N},\hat{N}+\frac{1}{2}\right]\hbar\omega = \hbar\omega[\hat{N},\hat{N}] + \frac{\hbar\omega}{2}[\hat{N},\hat{I}] = 0. \qquad (4.13.20)$$

Since, $\hat{N}$ and $\hat{H}$ commute, they must have a common set of eigenvectors. Let $|n\rangle$ be the nth joint eigenvector of these operators:

$$\hat{N}|n\rangle = n\,|n\rangle, \qquad (4.13.21)$$

and

$$\hat{H}|n\rangle = E_n\,|n\rangle, \qquad (4.13.22)$$

where n is a positive integer and $E_n, n = 1,2,3,\ldots$ are the energy eigenvalues. Using the definition of $\hat{N}$, along with equations (4.13.8) and (4.13.21), the energy eigenvalues for the oscillator are readily obtained to be $E_n = \hbar\omega\left(n+\frac{1}{2}\right)$.

Next, we compute the commutator of $\hat{a}$ and $\hat{a}^\dagger$ with $\hat{N}$. We have

$$[\hat{N}\hat{a}] = [\hat{a}^\dagger\hat{a},\hat{a}] = \hat{a}^\dagger[\hat{a},\hat{a}] + [\hat{a}^\dagger,\hat{a}]\hat{a} = -\hat{a}. \qquad (4.13.23)$$

Similarly,

$$[\hat{N},\hat{a}^\dagger] = \hat{a}^\dagger. \qquad (4.13.24)$$

Therefore, $\hat{N}\hat{a} = \hat{a}(\hat{N}-1)$ and $\hat{N}\hat{a}^\dagger = \hat{a}^\dagger(\hat{N}+1)$. As a result,

$$\begin{aligned}\hat{N}(\hat{a}|n\rangle) &= \hat{a}(\hat{N}-1)|n\rangle = (n-1)\hat{a}|n\rangle,\\ \hat{N}(\hat{a}^\dagger|n\rangle) &= \hat{a}^\dagger(\hat{N}+1)|n\rangle = (n+1)\hat{a}^\dagger|n\rangle.\end{aligned} \qquad (4.13.25)$$

These results say that, if $|n\rangle$ is an eigenstate of the number operator $\hat{N}$, then $\hat{a}|n\rangle$ and $\hat{a}^\dagger|n\rangle$ are also eigenstates of $\hat{N}$, but with eigenvalues $(n-1)$ and $(n+1)$, respectively. In other words, by acting on the state $|n\rangle$, the operator $\hat{a}$ decreases the number n by unity and generates a new eigenstate $|n-1\rangle$, that is, $\hat{a}|n\rangle = a_n|n-1\rangle$. Similarly, the operator $\hat{a}^\dagger$, when acting on $|n\rangle$, increases n by unity and generates a new eigenstate, $|n+1\rangle$ of $\hat{N}$, that is, $\hat{a}^\dagger|n\rangle = b_n|n-1\rangle$. Here, a_n and b_n are constants to be determined from the condition that the states $|n\rangle$ be normalized for all values of n.

Using $\hat{a}|n\rangle = a_n|n-1\rangle$, we have

$$(\langle n\hat{a}^\dagger|)\cdot(\hat{a}|n\rangle) = \langle n|\hat{a}^\dagger\hat{a}|n\rangle = \langle n|\hat{N}|n\rangle = n\langle n|n\rangle = n. \qquad (4.13.26)$$

On the other hand

$$(\langle n\hat{a}^\dagger|)\cdot(\hat{a}|n\rangle) = |a_n|^2\,\langle n-1|n-1\rangle = |a_n|^2. \tag{4.13.27}$$

Therefore, we must have

$$|a_n|^2 = n. \quad \Rightarrow \quad a_n = \sqrt{n}. \tag{4.13.28}$$

Similarly, we arrive at

$$|b_n|^2 = n+1. \quad \Rightarrow \quad b_n = \sqrt{n+1}. \tag{4.13.29}$$

Equation (4.13.28) shows that n is equal to the norm squared of the vector $\hat{a}|n\rangle$ and, hence, cannot be negative, that is, $n \geq 0$. Since, n is a positive integer, the energy spectrum of the one-dimensional harmonic oscillator is discrete and non-degenerate.

We can now apply the creation operator $\hat{a}^\dagger$ on $|0\rangle$ to generate all possible excited state energy eigenvectors. We have

$$|1\rangle = \hat{a}^\dagger|0\rangle, \tag{4.13.30}$$

$$|2\rangle = \frac{1}{\sqrt{2}}\,\hat{a}^\dagger|1\rangle = \frac{1}{\sqrt{2!}}\,\hat{a}^{\dagger 2}|0\rangle, \tag{4.13.31}$$

$$|3\rangle = \frac{1}{\sqrt{3}}\,\hat{a}^\dagger|2\rangle = \frac{1}{\sqrt{3!}}\,\hat{a}^{\dagger 3}|0\rangle, \tag{4.13.32}$$

$$\ldots\ldots\ldots\ldots\ldots\ldots, \tag{4.13.33}$$

$$\ldots\ldots\ldots\ldots\ldots\ldots, \tag{4.13.34}$$

$$|n\rangle = \frac{1}{\sqrt{n}}\,\hat{a}^\dagger|n-1\rangle = \frac{1}{\sqrt{n!}}\,\hat{a}^{\dagger n}|0\rangle. \tag{4.13.35}$$

Hence, to find any excited state eigenvector $|n\rangle$, we need to apply the creation operator n successive times to the vacuum state $|0\rangle$. Furthermore, since the energy spectrum of the Hamiltonian is non-degenerate, any two of the energy eigenvectors $|n'\rangle$ and $|n\rangle$ (corresponding to different eigenvalues) are orthogonal, and the sequence of the vectors $\{|0\rangle, |1\rangle, |2\rangle\, |3\rangle, \ldots, |n\rangle\}$ constitutes an orthonormal and complete basis:

$$\langle n'|n\rangle = \delta_{n'n}, \qquad \sum_{n=0}^{\infty} |n\rangle\langle n| = \hat{I}. \tag{4.13.36}$$

Note that the formalism of number operator for the harmonic oscillator is very useful and is frequently used in quantum optics and quantum field theory in general.

Homework Problems

1. Check whether the vectors $\vec{a} = (1,-2,1), \vec{b} = (0,3,1)$ and $\vec{c} = (0,0,5)$ in a usual three-dimensional Euclidean space are linearly independent or not.
2. The following functions are defined on the real x-axis.

   (a) $f(x) = x, g(x) = (x-1)^2, h(x) = (x+1)^2$,
   (b) $f(x) = \sin^2(x), g(x) = \cos^2(x), h(x) = \sin(2x)$,
   (c) $f(x) = \tan^2(x), g(x) = 1/2, h(x) = \sec^2(x)$,

   Which of these sets of functions is linearly dependent and which one is linearly independent and why?
3. Show that the vectors

$$\psi_1 = \begin{pmatrix} 1 \\ 1 \\ 0 \end{pmatrix}, \quad \psi_2 = \begin{pmatrix} 0 \\ 0 \\ 2 \end{pmatrix}, \quad \psi_3 = \begin{pmatrix} i \\ i \\ i \end{pmatrix}$$

   are linearly dependent.
4. In an orthonormal basis, consisting of three vectors $\{|\phi_i\rangle\}, i = 1,2,3$, two ket vectors $|\psi\rangle$ and $|\phi\rangle$ are given by the following expressions:

$$|\psi\rangle = 2i|\phi_1\rangle - 3i|\phi_2\rangle + |\phi_3\rangle, \quad |\phi\rangle = 2|\phi_1\rangle + 2i|\phi_2\rangle - 3i|\phi_3\rangle.$$

   (a) Calculate the norms $\|\psi\|$ and $\|\phi\|$. (b) Calculate the inner products $\langle\psi|\phi\rangle, \langle\phi|\psi\rangle$ and $\langle\psi+\phi|\psi+\phi\rangle$.
5. In an orthonormal basis consisting of the vectors $|\phi_1\rangle, |\phi_2\rangle$ and $|\phi_3\rangle$, we have two kets

$$|\psi\rangle = a\,|\phi_1\rangle - 3i|\phi_2\rangle + i|\phi_3\rangle, \quad |\phi\rangle = |\phi_1\rangle + i|\phi_2\rangle - i|\phi_3\rangle,$$

   where a is a constant. Find the value of a so that these kets are orthogonal.
6. Consider a potential well having an infinite wall at $x = 0$ and a wall of height V_0 at $x = L$. For the case $E < V_0$, obtain solutions to the Schrödinger equation inside the well ($0 \leq x \leq L$) and in the region beyond ($x > L$) that satisfy the appropriate boundary conditions at $x = 0$ and $x = \infty$. Enforce the proper matching conditions at $x = L$ to find an equation for the allowed energies of this system. Are there conditions for which no solution is possible? Explain.
7. In an orthonormal basis, consisting of three vectors $\{|\phi_i\rangle\}, i = 1,2,3$, two ket vectors $|\psi\rangle$ and $|\phi\rangle$ are given by the following expressions:

$$|\psi\rangle = 2i|\phi_1\rangle - 5i|\phi_2\rangle + |\phi_3\rangle, \quad |\phi\rangle = 2|\phi_1\rangle + 4i|\phi_2\rangle - 3i|\phi_3\rangle.$$

   Find the matrix representing $|\psi\rangle + |\phi\rangle$.

8. Let $|\psi\rangle = i|\phi_1\rangle - 2i|\phi_2\rangle + 2|\phi_3\rangle$, and $|\phi\rangle = |\phi_1\rangle + i|\phi_2\rangle - i|\phi_3\rangle$, where $\{|\phi_i\rangle\}, i = 1,2,3$ are three orthonormal vectors. Show that $|\psi\rangle$ and $|\phi\rangle$ satisfy the Schwartz inequality.
9. Let $|\psi\rangle = 2i|\phi_1\rangle - i|\phi_2\rangle + |\phi_3\rangle$, and $|\phi\rangle = |\phi_1\rangle + i|\phi_2\rangle - 3i|\phi_3\rangle$, where $\{|\phi_i\rangle\}, i = 1,2,3$ are three orthonormal vectors. Show that they satisfy the triangle inequality.
10. You are given a function $f(x) = x^n$. Determine the range of n for which $f(x)$ belongs to a Hilbert space on the interval $x \in (0,1)$. Here n is real but not necessarily positive.
11. Consider the kets $|\psi\rangle = 3i|\phi_1\rangle - 2i|\phi_2\rangle + 5|\phi_3\rangle$, and $|\phi\rangle = 2|\phi_1\rangle + i|\phi_2\rangle - i|\phi_3\rangle$, where $\{|\phi_i\rangle\}, i = 1,2,3$ three orthonormal vectors. Calculate $|\psi\rangle\langle\phi|$ and $|\phi\rangle\langle\psi|$ and check whether they are equal.
12. Let $|\psi\rangle = 2i|\phi_1\rangle - i|\phi_2\rangle + |\phi_3\rangle$, and $|\phi\rangle = |\phi_1\rangle + i|\phi_2\rangle - 3i|\phi_3\rangle$, where $\{|\phi_i\rangle\}, i = 1,2,3$ are three orthonormal vectors. Calculate $\mathrm{Tr}(|\psi\rangle\langle\phi|)$ and $\mathrm{Tr}(|\phi\rangle\langle\psi|)$
13. In a linear vector space V equipped with an orthonormal basis, $\{|\phi_1\rangle, |\phi_2\rangle, |\phi_3\rangle\}$, an operator $\hat{A}$ acts on the basis vectors and the results are as follows:

$$\hat{A}\,|\phi_1\rangle = 5\,|\phi_1\rangle + 3\,|\phi_2\rangle\,, \tag{4.13.37}$$

$$\hat{A}\,|\phi_2\rangle = 2\,|\phi_1\rangle - i\,|\phi_3\rangle\,, \tag{4.13.38}$$

$$\hat{A}\,|\phi_3\rangle = |\phi_1\rangle + 5\,|\phi_2\rangle\,. \tag{4.13.39}$$

    Compute the matrix corresponding to the operator $\hat{A}$ in V.
14. Show that the trace of an operator is independent of the basis in which it is expressed.
15. By using the ground state wave function and the machinery of the raising and lowering operators, compute the normalized fifth excited state wave function for the harmonic oscillator potential.
16. Using the commutation relation $[\hat{a}, \hat{a}^\dagger] = 1$, find the value of the commutators $[\hat{a}, \hat{N}^2]$ and $[\hat{a}^\dagger, \hat{N}^2]$.
17. Using the ladder operator formalism, show that

    (a) $\langle n'|\hat{x}|n\rangle = \sqrt{\frac{\hbar}{2m\omega}}\left(\sqrt{n'}\,\delta_{n'-1,n} + \sqrt{n}\,\delta_{n,n'-1}\right).$

    (b) $\langle n'|\hat{p}_x|n\rangle = i\sqrt{\frac{\hbar}{2m\omega}}\left(\sqrt{n'}\,\delta_{n'-1,n} - \sqrt{n}\,\delta_{n,n'-1}\right).$
18. Using the ladder operator formalism, show that

$$\langle n'|\hat{x}^3|n\rangle = 3\left(\frac{3\hbar}{2m\omega}\right)^{3/2}.$$

Chapter 5

Quantum Mechanics in Three Spatial Dimensions

So far we have discussed only one-dimensional problems. They serve as approximate models in several realistic situations and help us understand the basic features of quantum mechanics. However, in atomic, molecular and nuclear physics, we have to deal with problems in three spatial dimensions. Therefore, in what follows, we shall discuss the three-dimensional Schrödinger equation, its basic properties and the methods of its solution.

5.1 Three-dimensional Schrödinger Equation in Cartesian Coordinates

The Schrödinger equation, for a particle of mass m, in three spatial dimensions reads as

$$i\hbar \frac{\partial \psi(\vec{r},t)}{\partial t} = \hat{H}\psi(\vec{r},t), \tag{5.1.1}$$

where $\hat{H}$ is the Hamiltonian operator, given by

$$\hat{H} = \frac{\hat{\vec{p}}^2}{2m} + V(\vec{r}). \tag{5.1.2}$$

Note that the wave function and the potential energy are now functions of $\vec{r} = x\hat{i} + y\hat{j} + \hat{k}z$ and t. Here, in (5.1.2), $\hat{\vec{p}}$ is the three-dimensional momentum operator:

$$\hat{\vec{p}} = -i\hbar\vec{\nabla}, \quad \vec{\nabla} = \hat{i}\frac{\partial}{\partial x} + \hat{j}\frac{\partial}{\partial y} + \hat{k}\frac{\partial}{\partial z}. \tag{5.1.3}$$

With this $\hat{\vec{p}}$, the Hamiltonian operator takes the form

$$\hat{H} = -\frac{\hbar^2}{2m}\vec{\nabla}^2 + V(x,y,z), \tag{5.1.4}$$

where

$$\vec{\nabla}^2 = \frac{\partial^2}{\partial x^2} + \frac{\partial^2}{\partial y^2} + \frac{\partial^2}{\partial z^2}, \tag{5.1.5}$$

is the Laplacian (Laplace's operator) in Cartesian coordinates.

In the same way as in the one-dimensional case, the quantity $|\psi(\vec{r},t)|^2 dV$, $dV = dx\,dy\,dz$ is interpreted as the probability of finding the particle in an infinitesimal volume element $dV = dx\,dy\,dz$ around the point with position vector $\vec{r}$. Hence, as in the one-dimensional case, the normalization of the wave function reads

$$\int_{-\infty}^{+\infty} |\psi(\vec{r},t)|^2 dV = 1. \tag{5.1.6}$$

Note that sometimes we shall also use the symbols d^3r or d^3x for the infinitesimal volume element dV.

If the potential is time independent, the three-dimensional Schrödinger equation can also be solved by the method of separation of variables. The solution, in complete analogy with the one-dimensional case, allows us to write the stationary state solutions in the form

$$\psi(\vec{r},t) = \phi_n(\vec{r})e^{-\frac{i}{\hbar}Et}, \tag{5.1.7}$$

where E is the total energy. The function $\phi(\vec{r},t)$ satisfies the following time-independent three-dimensional Schrödinger equation

$$-\frac{\hbar^2}{2m}\vec{\nabla}^2\phi(\vec{r}) + V(\vec{r})\phi(\vec{r}) = E\phi(\vec{r}). \tag{5.1.8}$$

For the special case of a potential, $V(x,y,z)$, that can be written in the form

$$V(\vec{r}) = V_1(x) + V_2(y) + V_3(z), \tag{5.1.9}$$

the three-dimensional TISE reduces to a system of one-dimensional TISE. Indeed, if we write the solution in the form

$$\phi(\vec{r}) = X(x)Y(y)Z(z), \tag{5.1.10}$$

substitute it in (5.1.8) and divide throughout by $X(x)Y(y)Z(z)$, we obtain

$$\left[-\frac{\hbar^2}{2m}\frac{1}{X}\frac{d^2X}{dx^2} + V_1(x)\right] + \left[-\frac{\hbar^2}{2m}\frac{1}{Y}\frac{d^2Y}{dy^2} + V_2(y)\right] + \left[-\frac{\hbar^2}{2m}\frac{1}{Z}\frac{d^2Z}{dz^2} + V_3(z)\right] = E, \tag{5.1.11}$$

where E plays the role of the separation constant. Each term in the equation (5.1.11) depends on only one of the variables x, y, z and the sum of the three terms is a constant. This is possible only if each of these terms is a constant such that their sum is equal to E. In other words,

$$-\frac{\hbar^2}{2m}\frac{1}{X}\frac{d^2X}{dx^2}+V_1(x)=E_1, \tag{5.1.12}$$

$$-\frac{\hbar^2}{2m}\frac{1}{Y}\frac{d^2Y}{dy^2}+V_2(y)=E_2, \tag{5.1.13}$$

$$-\frac{\hbar^2}{2m}\frac{1}{Z}\frac{d^2Z}{dz^2}+V_3(z)=E_3, \tag{5.1.14}$$

where E_1, E_2 and E_3 are constants such that $E_1+E_2+E_3=E$. It is obvious that the solution of the aforementioned differential equations depends on the concrete form of the potentials. In what follows, we shall take up some important examples.

5.2 The Free Particle Solution in Cartesian Coordinates

Consider a particle of mass m moving freely in space in the absence of any external force field. In this case, $V(\vec{r})=0$ in the Schrödinger equation, and the system of equations (5.1.12)-(5.1.14) reduces to

$$-\frac{\hbar^2}{2m}\frac{d^2X}{dx^2}=E_1X, \tag{5.2.1}$$

$$-\frac{\hbar^2}{2m}\frac{d^2Y}{dy^2}=E_2Y, \tag{5.2.2}$$

$$-\frac{\hbar^2}{2m}\frac{d^2Z}{dz^2}=E_3Z. \tag{5.2.3}$$

The normalized solutions of these equations are

$$X(x)=\frac{1}{\sqrt{2\pi}}e^{ik_xx}, \tag{5.2.4}$$

$$Y(y)=\frac{1}{\sqrt{2\pi}}e^{ik_yy}, \tag{5.2.5}$$

$$Z(z)=\frac{1}{\sqrt{2\pi}}e^{ik_zz}, \tag{5.2.6}$$

where $k_j^2 = 2mE_j/\hbar^2, j = 1,2,3 = x,y,z$ and hence, $E_j = \hbar^2 k_j^2/2m$. As a result, the solution to the free time-independent Schrödinger equation (5.1.8) is given by

$$\psi(x,y,z) = \frac{1}{(2\pi)^{3/2}} e^{ik_1 x} e^{ik_2 y} e^{ik_3 z} = \frac{1}{(2\pi)^{3/2}} e^{i\vec{k}\cdot\vec{r}}, \tag{5.2.7}$$

where $\vec{k} = \hat{i}k_1 + \hat{j}k_2 + \hat{k}k_3 = \hat{i}k_x + \hat{j}k_y + \hat{k}k_z$ is the wave vector. The total energy of the particle E is given by the sum of the energy eigenvalues E_1, E_2 and E_3:

$$E = E_1 + E_2 + E_3 = \frac{\hbar^2}{2m}(k_1^2 + k_2^2 + k_3^2) = \frac{\hbar^2}{2m}\vec{k}^2. \tag{5.2.8}$$

We note here that the energy, E, depends on the magnitude of the wave vector $\vec{k}$ but not on its direction. Hence, different orientations of $\vec{k}$ satisfying the condition

$$|\vec{k}| = \sqrt{k_1^2 + k_2^2 + k_3^2} = \text{const.}, \tag{5.2.9}$$

lead to different eigenfunctions without changing the energy. Since there are infinite number of possible orientations of $\vec{k}$, the energy eigenvalue, E, is *infinitely degenerate*.

Thus, the solution to the time-dependent Schrödinger equation (5.1.1), for this special case of zero potential, is given by

$$\psi_{\vec{k}}(\vec{r},t) = \frac{1}{(2\pi)^{3/2}} e^{ik_1 x} e^{ik_2 y} e^{ik_3 z} e^{-i\frac{E}{\hbar}t} = \frac{1}{(2\pi)^{3/2}} e^{i(\vec{k}\cdot\vec{r} - \omega t)}, \tag{5.2.10}$$

where $\omega = E/\hbar$. Note that it is nothing but de Broglie's plane wave solution given by (1.4.7). The orthonormality condition, for the wave functions (5.2.10), reads

$$\langle \psi_{\vec{k}'}(\vec{r},t) | \psi_{\vec{k}}(\vec{r},t) \rangle = \frac{1}{(2\pi)^3} \int_{-\infty}^{+\infty} e^{i(\vec{k}-\vec{k}')\cdot\vec{r}} d^3 r = \delta(\vec{k} - \vec{k}'). \tag{5.2.11}$$

Once again, due to the infinite extension of the plane wave solutions, representation of free particles by such solutions leads to the same difficulties as mentioned in Chapter 3. Therefore, a free quantum particle is represented by a spatially localized wave whose amplitude is large and non-zero in a small region near the particle and tends to zero outside this region. Such a solution, as we know, is given by a wave packet, which is nothing but a superposition of an infinite (large) number of plane waves. Hence, a free particle is represented by the following three-dimensional wave packet:

$$\psi(\vec{r},t) = \frac{1}{(2\pi)^{3/2}} \int_{-\infty}^{+\infty} A(\vec{k},t) \psi_{\vec{k}}(\vec{r},t) d^3 k = \frac{1}{(2\pi)^{3/2}} \int_{-\infty}^{+\infty} A(\vec{k},t) e^{i(\vec{k}\cdot\vec{r} - \omega t)} d^3 k, \tag{5.2.12}$$

where

$$A(\vec{k},t) = \frac{1}{(2\pi)^{3/2}} \int_{-\infty}^{+\infty} \psi(\vec{r},t) e^{-i(\vec{k}\cdot\vec{r}-\omega t)} d^3r. \tag{5.2.13}$$

The position of the particle is given by the centre of the wave packet, which moves with the group velocity $v_g = \frac{\partial \omega}{\partial k}$ (see Fig.4 of Chapter 3).

5.3 The Infinite Rectangular Well Potential

Consider a spinless particle confined to move in an infinite rectangular potential well (rectangular box) given by

$$V(x,y,z) = \begin{cases} 0 & \text{for } 0 < x < a, 0 < y < b, 0 < z < c \\ +\infty & \text{elsewhere} \end{cases} \tag{5.3.1}$$

in three spatial dimensions. We want to find the solutions to the TISE for the given potential. In this case too, the variables separate because the potential can be written as $V(x,y,z) = V_1(x) + V_2(y) + V_3(z)$. If we write $\psi(x,y,z) = X(x)Y(y)Z(z)$, the original time-independent Schrödinger equation separates into three independent equations for $X(x)$, $Y(y)$ and $Z(z)$, each of which coincides with the TISE for the case of 1D asymmetric square well potential of Chapter 3. Therefore, the normalized solution of the three-dimensional Schrödinger equation, satisfying the required boundary conditions, can be written as

$$\psi_{n_1n_2n_3}(x,y,z) = \sqrt{\frac{8}{abc}} \sin\left(\frac{n_1\pi}{a}x\right) \sin\left(\frac{n_2\pi}{b}y\right) \sin\left(\frac{n_3\pi}{c}z\right), \; n_1,n_2,n_3 = 1,2,3,... \tag{5.3.2}$$

The corresponding energies are given by

$$E_{n_1n_2n_3} = \frac{\hbar^2\pi^2}{2m}\left(\frac{n_1^2}{a^2} + \frac{n_2^2}{b^2} + \frac{n_3^2}{c^2}\right), \tag{5.3.3}$$

where n_1, n_2 and n_3 can take all integer values starting from 1.

If $a = b = c = L$, the potential is called the infinite cubic well potential of side L. In this case, the wave functions are

$$\psi_{n_1n_2n_3}(x,y,z) = \sqrt{\frac{8}{L^3}} \sin\left(\frac{n_1\pi}{L}x\right) \sin\left(\frac{n_2\pi}{L}y\right) \sin\left(\frac{n_3\pi}{L}z\right), \; n_1,n_2,n_3 = 1,2,3,\ldots \tag{5.3.4}$$

and the corresponding energy eigenvalues are given by

$$E_{n_1n_2n_3} = \frac{\hbar^2\pi^2}{2mL^2}\left(n_1^2 + n_2^2 + n_3^2\right), \; n_1,n_2,n_3 = 1,2,3,\ldots \tag{5.3.5}$$

Note that most of the energy levels in the cubic well potential are degenerate. The ground state, with $n_1 = n_2 = n_3 = 1$ and energy

$$E_{111} = \frac{3\hbar^2\pi^2}{2mL^2}, \tag{5.3.6}$$

is not degenerate. The first excited state is characterized by three sets of quantum numbers $(n_1, n_2, n_3) = (2,1,1)$, $(n_1, n_2, n_3) = (1,2,1)$ and $(n_1, n_2, n_3) = (1,1,2)$ and its energy is given by

$$E_{211} = E_{121} = E_{112} = \frac{6\hbar^2\pi^2}{2mL^2}. \tag{5.3.7}$$

Since the same value energy corresponds to three distinct sets of quantum numbers n_1, n_2 and n_3, the first excited state is three-fold degenerate. The corresponding wave functions are as follows:

$$\psi_{211}(x,y,z) = \sqrt{\frac{8}{L^3}} \sin\left(\frac{2\pi}{L}x\right) \sin\left(\frac{\pi}{L}y\right) \sin\left(\frac{\pi}{L}z\right), \tag{5.3.8}$$

$$\psi_{121}(x,y,z) = \sqrt{\frac{8}{L^3}} \sin\left(\frac{\pi}{L}x\right) \sin\left(\frac{2\pi}{L}y\right) \sin\left(\frac{\pi}{L}z\right), \tag{5.3.9}$$

$$\psi_{112}(x,y,z) = \sqrt{\frac{8}{L^3}} \sin\left(\frac{\pi}{L}x\right) \sin\left(\frac{\pi}{L}y\right) \sin\left(\frac{2\pi}{L}z\right). \tag{5.3.10}$$

The second excited state is again characterized by three sets of quantum numbers $(n_1, n_2, n_3) = (2,2,1)$, $(n_1, n_2, n_3) = (2,1,2)$ and $(n_1, n_2, n_3) = (1,2,2)$, and it is also threefold degenerate with energy

$$E_{221} = E_{212} = E_{122} = \frac{9\hbar^2\pi^2}{2mL^2}. \tag{5.3.11}$$

The third excited state is three-fold degenerate too with $(n_1, n_2, n_3) = (3,1,1)$, $(n_1, n_2, n_3) = (1,3,1)$ and $(n_1, n_2, n_3) = (1,1,3)$ yielding the same value of energy. Similarly, one can determine the degeneracy of all other excited states.

It is worth mentioning here that degeneracy of energy levels is a consequence of some underlying symmetry of the potential (discussed in detail in Chapter 11). In the given case of cubic well potential, the symmetry is related to the equivalence of all the three spatial directions, which is absent in the case of rectangular well potential.

5.4 Schrödinger Equation in Spherical Coordinates

In most of the problems in atomic and molecular physics, the potential is spherically symmetric, that is, it depends only on the distance from the origin. In such cases, it is convenient to work with spherical polar coordinates (r, θ, φ) (shown in Figure 5.1). The transformation from the Cartesian system to the spherical system of coordinates is given by the following set of equations

$$x = r\sin\theta\cos\varphi, y = r\sin\theta\sin\varphi, z = r\cos\theta, \tag{5.4.1}$$

where

$$r = \sqrt{x^2+y^2+z^2}, \quad \theta = \cos^{-1}\left(\frac{z}{r}\right), \quad \varphi = \tan^{-1}\left(\frac{y}{x}\right). \tag{5.4.2}$$

In these expressions, r measures the radial distance from the origin, θ is the polar angle measured from the z-axis and φ is the azimuthal angle measured from the x-axis, as shown in the figure.

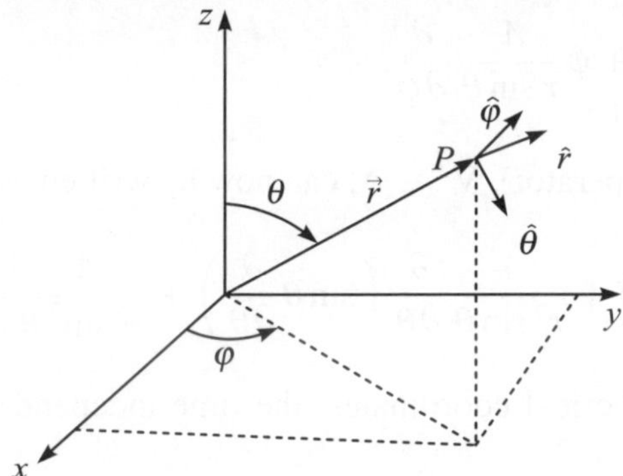

Figure 5.1 Spherical system of coordinates (r, θ, φ). $\hat{r}$, $\hat{\theta}$, and $\hat{\varphi}$ are the unit vectors along the r, θ and φ axes, respectively.

Using the transformation formulae, we obtain

$$\frac{\partial}{\partial x} = \sin\theta\cos\varphi\frac{\partial}{\partial r} + \frac{\cos\theta\cos\varphi}{r}\frac{\partial}{\partial\theta} - \frac{\sin\varphi}{r\sin\theta}\frac{\partial}{\partial\varphi}, \tag{5.4.3}$$

$$\frac{\partial}{\partial y} = \sin\theta\sin\varphi\frac{\partial}{\partial r} + \frac{\cos\theta\sin\varphi}{r}\frac{\partial}{\partial\theta} + \frac{\cos\varphi}{r\sin\theta}\frac{\partial}{\partial\varphi}, \tag{5.4.4}$$

$$\frac{\partial}{\partial z} = \cos\theta\frac{\partial}{\partial r} - \frac{\sin\theta}{r}\frac{\partial}{\partial\theta}. \tag{5.4.5}$$

The unit vectors of the spherical system of coordinates can also be calculated to be

$$\hat{r} = \left(\sin\theta\cos\varphi\hat{\imath} + \sin\theta\sin\varphi\hat{\jmath} + \cos\theta\hat{k}\right), \tag{5.4.6}$$

$$\hat{\theta} = \left(\cos\theta\cos\varphi\hat{\imath} + \cos\theta\sin\varphi\hat{\jmath} - \sin\theta\hat{k}\right), \tag{5.4.7}$$

$$\hat{\phi} = \left(-\sin\varphi\hat{\imath} + \cos\varphi\hat{\jmath}\right). \tag{5.4.8}$$

Taking these results into account, the gradient operator, $\vec{\nabla}$, can be written as

$$\begin{aligned}\vec{\nabla} &= \left(\sin\theta\cos\varphi\hat{\imath} + \sin\theta\sin\varphi\hat{\jmath} + \cos\theta\hat{k}\right)\frac{\partial}{\partial r}\\ &+ \frac{1}{r}\left(\cos\theta\cos\varphi\hat{\imath} + \cos\theta\sin\varphi\hat{\jmath} - \sin\theta\hat{k}\right)\frac{\partial}{\partial\theta}\\ &+ \frac{1}{r\sin\theta}\left(-\sin\varphi\hat{\imath} + \cos\varphi\hat{\jmath}\right)\frac{\partial}{\partial\varphi}.\end{aligned} \tag{5.4.9}$$

Or, using equations (5.4.6)–(5.4.8), we have

$$\vec{\nabla} = \hat{r}\frac{\partial}{\partial r} + \hat{\theta}\frac{1}{r}\frac{\partial}{\partial\theta} + \hat{\phi}\frac{1}{r\sin\theta}\frac{\partial}{\partial\varphi}. \tag{5.4.10}$$

The Laplacian (Laplace operator), $\vec{\nabla}^2 \equiv \Delta$, can now be written as

$$\vec{\nabla}^2 = \frac{1}{r^2}\frac{\partial}{\partial r}\left(r^2\frac{\partial}{\partial r}\right) + \frac{1}{r^2\sin\theta}\frac{\partial}{\partial\theta}\left(\sin\theta\frac{\partial}{\partial\theta}\right) + \frac{1}{r^2\sin^2\theta}\frac{\partial^2}{\partial\varphi^2}. \tag{5.4.11}$$

As a consequence, in spherical coordinates, the time-independent Schrödinger equation takes the form

$$-\frac{\hbar^2}{2m}\left[\frac{1}{r^2}\frac{\partial}{\partial r}\left(r^2\frac{\partial\phi}{\partial r}\right) + \frac{1}{r^2\sin\theta}\frac{\partial}{\partial\theta}\left(\sin\theta\frac{\partial\phi}{\partial\theta}\right) + \frac{1}{r^2\sin^2\theta}\frac{\partial^2\phi}{\partial\varphi^2}\right] + V(r)\phi = E\phi. \tag{5.4.12}$$

In the following sections,we shall discuss the solutions of this equation for some important cases.

5.5 Spherically Symmetric Potentials and Separation of Variables

If the potential is spherically symmetric, that is it is independent of the angles θ and ϕ and depends only on the radial distance r, the radial and angular variables in the Schrödinger equation can be separated. In such cases, we look for the solution in the form

$$\phi(r,\theta,\phi) = R(r)Y(\theta,\varphi). \tag{5.5.1}$$

Substitution of (5.5.1) into (5.4.12) with subsequent division throughout by RY gives

$$\left[\frac{1}{R}\frac{d}{dr}\left(r^2\frac{dR}{dr}\right) - \frac{2mr^2}{\hbar^2}(V-E)\right] + \frac{1}{Y}\left[\frac{1}{\sin\theta}\frac{\partial}{\partial\theta}\left(\sin\theta\frac{\partial Y}{\partial\theta}\right) + \frac{1}{\sin^2\theta}\frac{\partial^2 Y}{\partial\varphi^2}\right] = 0. \tag{5.5.2}$$

The first term in (5.5.2) is just a function of r, while the second term is a function of θ and φ only. Since the sum of these terms is zero, each of them must be equal to the same constant with opposite signs. We take this separation constant to be $\ell(\ell+1)$. The reason for this specific choice of the separation constant will be clear later, when we discuss the quantum mechanical theory of angular momentum. There, ℓ will represent the orbital quantum number and $\hbar\sqrt{\ell(\ell+1)}$, the value of the angular momentum of the particle in a given state with quantum number ℓ.

Thus, we have the system of differential equations, one each for the radial part $R(r)$ and the angular part $Y(\theta,\varphi)$ of the wave function:

$$\left[\frac{1}{R}\frac{d}{dr}\left(r^2\frac{dR}{dr}\right)-\frac{2mr^2}{\hbar^2}(V(r)-E)\right]=\ell(\ell+1), \tag{5.5.3}$$

$$\frac{1}{Y}\left[\frac{1}{\sin\theta}\frac{\partial}{\partial\theta}\left(\sin\theta\frac{\partial Y}{\partial\theta}\right)+\frac{1}{\sin^2\theta}\frac{\partial^2 Y}{\partial\varphi^2}\right]=-\ell(\ell+1). \tag{5.5.4}$$

We now proceed to discuss the solutions of these equations one-by-one.

5.6 Solution of the Angular Part of the Schrödinger Equation in Spherical Coordinates

Let us take the angular (Equation (5.5.4)) first. We have

$$\frac{1}{\sin\theta}\frac{\partial}{\partial\theta}\left(\sin\theta\frac{\partial Y}{\partial\theta}\right)+\frac{1}{\sin^2\theta}\frac{\partial^2 Y}{\partial\varphi^2}=-\ell(\ell+1)Y. \tag{5.6.1}$$

This can be rewritten in a more familiar form as

$$\sin\theta\frac{\partial}{\partial\theta}\left(\sin\theta\frac{\partial Y}{\partial\theta}\right)+\ell(\ell+1)\sin^2\theta Y+\frac{\partial^2 Y}{\partial\varphi^2}=0. \tag{5.6.2}$$

Separating the variables

$$Y(\theta,\varphi)=\vartheta(\theta)\Phi(\varphi), \tag{5.6.3}$$

substituting for $Y(\theta,\varphi)$ in (5.6.2) and dividing the resulting equation throughout by $\vartheta(\theta)\Phi(\varphi)$, we obtain

$$\frac{1}{\vartheta}\left[\sin\theta\frac{\partial}{\partial\theta}\left(\sin\theta\frac{d\vartheta}{\partial\theta}\right)\right]+\ell(\ell+1)\sin^2\theta+\frac{1}{\Phi}\frac{d^2\Phi}{d\varphi^2}=0. \tag{5.6.4}$$

The first term in (5.6.4) is a function of θ alone; whereas, the second term depends only on φ. Since the sum of theses terms is zero, each term must be equal to the same constant but with opposite signs. Taking this separation constant as m^2, we get

$$\sin\theta \frac{\partial}{\partial\theta}\left(\sin\theta \frac{d\vartheta}{\partial\theta}\right) + \ell(\ell+1)\sin^2\theta\,\vartheta = m^2 \tag{5.6.5}$$

$$\frac{1}{\Phi}\frac{d^2\Phi}{d\varphi^2} = -m^2 \Rightarrow \frac{d^2\Phi}{d\varphi^2} + m^2\Phi = 0. \tag{5.6.6}$$

Solving the equation (5.6.6) for Φ, we get

$$\Phi(\varphi) = e^{im\varphi}, \tag{5.6.7}$$

where m is a number and we have omitted the constant of integration, which can be absorbed in ϑ. Since when φ advances by 2π, we return to the same point in space, we have

$$\Phi(\varphi + 2\pi) = \Phi(\varphi) \quad \Rightarrow \quad e^{im(\varphi+2\pi)} = e^{im\varphi}. \tag{5.6.8}$$

Or, $\exp(2i\pi m) = 1$, which gives that m is an integer:

$$m = 0, \pm 1, \pm 2, \pm 3, ... \tag{5.6.9}$$

The ϑ equation can be reduced to the standard form of the Legendre equation by the change of variable $x = \cos\theta$. Its solutions are

$$\vartheta(\theta) = AP_\ell^m(x), \;\; x = \cos\theta, \tag{5.6.10}$$

where A is a constant and $P_\ell^m(x)$ are the associated Legendre polynomials. They are given by

$$P_\ell^m(x) = \left(1 - x^2\right)^{\frac{|m|}{2}} \frac{d^{|m|}}{dx^{|m|}} P_\ell(x), \tag{5.6.11}$$

where $P_\ell(x)$ are the Legendre polynomials defined by

$$P_\ell(x) = \frac{1}{2^l l!}\frac{d^\ell}{dx^\ell}\left(x^2 - 1\right)^\ell, \tag{5.6.12}$$

with ℓ as a non-negative integer. This formula is known as the *Rodriguez formula*. From (5.6.11), we get that if $|m| > \ell$, then $P_\ell^m(x) = 0$. This in turn says that for any given ℓ, there are $(2\ell + 1)$ possible values of m:

$$\ell = 0, 1, 2, ...; \;\; m = -\ell, (-\ell+1), (-\ell+2), (-\ell+3), ..., -1, 0, 1, ..., (\ell-1), \ell. \tag{5.6.13}$$

So, for a given ℓ there is a $(2\ell + 1)$-fold degeneracy with respect to the quantum number m. The normalized angular wave functions are given by

$$Y_\ell^m(\theta,\varphi) = \varepsilon\sqrt{\frac{(2\ell+1)\ (\ell-|m|)!}{4\pi\ (\ell+|m|)!}}\ P_\ell^m(\cos\theta)e^{im\varphi}, \tag{5.6.14}$$

where $\varepsilon = (-1)^m$ for $m \geq 0$ and $\varepsilon = 1$ for $m < 0$. The functions $Y_\ell^m(\theta,\varphi)$ are called *spherical harmonics*. The normalization condition for the spherical harmonics reads:

$$\int_0^\pi d\theta\ \sin\theta \int_0^{2\pi} d\varphi\ |Y_\ell^m(\theta,\varphi)|^2 = 1, \tag{5.6.15}$$

Thus, for a given value of ℓ, concrete expressions for $Y_\ell^m(\theta,\varphi)$ can be determined easily with the help of the equations (5.6.11)–(5.6.14).

5.7 Solution of the Radial Part of the Schrödinger Equation in Spherical Coordinates

Let us consider now the radial equation (5.5.3) for a given ℓ, i.e., for R_ℓ. It can be written as

$$\frac{d}{dr}\left(r^2\frac{dR_{n\ell}}{dr}\right) - \frac{2mr^2}{\hbar^2}(V(r)-E)R_{n\ell} = l(l+1)R_{n\ell}, \tag{5.7.1}$$

where we have introduced an additional subscript n for the radial wave function R_ℓ. It is usually done to identify the energy eigenvalues of the Hamiltonian: $\hat{H}\phi_{n\ell m} = E_n\,\phi_{n\ell m}$. n is called the principal quantum number and, as we shall see later, the orbital quantum number ℓ is related to the principal quantum number n in that, for a given n, ℓ can take values from 0 to $(n-1)$.

Equation (5.7.1) can be simplified further by changing the variables:

$$u_{n\ell}(r) = rR_{n\ell}(r). \tag{5.7.2}$$

We have

$$\frac{dR_{n\ell}}{dr} = \frac{(du_{n\ell}/dr)}{r} - \frac{u_{n\ell}}{r^2},\quad r^2\frac{dR_{n\ell}}{dr} = r\frac{du_{n\ell}}{dr} - u_{n\ell} \tag{5.7.3}$$

$$\frac{d}{dr}\left(r^2\frac{dR_{n\ell}}{dr}\right) = r(d^2u_{n\ell}/dr^2). \tag{5.7.4}$$

From (5.7.1) and (5.7.4), we get

$$\frac{d^2u_{n\ell}}{dr^2} + \frac{2m}{\hbar^2}\left[E - V(r) - \frac{\hbar^2\ell(\ell+1)}{2mr^2}\right]u_{n\ell}(r) = 0. \tag{5.7.5}$$

Equation (5.7.5) for the radial function $u_{n\ell}(r)$ (and hence for $R_{n\ell}(r)$) can be solved only if the potential, $V(r)$, is prescribed.

It is customary to introduce an effective potential, $V_{\rm eff}(r)$, by

$$V_{\rm eff}(r) = V(r) + \frac{\hbar^2 \ell(\ell+1)}{2mr^2}, \tag{5.7.6}$$

and rewrite the radial equation (5.7.5) as

$$\frac{d^2 u_{n\ell}}{dr^2} + \frac{2m}{\hbar^2}\left[E - V_{eff}(r)\right] u_{n\ell}(r) = 0. \tag{5.7.7}$$

This equation is similar to the one-dimensional Schrödinger equation with the difference that the effective potential $V_{\rm eff}$ has an extra term $\hbar^2\ell(\ell+1)/2mr^2$. This term is called the repulsive or centrifugal potential that tries to throw the particle away from the centre. Although the structure of this equation resembles the one-dimensional Schrödinger equation, it differs from the latter in the fact that the variable, r, cannot be negative: $r \in [0,\infty]$. Therefore, the radial wave function, $R_{n\ell}(r)$, must be finite everywhere from $r = 0$ to $r = \infty$. Consequently, the function $u_{n\ell}(r)$ must satisfy

$$\lim_{r\to 0} u_{n\ell}(r) = \lim_{r\to 0} rR_{n\ell}(r) = 0. \tag{5.7.8}$$

Note that for the bound states to exist, the potential $V(r)$ in (5.7.6), must be attractive because the part $\hbar^2\ell(\ell+1)/2mr^2$ in $V_{\rm eff}(r)$ is repulsive.

Once we solve the radial wave equation for a given $V(r)$, the full wave function will be given by

$$\phi_{n\ell m}(r,\theta,\phi) = R_{n\ell}(r)\, Y_\ell^m(\theta,\varphi). \tag{5.7.9}$$

Normalization of the wave function: The volume element $d\tau$ in spherical coordinates is given by $d\tau = r^2 \sin\theta \; dr\; d\theta\; d\varphi$. Hence, the normalization condition for the total wave function $\phi_{n\ell m}(r,\theta,\phi)$ reads as

$$\int_0^\infty dr \int_0^\pi d\theta \sin\theta \int_0^{2\pi} d\varphi\, r^2 \,|\phi_{n\ell m}(r,\theta,\phi)|^2 \tag{5.7.10}$$

$$= \int_0^\infty r^2\, |R_{n\ell}(r)|^2\, dr \int_0^\pi d\theta \sin\theta \int_0^{2\pi} d\varphi |Y_\ell^m(\theta,\varphi)|^2 = 1. \tag{5.7.11}$$

Since the spherical harmonics $Y_\ell^m(\theta,\varphi)$ are already normalized, we have

$$\int_0^\pi d\theta \int_0^{2\pi} d\varphi \;\sin\theta\; |Y_\ell^m(\theta,\varphi)|^2 = 1. \tag{5.7.12}$$

Therefore, to have the full wave function, $\phi_{n\ell m}(r,\theta,\varphi)$, normalized to unity, we have to simply normalize the radial wave function, $R_{n\ell}(r)$. As a result, we get

$$\int_0^\infty r^2\,|R_{n\ell}(r)|^2\,dr = 1. \tag{5.7.13}$$

Therefore, the stationary state wave functions of a particle, subject to a spherically symmetric potential $V(r)$, can be written as

$$\psi_{n\ell m}(r.\theta,\varphi) = A\ \phi_{n\ell m}(r.\theta,\varphi)\ e^{-\frac{i}{\hbar}E_{n\ell m}t}, \tag{5.7.14}$$

where the constant A is to be determined from the normalization condition, (5.7.13), for the radial wave function.

Note that the quantity $|\psi_{n\ell m}(r.\theta,\varphi)|^2\,d\tau$ represents the probability of finding the particle in the volume element $d\tau$, while the probability of finding the particle in a spherical shell enclosed between r and $r+dr$ is given by

$$\begin{aligned} P_{n\ell}(r)dr &= \left(\int_0^\pi \sin\theta d\theta \int_0^{2\pi} d\varphi\ |\psi_{n\ell m}(r.\theta,\varphi)|^2\right) r^2\ dr \\ &= |R_{n\ell}(r)|^2\ r^2\ dr \int_0^\pi\int_0^{2\pi} (Y_\ell^m(\theta,\varphi))^*\ Y_\ell^m(\theta,\varphi)\ \sin\theta\ d\theta\ d\varphi \\ &= |R_{n\ell}(r)|^2\ r^2\ dr. \end{aligned} \tag{5.7.15}$$

If we integrate $P_{n\ell}(r)$ from $r=0$ to $r=a$, where a is a real constant, we get the probability of finding the particle in a sphere of radius a centered at the origin $r=0$.

In what follows, we shall discuss the solutions of the radial Schrödinger equation (5.7.1) for some important spherically symmetric potentials.

5.8 The Free Particle Solution in Spherical Coordinates

Consider a particle of mass m moving freely in space. In this case, the potential $V(r)$ is zero and the stationary Schrödinger equation (5.4.12), reduces to

$$-\frac{\hbar^2}{2m}\left[\frac{1}{r^2}\frac{\partial}{\partial r}\left(r^2\frac{\partial\phi}{\partial r}\right) + \frac{1}{r^2\sin\theta}\frac{\partial}{\partial\theta}\left(\sin\theta\frac{\partial\phi}{\partial\theta}\right) + \frac{1}{r^2\sin^2\theta}\frac{\partial^2\phi}{\partial\varphi^2}\right] = E\phi. \tag{5.8.1}$$

The variables separate and the solution can be represented as $\phi_{k\ell m}(r,\theta,\phi) = R_{k\ell}(r)Y_\ell^m(\theta,\varphi)$. Note that in the given case of a free particle the energy, $E_k = \hbar^2k^2/2m$, takes continuous values and hence the radial wave function is characterized by the continuous index k.

The angular part of the wave function, $Y(\theta,\varphi)$, satisfies (5.6.2) and is given by the equation (5.6.14). The radial wave function, $R_{k\ell}(r)$, satisfies

$$\frac{d}{dr}\left(r^2\frac{dR_{k\ell}}{dr}\right)+k^2r^2R_{k\ell}=\ell(\ell+1)R_{k\ell}. \tag{5.8.2}$$

where $k^2 = 2mE_k/\hbar^2$. Introducing $\rho = kr$, we have

$$\frac{d^2R_{k\ell}(\rho)}{d\rho^2}+\frac{2}{\rho}\frac{dR_{k\ell}(\rho)}{d\rho}+\left(1-\frac{\ell(\ell+1)}{\rho^2}\right)R_{k\ell}(\rho)=0. \tag{5.8.3}$$

This is the spherical Bessel equation whose general solution, for any k (that is, E_k), is given by

$$R_{k\ell}(\rho)=A_\ell j_\ell(\rho)+B_\ell n_\ell(\rho), \ell=0,1,2,3,... \tag{5.8.4}$$

where $j_\ell(\rho)$ and $n_\ell(\rho)$ are the spherical Bessel functions and the spherical Neumann functions, respectively. They are given by

$$j_\ell(\rho)=(-\rho)^\ell\left(\frac{1}{\rho^\ell}\frac{d^\ell}{d\rho^\ell}\right)\frac{\sin\rho}{\rho}, n_\ell(\rho)=-(-\rho)^\ell\left(\frac{1}{\rho^\ell}\frac{d^\ell}{d\rho^\ell}\right)\frac{\cos\rho}{\rho}. \tag{5.8.5}$$

The asymptotic forms of these functions for $\rho\to 0$ and $\rho\to\infty$ are, respectively, given by

$$j_\ell(\rho)=\frac{2^\ell\ell!}{(2^\ell+1)!}\rho^\ell, n_\ell(\rho)=-\frac{(2\ell-1)!}{2^\ell\ell!}\frac{1}{\rho^{\ell+1}}(\rho\to 0), \tag{5.8.6}$$

$$j_\ell(\rho)=\frac{1}{\rho}\sin\left(\rho-\frac{l\pi}{2}\right), n_l(\rho)=-\frac{1}{\rho}\cos\left(\rho-\frac{\ell\pi}{2}\right)(\rho\to\infty). \tag{5.8.7}$$

Note that for $\rho\to 0$, the Neumann function blows up. Since the wave function has to be finite everywhere in space, the part of the solution containing the Neumann function must be dropped. As a result, we have

$$\phi_{k\ell m}(\rho,\theta,\varphi)=A_\ell j_\ell(kr)Y_\ell^m(\theta,\varphi), \tag{5.8.8}$$

where $k=\sqrt{2mE_k/\hbar^2}$. Since E_k is a continuous function of k, the energy spectrum is continuous and infinitely degenerate. This degeneracy corresponds to the spherical symmetry in the momentum space: all directions of $\vec{k}$ are equivalent.

Recall that the free particle solution in Cartesian coordinates is proportional to $e^{i\vec{k}\cdot\vec{r}}$ (see (5.2.7)), which can be expanded in terms of the spherical Bessel functions as

$$e^{i\vec{k}\cdot\vec{r}}=\sum_{\ell=0}^{\infty}\sum_{m=-\ell}^{\ell}c_{\ell m}\,j_\ell(kr)\,Y_\ell^m(\theta,\varphi), \tag{5.8.9}$$

where $c_{\ell m}$ are arbitrary constants. Thus, the solution for a free particle in spherical coordinates (*in terms of the spherical Bessel functions*) is completely equivalent to the free particle solution in Cartesian coordinates (*in terms of the plane waves*) with appropriate expansion coefficients $c_{\ell m}$. The problem then reduces to finding the expansion coefficients $c_{\ell m}$. For the particular case of propagation along the z-axis ($\vec{k}\|\hat{z}$), $m = 0$ and we get

$$e^{i\vec{k}\cdot\vec{r}} = e^{kr\cos\theta} = \sum_{\ell=0}^{\infty} i^{\ell}(2\ell+1)j_{\ell}(kr)P_{\ell}(\cos\theta), \tag{5.8.10}$$

where $P_{\ell}(\cos\theta)$ are the Legendre polynomials. For this given particular case, the coefficients $c_{\ell m}$ are thus given by $c_{\ell m} = i^{\ell}\,(2\ell+1)$.

Note that although the free particle solutions in the Cartesian and the spherical coordinates are equivalent, they do differ in physical content. While the plane wave solution describes a free particle of energy E_k with a well-specified linear momentum but undefined angular momentum, the solution in terms of the spherical Bessel functions describes a free particle with a well-defined angular momentum but gives no information about its linear momentum.

5.9 The Infinite Spherical Well Potential

Consider a particle of mass m moving in the following potential

$$V(r) = \begin{cases} 0, & \text{for } r \leq a \\ \infty, & \text{for } r > a, \end{cases} \tag{5.9.1}$$

where a is a positive constant with dimensions of length. This potential is called infinite spherical well potential. Using the radial Schrödinger equation, we want to determine the bound state energy spectrum and the corresponding normalized wave functions for the case when the orbital angular momentum of the particle is zero ($\ell = 0$). Also, we would like to compute the probability of finding the particle: (i) in a sphere of radius, say, $r = a/2$ and (ii) in the annular region between $r = \frac{a}{4}$ and $r = \frac{a}{2}$.

In the region $r < a$, the radial wave function $u(r) = rR(r)$ satisfies the following ordinary differential equation

$$\frac{d^2u}{dr^2} + \left[\frac{2m}{\hbar^2}E - \frac{l(l+1)}{r^2}\right]u(r) = 0, \tag{5.9.2}$$

which, for $\ell = 0$, reduces to

$$\frac{d^2u}{dr^2} + k^2u(r) = 0, k^2 = \frac{2mE}{\hbar^2}. \tag{5.9.3}$$

It has its general solution of the form

$$u(r) = A\cos(kr) + B\sin(kr), \tag{5.9.4}$$

where A and B are arbitrary constant coefficients. As a result, the function $R(r)$ is given by

$$R(r) = A\frac{\cos(kr)}{r} + B\frac{\sin(kr)}{r}. \tag{5.9.5}$$

The finiteness of the radial wave function $R(r)$ at $r = 0$ demands the coefficient A to be zero.

For $r > a$, the potential is infinite. Therefore, the wave function must vanish in this region, that is, $u(r) = 0$ for $r > a$. The continuity of the radial wave function, R, at the boundary $r = a$ requires that $R(a) = 0$. So, we have

$$R(a) = B\frac{\sin(ka)}{a} = 0. \tag{5.9.6}$$

The solution of (5.9.6) yields: $k_n a = n\pi$, where $n = 1, 2, 3, ...$ This relation leads to the discrete energy spectrum of the particle inside the well

$$E_n = \frac{n^2\pi^2\hbar^2}{2ma^2}, \qquad n = 1, 2, 3, ... \tag{5.9.7}$$

The normalization of the radial wave function $R_{n0}(r) = B\,[\sin(k_n r)/r]$ reads

$$1 = |B|^2 \int_0^a \frac{\sin^2\left(\frac{n\pi}{a}r\right)}{r^2} r^2 dr = \frac{|B|^2 a}{n\pi}\int_0^{n\pi} \sin^2\rho\, d\rho = \frac{|B|^2 a}{2}. \tag{5.9.8}$$

Hence, $B = \sqrt{2/a}$ and the normalized wave functions are

$$R_{n0}(r) = \sqrt{\frac{2}{a}}\frac{\sin\left(\frac{n\pi}{a}r\right)}{r}. \tag{5.9.9}$$

Using this solution, we can determine the probability of finding the particle inside the sphere of a given radius. For instance, the probability of finding the particle in a sphere of radius $a/2$ is given by

$$P = \int_0^{a/2} |R_{n0}|^2 r^2 dr = \frac{2}{a}\int_0^{a/2} \sin^2\left(\frac{n\pi}{a}r\right) dr = \frac{2}{n\pi}\int_0^{n\pi/2} \sin^2\rho\, d\rho. \tag{5.9.10}$$

Or,

$$P = \frac{1}{n\pi}\int_0^{n\pi/2} [1 - \cos 2\rho]\, d\rho = \frac{1}{n\pi}\left(\frac{n\pi}{2} - \frac{1}{2}\left[\sin 2\rho\right]_0^{(n\pi/2)}\right) = \frac{1}{2}. \tag{5.9.11}$$

Similarly, the probability of finding the particle in the annular region between $r = \frac{a}{4}$ and $r = \frac{a}{2}$ is given by

$$P = \frac{1}{n\pi}\left(\frac{n\pi}{4} + \frac{1}{2n\pi}\sin\left(\frac{n\pi}{2}\right)\right) = \begin{cases} \frac{1}{4} & \text{for even } n, \\ \frac{1}{4} + \frac{1}{2n\pi} & \text{for } n = 1,5,9,\ldots, \\ \frac{1}{4} - \frac{1}{2n\pi} & \text{for } n = 3,7,11,\ldots. \end{cases} \tag{5.9.12}$$

5.10 The Finite Spherical Well Potential

A particle of mass m is moving under the influence of the following potential

$$V(t) = \begin{cases} -V_0, & \text{for } r < a \\ 0, & \text{for } r \geq a, \end{cases} \tag{5.10.1}$$

where V_0 and a are positive constants with dimensions of energy and length respectively. We wish to find the bound state energy eigenvalues and the corresponding wave functions for $|E| < V_0$.

In the region $r \leq a$, the radial wave function satisfies the equation

$$\frac{d^2R_{n\ell}}{dr^2} + \frac{2}{r}\frac{dR_{n\ell}}{dr} + \frac{2m}{\hbar^2}\left[V_0 - |E|\right]R_{n\ell} = \frac{l(l+1)}{r^2}R_{n\ell}. \tag{5.10.2}$$

For $r > a$, the potential $V(r) = 0$, and we have

$$\frac{d^2R_{n\ell}}{dr^2} + \frac{2}{r}\frac{dR_{n\ell}}{dr} - \frac{2m|E|}{\hbar^2}R_{n\ell} = \frac{l(l+1)}{r^2}R_{n\ell}. \tag{5.10.3}$$

For bound states, $|E| < V_0$. Therefore, we introduce

$$k_1 = \sqrt{\frac{2m}{\hbar^2}(V_0 - |E|)}, \tag{5.10.4}$$

$$k_2 = \sqrt{\frac{2m}{\hbar^2}|E|}. \tag{5.10.5}$$

In terms of the function $u_{n\ell}(r) = rR_{n\ell}(r)$, these equations can then be written as

$$\frac{d^2u_{n\ell}}{dr^2} + \left[k_1^2 - \frac{l(l+1)}{r^2}\right]u_{n\ell}(r) = 0, (0 < r < a), \tag{5.10.6}$$

$$\frac{d^2u_{n\ell}}{dr^2} + \left[(ik_2)^2 - \frac{l(l+1)}{r^2}\right]u_{n\ell}(r) = 0, (r > a). \tag{5.10.7}$$

Equation (5.10.6) coincides with the equation satisfied by the free particle wave function with wave number k_1. Its solution is given by

$$u_{n\ell}(r) = A_\ell j_\ell(k_1 r) + B_\ell n_\ell(k_1 r), \tag{5.10.8}$$

where A_ℓ and B_ℓ are arbitrary constant coefficients. Once again due to the requirement of finiteness of the radial wave function, $R_{n\ell}(r)$, at $r = 0$, we omit the second term in (5.10.8). Therefore, the solution $R_{n\ell}^{(I)}(r)$ for $0 < r < a$ is given by

$$R_{n\ell}^{(I)}(r) = \frac{u_{n\ell}^{(I)}(r)}{r} = A_\ell j_\ell(k_1 r). \tag{5.10.9}$$

For $r > a$, the radial equation (5.10.7), has the solution

$$R_{n\ell}^{(II)}(r) = \frac{u_{n\ell}^{(II)}(r)}{r} = D_\ell h_\ell(ik_2), \tag{5.10.10}$$

where $h_\ell(ik_2) = j_\ell(ik_2 r) + in_\ell(ik_1 r)$ is the Hankel function that asymptotically behaves as $e^{-k_2 r}/r$ as $r \to +\infty$. Any other linear combination will diverge for $r \to +\infty$. Therefore, the radial wave function of the particle for the given potential, can be written as

$$R_{n\ell}(r) = \begin{cases} A_\ell j_\ell(k_1 r), & r \leq a \\ D_\ell h_\ell(ik_2 r), & r > a, \end{cases} \tag{5.10.11}$$

where A_ℓ and B_ℓ are to be determined from the boundary conditions.

The continuity of the wave function and its first derivative at $r = a$ leads to the transcendental equation

$$k_1 \frac{j'_\ell(k_1 a)}{j_\ell(k_1 a)} = k_2 \frac{h'_\ell(ik_2 a)}{h_\ell(ik_2 a)}, \tag{5.10.12}$$

for the determination of the energy eigenvalues. The solution is usually found numerically. It turns out that the roots of the equation (5.10.12) yield a discrete set of energy eigenvalues for the particle. The constants A_ℓ and D_ℓ are related through

$$A_\ell j_\ell(k_1 a) = D_\ell h_\ell(ik_2 a). \tag{5.10.13}$$

The full bound state wave functions are given by

$$R_{n\ell}(r) = A_\ell \begin{cases} j_\ell(k_1 r) Y_\ell^m(\theta, \varphi), & r \leq a \\ \frac{j_\ell(k_1 a)}{h_\ell(ik_2 a)} h_\ell(ik_2 r) Y_\ell^m(\theta, \varphi), & r > a, \end{cases} \tag{5.10.14}$$

where the constant A_ℓ is determined from the normalization of the radial wave function.

For the case $\ell = 0$, the radial wave function, $u_{n\ell}$, satisfies

$$\frac{d^2u_{n\ell}}{dr^2} + k_1^2 u_{n\ell}(r) = 0, (0 < r < a), \tag{5.10.15}$$

$$\frac{d^2u_{n\ell}}{dr^2} - k_2^2 u_{n\ell}(r) = 0, (r > a). \tag{5.10.16}$$

Since $j_0(k_1r) = \sin(k_1r)/(k_1r)$ and $h_0(ik_2r) = -(e^{-k_2r}/(k_2r)$, the solutions to these equations are given by

$$R_{n\ell}^{(I)}(r) = A\frac{\sin(k_1r)}{r}, (0 < r < a), \tag{5.10.17}$$

$$R_{n\ell}^{(II)}(r) = B\frac{e^{-k_2r}}{r}, (r > a), \tag{5.10.18}$$

where A and B are arbitrary constants. The continuity of the radial wave functions and their first derivatives at $r = a$ can be simultaneously satisfied by matching the logarithmic derivatives of the wave functions at $r = a$:

$$\frac{R_{n\ell}^{(I)\prime}(a)}{R_{n\ell}^{(I)}(a)} = \frac{R_{n\ell}^{(II)\prime}(a)}{R_{n\ell}^{(II)}(a)}. \tag{5.10.19}$$

Calculating the required derivatives, we get

$$k_1\cot(k_1a) - \frac{1}{a} = -k_2 - \frac{1}{a}. \tag{5.10.20}$$

From here, we arrive at the transcendental equation, whose roots determine the discrete values of the energy of the particle:

$$k_2 = -k_1\cot(k_1a). \tag{5.10.21}$$

Equation (5.10.21) can be solved graphically as follows. Let us introduce

$$\xi = k_1a, \eta = k_2a. \tag{5.10.22}$$

Multiplying (5.10.21) by a, we get

$$\xi\cot\xi = -\eta. \tag{5.10.23}$$

Furthermore, we have

$$\xi^2 + \eta^2 = a^2(k_1^2 + k_2^2) = \frac{2m}{\hbar^2}a^2(V_0 - |E|) + \frac{2m}{\hbar^2}a^2|E| = \frac{2mV_0a^2}{\hbar^2}, \tag{5.10.24}$$

which is the equation of a circle in the coordinates ξ and η with radius $\sqrt{2mV_0a^2/\hbar^2}$. The discrete energy levels are determined by the points of intersection of this circle with the curve $\eta = -\xi \cot \xi$.

In the limit $E \to 0$, we have

$$\sqrt{\frac{2m}{\hbar^2}V_0}\ \cot\left(\sqrt{\frac{2m}{\hbar^2}V_0}\right) = 0, \tag{5.10.25}$$

which yields

$$\sqrt{\frac{2m}{\hbar^2}V_0} = \frac{(2n+1)}{2}\pi, n = 0,1,2,3,.... \tag{5.10.26}$$

Therefore, for one, two and three bound states, we have $n = 0,1$ and $n = 2$ and the corresponding values of V_0 are

$$V_0^{(1)} = \frac{\pi^2\hbar^2}{8ma^2},\ V_0^{(2)} = \frac{9\pi^2\hbar^2}{8ma^2} \text{ and } V_0^{(3)} = \frac{25\pi^2\hbar^2}{8ma^2}. \tag{5.10.27}$$

Therefore, if $V_0a^2 < \pi^2\hbar^2/8m$, no bound state exists.

5.11 The Hydrogen Atom

A hydrogen atom consists of a proton (charge e and mass m_p) in the nucleus and an electron (charge $-e$ and mass m_e) orbiting around it, which is held in its orbit by the attractive Coulomb force. For simplicity, we shall ignore the spin degree of freedom in our treatment of this system.

Let $\vec{r}_e = (x_e, y_e, z_e)$ and $\vec{r}_p = (x_p, y_p, z_p)$ be the position vectors for the electron and the proton, respectively. Since the potential, V, depends only on the relative distance, r, between the electron and the proton, it is convenient to go over to the center of the mass system. Let $\vec{R} = (X,Y,Z)$ be the position vector of the centre of mass, defined by

$$\vec{R} = \frac{m_e\vec{r}_e + m_p\vec{r}_p}{m_e + m_p}, \tag{5.11.1}$$

and let $\vec{r} = (x,y,z) = \vec{r}_e - \vec{r}_p$ represent the relative position vector. The Schrödinger equation for the system is written as

$$i\hbar\frac{\partial\psi(\vec{r}_e,\vec{r}_p,t)}{\partial t} = \left[-\frac{\hbar^2}{2m_e}\vec{\nabla}_e^2 - \frac{\hbar^2}{2m_p}\vec{\nabla}_p^2 + V(r)\right]\psi(\vec{r}_e,\vec{r}_p,t). \tag{5.11.2}$$

It is easy to check that

$$\frac{1}{m_e}\vec{\nabla}_e^2 + \frac{1}{m_p}\vec{\nabla}_p^2 = \frac{1}{M}\vec{\nabla}_R^2 + \frac{1}{\mu}\vec{\nabla}_r^2, \tag{5.11.3}$$

$$M = m_e + m_p, \qquad \mu = \frac{m_e m_p}{m_e + m_p}, \tag{5.11.4}$$

where M and μ are the total and the so-called reduced mass, respectively. The operators $\vec{\nabla}_R^2$ and $\vec{\nabla}_r^2$ in (5.11.3) are given by

$$\vec{\nabla}_R^2 = \frac{\partial^2}{\partial X^2} + \frac{\partial^2}{\partial Y^2} + \frac{\partial^2}{\partial Z^2}, \tag{5.11.5}$$

and

$$\vec{\nabla}_r^2 = \frac{\partial^2}{\partial x^2} + \frac{\partial^2}{\partial y^2} + \frac{\partial^2}{\partial z^2}. \tag{5.11.6}$$

We look for the stationary state solutions of the Schrödinger equation (5.11.2) in the form

$$\psi(\vec{r}_e, \vec{r}_p, t) = \psi(\vec{r}_e, \vec{r}_p) e^{-i\frac{E_T}{\hbar}t}, \tag{5.11.7}$$

where E_T is the total energy of the system. Taking into account that, in the SI units, the Coulomb potential between the electron and proton is given by

$$V(r) = -\frac{e^2}{4\pi\varepsilon_0|\vec{r}_e - \vec{r}_p|}, \tag{5.11.8}$$

we get from (5.11.2) and (5.11.8) that $\psi(\vec{r}_e, \vec{r}_p)$ satisfies

$$\left[-\frac{\hbar^2}{2m_p}\vec{\nabla}_p^2 - \frac{\hbar^2}{2m_e}\vec{\nabla}_e^2 - \frac{e^2}{4\pi\varepsilon_0|\vec{r}_e - \vec{r}_p|}\right]\psi(\vec{r}_e, \vec{r}_p) = E_T\psi(\vec{r}_e, \vec{r}_p). \tag{5.11.9}$$

Equation (5.11.9) can be rewritten in the centre of the mass system as

$$\left[-\frac{\hbar^2}{2M}\vec{\nabla}_R^2 - \frac{\hbar^2}{2\mu}\vec{\nabla}_r^2 - \frac{e^2}{4\pi\varepsilon_0 r}\right]\psi(\vec{R}, \vec{r}) = E_T\psi(\vec{R}, \vec{r}). \tag{5.11.10}$$

Since the potential depends only on the relative coordinate r, we expect the variables to separate and look for the solution in the form

$$\psi(\vec{R}, \vec{r}) = \Phi(\vec{R})\phi(\vec{r}). \tag{5.11.11}$$

Inserting $\psi(\vec{R},\vec{r})$ from (5.11.11) into (5.11.10), we obtain

$$\left[-\frac{\hbar^2}{2M}\frac{1}{\Phi(\vec{R})}\vec{\nabla}_R^2\Phi(\vec{R})\right]-\left[\frac{\hbar^2}{2\mu}\frac{1}{\phi(\vec{r})}\vec{\nabla}_r^2\phi(\vec{r})+\frac{e^2}{4\pi\varepsilon_0 r}\right]=E_T. \tag{5.11.12}$$

The first term on the left-hand side of (5.11.12) is just a function of $\vec{R}$, whereas the second term depends only on $\vec{r}$. The sum of these terms equals a constant E_T. Since the vectors $\vec{R}$ and $\vec{r}$ are independent, for this equation to hold, each term on the left-hand side must be a constant. This leads to the following pair of equations

$$-\frac{\hbar^2}{2M}\vec{\nabla}_R^2\Phi(\vec{R})=E_R\Phi(\vec{R}), \tag{5.11.13}$$

$$-\frac{\hbar^2}{2\mu}\vec{\nabla}_r^2\phi(\vec{r})+\frac{e^2}{4\pi\varepsilon_0 r}\phi(\vec{r})=E_r\phi(\vec{r}), \tag{5.11.14}$$

where

$$E_T=E_R+E_r. \tag{5.11.15}$$

Note that (5.11.13) can be interpreted as the stationary Schrödinger equation of a free particle of mass M. Thus, we conclude that the centre of mass of the electron–proton pair in a hydrogen atom moves as a free particle of mass M. Consequently, the normalized solution of this equation is written as

$$\Phi(\vec{R})=\frac{1}{(2\pi)^{3/2}}e^{-i\vec{\kappa}\cdot\vec{R}}, \tag{5.11.16}$$

where $\vec{\kappa}$ is the wave vector associated with the free motion of the total mass M in the centre of mass frame and $E_R=\hbar^2\vec{\kappa}^2/2M$ the kinetic energy of M.

So far as (5.11.14) is concerned, it is nothing but the time-independent Schrödinger equation for a fictitious particle of mass μ moving in a central potential

$$V(r)=-\frac{e^2}{4\pi\varepsilon_0 r}. \tag{5.11.17}$$

Since the potential is spherically symmetric, it is convenient to solve this equation in spherical coordinates. Using the expression for $\vec{\nabla}_r^2$ in spherical polar coordinates, we can write (5.11.14) as

$$-\frac{\hbar^2}{2\mu}\left[\frac{1}{r^2}\frac{\partial}{\partial r}\left(r^2\frac{\partial\phi}{\partial r}\right)+\frac{1}{r^2\sin\theta}\frac{\partial}{\partial\theta}\left(\sin\theta\frac{\partial\phi}{\partial\theta}\right)+\frac{1}{r^2\sin^2\theta}\frac{\partial^2\phi}{\partial\varphi^2}\right]+\frac{e^2}{4\pi\varepsilon_0 r}\phi=E\phi, \tag{5.11.18}$$

where we have set $E_r \equiv E$ for convenience. Note that for $E > 0$, the Coulomb potential in (5.11.18) admits continuum energy states describing the scattering of electron on the proton. For $E < 0$, it admits the discrete set of bound states that represent the bound states of the hydrogen atom. Here, we shall take up the solution of the bound state problem only.

The ground state of hydrogen: We start with the simplest case of the ground state of the hydrogen atom with $n = 1$ and $\ell = 0$. This state, called the s state, possesses a complete spherical symmetry and the wave function corresponding to this state is given by

$$\phi_{n\ell m}(r,\theta,\varphi) = R_{10}(r)Y_0^0(\theta,\varphi) = \frac{1}{\sqrt{4\pi}}R_{10}(r), \tag{5.11.19}$$

where the radial wave function, $R_{10}(r)$, satisfies

$$-\frac{\hbar^2}{2\mu}\frac{1}{r^2}\frac{\partial}{\partial r}\left(r^2\frac{\partial R_{10}(r)}{\partial r}\right) + \frac{e^2}{4\pi\varepsilon_0 r}R_{10}(r) = ER_{10}(r). \tag{5.11.20}$$

Note that in (5.11.20), we have dropped the suffix r from E_r for convenience. Equation (5.11.20) can also be written as

$$\frac{1}{r^2}\frac{d}{dr}\left(r^2\frac{dR_{10}(r)}{dr}\right) + \frac{2\mu}{\hbar^2}\left[E + \frac{e^2}{4\pi\varepsilon_0 r}\right]R_{10}(r) = 0. \tag{5.11.21}$$

By introducing

$$\lambda = \frac{2\mu E}{\hbar^2}; \alpha = \frac{\mu e^2}{4\pi\varepsilon_0\hbar^2}, \tag{5.11.22}$$

we rewrite (5.11.21) in a more compact form as

$$\frac{d^2R_{10}(r)}{dr^2} + \frac{2}{r}\frac{dR_{10}(r)}{dr} + \left(\lambda + \frac{2\alpha}{r}\right)R_{10}(r) = 0. \tag{5.11.23}$$

Clearly, we need a solution for this equation satisfying the standard conditions, that is, the solution must be finite everywhere (including $r = 0$) and must tend to zero at spatial infinity, that is, for $r \to \infty$. We look for such a solution in the form

$$R_{10}(r) = e^{-\beta r}, \tag{5.11.24}$$

where β is a constant and we have omitted a constant factor on the right-hand side, which can be taken care of by normalization. From (5.11.23) and (5.11.24), we get

$$\beta^2 - \frac{2}{r}\beta + \left(\lambda + \frac{2\alpha}{r}\right) = 0 \Rightarrow (\beta^2 + \lambda) + (\alpha - \beta)\frac{2}{r} = 0. \tag{5.11.25}$$

Since (5.11.25) must be satisfied for any value of r, we have only the following two possibilities for the value of β

$$\beta^2 = -\lambda \text{ and } \beta = \alpha. \tag{5.11.26}$$

Using the expressions for α and λ from (5.11.22) in (5.11.26), we obtain the following formula for the ground state energy, E_g, of the hydrogen atom

$$E_g = -\frac{\mu^2 e^4}{(4\pi\varepsilon_0)^2\hbar^4} \times \frac{\hbar^2}{2\mu} = -\frac{\mu}{2\hbar^2}\left(\frac{e^2}{4\pi\varepsilon_0}\right)^2. \tag{5.11.27}$$

Inserting the values of the fundamental constants and the mass of the electron for μ, we get the value of the ground state energy of the hydrogen atom

$$E = -13.6\,\text{eV} \tag{5.11.28}$$

which coincides with the ground state energy, E_1, in the Bohr theory of hydrogen atom.

Clearly, the probability of finding the electron in the volume element $d\tau = r^2 \sin\theta\, dr d\theta\, d\varphi$ is given by

$$w d\tau = |N|^2 |R_{10}(r)|^2 |Y_0^0(\theta,\varphi)|^2 r^2 \sin\theta dr d\theta d\varphi, \tag{5.11.29}$$

where N is the normalization constant. Consequently, the probability of finding the electron at a distance in the interval $[r, r+dr]$ from the nucleus is obtained by integrating over the angles. Taking into account the normalization of the spherical harmonics, we get

$$w(r)dr = |N|^2 r^2 e^{-2\beta r} dr = |N|^2 r^2 e^{-2\beta r} dr. \tag{5.11.30}$$

Since β has the dimension of length inverse, we introduce a new constant a such that $\beta = 1/a$. Then,

$$w(r)dr = |N|^2 r^2 e^{-2r/a} dr. \tag{5.11.31}$$

The probability density $w(r) = |N|^2 r^2 e^{-2r/a_0}$ equals zero at $r = 0$ and tends to zero for $r \to \infty$. Therefore, in principle, there is a non-zero probability of finding the electron at any distance from the nucleus between $r \geq 0$ and $r = \infty$. Let us determine the distance at which this probability reaches its maximum value. We have

$$\frac{dw}{dr} = |N|^2 \left(2r - \frac{2r^2}{a_0}\right) e^{-2r/a} = 0, \Rightarrow r - \frac{r^2}{a} = 0. \tag{5.11.32}$$

Hence, the probability of finding the electron at a distance from the nucleus reaches it maximum value at $r_0 = a$. Now taking into account that $a = 1/\beta = 1/\alpha = 4\pi\varepsilon_0\hbar^2/\mu e^2$, we get

$$a = \frac{4\pi\varepsilon_0\hbar^2}{\mu e^2} = 0.529 \times 10^{-10}\text{m}. \tag{5.11.33}$$

If we recall the Bohr theory of hydrogen atom, we notice that a is nothing but the radius of the *first Bohr orbit*. This is called *Bohr radius*. In literature, it is usually written as a_0.

Let us now find the normalization constant N. The normalization condition reads

$$1 = |N|^2 \int_0^\infty e^{-2r/a_0} r^2 dr = |N|^2 \int_0^\infty e^{-2r/a_0} r^2 dr$$

$$= |N|^2 \frac{2a_0^3}{8} = |N|^2 \frac{a_0^3}{4}, \tag{5.11.34}$$

where we have used the standard integral

$$I = \int_0^\infty x^p e^{-bx} = \frac{p!}{b^{p+1}} = \frac{a_0^{p+1} p!}{2^{p+1}}, \tag{5.11.35}$$

for $p = 2$. As a result,

$$N = \frac{2}{(a_0^3)^{1/2}}. \tag{5.11.36}$$

The normalized ground state wave function of the hydrogen atom is given by

$$R_0(r) = \frac{2}{(a_0^3)^{1/2}} e^{-r/a_0} \frac{1}{\sqrt{4\pi}} = \frac{1}{\sqrt{\pi a_0^3}} e^{-r/a_0}. \tag{5.11.37}$$

Example 5.11.1: Calculate the average distance of the electron from the nucleus in the ground state of the hydrogen atom. Also, calculate the average values of the potential and kinetic energies in the ground state of the hydrogen atom. Using the uncertainty relation, discuss the stability of the hydrogen and hydrogenic atoms in the s state.

Solution: The average value $\langle r \rangle$ of the distance of the electron from the nucleus is given by

$$\langle r \rangle = \frac{\int r|\phi(\vec{r})|^2 d\tau}{\int |\phi(\vec{r})|^2 d\tau} = \frac{1}{(\pi a_0^3)} \int_0^\infty r e^{-2r/a_0} r^2 dr \int_0^{2\pi} d\varphi \int_0^\pi d\theta \sin\theta$$

$$= \frac{4\pi}{(\pi a_0^3)} \int_0^\infty e^{-2r/a_0} r^3 dr = \frac{4\pi}{\pi a_0^3} \frac{3! a_0^4}{(2)^4} = \frac{3}{2} a_0. \tag{5.11.38}$$

The average value of $1/r$ is

$$\left\langle \frac{1}{r} \right\rangle = \int \frac{1}{r} |\phi(\vec{r})|^2 d\tau = \frac{1}{(\pi a_0^3)} \int_0^\infty \frac{1}{r} e^{-2r/a_0} r^2 dr \int_0^{2\pi} d\varphi \int_0^\pi \sin\theta d\theta$$

$$= 4\pi \frac{1}{(\pi a_0^3)} \int_0^\infty e^{-2r/a_0} r dr = \frac{4}{a_0^3} \frac{1}{(2/a_0)^2} = \frac{1}{a_0}. \tag{5.11.39}$$

Therefore, the average value of the potential energy of the electron

$$\langle U \rangle = -\frac{e^2}{4\pi\varepsilon_0} \left\langle \frac{1}{r} \right\rangle = -\frac{e^2}{4\pi\varepsilon_0 a_0} = -\frac{e^2}{4\pi\varepsilon_0} \frac{\mu e^2}{4\pi\varepsilon_0 \hbar^2} = 2\left[-\frac{\mu}{2\hbar^2} \left(\frac{e^2}{4\pi\varepsilon_0} \right)^2 \right]. \tag{5.11.40}$$

Recollecting the expression for the ground state energy E_1, we get that $\langle U \rangle = 2E_1$. Now, the total energy E_1 is equal to sum of the average value of the kinetic energy $\langle T \rangle$ and the average value of the potential energy $\langle U \rangle$: $E_1 = \langle T \rangle + \langle U \rangle$. Therefore, we get

$$\langle T \rangle = E_1 - \langle U \rangle = -E_1 = \frac{\mu}{2\hbar^2} \left(\frac{e^2}{4\pi\varepsilon_0} \right)^2. \tag{5.11.41}$$

These results allow us to explain the stability of the hydrogen and hydrogenic atoms in the s state in the following way.

Let us assume that the electron in a hydrogen or hydrogenic atom remains at a distance $r = a_0$ on the average. Let the uncertainty in the momentum of the electron be Δp_r. Then in accordance with the aforementioned result, $\Delta p_r^2/2\mu$ has to be of the order of $\langle T \rangle$, that is,

$$\frac{(\Delta p_r)^2}{2\mu} = \langle T \rangle = \frac{\mu}{2\hbar^2} \left(\frac{e^2}{4\pi\varepsilon_0} \right)^2 = \frac{\hbar^2}{d\mu a_0^2}. \tag{5.11.42}$$

The uncertainty principle, on the other hand, says that

$$\Delta r \Delta p_r \geq \frac{\hbar}{2} \Rightarrow \Delta r \geq \frac{\hbar}{2} \frac{1}{\Delta p_r} = \frac{\hbar}{2} \times \frac{a_0}{\hbar} = \frac{a_0}{2}. \tag{5.11.43}$$

Taking the equality sign in this expression, we conclude that for hydrogen and hydrogenic atoms to be stable in the s state (with angular momentum zero), the uncertainty in position, that is, the radius of the sphere in which the electron is confined, cannot be less than $a_0/2$. In the opposite case, the law of conservation of energy will be violated.

***The general solution for the hydrogen atom*:** Let us go back to (5.11.18),

$$-\frac{\hbar^2}{2\mu} \left[\frac{1}{r^2} \frac{\partial}{\partial r} \left(r^2 \frac{\partial \phi}{\partial r} \right) + \frac{1}{r^2 \sin\theta} \frac{\partial}{\partial \theta} \left(\sin\theta \frac{\partial \phi}{\partial \theta} \right) + \frac{1}{r^2 \sin^2\theta} \frac{\partial^2 \phi}{\partial \varphi^2} \right] - \frac{e^2}{4\pi\varepsilon_0 r} \phi = E\phi, \tag{5.11.44}$$

and try to find its solution in the general case. Since the potential is spherically symmetric, the general solutions (with an arbitrary value of ℓ) are given by

$$\phi(r,\theta,\phi) = R_{n\ell}(r)\, Y_\ell^m(\theta,\varphi), \tag{5.11.45}$$

where $Y_\ell^m(\theta,\varphi)$ are the normalized spherical harmonics and the radial wave function $R_{n\ell}(r)$ satisfies

$$\frac{d}{dr}\left(r^2\frac{dR_{n\ell}(r)}{dr}\right) + \left[\lambda + \frac{2\alpha}{r}\right] r^2 R_{n\ell}(r) = \ell(\ell+1)R_{n\ell}(r), \tag{5.11.46}$$

where

$$\lambda = \frac{2\mu E}{\hbar^2}; \alpha = \frac{\mu e^2}{4\pi\varepsilon_0\hbar^2}. \tag{5.11.47}$$

Or,

$$\frac{d^2R_{n\ell}}{dr^2} + \frac{2}{r}\frac{dR_{n\ell}(r)}{dr} + \left(\lambda + \frac{2\alpha}{r} - \frac{\ell(\ell+1)}{r^2}\right) R_{n\ell}(r) = 0. \tag{5.11.48}$$

As mentioned earlier, we are interested in the bound state solutions for which $E < 0$ and hence, $\lambda < 0$. Since $\sqrt{|\lambda|}$ has dimensions of inverse of length, we introduce $r_0 = 1/\sqrt{-\lambda}$. Then (5.11.48) takes the form:

$$\frac{d^2R_{n\ell}}{dr^2} + \frac{2}{r}\frac{dR_{n\ell}(r)}{dr} + \left(-\frac{1}{r_0^2} + \frac{2\alpha}{r} - \frac{\ell(\ell+1)}{r^2}\right) R_{n\ell}(r) = 0. \tag{5.11.49}$$

Let us introduce the dimensionless independent variable

$$\rho = 2\frac{r}{r_0} = 2r\sqrt{-\lambda}. \tag{5.11.50}$$

Then, we have

$$\frac{d}{dr} = \frac{d\rho}{dr}\frac{d}{d\rho} = \frac{2}{r_0}\frac{d}{d\rho}, \tag{5.11.51}$$

$$\frac{d^2}{dr^2} = \frac{d\rho}{dr}\frac{d}{d\rho}\left(\frac{2}{r_0}\frac{d}{d\rho}\right) = \frac{4}{r_0^2}\frac{d^2}{d\rho^2}. \tag{5.11.52}$$

Using these results in (5.11.49) and multiplying throughout by $r_0^2/4$, we get

$$\frac{d^2R_{n\ell}(\rho)}{d\rho^2} + \frac{2}{\rho}\frac{dR_{n\ell}(\rho)}{d\rho} + \left(-\frac{1}{4} + \frac{\alpha}{\sqrt{-\lambda}}\frac{1}{\rho} - \frac{\ell(\ell+1)}{\rho^2}\right) R_{n\ell}(\rho) = 0. \tag{5.11.53}$$

For $\rho \to \infty$, (5.11.53) reduces to

$$\frac{d^2R_{n\ell}(\rho)}{d\rho^2} - \frac{1}{4}R_{n\ell}(\rho) = 0, \tag{5.11.54}$$

which has simple solutions $R_{n\ell}(r) = \exp(\pm\rho/2)$. Since the solution with positive exponent tends to ∞ as $r \to \infty$, it does not satisfy the standard conditions. Hence, it has to be omitted. Consequently, we look for the solution of (5.11.52) in the following form

$$R_{n\ell}(\rho) = e^{-\rho/2}u_{n\ell}(\rho), \tag{5.11.55}$$

where the function $u_{n\ell}(\rho)$ must obey the boundary conditions. Differentiating $R_{n\ell}(\rho)$ with respect to ρ, we obtain

$$\frac{dR_{n\ell}}{d\rho} = \left(\frac{du_{n\ell}}{d\rho} - \frac{1}{2}u_{n\ell}\right)e^{-\rho/2}, \tag{5.11.56}$$

$$\frac{d^2R_{n\ell}}{d\rho^2} = \left(\frac{d^2u_{n\ell}}{d\rho^2} - \frac{du_{n\ell}}{d\rho} + \frac{1}{4}\right)e^{-\rho/2}. \tag{5.11.57}$$

Equation (5.11.53), along with (5.11.55)–(5.11.57), leads to the following differential equation for the function $u_{n\ell}(\rho)$

$$\frac{d^2u_{n\ell}}{d\rho^2} + \left(\frac{2}{\rho} - 1\right)\frac{du_{n\ell}}{d\rho} + \left[\left(\frac{\alpha}{\sqrt{-\lambda}} - 1\right)\frac{1}{\rho} - \frac{\ell(\ell+1)}{\rho^2}\right]u_{n\ell} = 0. \tag{5.11.58}$$

The form of the equation (5.11.58) suggests that we look for the solution in the form

$$u_{n\ell}(\rho) = \rho^{\gamma}\sum_{j=0}^{\infty} c_j\rho^j, \tag{5.11.59}$$

where γ, $c_1, c_2, c_3, \ldots$ are constants to be determined. Note that, as in the case of a one-dimensional harmonic oscillator, in order to guarantee the boundedness of the solution for $\rho \to 0$, the series solution must start with ρ^{γ} instead of a constant. The value of γ will be determined from the requirement that the function $u_{n\ell}$ is finite everywhere. Differentiating the infinite sum (5.11.58) term by term, we get

$$\frac{du_{n\ell}}{d\rho} = \sum_{j=0}^{\infty} c_j(\gamma+j)\rho^{\gamma+j-1}, \tag{5.11.60}$$

where we have simply rewritten the resulting infinite sum by changing the dummy index of summation from j to $j+1$. Differentiating once again, we obtain

$$\frac{d^2u_{n\ell}}{d\rho^2} = \sum_{j=0}^{\infty} c_j(\gamma+j)(\gamma+j-1)\rho^{\gamma+j-2} \tag{5.11.61}$$

From (5.11.58)–(5.11.61), we arrive at

$$\sum_{j=0}^{\infty} c_j(\gamma+j)(\gamma+j-1)\rho^{\gamma+j-2} + \sum_{j=0}^{\infty} 2c_j(\gamma+j)\rho^{\gamma+j-2} - \sum_{j=0}^{\infty} c_j(\gamma+j)\rho^{\gamma+j-1}$$

$$+\sum_{j=0}^{\infty} c_j\left[\left(\frac{\alpha}{\sqrt{-\lambda}}-1\right)\frac{1}{\rho} - \frac{\ell(\ell+1)}{\rho^2}\right]\rho^{\gamma+j} = 0. \tag{5.11.62}$$

Or,

$$\sum_{j=0}^{\infty} c_j\left[(\gamma+j)(\gamma+j+1)-\ell(\ell+1)\right]\rho^{\gamma+j-2}$$

$$= \sum_{j=0}^{\infty} c_j\left[(\gamma+j+1)-\frac{\alpha}{\sqrt{-\lambda}}\right]\rho^{\gamma+j-1}. \tag{5.11.63}$$

Equation (5.11.63) must hold identically and, hence, the coefficients before identical powers of ρ, on both sides of the equation, must be equal. The lowest order term on the left-hand side contains $\rho^{\gamma-2}$ with the coefficient $(\gamma(\gamma+1)-\ell(\ell+1))c_0$. The lowest term on the left-hand side contains $\rho^{\gamma-1}$. Therefore,

$$(\gamma(\gamma+1)-\ell(\ell+1))c_0 = 0 \Rightarrow (\gamma(\gamma+1)-\ell(\ell+1)) = 0. \tag{5.11.64}$$

Equation (5.11.64) has two possible solutions

$$\gamma = \ell \;\text{ or }\; \gamma = -(\ell+1). \tag{5.11.65}$$

If we take the solution $\gamma = -(\ell+1)$, then the series in (5.11.59) would start with the term $c_0/\rho^{\ell+1}$ that goes to infinity for $\rho \to 0$. Therefore, we omit it and take $\gamma = \ell$. As a result, (5.11.63) reads

$$\sum_{j=0}^{\infty} c_j\left[(\ell+j)(\ell+j+1)-\ell(\ell+1)\right]\rho^{\ell+j-2}$$

$$= \sum_{j=0}^{\infty} c_j\left[(\ell+j+1)-\frac{\alpha}{\sqrt{-\lambda}}\right]\rho^{\ell+j-1}. \tag{5.11.66}$$

Since the coefficient for $j=0$ on the left-hand (LHS) side of (5.11.66) is zero, the series on the LHS starts with $j=1$. If we change the dummy index of summation j to $j+1$, we have

$$\text{LHS} = \sum_{j=0}^{\infty} c_{j+1} \left[(\ell+j+1)(\ell+j+2) - \ell(\ell+1)\right] \rho^{\ell+j-1}. \tag{5.11.67}$$

Consequently Equation (5.11.66) can be written as

$$\sum_{j=0}^{\infty} \left(c_{j+1} \left[(\ell+j+1)(\ell+j+2) - \ell(\ell+1)\right] - c_j \left[(\ell+j+1) - \frac{\alpha}{\sqrt{-\lambda}}\right] \right) \rho^{\ell+j-1} = 0. \tag{5.11.68}$$

From (5.11.68), we get the following recursion relation for the coefficients of the series in (5.11.59):

$$c_{j+1} = \frac{\left[(\ell+j+1) - \frac{\alpha}{\sqrt{-\lambda}}\right]}{(\ell+j+1)(\ell+j+2) - \ell(\ell+1)} c_j. \tag{5.11.69}$$

The recursion relation allows us to calculate all the coefficients of the series in (5.11.59) in terms of one coefficient, say c_0, with which the series starts. This coefficient is determined by the normalization condition. Thus, the series solution of (5.11.58) is given by (5.11.59) with coefficients determined by the recursion relation (5.11.69)) and the normalization of the radial wave function. The series in (5.11.59) is an infinite series and hence, we must check whether its behaviour as $\rho \to \infty$ is consistent with the finiteness of $u(\rho)$ or not. For this, let us look at the ratio c_{j+1}/c_j for large values of j (which obviously corresponds to large values of ρ):

$$\lim_{j\to\infty} \frac{c_{j+1}}{c_j} = \lim_{j\to\infty} \frac{\left[(\ell+j+1) - (\alpha/\sqrt{-\lambda})\right]}{(\ell+j+1)(\ell+j+2) - \ell(\ell+1)} = \frac{1}{j}. \tag{5.11.70}$$

On the other hand, the ratio a_{k+1}/a_k for the series

$$e^{\rho} = \sum_{k=0}^{\infty} \frac{\rho^k}{k!} a_k \tag{5.11.71}$$

is

$$\lim_{k\to\infty} \frac{a_{k+1}}{a_k} = \lim_{k\to\infty} \frac{k!}{(k+1)!} = \frac{1}{k+1} \sim \frac{1}{k}. \tag{5.11.72}$$

Hence, for large values of ρ, the series in (5.11.59) is proportional to e^{ρ}. That means, for $\rho \to \infty$,

$$u(\rho) \sim \rho^{\ell} e^{\rho} \quad \Rightarrow \quad R_{n\ell}(\rho) = e^{-\rho/2} u_{n\ell}(\rho) \sim \rho^{\ell} e^{\rho/2}, \tag{5.11.73}$$

and blows up. Therefore, if we want the required solutions for the radial wave function to satisfy the standard conditions, the series must be converted into a polynomial. That is, it must truncate at some appropriate term. This is possible only if, for some value $j = j_{\max}$, the numerator in (26) becomes zero, that is,

$$(n_r + \ell + 1) - \frac{\alpha}{\sqrt{-\lambda}} = 0, \tag{5.11.74}$$

where $n_r = j_{\max}$ is the maximum value of j for which $c_{n_r+1} = 0$. The number n_r is called the radial quantum number. Introducing a new quantum number, n, by the relation

$$n = n_r + \ell + 1, \tag{5.11.75}$$

we get that

$$\frac{\alpha}{\sqrt{-\lambda}} = n. \tag{5.11.76}$$

n is called the ***principal quantum number***. It allows us to write the recursion relation for the coefficients of the polynomial as

$$c_{j+1} = \frac{[(\ell + j + 1) - n]}{(\ell + j + 1)(\ell + j + 2) - \ell(\ell + 1)} c_j. \tag{5.11.77}$$

Further, we have

$$\sqrt{-\lambda} = \frac{\alpha}{n} = \frac{\mu e^2}{4\pi\varepsilon_0 n \hbar^2}. \tag{5.11.78}$$

Or,

$$-\frac{2\mu E}{\hbar^2} = \left(\frac{\mu e^2}{4\pi\varepsilon_0 n \hbar^2}\right)^2. \tag{5.11.79}$$

Therefore, the possible values of energy are

$$E_n = -\frac{\mu}{2\hbar^2}\left(\frac{e^2}{4\pi\varepsilon_0}\right)^2 \frac{1}{n^2}. \tag{5.11.80}$$

We see that the energy depends only on the *principal quantum number n*. Since the minimum value of ℓ is 0, it follows from (5.11.73) that the maximum value of ℓ is obtained when $n_r = 0$, i.e., $\ell_{\max} = n - 1$. Therefore, the possible values of ℓ, for a given n, are: $\ell = 0, 1, 2, 3, ..., n-1$.

We thus see that the energy states of a hydrogen atom can be characterized by three quantum numbers n_r, ℓ, and m. However, since n_r is determined by n and ℓ ($n_r = n - \ell - 1$), the energy states of a hydrogen atom are usually described by the triplet of quantum numbers n, ℓ, and m, as in other cases considered earlier.

Note that all the energy states of hydrogen, except the ground state with $n = 1$ and $\ell = 0$, are degenerate. The degree of degeneracy is determined as follows. For a given value of n, there are n possible values of ℓ $(0, 1, 2, 3, ..., n-1)$ and for every ℓ there are $2\ell + 1$ values of m from $-\ell$ to $+\ell$. Therefore, the degeneracy g is given by

$$g = \sum_{\ell=0}^{n-1} (2\ell + 1) = 1 + 3 + 5 + ... + (2n - 1). \tag{5.11.81}$$

This series is an arithmetic series with n terms and the common difference $d = 2$. Hence, the sum is given by

$$g = \sum_{\ell=0}^{n-1} (2\ell + 1) = \frac{n}{2}[2 \times 1 + (n-1) \times 2] = n^2. \tag{5.11.82}$$

In atomic physics, stationary states with different quantum numbers n are denoted by specific symbols. A symbol has the principal quantum number n as the coefficient before a letter which corresponds to different values of ℓ. For instance, the state with $n = 1$ and $\ell = 0$ is written as s state. For $n = 2$, the states are written as $2s$ and $2p$; for $n = 3$, they are written as $3s$, $3p$ and $3d$, and so on and so forth.

Let us write down the full form of the stationary state wave functions for the hydrogen atom. Note that

$$\sqrt{-\lambda} = \sqrt{-\frac{2\mu E_n}{\hbar^2}} = \left(\frac{1}{n^2}\left(\frac{\mu e^2}{4\pi\varepsilon_0 \hbar^2}\right)^2\right)^{1/2} = \frac{1}{na_0}, \tag{5.11.83}$$

where a_0 is the Bohr radius. Hence, $\rho = 2r/na_0$. Consequently, the stationary state energies and the corresponding wave functions of the hydrogen atom are, respectively,

$$E_n = -\frac{\mu}{2\hbar^2}\left(\frac{e^2}{4\pi\varepsilon_0}\right)^2 \frac{1}{n^2}, (n = 1, 2, 3, ...), \tag{5.11.84}$$

$$\psi_{n\ell m}(r, \theta, \varphi) = \phi_{n\ell m}(r, \theta, \varphi) e^{-\frac{i}{\hbar}E_n t} = R_{n\ell}(r) Y_\ell^m(\theta, \varphi) e^{-\frac{i}{\hbar}E_n t}, \tag{5.11.85}$$

$$R_{n\ell}(r) = e^{-r/na_0}\left(\rho^\ell \sum_{j=0}^{n_r} c_j \rho^j\right) = e^{-r/na_0}\left(\frac{2r}{na_0}\right)^\ell \sum_{j=0}^{n_r} c_j \left(\frac{2r}{na_0}\right)^j, \tag{5.11.86}$$

$$Y_\ell^m(\theta,\varphi) = (-1)^m \sqrt{\frac{(2\ell+1)(\ell-m)!}{4\pi(\ell+m)!}} P_\ell^m(\cos\theta)e^{im\varphi}, (m \geq 0), \tag{5.11.87}$$

$$c_{j+1} = \frac{[(\ell+j+1)-n]}{(\ell+j+1)(\ell+j+2)-\ell(\ell+1)} c_j, \tag{5.11.88}$$

where the associated Legendre polynomials, $P_\ell^m(x)$ and the Legendre polynomials, $P_\ell(x)$, of degree ℓ, are given by

$$P_\ell^m(x) = (1-x^2)^{|m|/2} \frac{\partial^{|m|}}{\partial x^{|m|}} P_\ell(x), x = \cos\theta, \tag{5.11.89}$$

$$P_\ell(x) = \frac{1}{2^\ell \ell!} \frac{\partial^\ell}{\partial x^\ell}(x^2-1)^\ell. \tag{5.11.90}$$

The constant c_0 is determined in each case from the normalization of the radial wave function. Note that the resulting wave functions $\psi_{n\ell m}(r,\theta,\varphi)$ are mutually orthogonal

$$\int \psi^*_{n\ell m}(r,\theta,\varphi)\psi_{n'\ell' m'}(r,\theta,\varphi)r^2 \sin\theta dr d\theta d\varphi = \delta_{nn'}\delta_{ll'}\delta_{mm'}, \tag{5.11.91}$$

which follows from the orthogonality of the spherical harmonics and from the fact that, for $n \neq n'$, they are eigenfunctions of the Hamiltonian with distinct eigenvalues.

To illustrate the procedure of calculations, let us now determine the analytical expressions for the wave functions of the ground state and the first excited state of the hydrogen atom.

Ground state: For the ground state of hydrogen, $n = 1$, $\ell = 0$ and $m = 0$. Therefore, the wave function is given by

$$\phi_{100}(r,\theta,\varphi) = R_{10}(r)Y_0^0(\theta,\varphi). \tag{5.11.92}$$

Now, the radial quantum number $n_r = n - \ell - 1 = 0$ and the recursion relation (5.11.77) gives $c_1 = 0$. Hence, the radial wave function is given by

$$R_{10}(r) = c_0 e^{-r/a_0}. \tag{5.11.93}$$

Normalizing the radial wave function, we get

$$1 = c_0^2 \int_0^\infty r^2 e^{-2r/a_0} dr = \frac{c_0^2 a_0^3}{4}, \Rightarrow c_0 = \frac{2}{\sqrt{a_0^3}}. \tag{5.11.94}$$

Therefore,

$$R_{10}(r) = \frac{2}{\sqrt{a_0^3}} e^{-r/a_0}. \tag{5.11.95}$$

Since $Y_0^0(\theta, \varphi) = 1/\sqrt{4\pi}$, the ground state wave function is

$$\phi_{100}(r, \theta, \varphi) = \frac{2}{\sqrt{a_0^3}} \frac{1}{\sqrt{4\pi}} e^{-r/a_0} = \frac{1}{\sqrt{\pi a_0^3}} e^{-r/a_0}. \tag{5.11.96}$$

First excited state: Here, $n = 2$ and ℓ can take two values: 0 and 1. For $\ell = 0$, we have $m = 0$. This state is described by the wave function

$$\phi_{200}(r, \theta, \varphi) = R_{20}(r) Y_0^0(\theta, \varphi), \tag{5.11.97}$$

and is called the $2s$ state. For $\ell = 1$, m can take three values $-1, 0$ and $+1$. This state is called the $2p$ state and it is 3-fold degenerate. The corresponding wave functions are

$$\phi_{21-1}(r, \theta, \varphi) = R_{21}(r) Y_1^{-1}(\theta, \varphi), (n = 2, \ell = 1, m = -1), \tag{5.11.98}$$

$$\phi_{210}(r, \theta, \varphi) = R_{21}(r) Y_1^0(\theta, \varphi), (n = 2, \ell = 1, m = 0), \tag{5.11.99}$$

$$\phi_{211}(r, \theta, \varphi) = R_{21}(r) Y_1^{+1}(\theta, \varphi), (n = 2, \ell = 1, m = +1). \tag{5.11.100}$$

Consider first the case: $n = 2$ and $\ell = 0$. We have $n_r = 1$. The recursion relation (5.11.77) now gives $c_1 = -c_0/2$ and $c_2 = 0$. Therefore,

$$R_{20}(r) = \left(c_0 + c_1 \frac{r}{a_0}\right) e^{-r/2a_0} = c_0 \left(1 - \frac{r}{2a_0}\right) e^{-r/2a_0}. \tag{5.11.101}$$

The normalization for $R_{20}(r)$ reads

$$1 = c_0^2 \int_0^\infty \left[1 - \frac{r}{a_0} + \frac{r^2}{4a_0^2}\right] r^2 e^{-r/a_0} dr = c_0^2 a_0^3 \left[2 - 6 + \frac{24}{4}\right]$$

$$= 2c_0^2 a_0^3, \Rightarrow c_0 = \frac{1}{\sqrt{2a_0^3}}. \tag{5.11.102}$$

As a result,

$$R_{20}(r) = \frac{1}{\sqrt{2a_0^3}} \left(1 - \frac{r}{2a_0}\right) e^{-r/2a_0}. \tag{5.11.103}$$

Once again, $Y_0^0(\theta,\varphi) = 1/\sqrt{4\pi}$, and hence, the $2s$ state wave function is given by

$$\phi_{200}(r,\theta,\varphi) = \frac{1}{2\sqrt{a_0^3}} \frac{1}{\sqrt{4\pi}} \left(1 - \frac{r}{2a_0}\right) e^{-r/2a_0}$$

$$= \frac{1}{\sqrt{8\pi a_0^3}} \left(1 - \frac{r}{2a_0}\right) e^{-r/2a_0}. \tag{5.11.104}$$

For $n = 2$ and $\ell = 1$, we have $n_r = 0$. The recursion relation (5.11.77) now gives $c_1 = 0$. Hence,

$$R_{21}(r) = c_0 \frac{r}{a_0} e^{-r/2a_0}. \tag{5.11.105}$$

Normalization gives

$$c_0 = \frac{1}{\sqrt{24a_0^3}}. \tag{5.11.106}$$

Hence, we get

$$R_{21}(r) = \frac{1}{\sqrt{24a_0^3}} \frac{r}{a_0} e^{-r/2a_0}. \tag{5.11.107}$$

Calculating the required spherical harmonics from (5.6.11), (5.6.12) and (5.6.14), we have

$$Y_1^{-1}(\theta,\varphi) = \sqrt{\frac{3}{8\pi}} \sin\theta e^{-i\varphi}, \tag{5.11.108}$$

$$Y_1^0(\theta,\varphi) = \sqrt{\frac{3}{4\pi}} \cos\theta, \tag{5.11.109}$$

$$Y_1^1(\theta,\varphi) = -\sqrt{\frac{3}{8\pi}} \sin\theta e^{+i\varphi}. \tag{5.11.110}$$

Consequently, the $2p$ state wave functions are given by

$$\phi_{21-1}(r,\theta,\varphi) = \frac{1}{8\sqrt{\pi a_0^3}} \frac{r}{a_0} e^{-r/2a_0} \sin\theta e^{-i\varphi}, \tag{5.11.111}$$

$$\phi_{210}(r,\theta,\varphi) = \frac{1}{4\sqrt{2\pi a_0^3}} \frac{r}{a_0} e^{-r/2a_0} \cos\theta, \tag{5.11.112}$$

$$\phi_{211}(r,\theta,\varphi) = -\frac{1}{8\sqrt{\pi a_0^3}}\frac{r}{a_0}e^{-r/2a_0}\sin\theta e^{i\varphi}. \tag{5.11.113}$$

For convenience in calculations, the first few radial wave functions, $R_{n\ell}(r)$, are presented in Table 1.

Table 5.1 The first few radial wave functions of hydrogen.

$$R_{10}(r) = \frac{2}{\sqrt{a_0^3}}e^{-r/a_0}$$

$$R_{20}(r) = \frac{1}{\sqrt{2a_0^3}}\left(1-\frac{r}{2a_0}\right)e^{-r/2a_0}$$

$$R_{21}(r) = \frac{1}{\sqrt{24a_0^3}}\frac{r}{a_0}e^{-r/2a_0}$$

$$R_{30}(r) = \frac{2}{\sqrt{27a_0^3}}\left[1-\frac{2r}{3a_0}+\frac{2}{27}\left(\frac{r}{a_0}\right)^2\right]e^{-r/3a_0}$$

$$R_{31}(r) = \frac{8}{27\sqrt{6a_0^3}}\frac{r}{a_0}\left(1-\frac{r}{6a_0}\right)e^{-r/3a_0}$$

$$R_{32}(r) = \frac{4}{81\sqrt{30a_0^3}}\left(\frac{r}{a_0}\right)^2 e^{-r/3a_0}$$

$$R_{40}(r) = \frac{1}{4\sqrt{a_0^3}}\left[1-\frac{3r}{4a_0}+\frac{1}{8}\left(\frac{r}{a_0}\right)^2-\frac{1}{192}\left(\frac{r}{a_0}\right)^3\right]e^{-r/4a_0}$$

$$R_{41}(r) = \frac{\sqrt{5}}{16\sqrt{3a_0^3}}\frac{r}{a_0}\left[1-\frac{r}{4a_0}+\frac{1}{80}\left(\frac{r}{a_0}\right)^2\right]e^{-r/4a_0}$$

$$R_{42}(r) = \frac{1}{64\sqrt{5a_0^3}}\left(\frac{r}{a_0}\right)^2\left[1-\frac{r}{12a_0}\right]e^{-r/4a_0}$$

$$R_{43}(r) = \frac{1}{768\sqrt{35a_0^3}}\left(\frac{r}{a_0}\right)^3 e^{-r/4a_0}$$

Laguerre polynomials and the radial wave function: The polynomials

$$\sum_{j=0}^{n_r} c_j\rho^j = \sum_{j=0}^{n_r} c_j\left(\frac{2r}{na_0}\right)^j, \tag{5.11.114}$$

in (5.11.86), whose coefficients are defined by the recursion relation (5.11.88), are known as *associated Laguerre polynomials* in mathematical physics. In our case, barring normalization, we can write

$$\sum_{j=0}^{n_r} c_j \rho^j = L_{n-\ell-1}^{2\ell+1}(\rho), \tag{5.11.115}$$

where $L_{q-p}^{p}(x)$ are expressed in terms of the qth Laguerre polynomials, $L_q(x)$, as

$$L_{q-p}^{p}(x) = (-1)^p \frac{d^p}{d\rho^p} L_q(x). \tag{5.11.116}$$

The qth Laguerre polynomial is given by

$$L_q(x) = e^x \frac{d^q}{dx^q}(x^q e^{-x}). \tag{5.11.117}$$

The first few Laguerre polynomials, $L_q(x)$, and the associated Laguerre polynomials, $L_{q-p}^{p}(x)$, are listed in Table 2 and Table 3, respectively.

Table 5.2 The first few Laguerre polynomials, $L_q(x)$.

$L_0(x) = 1$
$L_1(x) = -x + 1$
$L_2(x) = x^2 - 4x + 2$
$L_3(x) = -x^3 + 9x^2 - 18x + 6$
$L_4(x) = x^4 - 16x^3 + 72x^2 - 96x + 24$
$L_5(x) = -x^5 + 25x^4 - 200x^3 + 600x^2 - 600x + 120$
$L_6(x) = x^6 - 36x^5 + 450x^4 - 2400x^3 + 5400x^2 - 4320x + 720$

Table 5.3 The first few associated Laguerre polynomials, $L_{q-p}^{p}(x)$.

$$L_0^0 = 1,\ L_1^0 = -x+1,\ L_2^0 = x^2 - 4x + 2,$$

$$L_0^1 = 1,\ L_1^1 = -2x+4,\ L_2^1 = 3x^2 - 18x + 18$$

$$L_0^2 = 2,\ L_1^2 = -6x+18,\ L_2^2 = 12x^2 - 96x + 144$$

$$L_0^3 = 6,\ L_1^3 = -24x+96,\ L_2^3 = 60x^2 - 600x + 1200,$$

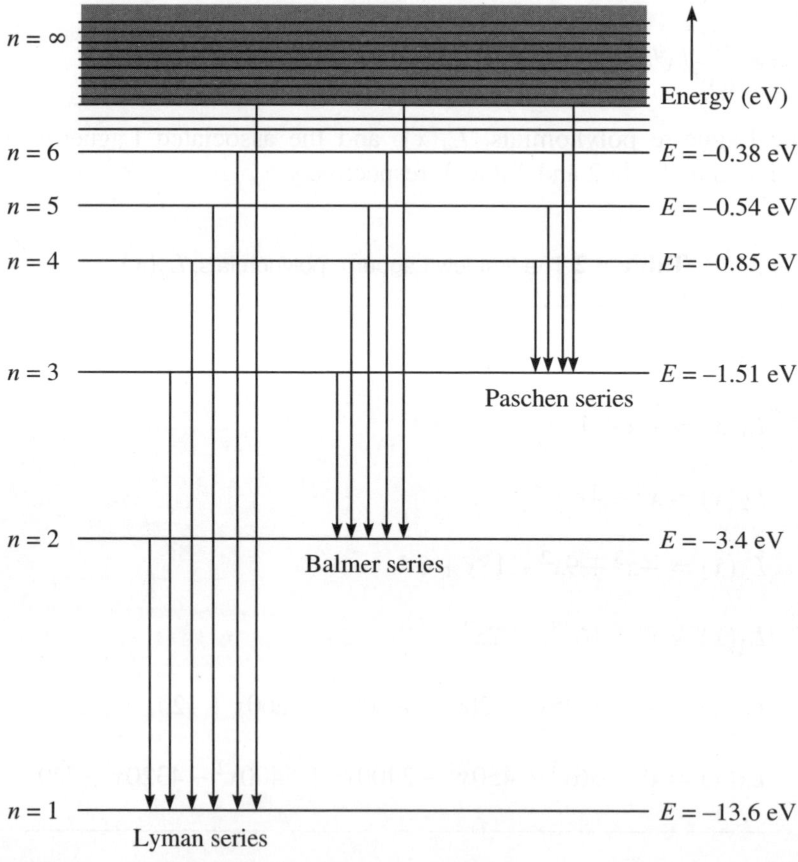

Figure 5.2 Energy levels and transitions between them for the hydrogen atom.

Using these Laguerre polynomials, the normalized wave function of the hydrogen atom can be written as:

$$\psi_{n\ell m}(r,\theta,\varphi) = \sqrt{\left(\frac{2}{na_0}\right)^3 \frac{(n-\ell-1)!}{2n\left[(n+\ell)!\right]^3}} e^{-r/na_0} \left(\frac{2r}{na_0}\right)^{\ell} \times \left[L_{n-\ell-1}^{2\ell+1}\left(\frac{2r}{na_0}\right)\right] Y_\ell^m(\theta,\varphi). \tag{5.11.118}$$

The spectrum of hydrogen: In principle, if the hydrogen atom is in one of the stationary states, it will reside there for ever. However, when perturbed (in fact, perturbations are always present), it may make a transition to another stationary state either by absorbing energy from the perturbation or by giving off energy under the action of the applied perturbation. In the first case, it makes a transition to an energetically higher state by absorbing electromagnetic radiation (usually), while in the latter case, it slides down to an energetically lower state by emitting electromagnetic radiation. The energy of the radiation, E_γ, is equal to the difference in energy of the stationary states involved in the transition:

$$E_\gamma = E_i - E_f = E_1\left(\frac{1}{n_i^2} - \frac{1}{n_f^2}\right), \tag{5.11.119}$$

where E_f and E_i are the energy of the final and the initial stationary states, respectively, and

$$E_1 = -\frac{\mu}{2\hbar^2}\left(\frac{e^2}{4\pi\varepsilon_0}\right)^2 \tag{5.11.120}$$

is the energy of the ground state ($n = 1$). The energy of a photon is proportional to the frequency, ν, of the emitted or absorbed radiation according to the formula $E_\gamma = h\nu$, where h is the Planck's constant. Also, the wavelength, λ, of the emitted or absorbed radiation is given by$\lambda = c/\nu$, where c is the speed of light in vacuum. Therefore,

$$\frac{1}{\lambda} = \frac{\mu}{4\pi\hbar^3 c}\left(\frac{e^2}{4\pi\varepsilon_0}\right)^2\left(\frac{1}{n_f^2} - \frac{1}{n_i^2}\right) = \mathscr{R}\left(\frac{1}{n_f^2} - \frac{1}{n_i^2}\right), \tag{5.11.121}$$

where $\mathscr{R} = 1.097 \times 10^7 \ (1/\mathrm{m})$ is the Rydberg constant. Equation (5.11.121) represents the well-known Rydberg formula for the spectrum of hydrogen.

The energy levels and the transitions between them have been depicted in Figure 5.2. The collection of spectral lines corresponding to the transitions from the higher energy states to the lower ones are named after the scientists who discovered them experimentally. The spectral lines corresponding to transitions to the ground state, $n_f = 1$, fall in the *ultraviolet* region of the electromagnetic spectrum and constitute the *Lyman series*. The spectral lines corresponding to transitions to the first excited state, $n_f = 2$, fall in the *visible* region of the electromagnetic spectrum and constitute the *Balmer series*, while the series consisting of spectral lines related to transitions to the 2*nd* excited state, with $n_f = 3$, falls in the *infrared* region and is known as *Paschen series*; and so on and so forth.

Example 5.11.2: At $t = 0$, the wave function of a hydrogen atom is given by

$$\psi(r,\theta,\varphi) = \frac{1}{\sqrt{2}}\phi_{300}(r,\theta,\varphi) + \frac{1}{\sqrt{3}}\phi_{311}(r,\theta,\varphi) + \frac{1}{\sqrt{6}}\phi_{322}(r,\theta,\varphi).$$

(a) What is the wave function at any $t > 0$? (b) If a measurement of energy is carried out in this state, what values would result and with what probabilities?

Solution: (a) Since $\hat{H}\phi_{n\ell m} = E_n\phi_{n\ell m}$,

$$\hat{H}\psi(r,\theta,\varphi,0) = E_3\psi(r,\theta,\varphi,0), \tag{5.11.122}$$

and the wave function at any $t > 0$ would be

$$\psi(r,\theta,\varphi,t) = \left[\frac{1}{\sqrt{2}}\phi_{300}(r,\theta,\varphi) + \frac{1}{\sqrt{3}}\phi_{311}(r,\theta,\varphi) + \frac{1}{\sqrt{6}}\phi_{322}(r,\theta,\varphi)\right] e^{-\frac{i}{\hbar}E_3 t}, \tag{5.11.123}$$

where $E_3 = E_1/9 = -13.6/9$ eV.

(b) Since the wave function is normalized: $\langle \psi(r,\theta,\varphi,0)\,|\,\psi(r,\theta,\varphi,0)\rangle = 1$, and it is an eigenfunction of the Hamiltonian, the measurement of energy will give E_3 with probability 1.

Example 5.11.3: Suppose we carry out the following transformation of the independent variable, r, and the radial wave function $R_{n\ell}$:

$$r = \frac{\lambda}{2}\rho^2, \quad R_{n\ell} = \frac{\chi(\rho)}{\rho},$$

in the radial equation for hydrogen, where λ is a constant. (a) Show that $\chi(\rho)$ is a solution of the radial equation of a two-dimensional harmonic oscillator with frequency $\omega = \sqrt{-2\lambda^2 E/\mu}$ and energy $2e^2\lambda/(4\pi\varepsilon_0)$, where E is the energy of the hydrogen atom. (b) Using the expression for the energy eigenvalues of the two-dimensional harmonic oscillator and comparing both the radial equations (for the hydrogen atom and for the two-dimensional oscillator) term by term to extract the correspondence between the parameters, determine the energy spectrum of the hydrogen atom. (c) Also, explicitly construct the normalized ground state wave function of the hydrogen atom.

Solution: The differential equation for the radial wave function $R_{n\ell}(r)$ of the hydrogen atom reads

$$\frac{d}{dr}\left(r^2\frac{dR_{n\ell}(r)}{dr}\right)+\left[\frac{2\mu E}{\hbar^2}r^2+\frac{2\mu e^2}{4\pi\varepsilon_0\hbar^2}r\right]R_{n\ell}(r)=\ell(\ell+1)R_{n\ell}(r). \tag{5.11.124}$$

The transformation of the independent variable gives

$$r=\frac{\lambda\rho^2}{2}\Rightarrow\frac{d}{dr}=\frac{1}{\lambda\rho}\frac{d}{d\rho}. \tag{5.11.125}$$

As a result, (5.11.122) takes the form

$$\left[\frac{1}{4\rho}\frac{d}{d\rho}\left(\rho^3\frac{d}{d\rho}\right)+\frac{\mu e^2\lambda}{4\pi\varepsilon_0\hbar^2}\rho^2+\frac{\mu E\lambda^2}{2\hbar^2}\rho^4-\ell(\ell+1)\right]\frac{\chi(\rho)}{\rho}=0. \tag{5.11.126}$$

It can be further simplified as

$$\begin{aligned}
&\frac{1}{4\rho}\left[\rho^3\left(-\frac{\chi}{\rho^2}+\frac{1}{\rho}\frac{d\chi}{d\rho}\right)\right]+\left[\frac{\mu e^2\lambda}{4\pi\varepsilon_0\hbar^2}\rho+\frac{\mu E\lambda^2}{2\hbar^2}\rho^3-\frac{\ell(\ell+1)}{\rho}\right]\chi(\rho)\\
&=\frac{1}{4}\left[\rho\frac{d^2\chi}{d\rho^2}+\frac{d\chi}{d\rho}-\frac{1}{\rho}\chi\right]+\left[\frac{\mu e^2\lambda}{4\pi\varepsilon_0\hbar^2}\rho+\frac{\mu E\lambda^2}{2\hbar^2}\rho^3-\frac{\ell(\ell+1)}{\rho}\right]\chi(\rho)\\
&=\frac{d^2\chi}{d\rho^2}+\frac{1}{\rho}\frac{d\chi}{d\rho}-\frac{1}{\rho^2}\chi+\left[\frac{\mu e^2\lambda}{\pi\varepsilon_0\hbar^2}+\frac{2\mu E\lambda^2}{\hbar^2}\rho^2-\frac{4\ell(\ell+1)}{\rho^2}\right]\chi(\rho)\\
&=\frac{d^2\chi}{d\rho^2}+\frac{1}{\rho}\frac{d\chi}{d\rho}+\left[\frac{\mu e^2\lambda}{\pi\varepsilon_0\hbar^2}+\frac{2\mu E\lambda^2}{\hbar^2}\rho^2\right]\chi-\left[\frac{1}{\rho^2}+\frac{4\ell(\ell+1)}{\rho^2}\right]\chi=0.
\end{aligned} \tag{5.11.127}$$

Finally, we obtain

$$\frac{d^2\chi}{d\rho^2}+\frac{1}{\rho}\frac{d\chi}{d\rho}-\left[\frac{(2\ell+1)^2}{\rho^2}-\frac{\mu e^2\lambda}{\pi\varepsilon_0\hbar^2}-\frac{2\mu E\lambda^2}{\hbar^2}\rho^2\right]\chi=0. \tag{5.11.128}$$

Let us now compare (5.11.126) with the equation satisfied by the radial wave function $R(\rho)$ of a two-dimensional harmonic oscillator:

$$\frac{d^2R(\rho)}{d\rho^2}+\frac{1}{\rho}\frac{dR(\rho)}{d\rho}-\left[\frac{m_\ell^2}{\rho^2}+\frac{m^2\omega^2}{\hbar^2}\rho^2-\frac{2\mu E}{\hbar^2}\right]R=0, \tag{5.11.129}$$

where $m_\ell=0,\pm1,\pm2,\pm3,\ldots$ is the magnetic quantum number and m is the mass. The correspondence among the parameters are given by

$$m_\ell = (2\ell+1), \qquad \omega = \sqrt{\frac{-2\lambda^2 E}{\mu}}, \qquad E' = \frac{e^2\lambda}{2\pi\varepsilon_0}, \tag{5.11.130}$$

where E is the energy of the hydrogen atom and E' is the energy of the oscillator. With this correspondence, $\chi(\rho)$ is indeed satisfies the radial equation of a two-dimensional harmonic oscillator with frequency $\omega = \sqrt{-2\lambda^2 E/\mu}$ and energy $2e^2\lambda/(4\pi\varepsilon_0)$, where E is the energy of the hydrogen atom.

(b) The energy levels of the two-dimensional harmonic oscillator is given by

$$E' = \hbar\omega(n'+1), \tag{5.11.131}$$

where

$$n' = |m_\ell| + 2n_\rho = 2(\ell+n_\rho)+1 = 2n-1, \tag{5.11.132}$$

n_ρ being the radial quantum number for the two-dimensional oscillator and $n = (\ell + n_\rho + 1)$ is the *principal quantum number* as introduced in the case of the hydrogen atom. Hence,

$$E' = \hbar\omega[(2n-1)+1] = 2\hbar\omega n. \tag{5.11.133}$$

As a result, we have

$$\frac{e^2\lambda}{2\pi\varepsilon_0} = 2n\hbar\sqrt{-\frac{2\lambda^2 E_n}{\mu}}. \tag{5.11.134}$$

The last equation gives the energy levels of the hydrogen atom as

$$E_n = -\frac{\mu}{2\hbar^2}\left(\frac{e^2}{4\pi\varepsilon_0}\right)^2 \frac{1}{n^2}, \qquad n = 1,2,3,\ldots \tag{5.11.135}$$

Equation (5.11.133) gives the same energy spectrum as obtained earlier(see (5.11.84)).

Degeneracy: Given a fixed value of $n \geq 1$, the quantum number ℓ will take $0,1,2,3,\ldots,n-1$ values, while the radial quantum number n_ρ will take values $n-1, n-2, n-3, \ldots, 0$. Therefore, for a given $n \geq 1$, the total number of energy states that correspond to the same energy E_n is

$$\sum_{\ell=1}^{(n-1)} (2\ell+1) = n^2. \tag{5.11.136}$$

(c) Note that the ground state energy of the hydrogen atom has $n = 1$ and $\ell = 0$. This means that it corresponds to the $m_\ell = 1$ and $n_\rho = 0$ state of the $2D$ harmonic oscillator.

The corresponding ground state wave function of the 2D oscillator in polar coordinates is given by

$$\chi(\rho) = N\rho\, e^{-\frac{m\omega}{2\hbar}\rho^2}, \tag{5.11.137}$$

where N is the normalization constant. As a consequence the ground state radial wave function of the hydrogen atom will be given by

$$R_{10}(\rho) = \frac{\chi(\rho)}{\rho} = N\, e^{-\frac{m\omega}{2\hbar}\rho^2}. \tag{5.11.138}$$

If we replace m by the reduced mass μ and replace E by the expression for the ground state energy of the hydrogen atom, we have

$$\omega = \sqrt{\frac{2\lambda^2}{\mu}\left(\frac{e^2}{4\pi\varepsilon_0}\right)^2 \frac{\mu}{2\hbar^2}} = \frac{e^2\lambda}{4\pi\varepsilon_0\hbar}, \tag{5.11.139}$$

Therefore, we get

$$\frac{\mu\omega}{2\hbar} = \frac{\lambda\mu e^2}{8\pi\varepsilon_0\hbar^2} = \frac{\lambda}{2a_0}, \tag{5.11.140}$$

where a_0 is the Bohr radius. The ground state radial wave function of the hydrogen atom can now be written as

$$R_{10}(r) = N\, e^{-\frac{\lambda}{2a_0}\rho^2} = N\, e^{-\frac{r}{a_0}}. \tag{5.11.141}$$

The normalization condition for this wave function is

$$1 = N^2 \int_0^\infty r^2\, e^{-2\frac{r}{a_0}}\, dr = \frac{a_0^3}{4} N^2 \tag{5.11.142}$$

The normalization constant is thus given by

$$N = \frac{2}{\sqrt{a_0^3}}. \tag{5.11.143}$$

The normalized ground state radial wave function of the hydrogen atom is

$$R_{10}(r) = \frac{2}{\sqrt{a_0^3}}\, e^{-\frac{r}{a_0}}. \tag{5.11.144}$$

5.12 The Isotropic Harmonic Oscillator in Spherical Coordinates

Consider a particle of effective mass μ moving in isotropic harmonic oscillator potential

$$V(r) = \frac{1}{2}\mu\,\omega^2 r^2,$$

where μ is the mass, ω is the angular frequency of the oscillator and r is the radial distance from the origin. Let us find the energy levels of the particle and determine the full stationary state wave functions.

We start with the Schrödinger equation in spherical coordinates

$$-\frac{\hbar^2}{2\mu}\left[\frac{1}{r^2}\frac{\partial}{\partial r}\left(r^2\frac{\partial\phi}{\partial r}\right)+\frac{1}{r^2\sin\theta}\frac{\partial}{\partial\theta}\left(\sin\theta\frac{\partial\phi}{\partial\theta}\right)+\frac{1}{r^2\sin^2\theta}\frac{\partial^2\phi}{\partial\varphi^2}\right]+\frac{1}{2}\mu\,\omega^2 r^2\phi = E\phi. \tag{5.12.1}$$

and look for the solution in the form

$$\phi(r,\theta,\phi) = R(r)Y(\theta,\varphi). \tag{5.12.2}$$

The angular solutions remain unchanged and are still given by the spherical harmonics $Y_{\ell,m}(\theta,\varphi)$, where ℓ is the orbital quantum number and m is the corresponding magnetic quantum number. Hence, we concentrate on solving the radial equation

$$\left[\frac{1}{r^2}\frac{d}{dr}\left(r^2\frac{d}{dr}\right)-\frac{2\mu}{\hbar^2}\left(E-\frac{1}{2}\mu\,\omega^2 r^2-\frac{\hbar^2\ell(\ell+1)}{2\mu r^2}\right)\right]R(r) = 0. \tag{5.12.3}$$

Substituting once again $R(r) = u(r)/r$, we arrive at

$$\frac{d^2u}{dr^2}+\frac{2\mu}{\hbar^2}\left[E-\frac{1}{2}\mu\,\omega^2 r^2-\frac{\hbar^2\ell(\ell+1)}{2\mu r^2}\right]u(r) = 0. \tag{5.12.4}$$

To determine the form of solution satisfying the standard conditions, we first examine the form of possible solutions for the limiting cases of $r \to 0$ and $r \to \infty$.

For $r \to 0$, neglecting $(2\mu E/\hbar^2)u(r)$ in comparison with the last term in (5.12.4), we get

$$\frac{d^2u}{dr^2}-\frac{\ell(\ell+1)}{r^2}u(r) = 0. \tag{5.12.5}$$

Let us look for $u(r)$ in the form r^s. We then obtain from (5.12.5)

$$s(s-1)-\ell(\ell+1) = 0, \tag{5.12.6}$$

which has two solutions $s = -\ell$ and $s = \ell + 1$. Since $u \sim r^{-\ell}$ blows up at $r = 0$, it is excluded due to the standard conditions, we conclude that in the vicinity of $r = 0$, we should have $u \sim r^{\ell+1}$.

For $r \to \infty$, the radial equation (5.12.4) reduces to

$$\frac{d^2u}{dr^2} + \left(\frac{2\mu E}{\hbar^2} - \frac{\mu^2\omega^2}{\hbar^2}r^2\right) u(r) = 0. \tag{5.12.7}$$

Recalling our experience with the Coulomb potential problem, we see that the form of this equation suggests

$$u(r) \sim Ae^{-\alpha r^2} v(r), \tag{5.12.8}$$

where $v(r)$ is some polynomial in r. Let us try the simplest $v(r) = 1$. Then $u(r) = Ae^{-\alpha r^2}$. Then we obtain from (5.12.7)

$$4\alpha^2 r^2 - 2\alpha + \frac{2\mu E}{\hbar^2} - \frac{\mu^2\omega^2}{\hbar^2}r^2 = 0. \tag{5.12.9}$$

From the above equation we obtain

$$\alpha = \frac{\mu\omega}{2\hbar}, \quad and \quad E = \frac{\hbar\omega}{2}. \tag{5.12.10}$$

Next, we put $v(r) = r$ and get that

$$\alpha = \frac{\mu\omega}{2\hbar}, \quad and \quad E = \frac{3\hbar\omega}{2}. \tag{5.12.11}$$

Similarly, for $v(r) = r^2$, we obtain $\alpha = \mu\omega/2\hbar$ and $E = 5\hbar\omega/2$, and so on and so forth. Taking into account all this, we come to the conclusion that we must look for the solution to the radial equation (5.12.4) in the form

$$u(r) = r^{\ell+1} e^{-\frac{\mu\omega}{2\hbar}r^2} v(r). \tag{5.12.12}$$

The above form of the solution tells us that we should change to the new variable

$$\rho = \sqrt{\frac{\mu\omega}{\hbar}}r. \tag{5.12.13}$$

In the new variable ρ, (5.12.4) reduces to

$$\frac{d^2u}{dr^2} + \left(\lambda - \rho^2 - \frac{\ell(\ell+1)}{\rho^2}\right) u = 0, \tag{5.12.14}$$

where $u = u(\rho)$ and $\lambda = 2E/\hbar\omega$. Consequently, we look for the solution of the equation (5.12.14) in the form

$$u(\rho) = \rho^{\ell+1} e^{-\rho^2/2} v(\rho). \tag{5.12.15}$$

Substitution of u from (5.12.15) into (5.12.14) yields the following ordinary second order differential equation with variable coefficients for the function $v(\rho)$:

$$\frac{d^2v}{d\rho^2} + \left(\frac{2(\ell+1)}{\rho} - 2\rho\right)\frac{dv}{d\rho} + (\lambda - 2\ell - 3)\, v = 0. \tag{5.12.16}$$

We look for the solution in terms of an infinite series

$$v(\rho) = \sum_{p=0}^{\infty} a_p \rho^p, \tag{5.12.17}$$

where a_p are constant expansion coefficients. Using it in (5.12.16), we obtain

$$\sum_{p=0}^{\infty} a_p \left[p(p-1)\,\rho^{p-2} + (2\ell+2)\, p\,\rho^{p-2} - 2\,p\,\rho^p + (\lambda - 2\ell - 3)\,\rho^p\right] = 0. \tag{5.12.18}$$

Let us replace p by $(p+2)$ in the first two terms. We then get

$$\sum_{p=-2}^{\infty} a_{p+2}\, \left[(p+1)(p+2)\,\rho^p + (2\ell+2)\,(p+2)\,\rho^p\right]$$
$$+ \sum_{k=0}^{\infty} a_p\, (\lambda - 2\ell - 3 - 2k)\, \rho^p = 0. \tag{5.12.19}$$

The last equation leads to

$$\sum_{p=0}^{\infty} \left[(p+1)(p+2)a_{p+2} + (2\ell+2)\,(p+2)a_{p+2} + (\lambda - 2\ell - 3 - 2p)\,a_p\right]\rho^p$$
$$+\ (2\ell+2)\,a_1\frac{1}{\rho} = 0. \tag{5.12.20}$$

For (5.12.20) to hold both the terms must separately be equal to zero. This leads to $a_1 = 0$ and the recursion relation for the expansion coefficients

$$a_{p+2} = \frac{(2\ell + 2p + 3 - \lambda)}{(p+1)(p+2) + (2\ell+2)(p+2)}\, a_p. \tag{5.12.21}$$

Since $a_1 = 0$, the recursion relation (5.12.21) tells us that all the coefficients corresponding to odd values of p in the series (5.12.17) are zero. As a result, we obtain the solution of the radial equation (5.12.14) as

$$u(\rho) = \rho^{\ell+1} e^{-\rho^2/2} v(\rho), \quad v(\rho) = \sum_p a_p \rho^p, \quad p = 0, 2, 4, \ldots. \tag{5.12.22}$$

For $\rho \to \infty$, the above solution diverges as e^{ρ^2} unless the infinite series is terminated at some term. Hence, for some $p = p_{\max} = k$, the coefficient a_{p+2} must vanish. Clearly, this can be achieved if

$$2\ell + 2k + 3 - \lambda = 0. \quad \Rightarrow \quad \lambda = \frac{2E}{\hbar\omega} = 2\ell + 2k + 3. \tag{5.12.23}$$

The above condition leads to the energy eigenvalues of the oscillator associated with a given value of ℓ:

$$E_{k,\ell} = \hbar\omega\left(k + \ell + \frac{3}{2}\right), \tag{5.12.24}$$

where k is any even positive integer or zero. If we introduce the quantum number $n = k + \ell$, the energy levels of the $3D$ isotropic oscillator can be written as

$$E_n = \hbar\omega\left(n + \frac{3}{2}\right), \tag{5.12.25}$$

where, in view of the fact that ℓ can take any positive integer value (including zero) and k is an even and positive integer (including zero), $n = k + \ell = 0, 1, 2, 3, \ldots$. Therefore, n can take any positive integer values or zero. In this case, we shall have $\lambda = 2n + 3$ and the recursion relation will read as

$$a_{p+2} = \frac{2(\ell + p - n)}{(p+1)(p+2) + (2\ell+2)(p+2)} a_p. \tag{5.12.26}$$

For a given $n = k + \ell$, there exists (except for normalization) a unique eigenfunction

$$\phi_{n\ell m}(r, \theta, \phi) = R_{n,\ell}(r) Y_{\ell m}(\theta, \varphi). \tag{5.12.27}$$

with

$$R_{n\ell}(r) = \beta^\ell r^\ell e^{-\beta r^2/2} v(\beta r), \quad v(\beta r) = \sum_p a_p (\beta r)^p, \quad p = 0, 2, 4, \ldots k, \tag{5.12.28}$$

where $\beta = \sqrt{m\omega/\hbar}$. Taking into account that for even n, ℓ can take $(n/2 + 1)$ values: $0, 2, 4, \ldots, n$, while for odd n, it can take $[(n-1)/2 + 1]$ values: $1, 3, 5, \ldots, n$, the degeneracy of the energy levels is calculated to be

$$g_{n_{even}} = \sum_{\ell=0,2,4,\ldots,n} (2\ell+1) = \frac{(n+1)(n+2)}{2}, \qquad \text{for even } n, \tag{5.12.29}$$

$$g_{n_{odd}} = \sum_{\ell=1,3,5,\ldots,n} (2\ell+1) = \frac{(n+1)(n+2)}{2}, \qquad \text{for odd } n. \tag{5.12.30}$$

For illustration, let us determine the energies and the wave functions of the three lowest lying states of the oscillator.

Ground state: The ground state corresponds to $n=0$ for which $\ell = k = 0$. Therefore, the ground state energy is given by $E_0 = (3/2)\hbar\omega$.

Since $v(\beta r) = a_0$, the corresponding wave function is

$$\phi_{000} = a_0 e^{-\frac{\mu\omega}{2\hbar}r^2} Y_{00}(\theta,\varphi), \tag{5.12.31}$$

where a_0 is determined from the normalization of the radial part of the wave function. Finally, the normalized ground state wave function is given by

$$\phi_{000} = \frac{2}{\pi^{1/4}}\left(\frac{m\omega}{\hbar}\right)^{3/4} e^{-\frac{m\omega}{2\hbar}r^2} Y_{00}(\theta,\varphi). \tag{5.12.32}$$

First excited state: It corresponds to $n=1$. Since k has to be even, we have $\ell = 1$ and $k=0$. Thus, the energy of the first excited state is $E_1 = (5/2)\hbar\omega$.

Once again $v(\beta r) = a_0$, and the radial wave function is

$$R_{11} = a_0 \left(\frac{m\omega}{\hbar}\right)^{1/2} r e^{-\frac{m\omega}{2\hbar}r^2}. \tag{5.12.33}$$

The normalization of $R_{11}(r)$ yields

$$a_0 = \sqrt{\frac{8}{3\sqrt{\pi}}}\left(\frac{m\omega}{\hbar}\right)^{3/4}. \tag{5.12.34}$$

Consequently, we have

$$\phi_{11m} = \sqrt{\frac{8}{3\sqrt{\pi}}}\left(\frac{m\omega}{\hbar}\right)^{5/4} r e^{-\frac{m\omega}{2\hbar}r^2} Y_{1m}(\theta,\varphi), \quad m = -1,0,1. \tag{5.12.35}$$

Second excited state: It corresponds to $n=2$. Since k has to be even, we have two pairs of k and ℓ: $(2,0)$ and $(0,2)$. Since $n=2$, the energy of the first excited state is $E_1 = (7/2)\hbar\omega$.

Case 1: $k = 2$ and $\ell = 0$. In this case

$$v(\beta r) = a_0 + a_2\beta^2 r^2, \quad a_2 = -\frac{2}{3}a_0. \tag{5.12.36}$$

The radial wave function is given by

$$R_{20} = a_0\left(1 - \frac{2}{3}\beta r^2\right)e^{-\frac{m\omega}{2\hbar}r^2}. \tag{5.12.37}$$

After normalizing $R_{20}(r)$ we obtain

$$\phi_{200} = \frac{\sqrt{6}}{\pi^{\frac{1}{4}}}\left(\frac{m\omega}{\hbar}\right)^{3/4}\left(1 - \frac{2m\omega}{3\hbar}r^2\right)e^{-\frac{m\omega}{2\hbar}r^2}Y_{00}(\theta,\varphi). \tag{5.12.38}$$

Case 2: $k = 0$ and $\ell = 2$. In this case again $v(\beta r) = a_0$ and the radial wave function is

$$R_{22} = a_0\frac{m\omega}{\hbar}r^2 e^{-\frac{m\omega}{2\hbar}r^2}. \tag{5.12.39}$$

After normalizing $R_{20}(r)$ we obtain

$$\phi_{22m} = \frac{4}{\sqrt{15\sqrt{\pi}}}\left(\frac{m\omega}{\hbar}\right)^{7/4}r^2 e^{-\frac{m\omega}{2\hbar}r^2}Y_{2m}(\theta,\varphi). \tag{5.12.40}$$

Homework Problems

1. Consider the case of a particle moving in the infinite rectangular well potential discussed in the chapter. What is the probability of finding the particle in the volume given by $0 < x < a$, $0 < y < b$, and $0 < z < c/3$?
2. A particle is in the second excited state of an infinite cubic potential well of side a. Determine the wave functions and the corresponding energy for this state. What is the degeneracy of this level?
3. An electron moves in an infinite cubic potential well of side $a = 0.5$ nm. What energy does the electron have in (a) the ground state and (b) the first excited state?
4. A particle of mass m is confined to move in an infinite two-dimensional potential well of side L. (a) Solve the corresponding two-dimensional TISE by the method of separation of variables and determine the wave functions and the corresponding energy levels. (b) Find the energies of the ground state and the first excited state. Are these states degenerate?
5. Find the stationary state wave functions and the corresponding energies for a particle of mass m moving in a three-dimensional isotropic harmonic oscillator potential

$$V(x,y,z) = \frac{1}{2}m\omega^2(x^2+y^2+z^2).$$

6. Assume that the nucleus of an atom can be regarded as a three-dimensional box of width 2×10^{-14} m. If a proton moves as a particle in this box, find (a) the ground state energy of the proton in MeV and (b) the energies of the first and second excited states. (c) What are the degeneracies of these states?
7. Use the Rodriguez formula to construct the first five Legendre polynomials.
8. Use (5.611), (5.6.12) and (5.6.14) to work out the spherical harmonics Y_{00}, Y_{10}, $Y_{1\,-1}$ and Y_{11}. Check that they are normalized to unity.
9. Using the Rodriguez formula, derive the orthonormality condition for the Legendre polynomials

$$\int_{-1}^{1} P_\ell(x)P_{\ell'}(x)\,dx = \frac{2}{2\ell+1}\delta_{\ell\ell'},$$

where $x=\cos\theta$.

10. Find the $\ell=0$ energy and normalized wave function of a particle of mass m that is subject to the following central potential

$$V(r) = \begin{cases} 0, & \text{for } a<r<b \\ \infty, & \text{for } r>a, \end{cases}$$

where a and b are positive constants. Write down the full stationary state solution.

11. Find the energy levels and the corresponding normalized wave functions for a particle of mass m subject to the following central potential

$$V(r) = -\alpha\delta(r-a),\ a>0,$$

where α is a positive constant. Discuss the existence of bound states in terms of the size of a.

12. Write down the TISE in plane polar coordinates for a two-dimensional isotropic harmonic oscillator

$$V(x,y) = \frac{1}{2}m\omega^2(x^2+y^2),$$

where ω is the angular frequency of the oscillator. Solve it and find the energy levels and the corresponding eigenfunctions. Discuss the degeneracy of the energy states.

13. (a) Calculate the most probable distance of the electron from the nucleus in the ground state of hydrogen. (b) Find the average distance of the electron from the nucleus in the ground state of hydrogen and compare it with the result of (a).

14. Calculate $\langle r^2 \rangle$ in the ground state of the hydrogen atom. Using this and the result of the Problem 13, calculate the uncertainty in the measurement of the distance of the electron from the nucleus in the ground state of hydrogen.

15. The normalized ground state wave function for the electron in the hydrogen atom is

$$\psi(r,\theta,\phi) = \frac{1}{\sqrt{\pi a_0^3}} e^{-r/a_0},$$

where r is the radial coordinate of the electron and a_0 is the Bohr radius. (a) Sketch the wave function as a function of r. (b) Show that the probability of finding the electron between r and $r+dr$ is given by

$$\frac{4}{a_0^3} r^2 e^{-2r/a_0} dr.$$

(c) Show that the wave function as given is normalized. (e) Find the probability of locating the electron between $a_0/2 \le r \le 3a_0/2$.

16. Calculate $\langle x \rangle$ and $\langle x^2 \rangle$ in the ground state of the hydrogen atom.

17. The radial part of the wave function for the hydrogen atom in the $2p$ state is given by

$$\psi(r,\theta,\phi) = Are^{-r/2a_0},$$

where A is a constant and a_0 is the Bohr radius. Using this expression, calculate the average value of r for an electron in this state.

18. An electron in a hydrogen atom is in the energy eigenstate

$$\psi_{2,1,-1}(r,\theta,\phi) = Nre^{-r/2a_0} Y_1^{-1}(\theta,\phi).$$

(a) Find the normalization constant N.

(b) What is the probability per unit volume of finding the electron at $r = 2a_0, \theta = 45°$ and $\phi = 60°$?

19. An electron in the Coulomb field of a proton is in a state described by the wave function

$$\psi = \frac{1}{6}[4\psi_{100} + 3\psi_{211} - \psi_{210} + \sqrt{10}\psi_{21-1}].$$

What is the expectation value of energy?

20. Calculate $\langle r \rangle$, $\langle r^2 \rangle$ and Δr in the state with $n = 2, \ell = 0$ and $m = 0$.

21. Calculate $\langle \psi_{n\ell}|r|\psi_{n\ell}\rangle, \langle \psi_{n\ell}|r^2|\psi_{n\ell}\rangle$ and $\langle \psi_{n\ell}|1/r|\psi_{n\ell}\rangle$ in the nth stationary state, $\psi_{n\ell}$, of the hydrogen atom.

22. Using the recursion relation, derive the radial wave functions R_{30}, R_{31} and R_{32}. Normalize them and write down the stationary state wave functions ψ_{300}, ψ_{320}, ψ_{32-2}, and ψ_{32-1}.
23. A hydrogen-like atom consists of a single electron orbiting the nucleus with Z protons. (a) Determine the energy spectrum of a hydrogen-like atom, that is, determine $E_n(Z)$. (b) Determine the Bohr radius as a function of Z and the modified expression for the Rydberg constant $\mathscr{R}$.
24. (a) Determine the quantum numbers ℓ and m_ℓ for the He^+ ion in the state corresponding to $n = 3$. (b) What is the energy of this state?
25. (a) Determine the quantum numbers ℓ and m_ℓ for the Li^{2+} ion in the states corresponding to $n = 1$ and $n = 2$. (b) Determine the energies of these states.
26. The wavelength for the $n = 3$ to $n = 2$ transition of the hydrogen atom is 656.3 nm. What is the wavelength of this same transition in singly ionized helium?
27. Calculate the uncertainty product $\Delta r \Delta p$ for the $1s$ electron of a hydrogen-like atom with atomic number Z.

Chapter 6

Quantum Mechanical Theory of Orbital Angular Momentum

6.1 The Angular Momentum Operators in Cartesian Coordinates

In classical mechanics, the angular momentum of a particle is given by $\vec{L} = \vec{r} \times \vec{p}$, where $\vec{r}$ and $\vec{p}$ are the position vector and momentum of the particle, respectively. The quantum mechanical operator for $\hat{\vec{L}}$, is obtained by replacing $\vec{r}$ and $\vec{p}$, with their respective operators, that is,

$$\hat{\vec{L}} = \hat{\vec{r}} \times \hat{\vec{p}}. \tag{6.1.1}$$

Using the expressions for $\hat{\vec{r}}$ and $\hat{\vec{p}}$ in the Cartesian system of coordinates, we have $\hat{\vec{L}} = \hat{i}\hat{L}_x + \hat{j}\hat{L}_y + \hat{k}\hat{L}_z$, where

$$\hat{L}_x = y\,\hat{p}_z - z\,\hat{p}_y = -i\hbar\left(y\frac{\partial}{\partial z} - z\frac{\partial}{\partial y}\right), \tag{6.1.2}$$

$$\hat{L}_y = z\,\hat{p}_x - x\,\hat{p}_z = -i\hbar\left(z\frac{\partial}{\partial x} - x\frac{\partial}{\partial y}\right), \tag{6.1.3}$$

$$\hat{L}_z = x\,\hat{p}_y - y\,\hat{p}_x = -i\hbar\left(x\frac{\partial}{\partial y} - y\frac{\partial}{\partial x}\right). \tag{6.1.4}$$

It is easy to check that each of these operators is hermitian (see Example 6.2.1 below). The operator corresponding to the square of the angular momentum is a scalar operator given by

$$\hat{L}^2 = \hat{\vec{L}}\cdot\hat{\vec{L}} = \hat{L}_x^2 + \hat{L}_y^2 + \hat{L}_z^2. \tag{6.1.5}$$

6.2 Commutation Relations, Measurement and Uncertainty

The algebra of the angular momentum operators is given by their commutation relations, which can be readily calculated with the help of the fundamental commutators $[\hat{x}_j, \hat{p}_k] = i\hbar\,\delta_{jk}$, $[\hat{x}_j, \hat{x}_k] = [\hat{p}_j, \hat{p}_k] = 0$, $j,k = 1,2,3$, and the following properties of commutators

$$[\hat{A} \pm \hat{B}, \hat{C}] = [\hat{A}, \hat{C}] \pm [\hat{B}, \hat{C}], \tag{6.2.1}$$

$$[\hat{A}, \hat{B}\hat{C}] = \hat{B}[\hat{A}, \hat{C}] + [\hat{A}, \hat{B}]\hat{C}, \tag{6.2.2}$$

where $\hat{A}$, $\hat{B}$, and $\hat{C}$ are arbitrary operators. For instance,

$$[\hat{L}_x, \hat{L}_y] = [y\,\hat{p}_z - z\,\hat{p}_y, z\,\hat{p}_x - x\,\hat{p}_z] = [y\,\hat{p}_z, z\,\hat{p}_x] - [z\,\hat{p}_y, z\,\hat{p}_x] - [y\,\hat{p}_z, x\,\hat{p}_z] + [z\,\hat{p}_y, x\,\hat{p}_z]. \tag{6.2.3}$$

Simplifying, we get

$$[y\,\hat{p}_z, z\,\hat{p}_x] = y\,[\hat{p}_z, z\,\hat{p}_x] + [y, z\,\hat{p}_x]\,\hat{p}_z = y\,z\,[\hat{p}_z, \hat{p}_x] + y\,[\hat{p}_z, z]\,\hat{p}_x + z\,[y, \hat{p}_x]\,\hat{p}_z$$
$$+ [y, z]\,\hat{p}_x\,\hat{p}_z = -i\hbar\,y\,\hat{p}_x, \tag{6.2.4}$$

$$[z\,\hat{p}_y, z\,\hat{p}_x] = z\,[\hat{p}_y, z\,\hat{p}_x] + [z, z\,\hat{p}_x]\,\hat{p}_y = z^2\,[\hat{p}_y, \hat{p}_x] + z\,[\hat{p}_y, z]\,\hat{p}_x + z\,[z, \hat{p}_x]\,\hat{p}_y$$
$$+ [z, z]\,\hat{p}_x\,\hat{p}_y = 0, \tag{6.2.5}$$

$$[y\,\hat{p}_z, x\,\hat{p}_z] = y\,[\hat{p}_z, x\,\hat{p}_z] + [y, x\,\hat{p}_z]\,\hat{p}_z = y\,x\,[\hat{p}_z, \hat{p}_z] + y\,[\hat{p}_z, x]\,\hat{p}_z + x\,[y, \hat{p}_z]\,\hat{p}_z$$
$$+ [y, x]\,\hat{p}_x^2 = 0, \tag{6.2.6}$$

$$[z\,\hat{p}_y, x\,\hat{p}_z] = z\,[\hat{p}_y, x\,\hat{p}_z] + [z, x\,\hat{p}_z]\,\hat{p}_y = z\,x\,[\hat{p}_y, \hat{p}_z] + z\,[\hat{p}_y, x]\,\hat{p}_z + x\,[z, \hat{p}_z]\,\hat{p}_y$$
$$+ [z, x]\,\hat{p}_z\,\hat{p}_y = i\hbar\,x\,\hat{p}_y. \tag{6.2.7}$$

Hence, we get

$$[\hat{L}_x, \hat{L}_y] = i\hbar\,(x\,\hat{p}_y - y\,\hat{p}_x) = i\hbar\hat{L}_z. \tag{6.2.8}$$

The other two commutators are calculated in a similar manner. The net result is

$$[\hat{L}_x, \hat{L}_y] = i\hbar\hat{L}_z, \quad [\hat{L}_y, \hat{L}_z] = i\hbar\hat{L}_x, \quad [\hat{L}_z, \hat{L}_x] = i\hbar\hat{L}_y. \tag{6.2.9}$$

The commutation relations (6.2.9) can be combined together into a single vector equation

$$i\hbar\hat{\vec{L}} = \hat{\vec{L}} \times \hat{\vec{L}}. \tag{6.2.10}$$

Equivalently, they can also be written as

$$\left[\hat{L}_j, \hat{L}_k\right] = i\hbar\, \varepsilon_{jk\ell}\, \hat{L}_\ell, \tag{6.2.11}$$

where summation over the repeated index ℓ from 1 to 3 is understood. Here, in (6.2.11), the symbol ε_{ijk} is called the Levi-Civita tensor density and it is defined as

$$\varepsilon_{ijk} = \begin{cases} 1 & \text{if } (ijk) \text{ is an even permutation of } (1\,2\,3) \\ -1 & \text{if } (ijk) \text{ is an odd permutation of } (1\,2\,3) \\ 0 & \text{otherwise.} \end{cases} \tag{6.2.12}$$

Measurement and uncertainty relation: We have earlier shown (Chapter 3) that any two hermitian and non-commuting operators, $\hat{A}$ and $\hat{B}$, satisfy the generalized uncertainty relation

$$\Delta A\ \Delta B \geq \frac{1}{2}\sqrt{|\langle[\hat{A},\hat{B}]\rangle|^2}. \tag{6.2.13}$$

Using this result for the angular momentum operators, $\hat{L}_x$, $\hat{L}_y$ and $\hat{L}_z$, we conclude that they must satisfy the following uncertainty relations

$$\Delta L_j\ \Delta L_k \geq \frac{1}{2}\sqrt{|\langle[\hat{L}_j,\hat{L}_k]\rangle|^2} = \frac{\hbar}{2}|\langle L_\ell\rangle|, \tag{6.2.14}$$

where $(jk\ell)$ are cyclic permutations of (123). It then follows that no two components of the angular momentum can be measured simultaneously accurately. At first glance, it might appear that only one of the three components of $\vec{L}$ can be determined or specified to characterize a given state of a particle. However, a careful analysis shows that, along with one of the components of $\vec{L}$, the square of the total angular momentum (hence, the absolute value of the angular momentum) can also be measured accurately in a given state of the particle. This is because of the fact that $\hat{L}^2$ commutes with each of the components $\hat{L}_x$, $\hat{L}_y$ and $\hat{L}_z$. For instance, we have

$$\begin{aligned}
[\hat{L}^2,\hat{L}_x] &= [(\hat{L}_x^2+\hat{L}_y^2+\hat{L}_z^2),\hat{L}_x] = [\hat{L}_y^2,\hat{L}_x]+[\hat{L}_z^2,\hat{L}_x] \\
&= \hat{L}_y[\hat{L}_y,\hat{L}_x]+[\hat{L}_y,\hat{L}_x]\hat{L}_y+\hat{L}_z[\hat{L}_z,\hat{L}_x]+[\hat{L}_z,\hat{L}_x]\hat{L}_z \\
&= -i\hbar\hat{L}_y\hat{L}_z - i\hbar\hat{L}_z\hat{L}_y + i\hbar\hat{L}_z\hat{L}_y + i\hbar\hat{L}_y\hat{L}_z = 0.
\end{aligned} \tag{6.2.15}$$

Similarly, we can prove that $\hat{L}_y$ and $\hat{L}_z$ also commute with $\hat{L}^2$. As a consequence, we can write

$$[\hat{L}^2,\hat{\vec{L}}] = 0. \tag{6.2.16}$$

Hence, in terms of the angular momentum of a particle, its quantum state can be characterized by the magnitude of the total angular momentum and any one of the Cartesian components of the angular momentum. Because of the isotropy of space, this component is always taken to be L_z. Consequently, the wave functions of the particle are characterized by two quantum numbers, the orbital quantum number, ℓ and the magnetic quantum number, m_ℓ or simply m. The meaning of these quantum numbers and their relationship with the eigenvalues of the operator $\hat{L}^2$ will follow from our later analysis.

Example 6.2.1: Show that the operator $\hat{L}_x$, is hermitian.

Solution: For the hermiticity of $\hat{L}_x$ we must have $\hat{L}_x^\dagger = \hat{L}_x$. Recalling that $\hat{x}_k^\dagger = x_k$, $\hat{p}_k^\dagger = \hat{p}_k$, $k = 1,2,3$, we get

$$\hat{L}_x^\dagger = (y\,\hat{p}_z - z\,\hat{p}_y)^\dagger = (y\,\hat{p}_z)^\dagger - (z\,\hat{p}_y)^\dagger = \hat{p}_z^\dagger\,y^\dagger - \hat{p}_y^\dagger\,\hat{z}^\dagger = \hat{p}_z\,y - \hat{p}_y\,z. \tag{6.2.17}$$

Using now that $[\hat{x}_j, \hat{p}_k] = 0$, if $j \neq k$, we arrive at the required result

$$\hat{L}_x^\dagger = (y\,\hat{p}_z - z\,\hat{p}_y)^\dagger = y\,\hat{p}_z - z\,\hat{p}_y = \hat{L}_x. \tag{6.2.18}$$

Example 6.2.2: Find the value of the commutators (a) $[\hat{x}, \hat{L}_x]$, (b) $[\hat{x}, \hat{L}_y]$, and $[\hat{p}_x, \hat{L}_y]$.

Solution:

(a) Using the expression for $\hat{L}_x$ in terms of the position and momentum operators, we have

$$\begin{aligned}[\hat{x}, \hat{L}_x] &= [\hat{x}, (\hat{y}\,\hat{p}_z - \hat{z}\,\hat{p}_y] = [\hat{x}, \hat{y}\,\hat{p}_z] - [\hat{x}, \hat{z}\,\hat{p}_y] \\ &= [\hat{x}, \hat{y}]\,\hat{p}_z + \hat{y}\,[\hat{x}, \hat{p}_z] - [\hat{x}, \hat{z}]\,\hat{p}_y - \hat{z}\,[\hat{x}, \hat{p}_z].\end{aligned} \tag{6.2.19}$$

Since $[\hat{x}_j, \hat{x}_k] = 0$, for all values of j and k from 1 to 3 and $[\hat{x}_j, \hat{p}_k] = 0$ for $j \neq k$, $[\hat{x}, \hat{L}_x] = 0$.

(b) Similarly,

$$\begin{aligned}[\hat{x}, \hat{L}_y] &= [\hat{x}, (\hat{z}\,\hat{p}_x - \hat{x}\,\hat{p}_z] = [\hat{x}, \hat{z}\,\hat{p}_x] - [\hat{x}, \hat{x}\,\hat{p}_z] \\ &= [\hat{x}, \hat{z}]\,\hat{p}_x + \hat{z}\,[\hat{x}, \hat{p}_x] - [\hat{x}, \hat{x}]\,\hat{p}_z - \hat{x}\,[\hat{x}, \hat{p}_z].\end{aligned} \tag{6.2.20}$$

Using $[\hat{x}_j, \hat{x}_k] = 0$, and $[\hat{x}_j, \hat{p}_k] = i\hbar\,\delta_{jk}$, we get that $[\hat{x}, \hat{L}_y] = i\hbar\,\hat{z} = i\hbar\,z$.

(c) In this case, we have

$$\begin{aligned}[\hat{p}_x, \hat{L}_y] &= [\hat{p}_x, (\hat{z}\,\hat{p}_x - \hat{x}\,\hat{p}_z] = [\hat{p}_x, \hat{z}\,\hat{p}_x] - [\hat{p}_x, \hat{x}\,\hat{p}_z] \\ &= [\hat{p}_x, \hat{z}]\,\hat{p}_x + \hat{z}\,[\hat{p}_x, \hat{p}_x] - [\hat{p}_x, \hat{x}]\,\hat{p}_z - \hat{x}\,[\hat{p}_x, \hat{p}_z].\end{aligned} \tag{6.2.21}$$

Using $[\hat{p}_j, \hat{p}_k] = 0$, for all values of j and k from 1 to 3 and $[\hat{x}_j, \hat{p}_k] = i\hbar\,\delta_{jk}$, we get that $[\hat{p}_x, \hat{L}_y] = i\hbar\,\hat{z} = i\hbar\,\hat{p}_z$.

6.3 The Eigenvalues of $\hat{L}^2$ and $\hat{L}_z$

We shall determine the possible eigenvalues of $\hat{L}^2$ and $\hat{L}_z$ by algebraic means. In other words, we shall determine their eigenvalues without solving the differential equations representing the corresponding eigenvalue problems for these operators. Our discussion will revolve around the commutation relations (6.2.9) and their consequences in the framework of linear algebra. In such an approach, the angular momentum is simply an observable represented by three hermitian operators $\hat{L}_x$, $\hat{L}_y$, $\hat{L}_z$ that satisfy the commutation relations (6.2.9). Since the entire discussion is based only on the commutation relations (6.2.9) of the angular momentum operators, the consequences hold good for any set of operators satisfying an identical set of commutation relations.

Since $\hat{L}^2$ and $\hat{L}_z$ commute, they have a common set of eigenfunctions. Let $\psi_{\lambda\,\mu}(\vec{r})$ be a common eigenfunction of $\hat{L}^2$ and $\hat{L}_z$, corresponding to the eigenvalues $\hbar^2\lambda$, and $\hbar\mu$, respectively. That is,

$$\hat{L}^2\psi(\vec{r}) = \hbar^2\lambda\,\psi_{\lambda\,\mu}(\vec{r}), \tag{6.3.1}$$

$$\hat{L}_z\psi(\vec{r}) = \hbar\mu\,\psi_{\lambda\,\mu}(\vec{r}). \tag{6.3.2}$$

Note that the dimensions of the angular momentum are those of $\hbar$, owing to which we have introduced the factors $\hbar^2$ and $\hbar$ before λ and μ, respectively, so that they are dimensionless.

Analogous to the case of one-dimensional harmonic oscillator discussed earlier, let us introduce the operators:

$$\hat{L}_\pm = \hat{L}_x \pm i\hat{L}_y. \tag{6.3.3}$$

Using the commutation relations (6.2.9), the commutator of $\hat{L}_z$ with $\hat{L}_\pm$ can be readily computed as

$$[\hat{L}_z,\hat{L}_+] = [\hat{L}_z,\hat{L}_x] + i[\hat{L}_z,\hat{L}_y] = i\hbar\hat{L}_y + i\,(-i)\,\hbar\hat{L}_x = \hbar(\hat{L}_x + i\hat{L}_y) = \hbar\hat{L}_+, \tag{6.3.4}$$

$$[\hat{L}_z,\hat{L}_-] = [\hat{L}_z,\hat{L}_x] - i[\hat{L}_z,\hat{L}_y] = i\hbar\hat{L}_y - \hbar\hat{L}_x = -\hbar(\hat{L}_x - i\hat{L}_y) = -\hbar\hat{L}_-. \tag{6.3.5}$$

Hence, the operators $\hat{L}_\pm$ do not commute with $\hat{L}_z$. However, they do commute with $\hat{L}^2$:

$$\begin{aligned}[\hat{L}^2,\hat{L}_\pm] &= [\hat{L}_x^2,\hat{L}_\pm] + [\hat{L}_y^2,\hat{L}_\pm] + [\hat{L}_z^2,\hat{L}_\pm] = [\hat{L}_x^2,\hat{L}_x \pm i\hat{L}_y] \\ &+ [\hat{L}_y^2,\hat{L}_x \pm i\hat{L}_y] + [\hat{L}_z^2,\hat{L}_x \pm i\hat{L}_y] = \pm i[\hat{L}_x^2,\hat{L}_y] + [\hat{L}_y^2,\hat{L}_x] \\ &+ [\hat{L}_z^2,\hat{L}_x] \pm i[\hat{L}_z^2,\hat{L}_y] = \mp\hbar\left(\hat{L}_x\hat{L}_z + \hat{L}_z\hat{L}_x\right) - i\hbar\left(\hat{L}_y\hat{L}_z + \hat{L}_z\hat{L}_y\right) \\ &+ i\hbar\left(\hat{L}_y\hat{L}_z + \hat{L}_z\hat{L}_y\right) \pm \hbar\left(\hat{L}_x\hat{L}_z + \hat{L}_z\hat{L}_x\right) = 0.\end{aligned} \tag{6.3.6}$$

Since $\hat{L}_{\pm}$ do not commute with $\hat{L}_z$, the eigenfunctions of $\hat{L}_z$ are not the eigenfunctions of $\hat{L}_{\pm}$. However, we have the following results

$$\hat{L}_z\left(\hat{L}_+\psi_{\lambda\,\mu}\right) = \hbar\hat{L}_+\psi_{\lambda\,\mu} + \hat{L}_+(\hat{L}_z\psi_{\lambda\,\mu}) = \hbar\left(\mu+1\right)\left(\hat{L}_+\psi_{\lambda\,\mu}\right), \tag{6.3.7}$$

$$\hat{L}_z\left(\hat{L}_-\psi_{\lambda\,\mu}\right) = -\hbar\hat{L}_-\psi_{\lambda\,\mu} + \hat{L}_-(\hat{L}_z\psi_{\lambda\,\mu}) = \hbar\left(\mu-1\right)\left(\hat{L}_-\psi_{\lambda\,\mu}\right), \tag{6.3.8}$$

which show that if $\psi_{\lambda\,\mu}$ is an eigenfunction of $\hat{L}_z$ with eigenvalue $\hbar\mu$, then $\hat{L}_+\psi_{\lambda\,\mu}$ is also an eigenfunction of $\hat{L}_z$ but with an eigenvalue $(\hbar\mu+\hbar)$. That is, the operator $\hat{L}_+$, by acting on the eigenfunction of $\hat{L}_z$ with a given eigenvalue, converts it into an eigenfunction of $\hat{L}_z$ with an eigenvalue raised by one unit of $\hbar$. Similarly, the operator $\hat{L}_-$, by acting on the eigenfunction of $\hat{L}_z$ with a given eigenvalue, converts it into an eigenfunction of $\hat{L}_z$ with an eigenvalue lowered by one unit of $\hbar$. Therefore, the operators $\hat{L}_+$ and $\hat{L}_-$ are called the raising (creation) and the lowering (annihilation) operators, respectively.

As a result, for a given value of λ (i.e., for a given eigenvalue of the $\hat{L}^2$ operator), we can construct a ladder of discrete eigenstates of $\hat{L}_z$ by repeatedly acting on a given eigenfunction of $\hat{L}_z$ with the raising operator $\hat{L}_+$, in which any two neighbouring states will differ in eigenvalue by one unit of $\hbar$. Note that if, starting with an eigenstate of $\hat{L}_z$ with a given eigenvalue, this process is continued indefinitely, we shall land up with a state in which the z component of the angular momentum will exceed the total angular momentum. Therefore, we conclude that there must exist an eigenstate, $\psi_{\lambda\,\mu_{\max}}$, of $\hat{L}_z$ with the highest possible eigenvalue, $\hbar\mu_{\max}$, such that

$$\hat{L}^2\psi_{\lambda\,\mu_{\max}} = \hbar^2\lambda\,\psi_{\lambda\,\mu_{\max}}, \quad \hat{L}_z\psi_{\lambda\,\mu_{\max}} = \hbar\mu_{\max}\,\psi_{\lambda\,\mu_{\max}} \;\; and \;\; \hat{L}_+\psi_{\lambda\,\mu_{\max}} = 0. \tag{6.3.9}$$

The next question is: How to find $\mu_{\max}$? To answer this question we notice that

$$\begin{aligned}\hat{L}_{\pm}\hat{L}_{\mp} &= (\hat{L}_x \pm i\hat{L}_y)(\hat{L}_x \mp i\hat{L}_y) = \hat{L}_x^2 + \hat{L}_y^2 \mp i(\hat{L}_x\hat{L}_y - \hat{L}_y\hat{L}_x)\\ &= \hat{L}^2 - \hat{L}_z^2 \mp i(i\hbar\hat{L}_z) = \hat{L}^2 - \hat{L}_z^2 \pm \hbar\hat{L}_z,\end{aligned} \tag{6.3.10}$$

and hence

$$\hat{L}^2 = \hat{L}_{\pm}\hat{L}_{\mp} + \hat{L}_z^2 \mp (\hbar\hat{L}_z). \tag{6.3.11}$$

That is, either $\hat{L}^2 = \hat{L}_+\hat{L}_- + \hat{L}_z^2 - (\hbar\hat{L}_z)$, or $\hat{L}^2 = \hat{L}_-\hat{L}_+ + \hat{L}_z^2 + (\hbar\hat{L}_z)$. Therefore, using the lower sign in (6.3.11), we obtain

$$\begin{aligned}\hat{L}^2\psi_{\lambda\,\mu_{\max}} &= \hat{L}_-\hat{L}_+\psi_{\lambda\,\mu_{\max}} + \hat{L}_z^2\psi_{\lambda\,\mu_{\max}} + (\hbar\hat{L}_z)\psi_{\lambda\,\mu_{\max}}\\ &= 0 + \hbar^2\mu_{\max}^2\psi_{\lambda\,\mu_{\max}} + \hbar^2\mu_{\max}\psi_{\lambda\,\mu_{\max}} = \hbar^2\mu_{\max}(\mu_{\max}+1)\psi_{\lambda\,\mu_{\max}},\end{aligned} \tag{6.3.12}$$

and hence

$$\lambda = \hbar^2 \mu_{\max}(\mu_{\max}+1). \tag{6.3.13}$$

This gives us the eigenvalues of the operator $\hat{L}^2$ in terms of the maximal eigenvalue of $\hat{L}_z$.

An argument similar to the one used in the case of $\hat{L}_+$, there must exist an eigenstate, $\psi_{\lambda\,\mu_{\min}}$, of $\hat{L}_z$ with the lowest possible eigenvalue, $\mu_{\min}$, such that

$$\hat{L}^2\psi_{\lambda\,\mu_{\min}} = \hbar^2\lambda\,\psi_{\lambda\,\mu_{\min}}, \quad \hat{L}_z\psi_{\mu_{\min}} = \hbar\mu_{\min}\,\psi_{\lambda\,\mu_{\min}} \text{ and } \hat{L}_-\psi_{\lambda\,\mu_{\min}} = 0. \tag{6.3.14}$$

Using the upper sign in (6.3.11), we have

$$\begin{aligned} \hat{L}^2\psi_{\lambda\,\mu_{\min}} &= \hat{L}_+\hat{L}_-\psi_{\lambda\,\mu_{\min}} + \hat{L}_z^2\psi_{\lambda\,\mu_{\min}} - (\hbar\hat{L}_z)\psi_{\lambda\,\mu_{\min}} \\ &= (0+\hbar^2\mu_{\min}^2 - \hbar^2\mu_{\min})\psi_{\lambda\,\mu_{\min}} = \hbar^2\mu_{\min}(\mu_{\min}-1)\psi_{\lambda\,\mu_{\min}}. \end{aligned} \tag{6.3.15}$$

Therefore,

$$\lambda = \hbar^2\mu_{\min}(\mu_{\min}-1). \tag{6.3.16}$$

From (6.3.13) and (6.3.16), we conclude

$$\mu_{\max}(\mu_{\max}+1) = \mu_{\min}(\mu_{\min}-1). \tag{6.3.17}$$

We get from (6.3.17) that either $\mu_{\min} = \mu_{\max}+1$ or $\mu_{\min} = -\mu_{\max}$. The first solution is unacceptable since, if so, the eigenvalue of the lowest eigenstate of $\hat{L}_z$ will be greater than the eigenvalue of the highest eigenstate. Thus, $\mu_{\max} = -\mu_{\max}$.

It is obvious now that if we start with $\psi_{\lambda\,\mu_{\max}}$ and apply $\hat{L}_-$ to it N integer number of times, we arrive at $\psi_{\lambda\,\mu_{\min}}$. Therefore,

$$\mu_{\max} = \mu_{\min} + N. \tag{6.3.18}$$

Taking into account that $\mu_{\min} = -\mu_{\max}$ in (6.3.18), we arrive at

$$\mu_{\max} = \frac{N}{2}. \tag{6.3.19}$$

It is customary to denote $\mu_{\max}$ by ℓ and μ by m (or, m_ℓ). The numbers ℓ and m are called the *orbital quantum number* and the *magnetic quantum number*, respectively. The eigenvalues of $\hat{L}^2$ and $\hat{L}_z$ can now be written as

$$\lambda_\ell = \hbar^2\,\ell(\ell+1), \qquad \mu_m = \hbar m, \tag{6.3.20}$$

where, for a given ℓ, m takes $(2\ell+1)$ values from $-\ell$ to ℓ and ℓ must be an integer or a half-integer depending on whether N is even or odd.

Thus, the joint eigenfunctions of $\hat{L}^2$ and $\hat{L}_z$ are characterized by two quantum numbers ℓ and m, where ℓ can take integer as well as half-integer values and, for a given ℓ, m can take $(2\ell+1)$ values from $-\ell$ to ℓ. They are denoted as ψ_ℓ^m. The eigenvalue equations for $\hat{L}^2$ and $\hat{L}_z$, respectively, are

$$\hat{L}^2\psi_\ell^m = \hbar^2\ell(\ell+1)\psi_\ell^m, \quad \hat{L}_z\psi_\ell^m = \hbar m\psi_\ell^m, \tag{6.3.21}$$

where $\ell = 0, 1/2, 1, 3/2, ...$ and $m = -\ell, -\ell+1, -\ell+2, -\ell+3, \ldots, 0, 1, 2, 3, \ldots, \ell-1, \ell$. Since for a given value of ℓ, there are $(2\ell+1)$ different values of m, the eigenvalue $\lambda_\ell = \hbar^2\ell(\ell+1)$ of $\hat{L}^2$ is said to be $(2\ell+1)$-fold degenerate.

6.4 The Angular Momentum Operators in Spherical Coordinates

Having determined the possible eigenvalues of $\hat{L}^2$ and $\hat{L}_z$, let us proceed to find the corresponding eigenfunctions. For the given purpose it is convenient to go over to the spherical system of coordinates (see Fig. 5.1). Using the chain rule for differentiation and the transformation equations (5.4.1) and (5.4.2), we obtain

$$\begin{aligned}\frac{\partial}{\partial\varphi} &= \frac{\partial x}{\partial\varphi}\frac{\partial}{\partial x} + \frac{\partial y}{\partial\varphi}\frac{\partial}{\partial y} + \frac{\partial z}{\partial\varphi}\frac{\partial}{\partial z}\\ &= -r\sin\theta\,\sin\varphi\frac{\partial}{\partial x} + r\sin\theta\,\cos\varphi\frac{\partial}{\partial y}\\ &= x\frac{\partial}{\partial y} - y\frac{\partial}{\partial x},\end{aligned} \tag{6.4.1}$$

and

$$\frac{\partial}{\partial\theta} = \cot\theta\left(x\frac{\partial}{\partial x} + y\frac{\partial}{\partial y}\right) - \tan\theta\, z\frac{\partial}{\partial z}. \tag{6.4.2}$$

If we now use the equations (5.4.3)-(5.4.5) along with the transformation equations (5.4.1) and (5.4.2), then the expressions for the x, y and z components of the angular momentum operator in spherical coordinates can be written as

$$\hat{L}_x = i\hbar\left(\sin\varphi\,\frac{\partial}{\partial\theta} + \cot\theta\,\cos\varphi\,\frac{\partial}{\partial\varphi}\right), \tag{6.4.3}$$

$$\hat{L}_y = -i\hbar\left(\cos\varphi\,\frac{\partial}{\partial\theta} - \cot\theta\sin\varphi\,\frac{\partial}{\partial\varphi}\right) \tag{6.4.4}$$

$$\hat{L}_z = -i\hbar\,\frac{\partial}{\partial\varphi}. \tag{6.4.5}$$

The expression for $\hat{L}^2$ is obtained by first writing the operators $\hat{L}_+$ and $\hat{L}_-$ in spherical coordinates and then using any one of the two expressions for $\hat{L}^2$, in terms of $\hat{L}_+$ and $\hat{L}_-$ and $\hat{L}_z$, given in (6.3.11). Hence, we start with the operator $\hat{L}_+$. We have

$$\hat{L}_+ = \hat{L}_x + i\hat{L}_y = \hbar\left[iz\frac{\partial}{\partial y} + z\frac{\partial}{\partial x} - (x+iy)\frac{\partial}{\partial z}\right] \tag{6.4.6}$$

Taking into account that $z = r\cos\theta$ and

$$x \pm iy = r\sin\theta(\cos\varphi \pm i\sin\varphi) = re^{\pm i\varphi}\sin\theta, \tag{6.4.7}$$

we get

$$\begin{aligned}\hat{L}_+ &= \hbar e^{i\varphi}\left(ire^{-i\varphi}\cos\theta\frac{\partial}{\partial y} + re^{-i\varphi}\cos\theta\frac{\partial}{\partial x} - r\sin\theta\frac{\partial}{\partial z}\right)\\ &= \hbar e^{i\varphi}\left[i(x-iy)\cot\theta\frac{\partial}{\partial x} + (x-iy)\cot\theta\frac{\partial}{\partial x} - r\sin\theta\frac{\partial}{\partial z}\right]\\ &= \hbar e^{i\varphi}\left[\cot\theta\left(x\frac{\partial}{\partial x} + y\frac{\partial}{\partial y}\right) - \tan\theta\, z\frac{\partial}{\partial z} + i\cot\theta\left(x\frac{\partial}{\partial y} - y\frac{\partial}{\partial x}\right)\right].\end{aligned} \tag{6.4.8}$$

With the help of (6.4.1) and (6.4.2), we arrive at

$$\hat{L}_+ = \hbar e^{i\varphi}\left(\frac{\partial}{\partial\theta} + i\cot\theta\frac{\partial}{\partial\varphi}\right). \tag{6.4.9}$$

A similar calculation leads to

$$\hat{L}_- = -\hbar e^{-i\varphi}\left(\frac{\partial}{\partial\theta} - i\cot\theta\frac{\partial}{\partial\varphi}\right). \tag{6.4.10}$$

Using now

$$\begin{aligned}\hat{L}_+\hat{L}_- &= -\hbar^2 e^{i\varphi}\left(\frac{\partial}{\partial\theta} + i\cot\theta\frac{\partial}{\partial\varphi}\right)\left\{e^{-i\varphi}\left(\frac{\partial}{\partial\theta} - i\cot\theta\frac{\partial}{\partial\varphi}\right)\right\}\\ &= -\hbar^2\left(\frac{\partial^2}{\partial\theta^2} + \cot\theta\frac{\partial}{\partial\theta} + \cot^2\theta\frac{\partial^2}{\partial\varphi^2} + i\frac{\partial}{\partial\varphi}\right),\end{aligned} \tag{6.4.11}$$

and

$$\hat{L}^2 = \hat{L}_+\hat{L}_- + \hat{L}_z^2 - \hbar\hat{L}_z, \tag{6.4.12}$$

we finally obtain the formula for $\hat{L}^2$ in spherical coordinates:

$$\hat{L}^2 = -\hbar^2 \left(\frac{\partial^2}{\partial \theta^2} + \cot\theta \frac{\partial}{\partial \theta} + \frac{1}{\sin^2\theta} \frac{\partial^2}{\partial \varphi^2} \right). \tag{6.4.13}$$

6.5 The Eigenfunctions of $\hat{L}^2$ and $\hat{L}_z$

We start with the eigenfunctions of $\hat{L}_z$. To find it, we must solve the following eigenvalue equation for $\hat{L}_z$:

$$-i\hbar \frac{\partial \Phi(\varphi)}{\partial \varphi} = \alpha \Phi(\varphi), \tag{6.5.1}$$

where $\Phi(\varphi)$ is the eigenfunction and α is the corresponding eigenvalue. In addition, we must require the fulfillment of the standard conditions of continuity, single-valuedness and boundedness by the resulting solutions. The solution is readily obtained as

$$\Phi(\varphi) = \Phi_0 e^{\frac{i}{\hbar}\alpha\varphi}, \tag{6.5.2}$$

where Φ_0 is a constant to be determined by the normalization condition. Clearly, these solutions are continuous and bounded in the entire range of variation of $\varphi (0 \leq \varphi \leq 2\pi)$. To check whether the solutions (6.5.2) are single-valued or not, we notice that the variable φ is cyclic and, hence, the solutions will be single-valued only if

$$e^{\frac{i}{\hbar}\alpha\varphi} = e^{\frac{i}{\hbar}\alpha(\varphi+2\pi)} \quad \Rightarrow \quad e^{\frac{i}{\hbar}2\pi\alpha} = 1. \tag{6.5.3}$$

This is possible only if

$$\alpha = \pm m\hbar, \tag{6.5.4}$$

where m is an integer including zero. Thus, $\alpha_m = m\hbar, m = 0, \pm1, \pm2, \pm3, ...$ are the eigenvalues of $\hat{L}_z$, which rightly happen to be the same as obtained earlier by algebraic means. However, there is an important difference: *the boundary conditions have eliminated the half-integer values for m.* It means that, if we measure L_z, we must get only those values which are integral multiples of $\hbar$. Once again, the characteristic number m is called the orbital magnetic quantum number.

Finally, using the normalization condition

$$|\Phi_0|^2 \int_0^{2\pi} d\varphi e^{i(m-m')\varphi} = 1 \tag{6.5.5}$$

we obtain $\Phi_0 = 1/\sqrt{2\pi}$. Hence, the normalized eigenfunctions of the operator $\hat{L}_z$ are

$$\Phi(\varphi) = \frac{1}{\sqrt{2\pi}} e^{im\varphi}, \qquad m = 0, \pm 1, \pm 2, \pm 3, ... \tag{6.5.6}$$

Eigen functions of $\hat{L}^2$

Using (6.4.13), the eigenvalue equation for $\hat{L}^2$ can be written as

$$-\hbar^2 \left\{ \frac{1}{\sin\theta} \frac{\partial}{\partial\theta} \left(\sin\theta \frac{\partial \psi_\ell^m(\theta,\varphi)}{\partial\theta} \right) + \frac{1}{\sin^2\theta} \frac{\partial^2 \psi_\ell^m(\theta,\varphi)}{\partial\varphi^2} \right\} = \hbar^2 \ell(\ell+1) \psi_\ell^m(\theta,\varphi). \tag{6.5.7}$$

By multiplying (6.5.7) throughout by $\sin^2\theta$, we can rewrite it as

$$\sin\theta \frac{\partial}{\partial\theta} \left(\sin\theta \frac{\partial \psi_\ell^m(\theta,\varphi)}{\partial\theta} \right) + \frac{\partial^2 \psi_\ell^m(\theta,\varphi)}{\partial\varphi^2} + \ell(\ell+1) \sin^2\theta \, \psi_\ell^m(\theta,\varphi) = 0. \tag{6.5.8}$$

Equation (6.5.8) can be solved by the method of separation of variables. So, we look for the solution in the form

$$\psi_\ell^m(\theta,\varphi) = \vartheta_\ell^m(\theta) \Phi(\varphi). \tag{6.5.9}$$

Using $\psi_\ell^m(\theta,\varphi)$, given by (6.5.9), in (6.5.8) and dividing the resulting equation throughout by $\vartheta_\ell^m \Phi$, we obtain

$$\left\{ \frac{1}{\vartheta_\ell^m} \left[\sin\theta \frac{d}{d\theta} \left(\sin\theta \frac{\partial \vartheta_\ell^m}{\partial\theta} \right) \right] + \ell(\ell+1) \sin^2\theta \right\} + \frac{1}{\Phi} \frac{\partial^2 \Phi}{\partial\varphi^2} = 0, \tag{6.5.10}$$

The first term in (6.5.10) is a function of θ, while the second term is a function of φ alone. Therefore, for this equation to be valid for any values of the independent variables θ and φ, each of the terms must be a constant such that the sum is equal to zero. That is,

$$\frac{1}{\vartheta_\ell^m} \left[\sin\theta \frac{d}{d\theta} \left(\sin\theta \frac{d\vartheta_\ell^m}{d\theta} \right) \right] + \ell(\ell+1) \sin^2\theta = m^2, \tag{6.5.11}$$

$$\frac{1}{\Phi} \frac{\partial^2 \Phi}{d\phi^2} = -m^2, \quad \Rightarrow \quad \frac{\partial^2 \Phi}{d\varphi^2} = -m^2 \Phi, \tag{6.5.12}$$

where m^2 is the real and positive separation constant.

Equation (6.5.12) coincides with the eigenvalue equation for the $\hat{L}_z$ operator considered earlier. The solutions of this equation, satisfying the standard conditions, are once again given by

$$\Phi(\varphi) = \frac{1}{\sqrt{2\pi}} e^{im\varphi}, \qquad m = 0, \pm 1, \pm 2, \pm 3, ... \tag{6.5.13}$$

The differential equation (6.5.11) for ϑ_ℓ^m can be written as

$$\frac{1}{\sin\theta}\frac{d}{d\theta}\left(\sin\theta\frac{d\vartheta_\ell^m}{d\theta}\right)+\left[\ell(\ell+1)-\frac{m^2}{\sin^2\theta}\right]\vartheta_\ell^m=0. \tag{6.5.14}$$

With the substitution $x=\cos\theta$, it can be reduced to

$$(1-x^2)\frac{d^2\vartheta_\ell^m(x)}{dx^2}-2x\frac{d\vartheta_\ell^m(x)}{dx}+\left[\ell(\ell+1)-\frac{m^2}{(1-x^2)}\right]\vartheta_\ell^m(x)=0, \tag{6.5.15}$$

which is known as the associated Legendre differential equation in the theory of special functions. Its solutions are given in terms of the associated Legendre polynomials $P_\ell^m(\cos\theta)$:

$$\vartheta_\ell^m(\theta)=C_m^\ell P_\ell^m(\cos\theta), \tag{6.5.16}$$

where C_m^ℓ are constants to be determined by the normalization condition. The associated Legendre polynomials $P_\ell^m(\cos\theta)$ are defined by

$$P_\ell^m(x)=(1-x^2)^{|m|/2}\frac{d^{|m|}}{dx^{|m|}}P_\ell(x), \tag{6.5.17}$$

where $x=\cos\theta$ and $P_\ell(x)$ is the Legendre polynomial of degree ℓ, which is calculated with the help of the formula

$$P_\ell(x)=\frac{1}{2^\ell\ell!}\frac{d^\ell}{dx^\ell}(x^2-1)^\ell. \tag{6.5.18}$$

In literature, the formulae (6.5.17) and (6.5.18) are known as Rodriguez formulae. Note that if we change m to $-m$ in (6.5.17), we get that

$$P_\ell^{-m}(x)=P_\ell^m(x). \tag{6.5.19}$$

From the normalization condition for $\psi_\ell^m(\theta,\phi)$, we get

$$\frac{|C_\ell^m|^2}{2\pi}\int_0^{2\pi}d\varphi\int_0^\pi d\theta\sin\theta\left(P_{\ell'}^{m'}(\cos\theta)\right)^*P_\ell^m(\cos\theta)=1. \tag{6.5.20}$$

Integrating over φ and using the following orthogonality condition

$$\int_0^\pi d\theta\sin\theta\left(P_{\ell'}^{m'}(\cos\theta)\right)^*P_\ell^m(\cos\theta)=\frac{2}{2\ell+1}\frac{(\ell+m)!}{(\ell-m)!}\delta_{\ell\ell'} \tag{6.5.21}$$

for the associated Legendre polynomials in (6.5.20), we arrive at

$$C_\ell^m = (-1)^m \sqrt{\frac{(2\ell+1)}{2}\frac{(\ell-m)!}{(\ell+m)!}} \qquad (m \geq 0). \tag{6.5.22}$$

The expression for ϑ_ℓ^m now reads

$$\vartheta_\ell^m = (-1)^m \sqrt{\frac{(2\ell+1)}{2}\frac{(\ell-m)!}{(\ell+m)!}} P_\ell^m(\cos\theta). \tag{6.5.23}$$

The full normalized eigenfunctions of $\hat{L}^2$ are now given by

$$\psi_\ell^m(\theta,\varphi) = \varepsilon\sqrt{\frac{(2\ell+1)(\ell-m)!}{4\pi(\ell+m)!}}\, P_\ell^m(\cos\theta)\, e^{im\varphi}, \tag{6.5.24}$$

where $\varepsilon = (-1)^m$ for $m \geq 0$ and $\varepsilon = 1$ for $m < 0$. Note that it is straightforward to check that $\psi_\ell^m(\theta,\varphi)$ are also eigenfunctions of $\hat{L}_z$ with eigenvalues $m\hbar$, where $m = 0, \pm 1, \pm 2, \pm 3$, and so on. If we refer to the theory of spherical functions, we recognize that the functions given by (6.5.24) are nothing but the normalized spherical harmonics, $Y_\ell^m(\theta,\varphi)$. Thus, the complete set of spherical harmonics, $Y_\ell^m(\theta,\varphi)$, constitutes the set of common eigenfunctions of $\hat{L}^2$ and $\hat{L}_z$:

$$\hat{L}^2 Y_\ell^m(\theta,\varphi) = \ell(\ell+1)\hbar^2 Y_\ell^m(\theta,\varphi), \tag{6.5.25}$$

$$\hat{L}_z Y_\ell^m(\theta,\varphi) = m\hbar Y_\ell^m(\theta,\varphi), \tag{6.5.26}$$

$$\int_0^{2\pi} d\varphi \int_0^{\pi} d\theta\, \sin\theta\, Y_{\ell'}^{m'}(\theta,\varphi)\, Y_\ell^m(\theta,\varphi) = \delta_{\ell'\ell}\delta_{m'm}, \tag{6.5.27}$$

where $\ell = 0, 1, 2, 3, ...$ and $m = -\ell, -\ell+1, \ldots, \ell-1, \ell$.

For later references it is useful to have analytical expressions for a few of the frequently used associated Legendre functions and spherical harmonics. We have presented them in Tables 6.1 and 6.2, respectively.

Comments: 1. Since the magnitude of the angular momentum, for a given ℓ, is given by $\sqrt{\ell(\ell+1)}\,\hbar$, different values of ℓ correspond to the states with correspondingly different values of the angular momentum. In atomic physics, these states are denoted by the letters $s, p, d, f, ...$ according to the following scheme:

$\ell =$	0	1	2	3	...
state	s	p	d	f	...

Table 6.1 The first few associated Legendre functions.

$P_0^0 = 1$	$P_2^0 = \frac{1}{2}(3\cos^2\theta - 1)$
$P_1^1 = \sin\theta$	$P_3^3 = 15\sin\theta(1-\cos^2\theta)$
$P_1^0 = \cos\theta$	$P_3^2 = 15\sin^2\theta\cos\theta$
$P_2^2 = 3\sin^2\theta$	$P_3^1 = \frac{3}{2}\sin\theta\,(5\cos^2\theta - 1)$
$P_2^1 = 3\sin\theta\,\cos\theta$	$P_3^0 = \frac{1}{2}(5\cos^3\theta - 3\cos\theta)$

Table 6.2 The first few spherical harmonics.

$Y_0^0 = \left(\frac{1}{4\pi}\right)^{1/2}$	$Y_2^{\pm 2} = \left(\frac{15}{32\pi}\right)^{1/2}\sin^2\theta\, e^{\pm 2i\varphi}$
$Y_1^0 = \left(\frac{3}{4\pi}\right)^{1/2}\cos\theta$	$Y_3^0 = \left(\frac{7}{16\pi}\right)^{1/2}(5\cos^3\theta - 3\cos\theta)$
$Y_1^{\pm 1} = \mp\left(\frac{3}{8\pi}\right)^{1/2}\sin\theta\, e^{\pm i\varphi}$	$Y_3^{\pm 1} = \mp\left(\frac{21}{64\pi}\right)^{1/2}\sin\theta\,(5\cos^2\theta - 1)\, e^{\pm i\varphi}$
$Y_2^0 = \left(\frac{5}{16\pi}\right)^{1/2}(3\cos^2\theta - 1)$	$Y_3^{\pm 2} = \left(\frac{105}{32\pi}\right)^{1/2}\sin^2\theta\,\cos\theta\, e^{\pm 2i\varphi}$
$Y_2^{\pm 1} = \mp\left(\frac{15}{8\pi}\right)^{1/2}\sin\theta\,\cos\theta\, e^{\pm i\varphi}$	$Y_3^{\pm 3} = \mp\left(\frac{35}{64\pi}\right)^{1/2}\sin^3\theta\, e^{\pm 3i\varphi}$

2. In the s state, $\ell = 0$ and the total angular momentum of the particle is zero. Consequently, the value of m_ℓ is also zero. This state is non-degenerate. In the state p, we have $\ell = 1$ and hence the quantum number m can take three values $-1, 0, +1$. Consequently, the projection of the angular momentum on the z-axis can have three values $-\hbar, 0, +\hbar$, respectively. So, the p-state is three-fold degenerate. Similarly, the d-state is 5-fold degenerate, the f-state is 7-fold degenerate, and so on.

Example 6.5.1: Determine the wave function that represents a non-trivial state of a system, such that it is an eigenstate of $\hat{L}^2$ with eigenvalue $\hbar^2\ell(\ell+1) = 2\hbar^2$ and also an eigenstate of $\hat{L}_x$ with eigenvalue 0.

Solution: It is given that an eigenvalue of $\hat{L}^2$ equals $2\hbar^2$. That is, $\hbar^2\ell(\ell+1) = 2\hbar^2$. It implies that $\ell = 1$. Since for $\ell = 1$, the magnetic quantum number m, takes three values -1,0 and

$+1$, the given state is three-fold degenerate. The required wave function, $\psi(\theta,\varphi)$, will be a linear combination of the spherical harmonics Y_1^m with $m=-1,0,1$. Therefore, we can write

$$\psi(\theta,\varphi)=AY_1^0+BY_1^1+CY_1^{-1}, \tag{6.5.28}$$

A,B and C are arbitrary constants. Since this wave function is also an eigenfunction of $\hat{L}_x=\frac{\hat{L}_++\hat{L}_-}{2}$ with eigenvalue 0, we have

$$\frac{\hat{L}_++\hat{L}_-}{2}\left(AY_1^0+BY_1^1+CY_1^{-1}\right)=0. \tag{6.5.29}$$

With the help of the formulae

$$\hat{L}_\pm Y_\ell^m=\hbar\sqrt{(\ell\mp m)(\ell\pm m+1)}Y_\ell^{m\pm1}, \tag{6.5.30}$$

Equation (6.5.30) yields

$$\left[BY_1^1+(A+C)Y_1^0+BY_1^{-1}\right]=0. \tag{6.5.31}$$

Since the spherical harmonics are linearly independent, each of the coefficients in (6.5.31) must be zero. We obtain $A=-C$ and $B=0$. Hence, the normalized wave function is

$$\psi(\theta,\varphi)=\frac{1}{\sqrt{2}}\left(Y_1^1-Y_1^{-1}\right). \tag{6.5.32}$$

Example 6.5.2: A rigid rotator consists of two particles, each of mass m, attached to the ends of a weightless rigid rod of length a, whose midpoint is fixed in space. The system can rotate about this midpoint. Let this rigid rotator be free from any external force field. Find the rotational energy states and the corresponding eigenfunctions of the rotator.

Solution: The energy of the rotator is purely kinetic

$$E=\frac{1}{2}I\omega^2,\quad I=2m\left(\frac{a}{2}\right)^2=\frac{ma^2}{2}, \tag{6.5.33}$$

where I is the moment of inertia of the rotator about its midpoint and ω is its angular velocity. Since $L=I\omega$ is the angular momentum about the axis passing through the midpoint of the rotator, the Hamiltonian is given by

$$H=\frac{1}{2I}(I^2\omega^2)=\frac{L^2}{2I}. \tag{6.5.34}$$

The time-independent Schrödinger equation reads

$$\hat{H}\phi = E\phi, \;\Rightarrow\; \hat{L}^2\phi = E'\phi, \tag{6.5.35}$$

where $E' = 2IE$. Equation (6.5.35) is nothing but the eigenvalue equation for the square of the angular momentum operator in spherical coordinates

$$\hat{L}^2\phi_\ell^m(\theta,\varphi) = E'\phi_\ell^m(\theta,\varphi), \tag{6.5.36}$$

which we have already solved. According to the earlier solution, the energy levels of the rotator are

$$E'_\ell = \hbar\sqrt{\ell(\ell+1)}, \;\Rightarrow\; E_\ell = \frac{\hbar\sqrt{\ell(\ell+1)}}{2I}, \quad \ell = 0,1,2,3,\ldots \tag{6.5.37}$$

The corresponding normalized eigenfunctions are given by the spherical harmonics $Y_\ell^m(\theta,\varphi)$:

$$\phi_\ell^m(\theta,\varphi) = Y_\ell^m(\theta,\varphi). \tag{6.5.38}$$

Note that the energy levels of the rigid rotator are $(2\ell+1)$-fold degenerate: for a given value of ℓ, there are $(2\ell+1)$ eigenfunctions: $Y_\ell^\ell, Y_\ell^{\ell-1}, Y_\ell^{\ell-2}, \ldots, Y_\ell^{-\ell}$, all corresponding to the same energy given by (6.5.38).

Example [1] 6.5.3: A particle is in the state with the wave function

$$|\psi(x,y,z)\rangle = A\left\{\frac{1}{\sqrt{8\pi}}\frac{x+iy}{r} + \frac{1}{\sqrt{16\pi}}\frac{2z^2-x^2-y^2}{r^2} + \sqrt{\frac{3}{4\pi}}\frac{z}{r}\right\}, \tag{6.5.39}$$

where A is an arbitrary constant. (a) Find A. (b) What is the average value of the orbital angular momentum in this state? (c) What is the average value of $\hat{L}_+$ in this state?

Solution: Using the expressions for x, y and z in spherical polar coordinates and the expressions for the spherical harmonics, we get

$$|\psi(\theta,\varphi)\rangle = A\left\{\sqrt{\frac{1}{3}}Y_1^0(\theta,\varphi) + \sqrt{\frac{1}{3}}Y_1^1(\theta,\varphi) + \sqrt{\frac{1}{5}}Y_2^0(\theta,\varphi)\right\}. \tag{6.5.40}$$

(a) The normalization condition for the wave function reads

$$\langle\psi|\psi\rangle = |A|^2\left\{\frac{1}{3}\langle Y_1^0|Y_1^0\rangle + \frac{1}{3}\langle Y_1^1|Y_1^1\rangle + \frac{1}{5}\langle Y_2^0|Y_2^0\rangle\right\} = 1. \tag{6.5.41}$$

[1] adapted from N. Zettili, Quantum Mechanics: Concepts and Applications, John Wiley, 2009.

Using the orthonormality of the spherical harmonics, we get the normalization constant A and the normalized wave function as

$$A = \sqrt{\frac{15}{13}}, \quad \Rightarrow \quad |\psi(\theta,\varphi)\rangle = \sqrt{\frac{5}{13}}Y_1^0(\theta,\varphi) + \sqrt{\frac{5}{13}}Y_1^1(\theta,\varphi) + \sqrt{\frac{3}{13}}Y_2^0(\theta,\varphi). \tag{6.5.42}$$

(b) The average value of the orbital angular momentum is given by

$$\langle L\rangle = \sqrt{\langle\psi|\hat{L}^2|\psi\rangle} = \sqrt{2\times\frac{5}{13}(2\hbar^2)+\frac{3}{13}(6\hbar^2)} = \sqrt{\frac{38}{13}}\hbar. \tag{6.5.43}$$

(c) The average value of $\hat{L}_+$ is

$$\langle\hat{L}_+\rangle = \langle\psi|\hat{L}_+|\psi\rangle = \frac{5}{13}\sqrt{2}\hbar + \frac{3}{13}\frac{3\sqrt{8}}{13}\hbar = \frac{11\sqrt{2}}{13}\hbar. \tag{6.5.44}$$

6.6 Space Quantization

Space quantization is essentially the quantization of the direction of the orbital angular momentum $\vec{L}$ in space with respect to the z-axis.

This can be understood as follows. We have seen that, although we can precisely determine the magnitudes of the total angular momentum $\vec{L}$ and its projection on the z-axis L_z, we cannot have any information about the other two components (L_x and L_y) of $\vec{L}$. Since the knowledge of all the three components of a vector is essential for specifying its direction, it follows that, in quantum mechanics, the direction of the angular momentum cannot be specified. One can then ask the question: 'Does this mean that $\vec{L}$ can have an arbitrary direction in space in a given quantum state in which it has precisely determined magnitude'? The answer is 'No'. Let us explain. We have established that both L and L_z are quantized. The discrete set of values that L can have are given by $\hbar\sqrt{\ell(\ell+1)}$, where ℓ is an integer including zero. On the other hand, the discrete values that L_z can have are given by $m\hbar$, where $m = -\ell, -\ell+1, \ldots, \ell-1, \ell$ is the orbital magnetic quantum number. It is because of this intrinsic relation between ℓ and m that $\vec{L}$ cannot be arbitrarily oriented in space. For a given value of the orbital magnetic quantum number m, it has to be inclined to the z-axis at a precisely defined angle. This angle, say θ, can take only discrete values given by

$$\theta = \cos^{-1}\left(\frac{m\hbar}{\hbar\sqrt{\ell(\ell+1)}}\right) = \cos^{-1}\left(\frac{m}{\sqrt{\ell(\ell+1)}}\right). \tag{6.6.1}$$

For instance, consider the state with $\ell = 1$. The magnitude of $\vec{L}$ in this state is $\sqrt{2}\hbar$. The values that m can have in this state are $-1,0$ and $+1$. Correspondingly, the angular momentum vector $\vec{L}$ can be inclined to the z-axis only at three discrete angles θ:

$$\theta_1 = \cos^{-1}\left(\frac{1}{\sqrt{2}}\right), \quad \theta_2 = \cos^{-1}(0), \quad \theta_3 = -\cos^{-1}\left(\frac{1}{\sqrt{2}}\right). \tag{6.6.2}$$

In other words, the direction of $\vec{L}$ in space is quantized. This can be graphically demonstrated as shown in Fig. 6.1.

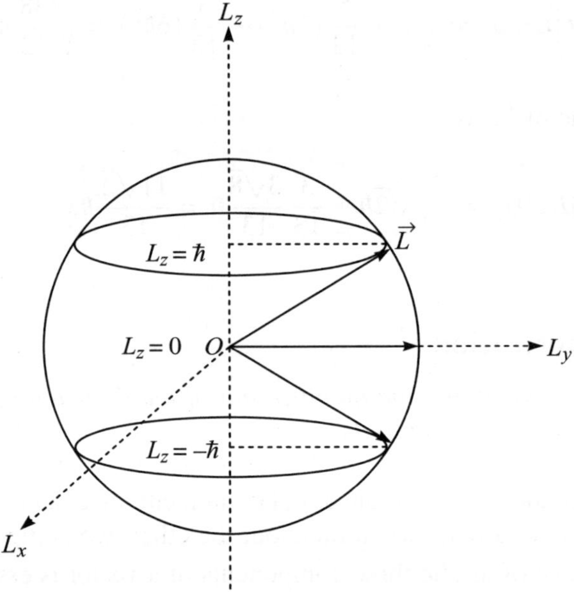

Figure 6.1 Graphical representation of the quantization of the direction of $\vec{L}$ for $\ell = 1$, where the radius of the sphere is equal to $L = \sqrt{2}\hbar$.

Note that in the Bohr theory of hydrogen atom too the angular momentum is quantized. However, the situation in quantum mechanics is radically different from the quantization of angular momentum in the Bohr theory. In the Bohr theory, all the three components of the angular momentum, $\vec{L}$, are strictly determined and hence, we can talk about the direction of $\vec{L}$ in space. In quantum mechanics, however, only one component, L_z, of $\vec{L}$ is determined, and hence, we can talk of the orientation of the angular momentum vector only with respect to a chosen axis; its overall orientation in space remains undefined.

In order to further clarify the concept of space quantization, consider once again the state with $\ell = 1$, that is, the p state. There are three eigenfunctions

$$\psi_+ = \sin\theta e^{i\phi}, \qquad \psi_0 = \cos\theta, \qquad \psi_- = \sin\theta e^{-i\phi} \tag{6.6.3}$$

of $\hat{L}^2$ with the same eigenvalue $2\hbar$. Since the state is three-fold degenerate, the wave function, ψ, of this state is given by the superposition of these three eigenfunctions,that is,

$$\psi = c_1\psi_+ + c_2\psi_0 + c_3\psi_-, \tag{6.6.4}$$

where c_1, c_2 and c_3 are arbitrary and in general, complex coefficients.

Because of the isotropy of space, all directions in space are equivalent unless we pick one of them up by imposing some specific physical condition. Therefore, if we want to know the projection of the angular momentum along a direction, we must somehow isolate it in space. For instance, this can be done by switching on a magnetic field parallel to this direction. If we now measure the projection of $\vec{L}$ on this direction and find the value $+\hbar$, then the state after the measurement will be described by the wave function ψ_+ and we shall have $|c_1|^2 = 1$, $|c_2|^2 = |c_3|^2 = 0$ and $\psi = \psi_+$, that is, L_z will have a definite value, while the other two components of $\vec{L}$ will be indeterminate. If we now decide to know the projection of $\vec{L}$ on some other axis, then we must switch off the magnetic field that was along the previous direction and switch it on along the newly chosen direction. As a result, the state preceding the measurement is destroyed and a new state comes into existence in which, once again, only one component of $\vec{L}$ can be specified.

Example 6.6.1: Find the angles between the angular momentum vector $\vec{L}$ and the z-axis, giving all possible orientations of $\vec{L}$ for $\ell = 3$.

Solution: In this case, $m = -3, -2, -1, 0, 1, 2, 3$. We know that the angle between the angular momentum vector $\vec{L}$ and the z-axis is given by

$$\theta_m = \cos^{-1}\left(\frac{m}{\sqrt{\ell(\ell+1)}}\right). \tag{6.6.5}$$

Therefore, we get

$$\theta_{-3} = \cos^{-1}\left(\frac{-3}{2\sqrt{3}}\right) = 150^\circ, \quad \theta_{-2} = \cos^{-1}\left(\frac{-2}{2\sqrt{3}}\right) = 125.26^\circ, \tag{6.6.6}$$

$$\theta_{-1} = \cos^{-1}\left(\frac{-1}{2\sqrt{3}}\right) = 106.78^\circ, \quad \theta_0 = \cos^{-1}(0) = 90^\circ, \tag{6.6.7}$$

$$\theta_1 = \cos^{1}\left(\frac{-1}{2\sqrt{3}}\right) = 73.23^\circ, \quad \theta_2 = \cos^{-1}\left(\frac{2}{2\sqrt{3}}\right) = 54.73^\circ, \tag{6.6.8}$$

$$\theta_3 = \cos^{-1}\left(\frac{3}{2\sqrt{3}}\right) = 30^\circ. \tag{6.6.9}$$

6.7 Matrix Representation of Angular Momentum Operators

We know that $\hat{L}^2$ and $\hat{L}_z$ commute. Therefore, it is convenient to take the complete set of spherical harmonics $\{Y_\ell^m(\theta,\varphi)$, which happens to be the common set of eigenfunctions of $\hat{L}^2$ and $\hat{L}_z$, as the basis set in the Hilbert space.

Obviously, $\hat{L}^2$ and $\hat{L}_z$ are diagonal in this basis

$$\int_0^{2\pi} d\varphi \int_0^{\pi} d\theta \, \sin\theta \, Y_{\ell'}^{m'} \, \hat{L}^2 Y_\ell^m = \hbar^2 \ell(\ell+1)\delta_{\ell'\ell}\delta_{m'm}, \tag{6.7.1}$$

$$\int_0^{2\pi} d\varphi \int_0^{\pi} d\theta \, \sin\theta \, Y_{\ell'}^{m'} \, \hat{L}_z Y_\ell^m = m\hbar\delta_{\ell'\ell}\delta_{m'm}. \tag{6.7.2}$$

The diagonal elements of the corresponding matrices L^2 and L_z are $\hbar^2\ell(\ell+1)$ and $m\hbar$, respectively. The operators $\hat{L}_+$ and $\hat{L}_-$ do not commute with $\hat{L}_z$. Therefore, they are represented by non-diagonal matrices in this basis. In order to determine these matrices, we shall have to first find the result of the action of the operators $\hat{L}_+$ and $\hat{L}_-$ on the basis functions $\{Y_\ell^m(\theta,\varphi)\}$.

What we know is that the operators $\hat{L}_+$ and $\hat{L}_-$, while acting on the eigenfunction of $\hat{L}_z$, change the value of m by unity. Therefore, we have

$$\hat{L}_\pm Y_\ell^m = C_{\ell m}^{\pm} \, Y_\ell^{m\pm1}, \tag{6.7.3}$$

where $C_{\ell m}^{\pm}$ are constants. They can be determined by requiring that the functions $Y_\ell^{m\pm1}$ be orthonormal. It is easy to check that $\hat{L}_+$ and $\hat{L}_-$ are hermitian conjugates of $\hat{L}_-$ and $\hat{L}_+$, respectively. So, we have

$$\int_0^{2\pi} d\varphi \int_0^{\pi} d\theta \, \sin\theta \, Y_{\ell'}^{*\,m'} \, (\hat{L}_\pm^\dagger \hat{L}_\pm) Y_\ell^m = \int_0^{2\pi} d\varphi \int_0^{\pi} d\theta \, \sin\theta \, Y_{\ell'}^{*\,m'} \, (\hat{L}_\mp \hat{L}_\pm) Y_\ell^m. \tag{6.7.4}$$

Alternatively,

$$\int_0^{2\pi} d\varphi \int_0^{\pi} d\theta \, \sin\theta \, Y_{\ell'}^{*\,m'} \, (\hat{L}_\pm^\dagger \hat{L}_\pm) Y_\ell^m = |C_{\ell m}^{\pm}|^2 \int_0^{2\pi} d\varphi \int_0^{\pi} d\theta \, \sin\theta \, Y_{\ell'}^{*\,(m'\pm1)} \, Y_\ell^{m\pm1}. \tag{6.7.5}$$

Therefore, on one hand, if the functions $Y_\ell^{m\pm1}$ are to be orthonormal, we have

$$\int_0^{2\pi} d\varphi \int_0^{\pi} d\theta \, \sin\theta \, Y_{\ell'}^{*\,m'} \, (\hat{L}_\pm^\dagger \hat{L}_\pm) Y_\ell^m = |C_{\ell m}^{\pm}|^2, \tag{6.7.6}$$

On the other hand, using the relations

$$\hat{L}_\mp \hat{L}_\pm = \hat{L}^2 - \hat{L}_z^2 \mp \hbar \hat{L}_z, \tag{6.7.7}$$

$$\hat{L}^2 Y_\ell^m = \ell(\ell+1)\hbar^2 Y_\ell^m, \tag{6.7.8}$$

$$\hat{L}_z Y_\ell^m = m\hbar Y_\ell^m, \tag{6.7.9}$$

we obtain

$$\int_0^{2\pi} d\varphi \int_0^{\pi} d\theta \sin\theta\, Y_{\ell'}^{*\,m'} (\hat{L}_\pm^\dagger \hat{L}_\pm) Y_\ell^m = \int_0^{2\pi} d\varphi \int_0^{\pi} d\theta \sin\theta\, Y_{\ell'}^{*\,m'} (\hat{L}^2 - \hat{L}_z^2 \mp \hbar \hat{L}_z) Y_\ell^m$$

$$= \left[\hbar^2 \ell(\ell+1) - \hbar^2 m^2 \mp \hbar^2 m\right] \int_0^{2\pi} d\varphi \int_0^{\pi} d\theta \sin\theta\, Y_{\ell'}^{*\,(m'\pm 1)} Y_\ell^{m\pm 1}$$

$$= \hbar^2 (l \mp m)(l \pm m + 1), \tag{6.7.10}$$

where, besides (6.7.1) and (6.7.2), we have used the orthonormality of the spherical harmonics given by (6.5.27). If we now compare (6.7.6) and (6.7.10) we get

$$C_{\ell m}^{\pm} = \hbar \sqrt{(l \mp m)(l \pm m + 1)}. \tag{6.7.11}$$

As a consequence, we have

$$\hat{L}_+ Y_\ell^m = \hbar \sqrt{(l-m)(l+m+1)}\, Y_\ell^{m+1}, \tag{6.7.12}$$

$$\hat{L}_- Y_\ell^m = \hbar \sqrt{(l+m)(l-m+1)}\, Y_\ell^{m-1}. \tag{6.7.13}$$

Since $\hat{L}_x = (\hat{L}_+ + \hat{L}_-)/2$ and $\hat{L}_y = (\hat{L}_+ - \hat{L}_-)/2i$, we get

$$\hat{L}_x Y_\ell^m = \frac{1}{2}[\hat{L}_+ + \hat{L}_-] Y_\ell^m$$

$$= \frac{\hbar}{2}\left[\sqrt{(l-m)(l+m+1)}\, Y_\ell^{m+1} + \sqrt{(l+m)(l-m+1)}\, Y_\ell^{m-1}\right] \tag{6.7.14}$$

$$\hat{L}_y Y_\ell^m = \frac{1}{2i}[\hat{L}_+ + \hat{L}_-] Y_\ell^m$$

$$= \frac{\hbar}{2i}\left[\sqrt{(l-m)(l+m+1)}\, Y_\ell^{m+1} - \sqrt{(l+m)(l-m+1)}\, Y_\ell^{m-1}\right]. \tag{6.7.15}$$

From (6.7.9), (6.7.12), (6.7.13) and the orthonormality condition for the spherical harmonics, (6.5.27), we find the matrix elements of $\hat{L}_+$, $\hat{L}_-$ and $\hat{L}_z$ as

$$\int_0^{2\pi} d\varphi \int_0^{\pi} d\theta \, \sin\theta \, Y_{\ell'}^{*\,m'} \, (\hat{L}_+ Y_\ell^m) = \hbar \sqrt{(l-m)(l+m+1)} \delta_{\ell'\ell} \delta_{m',m+1}, \tag{6.7.16}$$

$$\int_0^{2\pi} d\varphi \int_0^{\pi} d\theta \, \sin\theta \, Y_{\ell'}^{*\,m'} \, (\hat{L}_- Y_\ell^m) = \hbar \sqrt{(l+m)(l-m+1)} \delta_{\ell'\ell} \delta_{m',m-1}, \tag{6.7.17}$$

$$\int_0^{2\pi} d\varphi \int_0^{\pi} d\theta \, \sin\theta \, Y_{\ell'}^{*\,m'} \, (\hat{L}_z Y_\ell^m) = m\hbar \, \delta_{\ell'\ell} \delta_{m'm}. \tag{6.7.18}$$

With the help of the aforementioned calculated matrix elements, we can easily compute the matrices corresponding to the operators $\hat{L}^2$, $\hat{L}_z$, $\hat{L}_\pm$, $\hat{L}_x$ and $\hat{L}_y$ in a state with a definite value of the angular momentum,that is, for a given value of ℓ.

Example 6.7.1: Consider a particle in a superposition state with the wave function

$$|\psi(\theta,\varphi)\rangle = \sqrt{\frac{1}{5}} Y_1^{-1}(\theta,\varphi) + A Y_1^0 + \sqrt{\frac{1}{5}} Y_1^1(\theta,\varphi), \tag{6.7.19}$$

where A is an arbitrary constant and Y_ℓ^m are the spherical harmonics. (a) Find A so that ψ is normalized. (b) What is the probability that a measurement of L_z will yield a value $L_z = 0$? (c) Find the expectation values of $\hat{L}^2$ and L_+ in this state.

Solution:

(a) For the normalized wave function, we must have

$$\langle\psi|\psi\rangle = \int_0^{2\pi} d\varphi \int_0^{\pi} d\theta \, \psi^*(\theta,\varphi) \, \psi(\theta,\varphi) \sin\theta d\theta = 1. \tag{6.7.20}$$

Using the orthonormality condition for the spherical harmonics, we get

$$\langle\psi|\psi\rangle = \frac{2}{5} + A^2 = 1, \Rightarrow \quad A = \sqrt{\frac{3}{5}}. \tag{6.7.21}$$

(b) The normalized wave function is now given by

$$\psi(\theta,\varphi) = \sqrt{\frac{1}{5}} Y_1^{-1}(\theta,\varphi) + \sqrt{\frac{3}{5}} Y_1^0 + \sqrt{\frac{1}{5}} Y_1^1(\theta,\varphi), \tag{6.7.22}$$

and therefore the probability of finding the value $L_z = 0$ is

$$P = \frac{|\langle Y_1^0|\psi\rangle|^2}{\langle\psi|\psi\rangle} = \frac{3}{5}. \tag{6.7.23}$$

(c) From (6.7.8), we have

$$\hat{L}^2|\psi(\theta,\varphi)\rangle = \hat{L}^2\left[\sqrt{\frac{1}{5}}Y_1^{-1}(\theta,\varphi)+\sqrt{\frac{3}{5}}Y_1^0+\sqrt{\frac{1}{5}}Y_1^1(\theta,\varphi)\right] = 2\hbar^2|\psi(\theta,\varphi)\rangle. \tag{6.7.24}$$

The expectation value of $\hat{L}^2$ will be

$$\langle\hat{L}^2\rangle = \frac{\langle\psi|\hat{L}^2|\psi\rangle}{\langle\psi|\psi\rangle} = 2\hbar^2\frac{\langle\psi|\psi\rangle}{\langle\psi|\psi\rangle} = 2\hbar^2. \tag{6.7.25}$$

Using (6.7.12), we get

$$\hat{L}_+|\psi(\theta,\varphi)\rangle = \sqrt{\frac{6}{5}}Y_1^0+\sqrt{\frac{6}{5}}Y_1^1. \tag{6.7.26}$$

Therefore, the expectation value of $\hat{L}_+$ is given by

$$\langle\hat{L}_+\rangle = \frac{\langle\psi|\hat{L}_+|\psi\rangle}{\langle\psi|\psi\rangle} = \frac{\sqrt{6}}{5}\hbar+\frac{\sqrt{6}}{5}\hbar = \frac{2\sqrt{3}}{5}\hbar. \tag{6.7.27}$$

Example 6.7.2: Consider the case in which $\ell = 1$. Find the matrices representing the operators $\hat{L}^2$, $\hat{L}_z$, $\hat{L}_\pm$, $\hat{L}_x$ and $\hat{L}_y$. Show that the matrices L_x and L_y do not commute. Find their commutator.

Solution: For $\ell = 1$, we have $m = -1,0,1$ and the joint eigenfunctions of $\hat{L}^2$ and $\hat{L}_z$ are: $\left[Y_1^1, Y_1^0, Y_1^{-1}\right]$. Therefore, the matrix representing $\hat{L}^2$ is given by

$$L^2 = \begin{pmatrix} \langle Y_1^1,\hat{L}^2Y_1^1\rangle & \langle Y_1^1,\hat{L}^2Y_1^0\rangle & \langle Y_1^1,\hat{L}^2Y_1^{-1}\rangle \\ \langle Y_1^0,\hat{L}^2Y_1^1\rangle & \langle Y_1^0,\hat{L}^2Y_1^0\rangle & \langle Y_1^0,\hat{L}^2Y_1^{-1}\rangle \\ \langle Y_1^{-1},\hat{L}^2Y_1^1\rangle & \langle Y_1^{-1},\hat{L}^2Y_1^0\rangle & \langle Y_1^{-1},\hat{L}^2Y_1^{-1}\rangle \end{pmatrix} = 2\hbar^2\begin{pmatrix} 1 & 0 & 0 \\ 0 & 1 & 0 \\ 0 & 0 & 1 \end{pmatrix}, \tag{6.7.28}$$

where we have used (6.5.25) and (6.5.27) to get

$$\langle Y_l^k,\hat{L}^2Y_n^m\rangle = \hbar^2 n(n+1)\int_0^{2\pi}d\varphi\int_0^{\pi}d\theta\,\sin\theta\,Y_l^{*\,k}Y_n^m) = \hbar^2 n(n+1)\delta_{\ell n}\delta_{km}. \tag{6.7.29}$$

Similarly, by making use of (6.5.26) and (6.5.27), we obtain the matrix representing $\hat{L}_z$ in this basis:

$$L_z = \begin{pmatrix} \langle Y_1^1, \hat{L}_z Y_1^1 \rangle & \langle Y_1^1, \hat{L}_z Y_1^0 \rangle & \langle Y_1^1, \hat{L}_z Y_1^{-1} \rangle \\ \langle Y_1^0, \hat{L}_z Y_1^1 \rangle & \langle Y_1^0, \hat{L}_z Y_1^0 \rangle & \langle Y_1^0, \hat{L}_z Y_1^{-1} \rangle \\ \langle Y_1^{-1}, \hat{L}_z Y_1^1 \rangle & \langle Y_1^{-1}, \hat{L}_z Y_1^0 \rangle & \langle Y_1^{-1}, \hat{L}_z Y_1^{-1} \rangle \end{pmatrix} = \hbar \begin{pmatrix} 1 & 0 & 0 \\ 0 & 0 & 0 \\ 0 & 0 & -1 \end{pmatrix}. \tag{6.7.30}$$

The matrices, corresponding to $\hat{L}_+$ and $\hat{L}_-$ in this basis, are calculated to be

$$L_+ = \sqrt{2}\hbar \begin{pmatrix} 0 & 1 & 0 \\ 0 & 0 & 1 \\ 0 & 0 & 0 \end{pmatrix}, \quad L_- = \sqrt{2}\hbar \begin{pmatrix} 0 & 0 & 0 \\ 1 & 0 & 0 \\ 0 & 1 & 0 \end{pmatrix}. \tag{6.7.31}$$

Taking into account that $\hat{L}_x = (\hat{L}_+ + \hat{L}_-)/2$ and $\hat{L}_y = (\hat{L}_+ - \hat{L}_-)/2i$, we get

$$L_x = \frac{\hbar}{\sqrt{2}} \begin{pmatrix} 0 & 1 & 0 \\ 1 & 0 & 1 \\ 0 & 1 & 0 \end{pmatrix}, \quad L_y = \frac{\hbar}{\sqrt{2}} \begin{pmatrix} 0 & -i & 0 \\ i & 0 & -i \\ 0 & i & 0 \end{pmatrix}. \tag{6.7.32}$$

Now, we have to check whether L_x and L_y commute or not. We have

$$L_x L_y = \frac{\hbar^2}{2} \begin{pmatrix} 0 & 1 & 0 \\ 1 & 0 & 1 \\ 0 & 1 & 0 \end{pmatrix} \begin{pmatrix} 0 & -i & 0 \\ i & 0 & -i \\ 0 & i & 0 \end{pmatrix} = \frac{\hbar^2}{2} \begin{pmatrix} i & 0 & -i \\ 0 & 0 & 0 \\ i & 0 & -i \end{pmatrix}. \tag{6.7.33}$$

On the other hand,

$$L_y L_x = \frac{\hbar^2}{2} \begin{pmatrix} 0 & -i & 0 \\ i & 0 & -i \\ 0 & i & 0 \end{pmatrix} \begin{pmatrix} 0 & 1 & 0 \\ 1 & 0 & 1 \\ 0 & 1 & 0 \end{pmatrix} = \frac{\hbar^2}{2} \begin{pmatrix} -i & 0 & -i \\ 0 & 0 & 0 \\ i & 0 & i \end{pmatrix}. \tag{6.7.34}$$

Clearly, $L_x L_y \neq L_y L_x$ and hence the matrices L_x and L_y do not commute.

Their commutator is given by

$$\begin{aligned} L_x L_y - L_y L_x &= \frac{\hbar^2}{2} \begin{pmatrix} i & 0 & -i \\ 0 & 0 & 0 \\ i & 0 & -i \end{pmatrix} - \begin{pmatrix} -i & 0 & -i \\ 0 & 0 & 0 \\ i & 0 & i \end{pmatrix} \\ &= i\hbar^2 \begin{pmatrix} 1 & 0 & 0 \\ 0 & 0 & 0 \\ 0 & 0 & -1 \end{pmatrix} = i\hbar \hat{L}_z, \end{aligned} \tag{6.7.35}$$

as it should be.

Homework Problems

1. Calculate the commutators
   (a) $[\hat{x}, \hat{L}_x], [\hat{y}, \hat{L}_y]$ and $[\hat{z}, \hat{L}_z]$,
   (b) $[\hat{p}_x, \hat{L}_x], [\hat{p}_y, \hat{L}_y]$ and $[\hat{p}_z, \hat{L}_z]$.
   (c) $[\hat{x}, \hat{L}^2]$, and $[\hat{p}_x, \hat{L}^2]$.

2. Use the lowering operator, to find the angular dependence (without worrying about the normalization) of $Y_4^m(\theta,\phi)$ for $m = 3,2,1,0$. You are given $Y_4^4(\theta,\phi) = Ae^{4i\phi}\sin^4\theta$.

3. Use the expression for the spherical harmonics to construct $Y_l^l(\theta,\phi)$ and $Y_3^2(\theta,\phi)$. Check that they satisfy the angular equation for appropriate values of l and m.

4. A particle is in the state

$$|\psi\rangle = \frac{1}{\sqrt{5}}\, Y_2^0(\theta,\phi) + \frac{2}{\sqrt{5}}\, Y_2^{-1}(\theta,\phi) - \frac{2}{\sqrt{5}}\, Y_2^1(\theta,\phi),$$

   which is a superposition of the normalized eigenstates, $Y_\ell^m(\theta,\phi)$, of the $\hat{L}^2$ operator. Calculate the value of the total angular momentum of the particle in this state. Also, calculate the expectation value of the operator $\hat{L}_+\hat{L}_-$ in this state.

5. A particle moving in a two dimensional harmonic oscillator potential is in a state described by the wave function

$$\psi(\rho,\phi) = \frac{3}{\sqrt{\pi}\, a^3}\, \rho^2\, e^{-(\rho^2/2a^2)} \left(1 - \frac{4}{3}\sin^2\phi\right)\sin\phi,$$

   where ρ and ϕ are the planar polar coordinates and a is a constant. If the projection of the angular momentum on the axis perpendicular to the plane of oscillation is measured, what is (are) the possible value (values) that can be obtained?

6. Calculate numerical values for the total angular momentum $|\vec{L}|$, and L_z, for the $2p$ and $4d$ states of hydrogen.

7. Consider the case when $l = 1$. Find the matrices representing the operators $\hat{L}_x$ in the basis consisting of the eigenvectors of $\hat{L}_z$ and $\hat{L}^2$.

8. Calculate the expectation values (average values) of the operators $\hat{L}_x$, $\hat{L}_y$, $\hat{L}_x^2$, and $\hat{L}_y^2$, in the eigenstate Y_ℓ^m of $\hat{L}^2$.

9. A particle in a spherically symmetric potential is in a state described by the wave function

$$\psi(x,y,z) = C(xy + yz + zx)\, e^{-\alpha r^2},$$

where α is a constant. What is the probability that the measurement of the square of the angular momentum yields $6\hbar^2$?

10. Consider an electron for which $n = 4$, $\ell = 3$, and $m = 3$. Calculate the numerical value of (a) the orbital angular momentum and (b) the z component of the orbital angular momentum.

11. The Hamiltonian of a rotator is given by

$$\hat{H} = \frac{\hat{L}_x^2 + \hat{L}_y^2}{2I_1} + \frac{\hat{L}_z^2}{2I_2},$$

where I_1 and I_2 are the moments of inertia about the x-axis and the y-axis, respectively.

(a) Calculate the energy eigenvalues and the degeneracy of each of the corresponding energy levels.
(b) What are the energy eigenvalues for the various levels of $\ell = 3$?

12. The Hamiltonian of an axially symmetric rotator is given by

$$\hat{H} = \frac{\hat{L}_x^2 + \hat{L}_y^2 + \hat{L}_z^2}{2I},$$

where I is the moment inertia of the rotator. Find the energy levels and the corresponding degeneracies.

13. Calculate the orbital quantum number ℓ and the corresponding energy degeneracy for a rigid rotator for which the magnitude of the total angular momentum is $\sqrt{72}\,\hbar$.

14. A system is found in the state

$$\psi = Y_m^\ell(\theta, \varphi) = \sqrt{\frac{15}{8\pi}} \sin\theta \cos\theta \cos\varphi.$$

If the observable $\hat{L}_z$ is measured in this state, what are the values of L_z that the measurement will give and with what probabilities?

15. Determine the expectation value of $\hat{L}_x$ in the state of a system that is given in Problem 4.

16. Consider a particle of mass μ constrained to move on a circle of radius R. Show that the Hamiltonian of this system is

$$\hat{H} = \frac{\hat{L}^2}{2\mu R^2}.$$

Find the eigenvalues and the corresponding wave functions of the system. What is the degeneracy, if at all, of the eigenstates?

17. Consider the following 3×3 representation of the angular momentum operators

$$\hat{L}_x = \frac{1}{\sqrt{2}} \begin{pmatrix} 0 & 1 & 0 \\ 1 & 0 & 1 \\ 0 & 1 & 1 \end{pmatrix}, \; \hat{L}_y = \frac{1}{\sqrt{2}} \begin{pmatrix} 0 & -i & 0 \\ i & 0 & -i \\ 0 & i & 1 \end{pmatrix}, \; \hat{L}_z = \frac{1}{\sqrt{2}} \begin{pmatrix} 1 & 0 & 0 \\ i & 0 & 0 \\ 0 & 0 & -1 \end{pmatrix}.$$

of a system in a given state. (a) If $\hat{L}_z$ is measured, what values will obtain? (b) What are $\langle \hat{L}_x \rangle$ and $\langle \hat{L}_x^2 \rangle$ in the state in which $L_z = 1$? (c) What is the uncertainty ΔL_z in this state with $L_z = 1$?

Chapter 7

Simple Magnetic Field Effects

In this Chapter, we shall discuss a few important effects that arise when we consider quantum mechanical description of motion of a charged particle in an external magnetic field. We shall confine our discussions to the normal Zeeman effect in the presence of a constant magnetic field, Aharonov-Bohm effect arising from the gauge invariance of the electromagnetic fields, and the motion of free electrons in a constant magnetic field leading to what is known as Landau levels which are important for understanding the quantum Hall effects.

7.1 The Schrödinger Equation for a Spinless Charged Particle in an Electromagnetic Field

Consider a particle, without spin, of mass m and charge q in an electromagnetic field. Classically, the electromagnetic field is represented by the vector potential $\vec{A}(\vec{r},t)$ and the scalar potential $\Phi(\vec{r},t)$. In SI units, the expressions for the electric field $\vec{E}$ and the magnetic field $\vec{B}$ are

$$\vec{E} = -\vec{\nabla}\Phi - \frac{\partial \vec{A}}{\partial t}, \quad \vec{B} = \vec{\nabla} \times \vec{A}. \tag{7.1.1}$$

The Lagrangian of such a particle moving in an electromagnetic field is given by

$$L = \frac{1}{2}m\vec{v}^2 + q(\vec{v}\cdot\vec{A}) - q\Phi, \tag{7.1.2}$$

The generalized momentum $\vec{p}$ is given by

$$\vec{p} = \frac{\partial L}{\partial \vec{v}} = m\vec{v} + q\vec{A}. \tag{7.1.3}$$

The classical Hamiltonian is therefore given by

$$H = \vec{p}\cdot\vec{v} - L = m\vec{v}^2 + q\vec{A}\cdot\vec{v} - \frac{1}{2}m\vec{v}^2 - q(\vec{v}\cdot\vec{A}) + q\Phi = \frac{1}{2}m\vec{v}^2 + q\Phi. \tag{7.1.4}$$

Since

$$\vec{v} = \frac{\vec{p} - q\vec{A}}{m}, \tag{7.1.5}$$

we arrive at

$$H = \frac{(\vec{p} - q\vec{A})^2}{2m} + q\Phi. \tag{7.1.6}$$

Expanding the square, we obtain

$$\frac{1}{2m}\left(\vec{p} - q\vec{A}\right)^2 = \frac{1}{2m}\vec{p}^2 - \frac{q}{2m}\left(\vec{p}\cdot\vec{A} + \vec{A}\cdot\vec{p}\right) + \frac{q^2}{2m}\vec{A}^2. \tag{7.1.7}$$

The quantum mechanical Hamiltonian operator is obtained from the correspondence principle according to which we must replace $\vec{p}$ and $\vec{A}(\vec{r},t)$ by their corresponding operators. However, in doing so, we have to be careful about the product $\hat{\vec{p}}\cdot\vec{A}$ because $\hat{\vec{p}}$ is a differential operator and $\vec{A}$ is a function of coordinates. By taking into account that for any function $f(x,y,z)$,

$$\hat{\vec{p}}\cdot(\vec{A}f) = -i\hbar\vec{\nabla}\cdot(\vec{A}f) = -i\hbar f\,(\vec{\nabla}\cdot\vec{A}) - i\hbar\vec{A}\cdot(\vec{\nabla}f) = -i\hbar f\,(\vec{\nabla}\cdot\vec{A}) + (\vec{A}\cdot\vec{p})f, \tag{7.1.8}$$

we can write the Hamiltonian in the form

$$\hat{H} = -\frac{\hbar^2}{2m}\vec{\nabla}^2 + i\hbar\frac{q}{m}(\vec{A}\cdot\vec{\nabla}) + i\hbar\frac{q}{2m}(\vec{\nabla}\cdot\vec{A}) + \frac{q^2}{2m}\vec{A}^2 + q\Phi. \tag{7.1.9}$$

As we know, the electromagnetic potentials do not determine the electromagnetic fields uniquely. If we carry out the transformations

$$\vec{A}' = \vec{A} + \vec{\nabla}\lambda, \quad \Phi' = \Phi - \frac{\partial\lambda}{\partial t} \tag{7.1.10}$$

on the potentials, where $\lambda(\vec{r},t)$ is an arbitrary differentiable scalar function, the transformed fields $\vec{E}'$ and $\vec{B}'$ are identical with the old ones:

$$\vec{E}' = -\vec{\nabla}\Phi' - \frac{\partial\vec{A}'}{\partial t} = -\vec{\nabla}\Phi + \vec{\nabla}\left(\frac{\partial\lambda}{\partial t}\right) - \frac{\partial\vec{A}}{\partial t} - \frac{\partial}{\partial t}(\vec{\nabla}\lambda) = -\vec{\nabla}\Phi - \frac{\partial\vec{A}}{\partial t} = \vec{E}, \tag{7.1.11}$$

$$\vec{B}' = \vec{\nabla} \times \vec{A}' = \vec{\nabla} \times \vec{A} + \vec{\nabla} \times (\vec{\nabla}\lambda) = \vec{\nabla} \times \vec{A} = \vec{B}. \tag{7.1.12}$$

This property is known as the *gauge invariance of electrodynamics* and the aforementioned transformations of the potentials are known as *gauge transformations*. In order to fix this arbitrariness in the choice of the potentials, one of the following conditions

$$\vec{\nabla} \cdot \vec{A} + \frac{\partial \Phi}{\partial t} = 0, \tag{7.1.13}$$

$$\vec{\nabla} \cdot \vec{A} = 0, \tag{7.1.14}$$

is usually imposed on the potentials. When we do that, we say that we have chosen a gauge to work in. The first of these conditions is known as the Lorentz gauge, while the second is called the Coulomb gauge.

In what follows, we shall be working in the Coulomb gauge. In this gauge, the Hamiltonian takes the form

$$\hat{H} = -\frac{\hbar^2}{2m}\vec{\nabla}^2 + i\hbar\frac{q}{m}(\vec{A} \cdot \vec{\nabla}) + \frac{q^2}{2m}\vec{A}^2 + q\Phi. \tag{7.1.15}$$

The second term in (7.1.15), proportional to $(\vec{A} \cdot \vec{\nabla})$, is called the paramagnetic term, while the third term, proportional to $\vec{A}^2$, is known as the diamagnetic term. With the Hamiltonian given by (7.1.15), the Schrödinger equation for a charged but spinless particle in an electromagnetic field can be written as

$$i\hbar\frac{\partial \psi(\vec{r},t)}{\partial t} = \left[-\frac{\hbar^2}{2m}\vec{\nabla}^2 + i\hbar\frac{q}{m}(\vec{A}(\vec{r},t) \cdot \vec{\nabla}) + \frac{q^2}{2m}\vec{A}^2(\vec{r},t) + q\Phi(\vec{r},t)\right]\psi(\vec{r},t). \tag{7.1.16}$$

7.2 The Case of a Constant Magnetic Field

Let $\vec{B}$ be a spatially uniform and time-independent magnetic field. The vector potential for such a field is given by

$$\vec{A} = -\frac{1}{2}\vec{r} \times \vec{B}. \tag{7.2.1}$$

One can easily check that $\vec{\nabla} \times \vec{A}$ does produce the required constant magnetic field $\vec{B}$. If we use the aforementioned vector potential, we get the following expression for the paramagnetic term in the Hamiltonian:

$$i\hbar\frac{q}{m}(\vec{A} \cdot \vec{\nabla}) = \frac{q}{2m}(\vec{r} \times \vec{p}) \cdot \vec{B} = \frac{q}{2m}(\vec{B} \cdot \vec{L}), \tag{7.2.2}$$

where $\vec{L} = (\vec{r} \times \vec{p})$ is the angular momentum of the particle.

Using the well-known identities, $\vec{a}\cdot(\vec{b}\times\vec{c}) = \vec{b}\cdot(\vec{c}\times\vec{a}) = \vec{c}\cdot(\vec{a}\times\vec{b})$ and $\vec{a}\times(\vec{b}\times\vec{c}) = \vec{b}(\vec{a}\cdot\vec{c}) - \vec{c}(\vec{a}\cdot\vec{b})$, from vector algebra, we obtain the following expression for the diamagnetic term in the Hamiltonian:

$$\frac{q^2}{2m}\vec{A}^2 = \frac{q^2}{2m}\left([\vec{r}^2\,\vec{B}^2] - [\vec{r}\cdot\vec{B}]^2\right) = \frac{q^2\,B^2}{8m}\,(x^2+y^2), \tag{7.2.3}$$

where the direction of the magnetic field has been taken to be along the z-axis: $\vec{B} = B\,\hat{z}$.

Let us compare the order of magnitude of the paramagnetic and the diamagnetic terms for an electron without spin in an atom. With $x^2+y^2 \approx a_0^2$, where a_0 is the Bohr radius and the average value of the z component of the angular momentum is given by $\langle L_z\rangle \approx \hbar$, we get the ratio

$$\frac{(e^2/8m_e)\,(x^2+y^2)\,B^2}{(e^2/2m_e)\,\langle L_z\rangle\,B} = \frac{e}{4}\,\frac{a_0^2\,B}{\hbar} \approx 10^{-6}\,B\,\frac{1}{T}. \tag{7.2.4}$$

The realistic fields, usually achieved in laboratories, are of the order of 1.0 T. Thus, the quadratic term in $\vec{A}$ is negligible whenever $\langle L_z\rangle \neq 0$. Therefore, under laboratory conditions, diamagnetic effects are smaller than paramagnetic effects for electrons bound in atoms. However, there do, exist situations in which the diamagnetic and paramagnetic terms can be of comparable magnitude. For instance, this is the case for free electrons in a metal. The diamagnetic term is also important under conditions such as those prevailing on the surfaces of neutron stars: there, fields up to 10^8 T occur, which leads to a considerable change in the atomic structure.

Finally, let us compare the paramagnetic term with the Coulomb energy:

$$\frac{(e\,B\hbar/2m_e)}{m_e\,c^2\,\alpha^2/2} = \frac{e\hbar}{(m_e^2\,c^2\,\alpha^2)}\,B \approx 4.35\times10^{-6}\,B\,\frac{1}{T}, \tag{7.2.5}$$

where

$$\alpha = \frac{e^2}{4\pi\varepsilon_0\hbar c} \approx \frac{1}{137}, \tag{7.2.6}$$

is the fine structure constant. This means that under laboratory situations, the paramagnetic term is again very small compared to the Coulomb energy.

7.3 The Normal Zeeman Effect

If an atom is placed in an external magnetic field, the atomic energy levels are split. Because of this, when an emitting or absorbing atom is placed in a uniform external magnetic field, the observed spectrum changes. This was first seen experimentally by the Dutch physicist Pieter Zeeman and is hence called the *Zeeman effect*. Depending on the strength of the external magnetic field, the phenomenon manifests itself differently in terms of the number of spectral lines into which the original spectral line splits. For weak magnetic fields ($B \sim 0.8\,\mathrm{T}$ or less), it is divided into two classes: the *normal Zeeman effect* and the *anomalous Zeeman effect*. In the normal Zeeman effect, the splitting of spectral lines is determined by the orbital angular momentum alone, while in the anomalous case, it cannot be explained in terms of orbital angular momentum alone; one has to include the spin angular momentum. In fact, the generic nature of the anomalous Zeeman effect led to the discovery of spin angular momentum. For strong external fields ($B \geq 1.5$ T), the phenomenon manifests itself through what is called the *Paschen–Back effect*. In what follows, we shall deal only with the normal Zeeman effect.

Consider for concreteness, the case of a hydrogen-like atom, with a single electron placed in a uniform external magnetic field $\vec{B}$. As discussed earlier, keeping in mind the intensity of the magnetic field that can be achieved under realistic laboratory conditions, the stationary Schrödinger for a spinless electron, moving in a Coulomb field in the presence of a constant magnetic field, takes the form

$$\vec{\nabla}^2\,\psi(\vec{r}) - \frac{ieB}{\hbar}\frac{\partial\psi(\vec{r})}{\partial\varphi} + \frac{2m_e}{\hbar^2}\left(E - U(\vec{r})\right)\psi(\vec{r}) = 0, \tag{7.3.1}$$

where $U = -e\Phi(\vec{r})$ is the potential energy of the electron in the force field of the nuclear proton.

Since we are dealing with the motion of a particle in the central force field, we look for the solution in the form $\psi = \psi(r,\theta,\varphi) = R(r)\Theta(\theta)\phi(\varphi)$. Now, let us recollect that for the central potentials, the function $\phi(\varphi) = e^{im\varphi}$, where m is the magnetic quantum number, and hence, $\psi(r,\theta,\varphi) = R(r)\Theta(\theta)e^{im\varphi}$. In this case,

$$-\frac{ieB}{\hbar}\frac{\partial\psi}{\partial\varphi} = \frac{eB}{\hbar}m\,\psi(r,\theta,\varphi), \tag{7.3.2}$$

and we obtain

$$\Delta_{r,\theta,\varphi}\psi(r,\theta,\varphi) + \frac{2m_e}{\hbar^2}\left(E + \frac{eB\hbar}{2m_e}m - U(\vec{r})\right)\psi(r,\theta,\varphi) = 0. \tag{7.3.3}$$

Equation (7.3.3) can be rewritten in a more familiar form as

$$\Delta_{r,\theta,\varphi}\psi(r,\theta,\varphi) + \frac{2m_e}{\hbar^2}\left(E' - U(\vec{r})\right)\psi(r,\theta,\varphi) = 0, \tag{7.3.4}$$

where

$$E' = E + \frac{eB\hbar}{2m_e}m. \tag{7.3.5}$$

For a hydrogen-like atom, for instance,

$$U(\vec{r}) = -\frac{Ze^2}{4\pi\varepsilon_0\,r}. \tag{7.3.6}$$

The resulting equation coincides with the corresponding stationary Schrödinger equation for the hydrogen-like atom with a modified stationary state energy E'. The solution of this equation, as we know, will lead to a series of energy eigenvalues $E'_1, E'_2, E'_3, \ldots, E'_k, \ldots$. According to (7.3.5),

$$E_k = E'_k - \frac{eB\hbar}{2m_e} m = E'_k - \hbar\omega_\ell m, \tag{7.3.7}$$

where

$$\omega_\ell = \frac{eB}{2m_e}, \tag{7.3.8}$$

is the *Larmor frequency*.

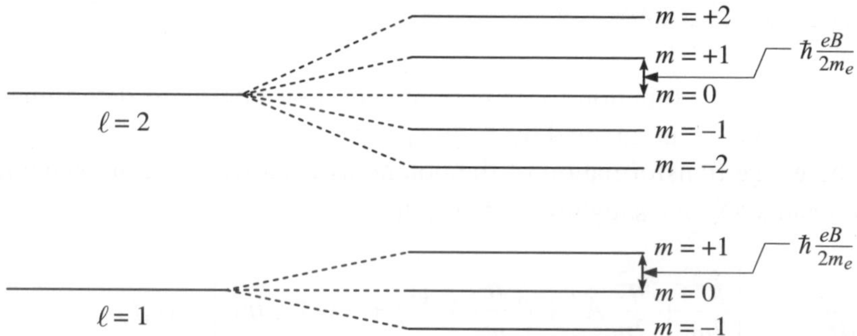

Figure 7.1 Zeeman effect in an external magnetic field.

Thus, the an energy level E_k in the absence of the magnetic field differs from the corresponding energy level E' by $-\frac{eB\hbar}{2m_e}m$. Since m can take all integer values from $-\ell$ to $+\ell$, each of the energy levels E, which is $(2\ell+1)$-fold degenerate in the absence of the magnetic field, splits up into $2\ell+1$ discrete energy levels. So far as the wave functions are concerned, they remain the same as in the absence of the magnetic field. Thus, we can say that the magnetic field removes degeneracy with respect to the magnetic quantum number m by displacing the $(2\ell+1)$ coinciding sub-levels with respect to each other in such a way that the distance between any two discrete sub-levels equals $\hbar\omega_\ell$. Evidently, this distance between the split sub-levels is proportional to the intensity of the magnetic field and is independent of the quantum numbers n and ℓ.

This equidistant splitting caused by the magnetic field is called the *normal Zeeman effect* (see Fig.7.1). Since m takes integer values, each degenerate energy level must split into an odd number of discrete levels. However, in the hydrogen atom the splitting leads to an even number of levels, as if the angular momentum were half integrals. This departure from the general rule (7.3.7) is due to the spin angular momentum of the electron (see Chapter 8) which takes half-integer values.

7.4 Transformation of the Wave Function under Gauge Transformation

We have seen in the previous section that Maxwell's electrodynamics is invariant under the local gauge transformation of the potentials given by (7.1.10). In view of this, we want to find out whether this has anything to do with the wave function of our charged particle and the differential equation satisfied by the transformed (new) wave function.

Since the electric and magnetic fields do not change under gauge transformation and they are the measurable physical entities (not the potentials), the wave function of the particle, interacting with the electromagnetic field, should not undergo any significant changes that could affect the physical content of the theory. It then follows that, under the gauge transformation of the electromagnetic potentials, the wave function of the particle can at most acquire a phase factor. Accordingly, we look for the transformed wave function $\psi'(\vec{r},t)$ in the following form:

$$\psi'(\vec{r},t) = \psi(\vec{r},t)\, e^{i\Lambda(\vec{r},t)}, \tag{7.4.1}$$

where Λ is an arbitrary function to be determined. Our goal is to determine Λ and the differential equation that is satisfied by $\psi'(\vec{r},t)$.

Before the gauge transformation of the potentials is carried out, the wave function, in the Coulomb gauge ($\vec{\nabla}\cdot\vec{A}$), satisfies the Schrödinger equation

$$i\hbar\frac{\partial\psi(\vec{r},t)}{\partial t} = \left[\frac{\hat{\vec{p}}^2}{2m} + \frac{q^2}{2m}\vec{A}^2 + i\hbar\frac{q}{m}(\vec{A}\cdot\vec{\nabla}) + V(\vec{r}) + q\Phi\right]\psi(\vec{r},t). \tag{7.4.2}$$

To determine the equation satisfied by $\psi'(\vec{r},t)$, we proceed as follows. The time and space derivatives of $\psi'(\vec{r},t)$ are

$$i\hbar\frac{\partial\psi}{\partial t} = \left(i\hbar\frac{\partial\psi'}{\partial t} + \hbar\frac{\partial\Lambda}{\partial t}\psi'\right)e^{-i\Lambda(\vec{r},t)}, \tag{7.4.3}$$

$$i\hbar\frac{q}{m}(\vec{A}\cdot\vec{\nabla})\psi = i\hbar\frac{q}{m}\left[(\vec{A}\cdot\vec{\nabla})\psi' - i(\vec{A}\cdot\vec{\nabla}\Lambda)\psi'\right]e^{-i\Lambda(\vec{r},t)}, \tag{7.4.4}$$

$$\begin{aligned}-\frac{\hbar^2}{2m}\vec{\nabla}^2\psi = \Bigg[&-\frac{\hbar^2}{2m}\vec{\nabla}^2\psi' + i\frac{\hbar^2}{m}(\vec{\nabla}\Lambda\cdot\vec{\nabla}\psi') + i\frac{\hbar^2}{2m}(\vec{\nabla}^2\Lambda)\,\psi' \\ &+ \frac{\hbar^2}{2m}(\vec{\nabla}\Lambda)^2\psi'\Bigg]e^{-i\Lambda(\vec{r},t)},\end{aligned} \tag{7.4.5}$$

Substituting for ψ in (7.4.2) from (7.4.1) and using (7.4.3)–(7.4.5), we obtain

$$i\hbar\frac{\partial\psi'}{\partial t} = -\frac{\hbar^2}{2m}\vec{\nabla}^2\psi' + \left[\frac{q^2}{2m}\vec{A}^2 + \frac{\hbar^2}{2m}(\vec{\nabla}\Lambda)^2 + \frac{q}{m}\hbar(\vec{A}\cdot\vec{\nabla}\Lambda)\right]\psi' + i\frac{\hbar^2}{2m}(\vec{\nabla}^2\Lambda)\psi'$$

$$+ i\frac{\hbar^2}{m}(\vec{\nabla}\Lambda\cdot\vec{\nabla})\psi' + i\hbar\frac{q}{m}(\vec{A}\cdot\vec{\nabla})\psi' + V(\vec{r})\psi' + \left(q\Phi - \hbar\frac{\partial\Lambda}{\partial t}\right)\psi'. \quad (7.4.6)$$

Note that if $\Lambda = \frac{q}{\hbar}\lambda(\vec{r},t)$, then the transformed wave function takes the form

$$\psi'(\vec{r},t) = \psi(\vec{r},t)\,e^{i\frac{q}{\hbar}\lambda(\vec{r},t)}. \quad (7.4.7)$$

Consequently, in the Coulomb gauge ($\vec{\nabla}\cdot\vec{A} = 0$), (7.4.6) can be rewritten as

$$i\hbar\frac{\partial\psi'}{\partial t} = \left[\frac{\vec{p}^{\,2}}{2m} + \frac{q^2}{2m}\vec{A}'^2 + i\hbar\frac{q}{m}(\vec{A}'\cdot\vec{\nabla}) + V'(\vec{r}) + q\Phi'\right]\psi'(\vec{r},t), \quad (7.4.8)$$

where the primed potentials are the gauge-transformed potentials:

$$\vec{A}' = \vec{A} + \vec{\nabla}\lambda, \quad \Phi' = \Phi - \frac{\partial\lambda}{\partial t}. \quad (7.4.9)$$

and we have taken into account that $\vec{\nabla}^2\Lambda = 0$ due to the Coulomb gauge condition.

It follows from these discussions that if, along with the gauge transformation of the electromagnetic potentials, we transform the wave function by multiplying it with a phase factor $e^{i\frac{q}{\hbar}\lambda(\vec{r},t)}$ that contains the same function $\lambda(\vec{r},t)$ which is used for the gauge transformation, the new wave function satisfies the same Schrödinger equation (7.4.2) (which is satisfied by the wave function prior to the gauge transformation) but with the gauge-transformed potentials.

Thus, the gauge transformation induces an additional space-and time-dependent phase factor into the wave function. However, the change in gauge has no observable physical consequences, since the probability density, $|\psi|^2$, and the average values of the observables do not change.

7.5 The Aharonov–Bohm Effect

The form invariance of the Schrödinger equation for a particle under the gauge transformation of the potentials and the induction of an additional phase into the wave function of a charged particle in an external electromagnetic field lead to an observable physical phenomenon which is known as *Aharonov-Bohm effect.*

Consider a charged particle, of mass m and charge q, travelling along a path L_1 from a point P_1 to another point P_2 in a region in which the magnetic field is identically equal to zero: $\vec{B} = 0$. However, the vanishing of the magnetic field does not imply that the vector potential $\vec{A}$ is zero. For instance, in the case of an infinitely long solenoid, the magnetic field is confined inside the solenoid and is identically equal to zero outside. However, in the region outside the solenoid, $\vec{A}$ can be expressed as the gradient of a scalar field $\lambda(\vec{r})$:

$$\vec{A} = \vec{\nabla}\lambda, \tag{7.5.1}$$

so that

$$\lambda = \int_{P_1}^{P_2} \vec{A}\cdot d\vec{\ell}. \tag{7.5.2}$$

Now the wave function in the field-free region can be obtained by solving either (7.4.2) or from the gauge transformed equation (7.4.8). Note that if we choose $-\lambda(\vec{r},t)$ as the function to gauge transform the potentials, the vector potential can be made to vanish in our field-free region

$$\vec{A}' = \vec{A} - \vec{\nabla}\lambda = 0. \tag{7.5.3}$$

Furthermore, since there is no electric field present here, we can set $\Phi = \Phi' = 0$ and the wave function can be found by solving either (7.4.2) in which the scalar potential is absent or from the gauge-transformed equation

$$i\hbar\frac{\partial\psi'}{\partial t} = \left[\frac{\hat{\vec{p}}^2}{2m} + V'(\vec{r})\right]\psi'(\vec{r},t), \tag{7.5.4}$$

in which both the potentials (the vector as well as the scalar potentials) do not occur. The relationship between these wave functions is obtained by replacing λ in (7.4.7) with $-\lambda$:

$$\psi(\vec{r},t) = \psi'(\vec{r},t)\exp\left[-i\frac{q}{\hbar}\int_L \vec{A}\cdot d\vec{\ell}\right], \tag{7.5.5}$$

where $\psi'(\vec{r},t)$ is the wave function in the potential $V(\vec{r})$ with magnetic field $\vec{B}$ identically equal to zero in all space.

Let us go back to our charged particle moving along the path L_1. In view of the aforementioned result, in traversing the path from P_1 to P_2, the wave function of the particle will acquire a phase

$$\varphi_1 = \frac{q}{\hbar}\int_{P_1}^{P_2}\vec{A}\cdot d\vec{\ell}. \tag{7.5.6}$$

If we consider another path L_2 with the same end points but traversed in the opposite direction, the phase accumulated by the wave function will be

$$\varphi_2 = -\frac{q}{\hbar}\int_{P_2}^{P_1}\vec{A}\cdot d\vec{\ell}, \tag{7.5.7}$$

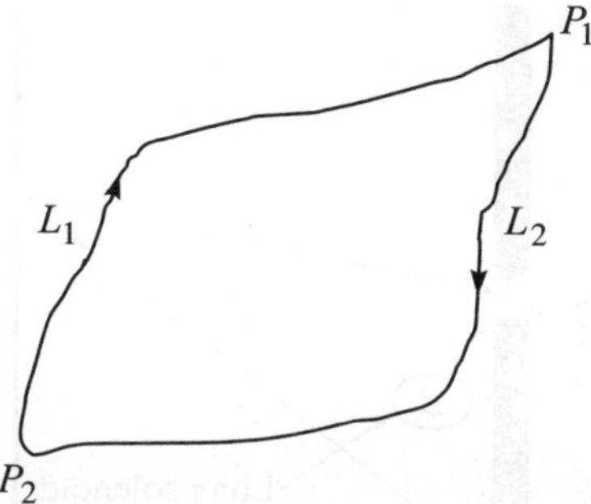

Figure 7.2 Closed path traversed by the particle in the field-free region.

where the minus sign is due to the reversed direction traversed along L_2. Therefore, the total change in phase of the wave function traversing the closed path, from P_1 to P_2 and back to P_1 (see Fig.7.2), will be given by

$$\Delta\varphi = \varphi_1 + \varphi_2 = \frac{q}{\hbar}\oint_L \vec{A}\cdot d\vec{\ell} = \frac{q}{\hbar}\int_S (\vec{\nabla}\times\vec{A})\cdot d\vec{a} = \frac{q}{\hbar}\int_S \vec{B}\cdot d\vec{a}, \tag{7.5.8}$$

where the line integral runs over the loop from P_1 to P_2 and back to P_1 and the surface integral, according to Stokes' theorem, is over the surface S enclosed by this loop. Thus, $\Delta\varphi$ is proportional to the flux of the magnetic field, through the surface S. Hence, in the absence of the magnetic field the total change in the phase of the wave function is zero.

If, however, the loop encloses a region of non-zero magnetic field, confined within a small region inside, the flux through the surface enclosed by the loop will be non-zero and, hence, even if there is no magnetic field along the paths L_1 and L_2, the net change in the phase of the wave function will be non-zero.

To show the physical consequence of this magnetic flux dependent phase shift, consider the double slit electron interference experiment, shown in Fig.7.3, in which the magnetic field is confined to the interior of the 'infinitely' long solenoid[1], perpendicular to the plane of the figure and depicted by a circle.

The electrons emitted by the electron gun are incident on the wall with two slits 1 and 2 as shown in Fig.7.3. A current carrying 'infinitely' long solenoid is placed behind the wall containing the slits and screened by the part of the wall between the slits. Because of this screening, the electrons are restricted to the region $\vec{B} = 0$. Let $\psi_1(\vec{r})$ be the wave function when only slit 1 is open. Then, according to (7.5.5), we have

$$\psi_1(\vec{r}) = \psi_1^{(0)}(\vec{r},t)\,\exp\left[i\frac{e}{\hbar}\int_{P\to 1}^{Q} \vec{A}\cdot d\vec{\ell}\right], \tag{7.5.9}$$

[1]F. Munley, Am. J. Phys., v. 53, p. 779, 1985.

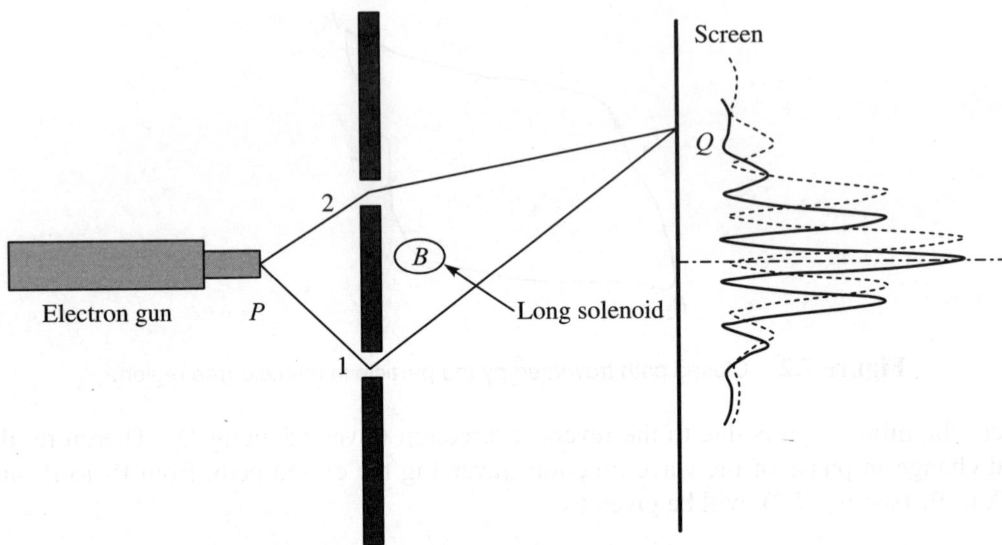

Figure 7.3 The electron interference experiment in which the electron cannot penetrate into the region of the magnetic field.

where $\psi_1^{(0)}(\vec{r},t)$ is the solution of the field-free Schrödinger equation and the line integral runs from P through slit 1 to the point Q on the screen. Similarly, the wave function $\psi_2(\vec{r})$, when only slit 2 is open, is given by

$$\psi_2(\vec{r}) = \psi_2^{(0)}(\vec{r},t)\exp\left[i\frac{e}{\hbar}\int_{P\to 2}^{Q}\vec{A}\cdot d\vec{\ell}\right], \tag{7.5.10}$$

where $\psi_2^{(0)}(\vec{r},t)$ is the solution of the field-free Schrödinger equation and the line integral runs from P through slit 2 to the point Q on the screen. Now, the wave function $\psi_{12}(\vec{r})$, when both the slits are open is given by the superposition of the wave functions $\psi_1(\vec{r})$ and $\psi_2(\vec{r},t)$:

$$\psi_{12}(\vec{r}) = \psi_1^{(0)}(\vec{r},t)\exp\left[i\frac{e}{\hbar}\int_{P\to 1}^{Q}\vec{A}\cdot d\vec{\ell}\right] + \psi_2^{(0)}(\vec{r},t)\exp\left[i\frac{e}{\hbar}\int_{P\to 2}^{Q}\vec{A}\cdot d\vec{\ell}\right]. \tag{7.5.11}$$

Using (7.5.8), $\psi_{12}(\vec{r})$ can be written as

$$\psi_{12}(\vec{r}) = \left(\psi_1^{(0)}(\vec{r},t)\exp\left[i\frac{e}{\hbar}\phi_B\right] + \psi_2^{(0)}(\vec{r},t)\right)\exp\left[i\frac{e}{\hbar}\int_{P\to 2}^{Q}\vec{A}\cdot d\vec{\ell}\right], \tag{7.5.12}$$

where ϕ_B is the flux of the magnetic field through the surface, S, enclosed by the loop P $\to$ 1 $\to$ Q$\to$ 2 $\to$ P:

$$\phi_B = \int_S \vec{B}\cdot d\vec{a}. \tag{7.5.13}$$

Thus, the relative phase between the wave functions $\psi_1(\vec{r})$ and $\psi_2(\vec{r},t)$ changes in the presence of a narrowly confined region of a magnetic field, behind the wall containing the slits. As a consequence, the position of the resulting interference pattern on the screen shifts relative to the interference pattern in the absence of the magnetic field, although the electrons cannot penetrate into the region of the magnetic field of the solenoid. This phenomenon was predicted by Aharonov and Bohm in 1959 and is known as *Aharonov–Bohm effect*. The prediction was experimentally confirmed within a few months by Robert G. Chambers in 1960.

7.6 Free Electrons in a Magnetic Field: Landau Levels

Let us consider the problem of a free (unbound) electron interacting with a static and uniform magnetic field. This study leads to the concept of Landau levels, which is crucial for the understating of phenomena, like Landau diamagnetism, de Haas–von Alphen effect and the integer quantum Hall effect.

Consider a spinless electron moving freely in the xy-plane and subject to a static magnetic field, $\vec{B} = B_0\,\hat{k}$, directed along the z-axis, where B_0 is constant. We want to solve the stationary state Schrödinger equation for the electron and determine energy eigenvalues and the corresponding wave functions. To accomplish this task, we first need to specify a gauge potential $\vec{A}$ such that $\vec{\nabla}\times\vec{A} = B_0\,\hat{k}$. Clearly, this can be done in several ways. In the given problem, though, it is convenient to work in the so-called Landau gauge, in which

$$\vec{A} = B_0 x\,\hat{j}. \tag{7.6.1}$$

It is trivial to check that $\vec{\nabla}\cdot\vec{A} = 0$ and $\vec{\nabla}\times\vec{A} = B_0\,\hat{k}$. It is also worth mentioning here that the given magnetic field $\vec{B}$ is invariant under translations and rotations in the xy-plane. However, our choice of $\vec{A}$ is not; it breaks the rotational symmetry and the translational symmetry in the x direction. But, there is nothing to worry about–although the intermediate calculations will not be manifestly invariant, the physics will be invariant under all symmetries. This kind of compromise is inevitable while dealing with magnetic fields.

The Hamiltonian for our electron in the Landau gauge takes the form

$$\hat{H} = \frac{1}{2m}\left(\hat{\vec{p}} + e\vec{A}\right)^2 = \frac{1}{2m}\left(\hat{p}_x^2 + \hat{p}_y^2 + 2eB_0 x\hat{p}_y + e^2B_0^2\right). \tag{7.6.2}$$

Note that the Hamiltonian commutes with $\hat{p}_y$; they both share a common set of eigenfunctions. Since the eigenfunctions of $\hat{p}_y$ are plane waves propagating along the y direction, we look for the solution of the stationary Schrödinger equation

$$\frac{1}{2m}\left(\hat{p}_x^2 + \hat{p}_y^2 + 2eB_0 x\hat{p}_y + e^2B_0^2\right)\psi(x,y) = E\,\psi(x,y) \tag{7.6.3}$$

in the following form

$$\psi(x,y) = e^{iky}\,\phi_k(x), \tag{7.6.4}$$

where $k_y \equiv k = p_y/\hbar$. Since

$$\hat{p}_y \psi(x,y) = -i\hbar \frac{\partial}{\partial y}\left(e^{ik_y y}\,\phi_k(x)\right) = (\hbar k)\,e^{iky}\,\phi_k(x) = \hbar k\,\psi(x,y), \tag{7.6.5}$$

the Schrödinger equation (7.6.3) takes the form

$$\left(-\frac{\hbar^2}{2m}\frac{\partial^2}{\partial x^2} + \frac{e^2B_0^2}{2m}x^2 + 2\,eB_0\hbar k\,x + \frac{\hbar^2k^2}{2m}\right)\,\phi_k(x) = E\,\phi_k(x), \tag{7.6.6}$$

which can be re-written as

$$\left[-\frac{\hbar^2}{2m}\frac{\partial^2}{\partial x^2} + \frac{e^2B_0^2}{2m}\left(x + \frac{\hbar k}{eB_0}\right)^2\right]\,\phi_k(x) = E\,\phi_k(x), \tag{7.6.7}$$

or,

$$\left[-\frac{\hbar^2}{2m}\frac{\partial^2}{\partial x^2} + \frac{1}{2}\omega_c^2\left(x + k\ell_B^2\right)^2\right]\,\phi_k(x) = E\,\phi_k(x), \tag{7.6.8}$$

where

$$\omega_c = \frac{eB}{m} \quad \text{and} \quad \ell_B = \sqrt{\frac{\hbar}{eB}}, \tag{7.6.9}$$

are the electron's cyclotron frequency and the magnetic length, respectively. Note that the cyclotron frequency, ω_c, is just the classical frequency of the orbital motion of the electron in a magnetic field and ℓ_B is a characteristic length scale that governs any quantum phenomena in a magnetic field.

Equation (7.6.8) is nothing but the Schrödinger equation of a harmonic oscillator in the x direction, which is centred at $x_k = -k\ell_B^2$. Using this fact, we can now straightaway write down the energy eigenvalues

$$E_n = \hbar\omega_c\left(n + \frac{1}{2}\right), \tag{7.6.10}$$

where n can take any integer values including 0 and $k \in R$. These energy levels of a free electron in a constant magnetic field are equally spaced and each of them is proportional to the magnitude of the magnetic field. They are known as Landau levels.

The corresponding wave functions can be written as

$$\psi_{n,k}(x,y) = C\,e^{iky}\,H_n(x+x_k)\,e^{-(x+x_k)^2/2\ell_B^2}, \tag{7.6.11}$$

where $x_k = k\ell_B^2$ is the classical centre of the electron's orbit and C is a constant to be determined by normalization. Here, $H_n(x+x_k)$ is the the usual hermite polynomial of degree n. We notice that the wave functions depend on two quantum numbers n and k but the energy levels depend only on n. Since we can have many different $k_y = k$ all with same E_n, each of the Landau levels is highly degenerate. In order to understand this degeneracy consider a finite region in the xy-plane with extensions L_x and L_y along the x and y directions. Since there is translational invariance along the y direction, we assume periodic boundary condition: $\psi(x, y+L_y) = \psi(x,y)$. Then the allowed values of $k-$ vectors are

$$k = \frac{2\pi}{L_y} n_y, \quad n_y = 0, 1, 2, 3, \ldots$$

Hence the allowed values of x_k are separated by

$$\Delta x = \ell_B^2 \cdot \Delta k = \frac{2\pi\,\ell_B^2}{L_y}.$$

So, the total number of states in the region is

$$\mathcal{N} = \frac{L_x L_y}{2\pi\,\ell_B^2} = \frac{eB}{2\pi\hbar} A,$$

where $A = L_x L_y$ is the area of the region under consideration. It means that each Landau level has $1/2\pi\ell_B^2$ states per unit area.

Finally, note that although we treated x and y asymmetrically for convenience of calculation, in reality, we cannot distinguish between the two due to the symmetry of the original problem with magnetic field in the z direction.

Homework Problems

1. Consider a particle of mass m attached to a rigid massless rod of fixed length R whose other end is fixed at the origin. The rod is free to rotate about the origin. The particle has no internal spin degree of freedom, but carries an electric charge $+e$. It is placed in a uniform magnetic field $\vec{B}$. Using the principle of minimal substitution, write down the Schrödinger equation for this charged rigid rotator.
2. The Schrödinger equation in the first question can be solved exactly if the magnetic field is weak. Compute the energy levels of the system , assuming that the magnetic

field is weak (i.e., assume that the term in the Hamiltonian that is quadratic in $\vec{B}$ can be neglected).

3. For $\vec{B} = B_0 \hat{k}$, where B_0 is a constant, show that the resulting Schrödinger equation, in the previous Problem 2, is exactly solvable. Find the solutions.

Chapter 8

Quantum Mechanical Theory of the Spin Angular Momentum

8.1 Spin

Spin angular momentum or simply *spin* is a fundamental property of all particles, irrespective of whether they are elementary or composite. It belongs to an internal degree of freedom (completely independent of the spatial degrees of freedom) and *manifests itself as some intrinsic angular momentum of the particle*. It was introduced in quantum mechanics as an attempt to explain the experimentally observed fine structures of the spectral lines in the emission spectra of alkali metals and the peculiarities involved in the anomalous (complex) Zeeman effect that showed the unusual splitting pattern of atomic energy levels in the presence of a weak external magnetic field. Note that all efforts, prior to the conjecture about spin, to explain the aforementioned experimental results on the basis of the Schrödinger equation without spin had miserably failed.

An atom of any of the alkali metals has an almost inert core, consisting of the nucleus and $(Z-1)$ inner electrons, together with a single outer electron. The transitions of the outer electron between energy levels are responsible for the aforementioned spectral lines. Therefore, any additional property required to be postulated for the explanation of the fine structures of the spectral lines or anomalous Zeeman effect, had to be attributed to the valence electron. It is because of this reason that Uhlenbeck and Goudsmit put forward their conjecture about electron's spin. They assumed that, similar to Earth's spinning motion about its axis, an electron, in addition to its orbital motion about the nucleus, also possessed a spinning motion about its axis of symmetry. The angular momentum related to this spinning motion was given the name 'spin'. Uhlenbeck and Goudsmit also assumed that, analogous to the magnetic dipole moment related to the orbital angular momentum, an electron possessed an *intrinsic* magnetic dipole moment associated with the spin angular momentum, whose interaction with the external magnetic field was the key for resolving the discrepancies. Later theoretical and experimental developments did confirm the existence of spin.

It is worth mentioning here that, conceptually, the spinning motion of an electron, proposed by Uhlenbeck and Goudsmit, was highly questionable in view of the fact that an electron was a point particle and the classical notion of angular momentum of a rigid body did not apply. However, as we know now, the theory constructed on the basis of such an ad hoc assumption did succeed in explaining the experimental results to a great degree of accuracy.

8.2 Spin Operators and their Commutation Relations

Spin is denoted by a vector $\vec{S}$. As required by the rules of quantum mechanics, it is represented by an operator $\hat{\vec{S}}$ with Cartesian components $\hat{S}_x$, $\hat{S}_y$ and $\hat{S}_z$. Since it is a kind of angular momentum, the operators $\hat{S}_x$, $\hat{S}_y$ and $\hat{S}_z$ must satisfy the same set of commutation relations that is satisfied by the Cartesian components of the orbital angular momentum. Consequently, we have

$$[\hat{S}_x, \hat{S}_y] = i\hbar\hat{S}_z, \tag{8.2.1}$$

$$[\hat{S}_y, \hat{S}_z] = i\hbar\hat{S}_x, \tag{8.2.2}$$

$$[\hat{S}_z, \hat{S}_x] = i\hbar\hat{S}_y. \tag{8.2.3}$$

Given the commutation relations (8.2.1)–(8.2.3), it is straightforward to check that $\hat{S}^2$ commutes with each of the operators $\hat{S}_x$, $\hat{S}_y$ and $\hat{S}_z$, that is,

$$[\hat{S}^2, \hat{S}_x] = 0, \quad [\hat{S}^2, \hat{S}_y] = 0, \quad [\hat{S}^2, \hat{S}_z] = 0. \tag{8.2.4}$$

From (8.2.1)–(8.2.4), it follows that only S^2, and hence the magnitude of the total spin, and the projection of spin on a given axis, say S_z, can be specified simultaneously in a given state of the particle. The other two components S_x and S_y of $\vec{S}$ cannot be specified at all. Thus, the orientation of spin in space, in general, cannot be defined. It can be defined only with respect to a chosen axis. Also, similar to the case of orbital angular momentum, the direction of spin in space is quantized and we have the second example of space quantization.

Further, since $[\hat{S}^2, \hat{S}_z] = 0$, the operators $\hat{S}^2$ and $\hat{S}_z$ can have a common set of eigenvectors, $|s, m_s\rangle$, characterized by two quantum numbers s and m_s. The quantum number s is called the *spin quantum number* and takes integers as well as half-integer values. On the other hand, the quantum number m_s is called the *spin magnetic quantum number* and takes $(2s+1)$ values from $-s$ to s. Similar to the case of orbital angular momentum, the eigenvectors $|s, m_s\rangle$ satisfy

$$\hat{S}^2|s, m_s\rangle = \hbar^2 s(s+1)|s, m_s\rangle, \tag{8.2.5}$$

$$\hat{S}_z|s, m_s\rangle = \hbar m_s|s, m_s\rangle, \tag{8.2.6}$$

and

$$\hat{S}_{\pm}|s,m_s\rangle = \hbar\sqrt{s(s+1)-m_s(m_s\pm 1)}|s,m_s\rangle, \tag{8.2.7}$$

where $\hat{S}_{\pm} = \hat{S}_x \pm i\hat{S}_y$ are the raising and lowering operators for spin. Also, in a given state with quantum number s, the magnitude of spin is given by $S = \sqrt{s(s+1)}\hbar$. The z component of spin is quantized, $S_z = m_s\hbar$, and takes $(2s+1)$ different values.

Now, the magnetic dipole moment associated with spin is given by

$$\vec{\mu}_s = -\frac{e}{m_e}\vec{S}, \tag{8.2.8}$$

where e is the magnitude of the electronic charge and m_e is the mass of the electron. As a consequence, the Hamiltonian for an electron, *with spin*, in an external magnetic field $\vec{B}$ along the positive z direction, will have a potential energy term

$$\Delta W = -\vec{\mu}_s\cdot\vec{B} = \frac{eB}{m_e}S_z = \frac{e\hbar B}{m_e}m_s. \tag{8.2.9}$$

As a consequence, the original energy levels will be shifted by $(e\hbar B/m_e)\,m_s$. Since m_s takes $(2s+1)$ values, the original degenerate energy level will split into $(2s+1)$ distinct levels.

Note that unlike the case of orbital angular momentum where the boundary conditions on the eigenfunctions allowed one to exclude the half-integer values of ℓ, for the spin angular momentum, there is no such restriction and s can take both the integer and the half-integer values in the units of $\hbar$. We shall see in the next chapter that nature supports both kinds of particles: particles with integer spin, called *bosons*, and particles with half-integer spin, called *fermions*. For instance, photons ($s=1$), π-mesons ($s=0$), gravitons ($s=2$) and so on are bosons, while electrons ($s=\frac{1}{2}$), protons ($s=\frac{1}{2}$), neutrons ($s=\frac{1}{2}$), delta particles ($s=\frac{3}{2}$) and so on are fermions.

8.3 Spin and Pauli Matrices

Let us consider the famous Stern–Gerlach experiment (schematically shown in Figure 8.1) in which a beam of silver atoms in their *ground state* was made to pass through a region of inhomogeneous magnetic field in the direction perpendicular to the direction of the field. At the exit from this region, the beam was collected on a screen. The results of the experiment showed that the original beam split into two after passing through the region of the inhomogeneous magnetic field. It was evident from the pair of spots that appeared on the screen, symmetrically placed on either side of the central spot that would have occurred in the absence of the magnetic field.

It was argued that the splitting of the beam could take place only if silver atoms had some magnetic moment subject to a force from the applied magnetic field whose direction depended on the relative orientation of the magnetic field and the magnetic moment of the atom. In this context, it is important to note that a silver atom has 47 electrons out of which 46 constitute the spherically symmetric charge distribution around the nucleus: they fill all the sub-shells for $n = 1, n = 2$, and $n = 3$, and the $4d$ sub-shell and contribute nothing to the orbital angular momentum of the atom. The 47th electron is in the $5s$ state and it cannot have any orbital angular momentum too. Thus, a silver atom in its ground state does not have any orbital angular momentum and hence there is no magnetic moment associated with it, which can lead to the splitting of the atomic beam during passage through an inhomogeneous magnetic field. Therefore, it was argued that the splitting of the beam into two could happen only if silver atoms had some kind of an intrinsic angular momentum (not at all related to orbital motion) and a magnetic moment associated with it. But then, this intrinsic angular momentum had to be attributed to the valence electron in the $5s$ sub-shell because of the reasons stated earlier. We shall keep this fact in mind while discussing the consequences of the Stern–Gerlach experiment.

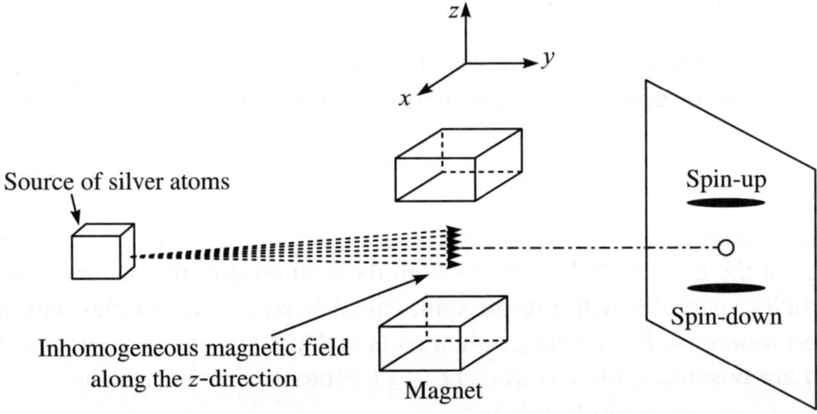

Figure 8.1 Schematic representation of the Stern–Gerlach experiment.

Since the beam split into two, it follows from the theory discussed in the previous section that

$$2s + 1 = 1 \quad \Rightarrow \quad s = \frac{1}{2}. \tag{8.3.1}$$

Therefore, for an electron, the spin quantum number $s = \frac{1}{2}$ and the spin magnetic quantum number m_s can take only two values: $+\frac{1}{2}$ and $-\frac{1}{2}$. It means that the eigenvalue of $\hat{S}^2$ is equal to $(3/4)\,\hbar^2$, and the projection of spin on the z-axis can have two values $+\hbar/2$ and $-\hbar/2$. Given these facts, we shall now try to construct the quantum mechanical theory of an electron by taking into account its spin properties.

We know by now that spin is a purely quantum mechanical property of a particle with no classical analogue at all. Also, the spin degree of freedom is completely independent of the spatial degrees of freedom. Therefore, new concepts and mathematical tools are required to incorporate this novel aspect of motion into the formalism of quantum mechanics. Let us see how it can be accomplished.

Since the state of an electron is characterized by two values, $+\hbar/2$ and $-\hbar/2$, of the projection of its spin on the z-axis, the wave function of the electron must consist of two components: $\psi_+(\vec{r},t)$, corresponding to $S_z = +\hbar/2$, and $\psi_-(\vec{r},t)$corresponding to $S_z = -\hbar/2$. It is convenient to write it as a column vector:

$$\psi(\vec{r},t) = \begin{pmatrix} \psi^+(\vec{r},t) \\ \psi^-(\vec{r},t) \end{pmatrix}. \tag{8.3.2}$$

Note that if only $\psi_+(\vec{r},t)$ is non-zero, it corresponds to the case when the projection of electron's spin is along the positive z-direction, and if only $\psi_-(\vec{r},t)$ is non-zero, the projection of the electron's spin is along the negative z-direction. A general state, ψ, is a superposition of these two states.

After defining the state of an electron with spin, we must now determine the operator $\hat{\vec{S}}$ corresponding to the dynamical variable $\vec{S}$. Since, $\hat{\vec{S}}$ acts on vectors belonging to a two-dimensional Euclidean space, it must be represented by a 2×2 matrix. Evidently, its Cartesian components, $\hat{S}_x, \hat{S}_x$, and $\hat{S}_z$ will also be a 2×2 matrix.

Following Pauli, let us introduce a new vector matrix $\hat{\vec{\sigma}} \equiv \hat{\sigma}_x \hat{\imath} + \hat{\sigma}_y \hat{\jmath} + \hat{\sigma}_z \hat{k}$ by the formula

$$\vec{S} = \frac{\hbar}{2}\hat{\vec{\sigma}}. \tag{8.3.3}$$

Replacing the Cartesian components of $\vec{S}$ in the commutation relations (8.1.1)–(8.1.3) with the corresponding Cartesian components of $\hat{\vec{\sigma}}$, we get the commutation relations satisfied by the σ matrices:

$$[\hat{\sigma}_x, \hat{\sigma}_y] = 2i\hat{\sigma}_z, \tag{8.3.4}$$

$$[\hat{\sigma}_y, \hat{\sigma}_z] = 2i\hat{\sigma}_x, \tag{8.3.5}$$

$$[\hat{\sigma}_z, \hat{\sigma}_x] = 2i\hat{\sigma}_y. \tag{8.3.6}$$

Let the axis for the projection of spin be the z-axis in an arbitrarily oriented Cartesian system of coordinates. Then the eigenvalues of S_z will be $+\hbar/2$ and $-\hbar/2$. Therefore, the eigenvalues of $\hat{\sigma}_z$ will be $+1$ and -1. It means that the operator $\hat{\sigma}_z$ must be represented by a diagonal matrix with diagonal elements $+1$ and -1, that is,

$$\hat{\sigma}_z = \begin{pmatrix} 1 & 0 \\ 0 & -1 \end{pmatrix} \tag{8.3.7}$$

and $\hat{\sigma}_z^2$ will be a 2×2 unit matrix. This is usually called the S_z-representation for the sigma matrices. It then follows from the isotropy of space (equivalence of all the directions in space) that the matrices $\hat{\sigma}_x^2$ and $\hat{\sigma}_y^2$ will also be 2×2 unit matrices with eigenvalues 1, that is,

$$\hat{\sigma}_x^2 = \hat{\sigma}_y^2 = \hat{\sigma}_z^2 = I = \begin{pmatrix} 1 & 0 \\ 0 & 1 \end{pmatrix}. \tag{8.3.8}$$

Since $\hat{\sigma}_y^2$ commutes with $\hat{\sigma}_z$, we have

$$\begin{aligned} \hat{\sigma}_y^2\hat{\sigma}_z - \hat{\sigma}_z\hat{\sigma}_y^2 &= \hat{\sigma}_y\hat{\sigma}_y\hat{\sigma}_z - \hat{\sigma}_y\hat{\sigma}_z\hat{\sigma}_y + \hat{\sigma}_y\hat{\sigma}_z\hat{\sigma}_y - \hat{\sigma}_z\hat{\sigma}_y\hat{\sigma}_y \\ &= \hat{\sigma}_y(\hat{\sigma}_y\hat{\sigma}_z - \hat{\sigma}_z\hat{\sigma}_y) + (\hat{\sigma}_y\hat{\sigma}_z - \hat{\sigma}_z\hat{\sigma}_y)\hat{\sigma}_y = 0. \end{aligned} \tag{8.3.9}$$

Taking into account the commutation relations of σ-matrices, we obtain

$$2i(\hat{\sigma}_y\hat{\sigma}_x + \hat{\sigma}_x\hat{\sigma}_y) = 0. \quad \Rightarrow \quad \hat{\sigma}_x\hat{\sigma}_y + \hat{\sigma}_y\hat{\sigma}_x = 0. \tag{8.3.10}$$

This means that the matrices $\hat{\sigma}_x$ and $\hat{\sigma}_y$ anti-commute. Similarly, one can prove that all the σ-matrices anti-commute with each other. This property along with the commutation relations leads to the following useful formulae

$$\hat{\sigma}_x\hat{\sigma}_y = -\hat{\sigma}_y\hat{\sigma}_x = i\hat{\sigma}_z, \tag{8.3.11}$$

$$\hat{\sigma}_y\hat{\sigma}_z = -\hat{\sigma}_z\hat{\sigma}_y = i\hat{\sigma}_x, \tag{8.3.12}$$

$$\hat{\sigma}_z\hat{\sigma}_x = -\hat{\sigma}_x\hat{\sigma}_z = i\hat{\sigma}_y. \tag{8.3.13}$$

If we multiply the first of the aforementioned relations by $\hat{\sigma}_z$ from the right, we arrive at the identity

$$\hat{\sigma}_x\hat{\sigma}_y\hat{\sigma}_z = iI \tag{8.3.14}$$

which will be useful later.

Let us determine the concrete expressions for the sigma matrices. The general form of $\hat{\sigma}_x$ can be written as

$$\hat{\sigma}_x = \begin{pmatrix} a_1 & a_2 \\ a_3 & a_4 \end{pmatrix}, \tag{8.3.15}$$

where the matrix elements a_1, a_2, a_3 and a_4 are, in general, complex and have to be determined using the basic properties of the sigma matrices.

First of all, $\hat{\sigma}_x$ must be hermitian, that is, $\hat{\sigma}_x = \hat{\sigma}_x^\dagger$. It gives $a_1^* = a_1, a_4^* = a_4$ and $a_2^* = a_3, a_3^* = a_2$. That is, the diagonal elements are real and the off-diagonal elements are complex conjugate to each other. Since $\hat{\sigma}_x$ and $\hat{\sigma}_z$ anti-commute, that is, $\hat{\sigma}_x\hat{\sigma}_z = -\hat{\sigma}_z\hat{\sigma}_x$, we have

$$\begin{pmatrix} a_1 & -a_2 \\ a_3 & -a_4 \end{pmatrix} = \begin{pmatrix} -a_1 & -a_2 \\ a_3 & a_4 \end{pmatrix}. \tag{8.3.16}$$

Therefore, $a_1 = 0$ and $a_4 = 0$. Using the property that $\hat{\sigma}_x^2 = I$, we get

$$\begin{pmatrix} a_2a_3 & 0 \\ 0 & a_3a_2 \end{pmatrix} = \begin{pmatrix} 1 & 0 \\ 0 & 1 \end{pmatrix} \quad \Rightarrow \quad a_2a_3 = a_3a_2 = 1. \tag{8.3.17}$$

Since $a_2 = a_3^\dagger$ and $a_3 = a_2^\dagger$, we conclude that $|a_2|^2 = |a_3|^2 = 1$. Therefore, $a_2 = e^{i\alpha}$ and $a_3 = e^{-i\alpha}$, where α is an arbitrary real constant. Since, without any loss of generality, we can put α equal to zero, we have

$$\hat{\sigma}_x = \begin{pmatrix} 0 & 1 \\ 1 & 0 \end{pmatrix}. \tag{8.3.18}$$

Now using the relation $i\hat{\sigma}_x = \hat{\sigma}_z\hat{\sigma}_x$, we obtain

$$\hat{\sigma}_y = \begin{pmatrix} 0 & -i \\ i & 0 \end{pmatrix}. \tag{8.3.19}$$

The matrices

$$\hat{\sigma}_x = \begin{pmatrix} 0 & 1 \\ 1 & 0 \end{pmatrix}, \quad \hat{\sigma}_y = \begin{pmatrix} 0 & -i \\ i & 0 \end{pmatrix}, \quad \hat{\sigma}_z = \begin{pmatrix} 1 & 0 \\ 0 & -1 \end{pmatrix}, \tag{8.3.20}$$

are called Pauli matrices in the S_z representation and along with the unit matrix

$$I = \begin{pmatrix} 1 & 0 \\ 0 & 1 \end{pmatrix} \tag{8.3.21}$$

form the basis in the space of 2×2 matrices. Any 2×2 matrix can be expanded as a linear combination of these matrices.

Now consider $\hat{\sigma}_z$. Its eigenvalues are ± 1. In the state corresponding to the eigenvalue $+1$, the spin of the electron points along the $+z$-axis and we call it *spin-up state*. Similarly, in the state corresponding to the eigenvalue -1, the spin of the electron points along the $-z$ direction and it is called the *spin-down* state. The eigenfunctions of $\hat{\sigma}_z$ with eigenvalues $+1$ and -1, respectively, are readily computed as

$$\chi_z^+ = \begin{pmatrix} 1 \\ 0 \end{pmatrix}, \quad \chi_z^- = \begin{pmatrix} 0 \\ 1 \end{pmatrix}, \tag{8.3.22}$$

Let us check whether these spin functions are eigenfunctions of $\hat{\sigma}_x$ and $\hat{\sigma}_y$ or not. We have

$$\hat{\sigma}_x\chi_z^+ = \begin{pmatrix} 0 & 1 \\ 1 & 0 \end{pmatrix}\begin{pmatrix} 1 \\ 0 \end{pmatrix} = \begin{pmatrix} 0 \\ 1 \end{pmatrix} = \chi_z^-, \tag{8.3.23}$$

$$\hat{\sigma}_x\chi_z^- = \begin{pmatrix} 0 & 1 \\ 1 & 0 \end{pmatrix}\begin{pmatrix} 0 \\ 1 \end{pmatrix} = \begin{pmatrix} 1 \\ 0 \end{pmatrix} = \chi_z^+, \tag{8.3.24}$$

$$\hat{\sigma}_y\chi_z^+ = \begin{pmatrix} 0 & -i \\ i & 0 \end{pmatrix}\begin{pmatrix} 1 \\ 0 \end{pmatrix} = \begin{pmatrix} 0 \\ i \end{pmatrix} = i\chi_z^-, \tag{8.3.25}$$

$$\hat{\sigma}_y\chi_z^- = \begin{pmatrix} 0 & -i \\ i & 0 \end{pmatrix}\begin{pmatrix} 0 \\ 1 \end{pmatrix} = \begin{pmatrix} -i \\ 0 \end{pmatrix} = -i\chi_z^+ \tag{8.3.26}$$

These results tell us that, in the states described by the eigenfunctions χ_z^+ and χ_z^- of $\hat{\sigma}_z$, only S_z has definite values equal to $\pm\hbar/2$. The projections of spin on the x-and y-axes are completely unknown. This is nothing but the consequence of the fact that the matrices $\hat{\sigma}_x$, $\hat{\sigma}_y$ and $\hat{\sigma}_z$ and hence the spin matrices, S_x, S_y and S_z do not commute with each other.

Example 8.3.1: (i) Find the eigenvalues and the eigenvectors (eigenfunctions) of the matrices $\hat{\sigma}_x$ and $\hat{\sigma}_y$. (ii) Show that irrespective of the direction of a chosen axis, the projection of spin on the axis can take only two values equal to $\pm\hbar/2$.

Solution:

(i) The characteristic equation for $\hat{\sigma}_x$ reads

$$\begin{vmatrix} 0-\lambda & 1 \\ 1 & 0-\lambda \end{vmatrix} = \lambda^2 - 1 = 0. \tag{8.3.27}$$

Therefore, the eigenvalues of $\hat{\sigma}_x$ are $\lambda = \pm 1$. Let

$$\begin{pmatrix} a \\ b \end{pmatrix} \tag{8.3.28}$$

be the eigenvector of $\hat{\sigma}_x$. Then for $\lambda_1 = \pm 1$, we have

$$\begin{pmatrix} 0 & 1 \\ 1 & 0 \end{pmatrix}\begin{pmatrix} a \\ b \end{pmatrix} = \pm\begin{pmatrix} a \\ b \end{pmatrix} \Rightarrow b = \pm a. \tag{8.3.29}$$

Thus, the eigenvectors of $\hat{\sigma}_x$ corresponding to the eigenvalues $\lambda_1 = +1$ and $\lambda_2 = -1$ are

$$\chi_x^+ = \frac{1}{\sqrt{2}}\begin{pmatrix} 1 \\ 1 \end{pmatrix}, \quad \chi_x^- = \frac{1}{\sqrt{2}}\begin{pmatrix} 1 \\ -1 \end{pmatrix}, \tag{8.3.30}$$

respectively. Similarly, the eigenvalues of $\hat{\sigma}_y$ are $\lambda = \pm 1$ and the corresponding normalized eigenvectors are given by

$$\chi_y^+ = \frac{1}{\sqrt{2}}\begin{pmatrix} 1 \\ i \end{pmatrix}, \quad \chi_y^- = \frac{1}{\sqrt{2}}\begin{pmatrix} 1 \\ -i \end{pmatrix}. \tag{8.3.31}$$

(ii) Since $\hat{\vec{\sigma}} = \hat{i}\hat{\sigma}_x + \hat{j}\hat{\sigma}_y + \hat{k}\hat{\sigma}_z$, the projection of $\hat{\vec{\sigma}}$ on an arbitrarily oriented axis with directional cosines, ℓ, m and n, is given by the matrix $M = \ell\sigma_x + m\sigma_y + n\sigma_z$, or

$$M = \begin{pmatrix} n & \ell - im \\ \ell + im & -n \end{pmatrix}. \tag{8.3.32}$$

The characteristic equation for the eigenvalues λ of M reads

$$\begin{vmatrix} n-\lambda & \ell - im \\ \ell + im & -(n+\lambda) \end{vmatrix} = \lambda^2 - (\ell^2 + m^2 + n^2) = \lambda^2 - 1 = 0, \tag{8.3.33}$$

where we have used the property of the directional cosines, namely, $\ell^2 + m^2 + n^2 = 1$. From (8.2.33), we get the eigenvalues of M to be ± 1. It means that the eigenvalues of the operator corresponding to the projection of spin $\vec{S}$ on this axis will be $S_{\ell mn} = \pm\hbar/2$. Thus, irrespective of the direction of the chosen axis, the projection of spin on that axis can take only two values $\pm\hbar/2$. Thus, we see that Pauli matrices $(\sigma_x, \sigma_y, \sigma_z)$ and the related spin matrices (S_x, S_y, S_z) satisfy all the requirements of quantum mechanics and are consistent with the experimental results.

Example 8.3.2: Find the eigenvalues and eigenstates of the spin operator $\hat{\vec{S}}$ of an electron in the direction of a unit vector $\hat{n}$ that lies in the xy plane making an angle θ with the x-axis.

Solution: The projection of the spin operator $\hat{\vec{S}}$ on $\hat{n}$ will be $S_n = \frac{\hbar}{2}\hat{\sigma}_n$, where

$$\hat{\sigma}_n = \begin{pmatrix} 0 & \cos\theta \\ \cos\theta & 0 \end{pmatrix} + \begin{pmatrix} 0 & -i\sin\theta \\ i\sin\theta & 0 \end{pmatrix} = \begin{pmatrix} 0 & e^{-i\theta} \\ e^{i\theta} & 0 \end{pmatrix}. \tag{8.3.34}$$

The requirement of non-trivial solutions to the eigenvalue equation for $\hat{\sigma}_n$ yields

$$\begin{vmatrix} -\lambda & e^{-i\theta} \\ e^{i\theta} & -\lambda \end{vmatrix} = 0, \;\Rightarrow\; \lambda = \pm 1. \tag{8.3.35}$$

Hence, the eigenvalues of the operator S_n are $\pm\frac{\hbar}{2}$.
For the eigenvectors of S_n, We have

$$\begin{pmatrix} 0 & e^{-i\theta} \\ e^{i\theta} & 0 \end{pmatrix}\begin{pmatrix} a \\ b \end{pmatrix} = \begin{pmatrix} be^{-i\theta} \\ ae^{i\theta} \end{pmatrix} = \pm\begin{pmatrix} a \\ b \end{pmatrix} \;\Rightarrow\; a = e^{-i\theta/2}, \; b = \pm e^{i\theta/2}, \tag{8.3.36}$$

where the $\pm$ signs in b correspond to $\pm\hbar/2$, respectively. The normalized eigenvectors of S_n, corresponding to the eigenvalues $\pm\hbar/2$, are

$$\chi_n^+ = \frac{1}{\sqrt{2}}\begin{pmatrix} e^{-i\theta/2} \\ e^{i\theta/2} \end{pmatrix}, \quad \chi_n^- = \frac{1}{\sqrt{2}}\begin{pmatrix} e^{-i\theta/2} \\ -e^{i\theta/2} \end{pmatrix}. \tag{8.3.37}$$

Before proceeding further, let us summarize the basic properties of sigma matrices and write down some useful formulae that are easily derived from them. We have

$$I.\ \sigma_k^\dagger = \hat{\sigma}_k,\ \ \mathrm{Tr}(\hat{\sigma}_k) = 0,\ \ \det(\hat{\sigma}_k) = -1 \qquad (k = x,y,z), \tag{8.3.38}$$

$$II.\ \hat{\sigma}_k^2 = \hat{I}, \qquad (k = x,y,z), \tag{8.3.39}$$

$$III.\ \hat{\sigma}_j\hat{\sigma}_k = i\varepsilon_{jk\ell}\hat{\sigma}_\ell \qquad (j \neq k), \tag{8.3.40}$$

$$IV.\ \hat{\sigma}_j\hat{\sigma}_k + \hat{\sigma}_k\hat{\sigma}_j = 0 \qquad (j \neq k), \tag{8.3.41}$$

$$V.\ [\hat{\sigma}_j, \hat{\sigma}_k] = 2i\varepsilon_{jkl}\hat{\sigma}_l \qquad (j,k,l = x,y,z), \tag{8.3.42}$$

where $\hat{I}$ the 2×2 unit matrix, ε_{jkl} is the Levi-Civita tensor density and the summation from 1 to 3 over the repeated index, ℓ, is implied. The properties in (8.2.39) and (8.2.41) can be combined together as

$$[\hat{\sigma}_j, \hat{\sigma}_k]_+ = 2\hat{I}\delta_{jk}, \tag{8.3.43}$$

where $[\hat{\sigma}_j, \hat{\sigma}_k]_+$ stands for the anti-commutator of $\hat{\sigma}_j$ and $\hat{\sigma}_k$ and δ_{jk} is the Kronecker delta. On the other hand, the properties in (8.2.39) and (8.2.40) can be combined together into a single formula

$$\hat{\sigma}_j\hat{\sigma}_k = \delta_{jk} + i\varepsilon_{jk\ell}\hat{\sigma}_\ell. \tag{8.3.44}$$

where once again summation from 1 to 3 over ℓ is implied.

Because the spin and spatial degrees of freedom are completely independent, the spin operators $\hat{S}_x, \hat{S}_y$ and $\hat{S}_z$ commute with the position operator $\hat{\vec{r}}$, the momentum operator $\hat{\vec{p}}$ and the angular momentum operator $\hat{\vec{L}}$:

$$VI.\ [\hat{S}_j, \hat{r}_k] = 0,\ \ [\hat{S}_j, \hat{p}_k] = 0,\ \ [\hat{S}_j, \hat{L}_k] = 0, \qquad (j,k = x,y,z). \tag{8.3.45}$$

The next question that we may ask is: "*How do we write the wave function of an electron by taking into account its spin properties*"? The answer to this question is as follows. Because of the independence of the spatial and the spin degrees of freedom, the total wave function of the particle is the product of the spatio-temporal part, $\psi(\vec{r},t)$, and the spin part, $\chi(S)$:

$$\psi(\vec{r},t;S) = \psi(\vec{r},t)\chi(S). \tag{8.3.46}$$

Or, taking into account our earlier notation given in (8.2.2), the same can be written as

$$\psi(\vec{r},t;S) = \begin{pmatrix} \psi^+ \\ \psi^- \end{pmatrix} = \psi(\vec{r},t) \begin{pmatrix} \chi^+ \\ \chi^- \end{pmatrix}, \tag{8.3.47}$$

where the spin functions χ^+ and χ^- correspond to spin-up and spin-down cases, respectively.

In Dirac notation, it can be written as

$$|\psi(\vec{r},t;S)\rangle = |\psi(\vec{r},t)\rangle \otimes |s,m_s\rangle, \tag{8.3.48}$$

where s is the total spin and m_s is the spin-magnetic quantum number. The spin functions, $|s,m_s\rangle$, are given by

$$\chi^+ = \left|\frac{1}{2},\frac{1}{2}\right\rangle, \quad \chi^- = \left|\frac{1}{2},-\frac{1}{2}\right\rangle. \tag{8.3.49}$$

Example 8.3.3: An electron is in the spin state

$$\chi = A\begin{pmatrix} i \\ 2 \end{pmatrix}. \tag{8.3.50}$$

(a) Find the constant A. (b) If a measurement of S_z is made on the electron, what is the probability of getting the value $S_z = -\frac{\hbar}{2}$? (c) If, instead, a measurement of S_y is carried out, what is the probability of getting the value $S_y = \frac{\hbar}{2}$?

Solution:

(a) The normalization of the wave function reads

$$|A|^2 \begin{pmatrix} -i & 2 \end{pmatrix} \begin{pmatrix} i \\ 2 \end{pmatrix} = 1, \;\Rightarrow\; 5|A|^2 = 1, \;\Rightarrow\; A = \frac{1}{\sqrt{5}}. \tag{8.3.51}$$

(b) To find the answer, we have to expand χ in terms of the complete set of eigenvectors of $\hat{\sigma}_z$, that is, $\chi = B\chi_z^+ + C\chi_z^-$, where B and C are arbitrary complex constants. A little bit of algebra gives, $B = i/\sqrt{5}$ and $C = 2/\sqrt{5}$. Therefore,

$$\chi = \frac{i}{\sqrt{5}}\chi_z^+ + \frac{2}{\sqrt{5}}\chi_z^- = \frac{i}{\sqrt{5}}\begin{pmatrix} 1 \\ 0 \end{pmatrix} + \frac{2}{\sqrt{5}}\begin{pmatrix} 0 \\ 1 \end{pmatrix}. \tag{8.3.52}$$

The probability, $P_z(-\hbar/2)$, of obtaining $S_z = -\frac{\hbar}{2}$ is therefore given by

$$P_z(-\hbar/2) = |\,(\chi_z^-)^*\chi|^2 = |C|^2 = \frac{4}{5}. \tag{8.3.53}$$

(c) In this case, we have to express χ in terms of the complete set of eigenvectors of $\hat{\sigma}_y$, that is, $\chi = D\chi_y^+ + E\chi_y^-$, where D and E are arbitrary complex constants. We get

$$\chi = -\frac{i}{\sqrt{10}}\left\{\frac{1}{\sqrt{2}}\begin{pmatrix} 1 \\ i \end{pmatrix}\right\} + \frac{3i}{\sqrt{10}}\left\{\frac{1}{\sqrt{2}}\begin{pmatrix} 1 \\ -i \end{pmatrix}\right\}. \tag{8.3.54}$$

The probability, $P_y(\hbar/2)$, of obtaining $S_y = \frac{\hbar}{2}$ is therefore given by

$$P_y(\hbar/2) = |\langle\chi_y^+|\chi\rangle|^2 = |D|^2 = \frac{1}{10}. \tag{8.3.55}$$

Example 8.3.4: Consider the spin state of an electron in the previous problem. (a) Find the expectation values of $\hat{S}_x, \hat{S}_y$ and $\hat{S}_z$. (b) Find the uncertainties in the measurements of the observables $\hat{S}_x, \hat{S}_y$ and $\hat{S}_z$.

Solution:

(a) The normalized spin wave function is given by

$$\chi = \frac{1}{\sqrt{5}}\begin{pmatrix} i \\ 2 \end{pmatrix}. \tag{8.3.56}$$

Therefore, the expectation values of $\hat{S}_x, \hat{S}_y$ and $\hat{S}_z$ are given by

$$\langle\hat{S}_x\rangle = \chi^\dagger\hat{S}_x\chi = \frac{\hbar}{2}\times\frac{1}{5}\begin{pmatrix} -i & 2 \end{pmatrix}\begin{pmatrix} 0 & 1 \\ 1 & 0 \end{pmatrix}\begin{pmatrix} i \\ 2 \end{pmatrix} = 0, \tag{8.3.57}$$

$$\langle\hat{S}_y\rangle = \chi^\dagger\hat{S}_y\chi = \frac{\hbar}{2}\times\frac{1}{5}\begin{pmatrix} -i & 2 \end{pmatrix}\begin{pmatrix} 0 & -i \\ i & 0 \end{pmatrix}\begin{pmatrix} i \\ 2 \end{pmatrix} = -\frac{2}{5}\hbar, \tag{8.3.58}$$

$$\langle\hat{S}_z\rangle = \chi^\dagger\hat{S}_y\chi = \frac{\hbar}{2}\times\frac{1}{5}\begin{pmatrix} -i & 2 \end{pmatrix}\begin{pmatrix} 1 & 0 \\ 0 & -1 \end{pmatrix}\begin{pmatrix} i \\ 2 \end{pmatrix} = -\frac{3}{10}\hbar. \tag{8.3.59}$$

(b) The average values of $\hat{S}_x^2$ $\hat{S}_y^2$ and $\hat{S}_z^2$ are given by

$$\langle\hat{S}_x^2\rangle = \langle\hat{S}_y^2\rangle = \langle\hat{S}_x^2\rangle = \frac{\hbar^2}{4}. \tag{8.3.60}$$

Therefore, the uncertainties in the measurements of $\hat{S}_x, \hat{S}_y$ and $\hat{S}_z$ are given by

$$\Delta S_x = \sqrt{\langle\hat{S}_x^2\rangle - \langle\hat{S}_x\rangle^2} = \sqrt{\frac{\hbar^2}{4} - 0} = \frac{\hbar}{2}, \tag{8.3.61}$$

$$\Delta S_y = \sqrt{\langle \hat{S}_y^2 \rangle - \langle \hat{S}_y \rangle^2} = \sqrt{\frac{\hbar^2}{4} - \frac{4}{25}\hbar^2} = \frac{3}{10}\hbar, \tag{8.3.62}$$

$$\Delta S_z = \sqrt{\langle \hat{S}_z^2 \rangle - \langle \hat{S}_z \rangle^2} = \sqrt{\frac{\hbar^2}{4} - \frac{9}{100}\hbar^2} = \frac{2}{5}\hbar. \tag{8.3.63}$$

Example 8.3.5: When $\hat{S}_x$ is measured on a spin $\frac{1}{2}$ particle, the result is $\frac{\hbar}{2}$. Immediately after that the projection of spin in the direction of a unit vector $\hat{n}$, which lies in the xy plane making an angle θ with the x-axis, is measured on the particle. What is the probability of getting the value $\frac{\hbar}{2}$?

Solution: After the first measurement, the particle must be in the eigenstate of $\hat{S}_x$ corresponding to the eigenvalue $\frac{\hbar}{2}$, that is, in the state

$$\chi_x^{(+)} = \frac{1}{\sqrt{2}} \begin{pmatrix} 1 \\ 1 \end{pmatrix}. \tag{8.3.64}$$

In order to answer the question, we must first expand $\chi_x^{(+)}$ into a linear combination of the eigenvectors of $\hat{S}_x \cos\theta + \hat{S}_y \sin\theta$. So, we write

$$\frac{1}{\sqrt{2}} \begin{pmatrix} 1 \\ 1 \end{pmatrix} = \alpha \frac{1}{\sqrt{2}} \begin{pmatrix} e^{-i\theta/2} \\ e^{i\theta/2} \end{pmatrix} + \beta \frac{1}{\sqrt{2}} \begin{pmatrix} e^{-i\theta/2} \\ -e^{i\theta/2} \end{pmatrix}, \tag{8.3.65}$$

and determine the coefficients α and β. From this matrix equation, we get the following set of algebraic equations:

$$\alpha + \beta = e^{i\theta/2}, \tag{8.3.66}$$

$$\alpha - \beta = e^{-i\theta/2}. \tag{8.3.67}$$

From (8.2.65) and (8.2.66), we obtain

$$\alpha = \cos\left(\frac{\theta}{2}\right), \quad \beta = i \sin\left(\frac{\theta}{2}\right). \tag{8.3.68}$$

Therefore, the probability of getting the value $\frac{\hbar}{2}$ for S_n is

$$P(+\hbar/2) = \cos^2\left(\frac{\theta}{2}\right). \tag{8.3.69}$$

8.4 Spin Precession in a Uniform External Magnetic Field

Let us consider what happens to a spin $\frac{1}{2}$ charged particle (charge q and mass m) when it is placed in an external magnetic field, $\vec{B}$ such that it is at rest at a fixed location. We know from electrodynamics that a spinning charged particle is equivalent to a magnetic dipole whose magnetic dipole moment, $\vec{\mu}$, is proportional to its spin angular momentum, $\vec{S}$:

$$\vec{\mu} = \gamma\vec{S}, \tag{8.4.1}$$

where γ is called the gyromagnetic ration[1]. We also know that a magnetic dipole experiences a torque $\vec{\tau} = \vec{\mu} \times \vec{B}$, in an external magnetic field $\vec{B}$, which tries to align it along the applied field. The magnetic potential energy of such a dipole in the external field is given by

$$W_m = -\vec{\mu}\cdot\vec{B} = -\gamma\vec{S}\cdot\vec{B}. \tag{8.4.2}$$

So, the Hamiltonian, $\hat{H}_S$, representing the potential energy of interaction is given by

$$\hat{H}_S = -\gamma\hat{\vec{S}}\cdot\vec{B}. \tag{8.4.3}$$

Therefore, in general, the time evolution of the wave function of a spin $\frac{1}{2}$ particle in the presence of an electromagnetic field will be governed by the time dependent Schrödinger equation (see Chapter 7)

$$i\hbar\frac{\partial\psi}{\partial t} = \frac{1}{2m}\left(\hat{\vec{P}} - q\vec{A}\right)^2\psi + q\Phi\psi + \hat{H}_S\psi, \tag{8.4.4}$$

Equation (8.4.4), which takes into account the contribution from the spin of the particle, is also known as *Pauli equation.*

Since the spinning charged particle is at rest at a fixed location in the magnetic field and its spin is the only degree of freedom, (8.4.4) yields the following equation for the time evolution of the spin wave function $\chi(t)$:

$$i\hbar\frac{\partial\chi(t)}{\partial t} = \hat{H}_S\chi(t). \tag{8.4.5}$$

If the magnetic field is uniform and directed along the z axis , i.e., $\vec{B} = \hat{z}B_0$, then $\hat{H} = -\gamma\vec{S}\cdot\vec{B} = -\gamma B_0\hat{S}_z$. So, the Hamiltonian commutes with $\hat{S}_z$ and the eigenstates of $\hat{S}_z$ are also the eigenstates of the Hamiltonian. Hence, we have

$$\hat{H}\chi_z^+ = -\gamma B_0\hat{S}_z\chi_z^+ = -\frac{\gamma B_0\hbar}{2}\chi_z^+ \equiv E^+\chi_z^+, \tag{8.4.6}$$

$$\hat{H}\chi_z^- = -\gamma B_0\hat{S}_z\chi_z^- = \frac{\gamma B_0\hbar}{2}\chi_z^- \equiv E^-\chi_z^-. \tag{8.4.7}$$

[1] See, for instance, D.J. Griffiths, *Introduction to Electrodynamics*, Prentice Hall, 1999, 3rd ed., page 252.

where $E^+ = -(\gamma B_0 \hbar)/2$ and $E^- = (\gamma B_0 \hbar)/2$. Thus, the general spin wave function of the particle at $t = 0$ will be given by the linear superposition $\chi(0) = a_0 \chi_z^+ + b_0 \chi_z^-$, where a_0 and b_0 are arbitrary constants. Normalization requires a_0 and b_0 to satisfy the condition $a_0^2 + b_0^2 = 1$. Without any loss of generality, this condition can be fulfilled by choosing $a_0 = \cos(\theta/2)$ and $b_0 = \sin(\theta/2)$, where θ is constant whose physical meaning will be clarified later.

To study the time-evolution of the spin state of the particle in the external magnetic field, we use (8.4.5). Using the expressions for χ_z^+ and χ_z^-, we get

$$i\hbar \begin{pmatrix} \dot{a} \\ \dot{b} \end{pmatrix} = -\frac{\gamma B_0 \hbar}{2} \begin{pmatrix} 1 & 0 \\ 0 & -1 \end{pmatrix} \begin{pmatrix} a \\ b \end{pmatrix} = \frac{\hbar \omega_L}{2} \begin{pmatrix} -a \\ b \end{pmatrix}, \tag{8.4.8}$$

where dot stands for the ordinary time derivative and $\omega_L = \gamma B_0$ is the Larmor frequency. We thus get the following set of equations

$$\dot{a} = i(\omega_L/2)\, a, \quad \dot{b} = -i(\omega_L/2)\, b. \tag{8.4.9}$$

The solutions of the above equations consistent with the initial condition lead to the following time-dependent spin function

$$\chi(t) = \begin{pmatrix} \cos(\theta/2)\, e^{i\gamma B_0 t/2} \\ \sin(\theta/2)\, e^{-i\gamma B_0 t/2} \end{pmatrix}. \tag{8.4.10}$$

In order to see the physical meaning of the constant θ, let us calculate the average values of the spin components S_x, S_y and S_z. We have

$$\langle S_x \rangle = \chi^\dagger(t) \hat{S}_x \chi(t) = \frac{\hbar}{2} \sin\theta \cos(\gamma B_0 t), \tag{8.4.11}$$

$$\langle S_y \rangle = \chi^\dagger(t) \hat{S}_y \chi(t) = -\frac{\hbar}{2} \sin\theta \sin(\gamma B_0 t), \tag{8.4.12}$$

$$\langle S_z \rangle = \chi^\dagger(t) \hat{S}_z \chi(t) = \frac{\hbar}{2} \cos\theta. \tag{8.4.13}$$

On the basis of (8.4.11)–(8.4.13), we conclude that the spin component along the field direction is conserved. The average spin $\langle \vec{S} \rangle$ is tilted at a constant angle θ to the field direction (that is the z-axis) and precesses about the field direction at the Larmor frequency ω_L.

Example 8.4.1: Consider a spin $\frac{1}{2}$ particle at rest in a uniform magnetic field pointing in the $+z$ direction. It is initially in the spin state χ_x^+ at time $t = 0$. (a) What will be the state at some later time t? (b) If we measure S_x at time t, what are the probabilities of finding $\pm\frac{\hbar}{2}$? (c) What are the probabilities of finding $\pm\frac{\hbar}{2}$, if we measure S_z at time t?

Solution: At $t = 0$, we have

$$\chi(0) = \chi_x^+ = \frac{1}{\sqrt{2}}\begin{pmatrix} 1 \\ 1 \end{pmatrix} = \frac{1}{\sqrt{2}}\begin{pmatrix} 1 \\ 0 \end{pmatrix} + \frac{1}{\sqrt{2}}\begin{pmatrix} 0 \\ 1 \end{pmatrix} = \frac{1}{\sqrt{2}}\chi_z^+ + \frac{1}{\sqrt{2}}\chi_z^-. \quad (8.4.14)$$

(a) As discussed earlier, the state at any $t > 0$ will be

$$\chi(t) = \frac{1}{\sqrt{2}} e^{i\frac{\gamma B_0 t}{2}} \chi_z^+ + \frac{1}{\sqrt{2}} e^{-i\frac{\gamma B_0 t}{2}} \chi_z^-. \quad (8.4.15)$$

(b) Since $\chi(t)$ is normalized to unity, according to the measurement postulate of quantum mechanics, the probability of obtaining $\frac{\hbar}{2}$ for S_x at any $t > 0$ is given by

$$P(S_x = \hbar/2) = \left|\langle \chi_x^+ | \chi(t) \rangle\right|^2. \quad (8.4.16)$$

Using (8.4.14) and (8.4.15), we obtain

$$P(S_x = \hbar/2) = \left| \frac{1}{2}\left(e^{i\frac{\gamma B_0 t}{2}} \langle \chi_z^+ | \chi_z^+ \rangle + e^{-i\frac{\gamma B_0 t}{2}} \langle \chi_z^- | \chi_z^- \rangle \right) \right.$$
$$\left. + \frac{1}{2}\left(e^{-i\frac{\gamma B_0 t}{2}} \langle \chi_z^+ | \chi_z^- \rangle + e^{-i\frac{\gamma B_0 t}{2}} \langle \chi_z^- | \chi_z^+ \rangle \right) \right|^2. \quad (8.4.17)$$

Taking into account the orthonormality of the eigenvectors of $\hat{\sigma}_z$, we get

$$P(S_x = \hbar/2) = \left| \frac{1}{2}\left[e^{i\frac{\gamma B_0 t}{2}} \langle \chi_z^+ | \chi_z^+ \rangle + e^{-i\frac{\gamma B_0 t}{2}} \langle \chi_z^- | \chi_z^- \rangle \right] \right|^2 = \cos^2\left(\frac{\gamma B_0 t}{2}\right). \quad (8.4.18)$$

By taking into account that

$$\chi(0) = \chi_x^- = \frac{1}{\sqrt{2}}\begin{pmatrix} 1 \\ -1 \end{pmatrix} = \frac{1}{\sqrt{2}}\begin{pmatrix} 1 \\ 0 \end{pmatrix} - \frac{1}{\sqrt{2}}\begin{pmatrix} 0 \\ 1 \end{pmatrix} = \frac{1}{\sqrt{2}}\chi_z^+ - \frac{1}{\sqrt{2}}\chi_z^-, \quad (8.4.19)$$

the probability for finding the value $S_x = -\frac{\hbar}{2}$ is computed in an analogous manner. The result is

$$P(S_x = -\hbar/2) = \left| i\frac{1}{2i}\left[e^{i\frac{\gamma B_0 t}{2}} \langle \chi_z^+ | \chi_z^+ \rangle - e^{-i\frac{\gamma B_0 t}{2}} \langle \chi_z^- | \chi_z^- \rangle \right] \right|^2 = \sin^2\left(\frac{\gamma B_0 t}{2}\right). \quad (8.4.20)$$

(c) Similarly, if a measurement of S_z is carried out at $t > 0$, the probabilities of obtaining the values $S_z = \pm\frac{\hbar}{2}$ are given by

$$P(S_z = \hbar/2) = \left|\langle \chi_z^+ | \chi(t)\rangle\right|^2 = \left|\frac{1}{\sqrt{2}} e^{i\frac{\gamma B_0 t}{2}} \langle \chi_z^+ | \chi_z^+ \rangle\right|^2 = \frac{1}{2}, \tag{8.4.21}$$

$$P(S_z = \hbar/2) = \left|\langle \chi_z^- | \chi(t)\rangle\right|^2 = \left|\frac{1}{\sqrt{2}} e^{-i\frac{\gamma B_0 t}{2}} \langle \chi_z^- | \chi_z^- \rangle\right|^2 = \frac{1}{2}. \tag{8.4.22}$$

Homework Problems

1. Let $\vec{a}$ and $\vec{b}$ be any two vectors that commute with $\hat{\vec{\sigma}} = \hat{i}\hat{\sigma}_x + \hat{j}\hat{\sigma}_y + \hat{k}\hat{\sigma}_z$. Show that $(\hat{\vec{\sigma}} \cdot \vec{a})(\hat{\vec{\sigma}} \cdot \vec{b}) = (\vec{a} \cdot \vec{b})\hat{I} + i\hat{\vec{\sigma}} \cdot (\vec{a} \times \vec{b})$.
2. Find the eigenvalues and eigenstates of the spin operator $\hat{\vec{S}}$ of an electron in the direction of a unit vector $\hat{n} = \hat{j}\cos\alpha + \hat{k}\sin\alpha$ that lies in the yz plane making an angle α with the y-axis.
3. The Hamiltonian of a system is $\hat{H} = \mathscr{E}\,(\hat{\vec{\sigma}} \cdot \hat{n})$, where $\mathscr{E}$ is a constant with the dimensions of energy. Here, $\hat{n} = \hat{i}\sin\theta\,\cos\phi + \hat{j}\sin\theta\,\sin\phi + \hat{k}\cos\theta$ is an arbitrary unit vector in three dimensions and $\hat{\vec{\sigma}} = \hat{i}\,\hat{\sigma}_x + \hat{j}\,\hat{\sigma}_x + \hat{k}\,\hat{\sigma}_z$. Find the energy eigenvalues and normalized eigenvectors of $\hat{H}$.
4. An electron in a hydrogen atom occupies a combined spin and position state, given by the wave function

$$\psi(r,\theta,\phi) = R_{32}\left(\sqrt{\frac{1}{5}} Y_2^0\, \chi_+ + \sqrt{\frac{2}{5}}\,(Y_2^{-1} - Y_2^1)\,\chi_-\right).$$

(a) If $\hat{L}^2$ and $\hat{L}_z$ are measured in this state, what values will result and with what probabilities? (b) If $\hat{S}^2$ and $\hat{S}_z$ are measured in this state, what values will result and with what probabilities?

5. An electron is in the spin state

$$\chi = A\begin{pmatrix} 2+3i \\ 6 \end{pmatrix}$$

(a) Determine the normalization constant A. (b) If S_z is measured, what is the probability of getting the value $\hbar/2$? (c) Calculate the average values of $\hat{S}_x$, $\hat{S}_y$, $\hat{S}_z$, $\hat{S}_x^2$, $\hat{S}_y^2$ and $\hat{S}_z^2$. (d) Show that the uncertainties in the measurements of $\hat{S}_x$ and $\hat{S}_y$ satisfy the Heisenberg uncertainty relation.

6. Consider a spin $\frac{1}{2}$ particle at rest in a spatially uniform but time-dependent magnetic field $\vec{B} = B_0\,\sin(\omega t)\,\hat{z}$, where B_0 and ω are constants. At $t = 0$, it is in the spin state χ_y^+. (a) What will be the state at some later time t? (b) If S_y is measured at a later time t, what are the probabilities of finding $\pm\frac{\hbar}{2}$?

Chapter 9

Addition of Angular Momenta

9.1 General Theory and the Clebsch–Gordan Coefficients

In many problems of interest it is necessary to add angular momenta. For instance, one is required to add the orbital angular momentum, $\hat{\vec{L}}$, and the spin angular momentum, $\hat{\vec{S}}$, while studying spin-orbit coupling in atoms. Then, there are problems related to the studies of multi-electron atoms where one has to add two or more orbital angular momenta. Therefore, it is important to discuss the procedure of addition of angular momenta in quantum mechanics. In view of this, in what follows, we shall discuss the general algebraic method for the addition of any two angular momenta.

Note that, in this Chapter, we shall write the eigenfunctions of $\hat{L}^2$ in the bra–ket notation as: $|\ell,m\rangle$. Thus, $|\ell,m\rangle$ is an eigenvector (or eigenket) of $\hat{L}^2$ with two quantum numbers ℓ and m. If $\hat{\vec{L}}$ happens to be orbital angular momentum, then ℓ represents the orbital quantum number and m stands for the orbital magnetic quantum number. On the other hand, if $\hat{\vec{L}}$ happens to be the spin angular momentum ($\hat{\vec{L}} = \hat{\vec{S}}$), then ℓ is spin quantum number i.e., $\ell = s$ and m equals the spin magnetic quantum number i.e., $m = m_s$.

Let us, without specifying the nature, consider the addition of two angular momenta $\hat{\vec{L}}_1$ and $\hat{\vec{L}}_2$: $\hat{\vec{J}} = \hat{\vec{L}}_1 + \hat{\vec{L}}_2$. Individually, $\hat{\vec{L}}_1$ and $\hat{\vec{L}}_2$ satisfy the following quantum mechanical commutation relations (see Chapter 6):

$$[\hat{L}_{1i}, \hat{L}_{1j}] = i\hbar \sum_k \varepsilon_{ijk} \hat{L}_{1k}, \tag{9.1.1}$$

$$[\hat{L}_{2i}, \hat{L}_{2j}] = i\hbar \sum_k \varepsilon_{ijk} \hat{L}_{2k}, \tag{9.1.2}$$

where the indices i, j and k take values from 1 to 3. Note that, it is assumed here that $\hat{\vec{L}}_1$ and $\hat{\vec{L}}_2$ either correspond to different degrees of freedom, or correspond to the same degree of freedom but belong to different particles.

In view of the preceding assumption, the operators $\hat{\vec{L}}_1$ and $\hat{\vec{L}}_2$ act in different vector spaces: $\hat{\vec{L}}_1$ acts in the $(2\ell_1+1)$ dimensional space spanned by the kets $\{|\ell_1,m_1\rangle\}$, while $\hat{\vec{L}}_2$ acts in the $(2\ell_2+1)$ dimensional space spanned by the kets $\{|\ell_2,m_2\rangle\}$). Hence, they commute and can have a common set of eigenvectors. Let us write these common eigenvectors as

$$|\ell_1,m_1;\ell_2,m_2\rangle = |\ell_1,m_1\rangle \otimes |\ell_2,m_2\rangle, \tag{9.1.3}$$

where $\ell_i, i=1,2$ and $m_i, i=1,2$ are the individual quantum numbers and $\otimes$ stands for the direct (tensorial) product. Then according to the earlier discussions

$$\hat{\vec{L}}_1^2|\ell_1,m_1;\ell_2,m_2\rangle = \hbar^2\ell_1(\ell_1+1)|\ell_1,m_1;\ell_2,m_2\rangle, \tag{9.1.4}$$

$$\hat{L}_{1z}|\ell_1,m_1;\ell_2,m_2\rangle = \hbar m_1|\ell_1,m_1;\ell_2,m_2\rangle \tag{9.1.5}$$

$$\hat{\vec{L}}_2^2|\ell_1,m_1;\ell_2,m_2\rangle = \hbar^2\ell_2(\ell_2+1)|\ell_1,m_1;\ell_2,m_2\rangle, \tag{9.1.6}$$

$$\hat{L}_{2z}|\ell_1,m_1;\ell_2,m_2\rangle = \hbar m_2|\ell_1,m_1;\ell_2,m_2\rangle \tag{9.1.7}$$

Let us show that the total angular momentum operators $\hat{J}_i = \hat{L}_{1i} + \hat{L}_{2i}, (i=1,2,3)$ also obey the usual angular momentum commutation relations, i.e.,

$$[\hat{J}_i,\hat{J}_j] = i\hbar\sum_k \varepsilon_{ijk}\hat{J}_k, \tag{9.1.8}$$

where, once again, each of the indices i, j and k takes three values 1, 2 and 3. We have

$$\begin{aligned}
[\hat{J}_i,\hat{J}_j] &= [\hat{L}_{1i}+\hat{L}_{2i},\hat{L}_{1j}+\hat{L}_{2j}] = [\hat{L}_{1i},\hat{L}_{1j}]+[\hat{L}_{1i},\hat{L}_{2j}]+[\hat{L}_{2i},\hat{L}_{1j}]+[\hat{L}_{2i},\hat{L}_{2j}] \\
&= i\hbar\sum_k \varepsilon_{ijk}\hat{L}_{1k} + i\hbar\sum_r \varepsilon_{ijk}\hat{L}_{2k} \\
&= i\hbar\sum_k \varepsilon_{ijk}(\hat{L}_{1k}+\hat{L}_{2k}) \\
&= i\hbar\sum_k \varepsilon_{ijk}\hat{J}_k,
\end{aligned} \tag{9.1.9}$$

where we have taken into account that $[\hat{L}_{1i}, \hat{L}_{2j}] = 0$ and $[\hat{L}_{2i}, \hat{L}_{2j}] = 0$.

Our main task is the following. Given the values of the individual angular momenta $|\vec{L}_1|$ and $|\vec{L}_2|$ (i.e., the quantum numbers ℓ_1 and ℓ_2), find the values that the total angular momentum $|\vec{J}|$ (i.e., the quantum number j corresponding to it) can take and given the values of the individual magnetic quantum numbers m_1 and m_2 find the values that the total magnetic quantum number m_j of the combined system can take. To accomplish this programme, we proceed as follows.

Since the total angular momentum operators $\hat{J}_i, (i = 1,2,3)$, satisfy the usual angular momentum commutation relations, using the results of the earlier chapters, we can easily show that

$$[\hat{J}^2, \hat{J}_z] = 0, \quad [\hat{J}^2, \hat{J}_\pm] = 0, \quad , [\hat{J}_+, \hat{J}_-] = 2\hbar \hat{J}_z, \quad , [\hat{J}_z, \hat{J}_\pm] = \pm\hbar \hat{J}_\pm, \tag{9.1.10}$$

where $\hat{J}_+ = (\hat{J}_x + i\hat{J}_y)/2$ and $\hat{J}_- = (\hat{J}_x - i\hat{J}_y)/2i$ are the total angular momentum raising and lowering operators, respectively. Further,the Hilbert space in which the total angular momentum operator $\hat{J}^2$ acts is the product space $\mathscr{H} = \mathscr{H}_1 \otimes \mathscr{H}_2$, spanned by the kets $|\ell_1, \ell_2, m_1, m_2\rangle = |\ell_1, m_1\rangle \otimes |\ell_2, m_2\rangle$. Since the kets $\{|j_1, m_1\rangle\}$ and $\{|j_2, m_2\rangle\}$ individually form a complete orthonormal basis (being the eigenvectors of hermitian operators), the kets $\{|\ell_1, \ell_2, m_1, m_2\rangle\}$ also form a complete and orthonormal basis:

$$\begin{aligned}\langle \ell_1, \ell_2; m_1, m_2 | \ell_1', \ell_2', m_1', m_2'\rangle &= \langle \ell_1, m_1 | \ell_1', m_1'\rangle \langle \ell_2, m_2 | \ell_2', m_2'\rangle \\ &= \delta_{\ell_1 \ell_1'} \delta_{\ell_2 \ell_2'} \delta_{m_1 m_1'} \delta_{m_2 m_2'},\end{aligned} \tag{9.1.11}$$

$$\begin{aligned}&\sum_{m_1 m_2} |\ell_1, \ell_2; m_1, m_2\rangle \langle \ell_1, \ell_2; m_1, m_2| \\ &= \left(\sum_{m_1=-\ell_1}^{\ell_1} |\ell_1, m_1\rangle \langle \ell_1, m_1| \right) \left(\sum_{m_2=-\ell_2}^{\ell_2} |\ell_2, m_2\rangle \langle \ell_2, m_2| \right) = \hat{I}^2 = \hat{I}.\end{aligned} \tag{9.1.12}$$

It is straightforward to prove that $[\hat{J}^2, \hat{L}_1^2] = 0$, $[\hat{J}^2, \hat{L}_2^2] = 0$, $[\hat{J}_z, \hat{L}_1^2] = 0$, $[\hat{J}_z, \hat{L}_2^2] = 0$, but $[\hat{J}^2, \hat{L}_{1z}] \neq 0$,and $[\hat{J}^2, \hat{L}_{2z}] \neq 0$. Therefore, the maximal set of commuting operators for the system is given by $\hat{J}^2, \hat{J}_z, \hat{L}_1^2$ and $\hat{L}_2^2$,. They can be simultaneously diagonalized and their joint eigenfunctions are characterized by four quantum numbers j (quantum number representing the total angular momentum), m_j (magnetic quantum number characterizing the projection of the total angular momentum on z-axis), ℓ_1 (orbital angular momentum of particle 1) and ℓ_2 (orbital angular momentum of particle 2).

Let $|\ell_1, \ell_2, j, m\rangle$ be the simultaneous eigenfunctions of $\hat{J}^2$ and $\hat{J}_z$. Since ℓ_1 and ℓ_2 are fixed, we shall write these vectors as $|j, m\rangle$. Clearly, for every j, the number m has $(2j+1)$ allowed values:$m = -j, -j+1, ..., j-1, j$. The above completeness and orthonormality conditions can now be re-written as

$$\sum_j \sum_{m=-j}^{j} |j, m\rangle \langle j, m| = \hat{I}, \tag{9.1.13}$$

$$\langle j', m' | j, m\rangle = \delta_{jj'} \delta_{mm'}. \tag{9.1.14}$$

Also, it is not difficult to show that

$$\hat{J}^2 |j, m\rangle = j(j+1)\hbar^2 |j, m\rangle, \tag{9.1.15}$$

$$\hat{J}_z|j,m\rangle = m\hbar|j,m\rangle, \tag{9.1.16}$$

$$\hat{J}_+|j,m\rangle = \sqrt{(j-m)(j+m+1)}\hbar\,|j,m+1\rangle, \tag{9.1.17}$$

$$\hat{J}_-|j,m\rangle = \sqrt{(j+m)(j-m+1)}\hbar\,|j,m-1\rangle, \tag{9.1.18}$$

$$\hat{J}_z|j,m\rangle = m\hbar\,|j,m\rangle, \tag{9.1.19}$$

$$\hat{J}_+|j,m=j\rangle = 0,\ \ \hat{J}_-|j,m=-j\rangle = 0. \tag{9.1.20}$$

Since $\hat{J}^2$ and $\hat{J}_z$ are hermitian, the vectors $|j,m\rangle$ also constitute an orthonormal and complete basis in $\mathscr{H}$.

In order to achieve the goal, stated above, we have to find the linear combination of $|\ell_1,\ell_2;m_1,m_2\rangle$, with fixed ℓ_1 and ℓ_2, which are eigenfunctions of $\hat{J}_z = \hat{J}_{1z}+\hat{J}_{2z}$ with eigenvalues m and also eigenstates of and $\hat{J}^2 = (\hat{\vec{L}}_1+\hat{\vec{L}}_2)^2$ with eigenvalues $j(j+1)$. It then follows from linear algebra that this goal can be achieved, if we succeed in finding a unitary matrix that relates the bases $\{|\ell_1,\ell_2;m_1,m_2\rangle\}$ and $\{|j,m\rangle\}$. For this purpose let us expand the besis vector $|j,m\rangle$ in terms of the basis $\{|\ell_1,\ell_2;m_1,m_2\rangle\}$ as

$$|j,m\rangle = \sum_{m_1=-\ell_1}^{\ell_1}\ \sum_{m_2=-\ell_2}^{\ell_2} C^{m_1\,m_2\,m}_{\ell_1\,\ell_2\,j}\,|\ell_1,\ell_2;m_1,m_2\rangle, \tag{9.1.21}$$

where the coefficients of expansion

$$C^{m_1\,m_2\,m}_{\ell_1\,\ell_2\,j} = \langle \ell_1,\ell_2;m_1,m_2|j,m\rangle \tag{9.1.22}$$

are called the *Clebsch-Gordan* (CG) *coefficients*. Therefore, the solution of the problem of addition of two angular momenta reduces to the determination of the Clebsch-Gordan coefficients which are nothing but the elements of the unitary matrix that effects the transition from the basis $\{|\ell_1,\ell_2;m_1,m_2\rangle\}$ to the basis $\{|j,m\rangle\}$.

By convention, Clebsch-Gordan coefficients are taken to be real, i.e.,

$$\langle \ell_1,\ell_2;m_1,m_2|j,m\rangle = \langle \ell_1,\ell_2;m_1,m_2|j,m\rangle^\dagger = \langle j,m|\ell_1,\ell_2;m_1,m_2\rangle. \tag{9.1.23}$$

Also, using (9.1.12), we get

$$\sum_{m_1 m_2} \langle j',m'|\ell_1,\ell_2;m_1,m_2\rangle\langle \ell_1,\ell_2;m_1,m_2|j,m\rangle = \delta_{j'j}\delta_{m'm}. \tag{9.1.24}$$

Since the Clebsch-Gordan coefficients are real, we can write this equation as

$$\sum_{m_1 m_2} \langle \ell_1,\ell_2;m_1,m_2|j',m'\rangle\langle \ell_1,\ell_2;m_1,m_2|j,m\rangle = \delta_{j'j}\delta_{m'm}. \tag{9.1.25}$$

The last equation leads to

$$\sum_{m_1 m_2} \langle \ell_1, \ell_2; m_1, m_2 | j, m \rangle^2 = 1. \tag{9.1.26}$$

Similarly, we can derive the following relation

$$\sum_j \sum_{m=-j}^{j} \langle \ell_1, \ell_2, m_1', m_2' | j, m \rangle \langle \ell_1, \ell_2; m_1, m_2 | j, m \rangle = \delta_{m_1' m_1} \delta_{m_2' m_2}, \tag{9.1.27}$$

which yields

$$\sum_j \sum_{m=-j}^{j} \langle j_1, j_2, m_1, m_2 | j, m \rangle^2 = 1. \tag{9.1.28}$$

Our next step is to find the eigenvalues of the operator $\hat{\vec{J}}^2$ in terms of the eigenvalues of the operators $\hat{L}_1^2$ and $\hat{L}_2^2$ and the eigenvalues of the operator $\hat{J}_z$ in terms of the eigenvalues of the operators $\hat{L}_{1z}$ and $\hat{L}_{2z}$ so that we could express j in terms of ℓ_1 and ℓ_2 and m in terms of m_1 and m_2.

Constraints on the indices of CG coefficients: (A) Since $\hat{J}_z = \hat{L}_{1z} + \hat{L}_{2z}$, we have

$$\langle \ell_1, \ell_2; m_1, m_2 | \hat{J}_z - \hat{L}_{1z} - \hat{L}_{2z} | j, m \rangle = 0. \tag{9.1.29}$$

Using the following relations

$$\hat{J}_z | j, m \rangle = m\hbar \, | j, m \rangle, \tag{9.1.30}$$

$$\langle \ell_1, \ell_2; m_1, m_2 | \hat{L}_{1z} = m_1 \hbar \langle \ell_1, \ell_2; m_1, m_2 |, \tag{9.1.31}$$

$$\langle \ell_1, \ell_2; m_1, m_2 | \hat{L}_{2z} = m_2 \hbar \langle \ell_1, \ell_2; m_1, m_2 |, \tag{9.1.32}$$

we obtain

$$(m - m_1 - m_2) \langle \ell_1, \ell_2; m_1, m_2 | j, m \rangle = 0. \tag{9.1.33}$$

Therefore, for $\langle \ell_1, \ell_2; m_1, m_2 | j, m \rangle$ to be nonzero, we must have $m = m_1 + m_2$. *This is the first constraint for the Clebsch-Gordan coefficients.*

(B) Further, the maximum values of m_1 and m_2 are ℓ_1 and ℓ_2, respectively, and hence the maximum value of m is $m_{\max} = \ell_1 + \ell_2$. However, $|m| \leq j$, and therefore, $j_{\max} = \ell_1 + \ell_2$.

We have to now find $j_{\min}$, i.e., the minimum possible value of j. Since the dimension of the product space is $N = (2\ell_1 + 1) \times (2\ell_2 + 1)$, there are $(2\ell_1 + 1) \times (2\ell_2 + 1)$ number of basis vectors $|j, m\rangle$ in this space. On the other hand, for each value of j there are $(2j + 1)$ basis vectors $|j, m\rangle$, and hence

$$\sum_{j=j\mathrm{min}}^{j\mathrm{max}} (2j+1) = (2\ell_1+1)(2\ell_2+1). \tag{9.1.34}$$

Since $j_{\max} = \ell_1 + \ell_2$, the left hand-side of (9.1.34) has $j_{\max} - j_{\min} + 1 = \ell_1 + \ell_2 + 1 - j_{\min}$ terms. Therefore, we have

$$S \equiv \sum_{j\mathrm{min}}^{j\mathrm{max}} (2j+1) = (2j_{\min}+1) + (2j_{\min}+3) + (2j_{\min}+5) + ... + (2(\ell_1+\ell_2)+1). \tag{9.1.35}$$

We can write the above series in two equivalent ways

$$S = (2j_{\min}+1) + 2(j_{\min}+3) + (2j_{\min}+5) + ... + (2(\ell_1+\ell_2)+1),$$
$$S = (2(\ell_1+\ell_2)+1) + (2(\ell_1+\ell_2)-1) + (2(\ell_1+\ell_2)-3) + ... + (2j_{\min}+1). \tag{9.1.36}$$

Adding up the aforementioned two equations term by term, we obtain

$$S = ([(\ell_1+\ell_2+1)+j_{\min}] + [(\ell_1+\ell_2+1)+j_{\min}] + [(\ell_1+\ell_2+1)+j_{\min}] + ... + [(\ell_1+\ell_2+1)+j_{\min}]). \tag{9.1.37}$$

Since S in (9.1.37) contains $(j_{\max} - j_{\min} + 1) = (\ell_1 + \ell_2 + 1 - j_{\min})$ terms, using (9.1.34), we arrive at

$$(\ell_1+\ell_2+1-j_{\min})[(\ell_1+\ell_2+1)+j_{\min}] = (2\ell_1+1)(2\ell_2+1). \tag{9.1.38}$$

Or,

$$j_{\min}^2 = (\ell_1-\ell_2)^2. \tag{9.1.39}$$

Hence $j_{\min} = |\ell_1 - \ell_2|$ and we have the following range of variation of j:

$$|\ell_1-\ell_2| \le j \le (\ell_1+\ell_2). \tag{9.1.40}$$

So, j changes from $|\ell_1 - \ell_2|$ to $(\ell_1 + \ell_2)$ in integer steps:

$$j = |\ell_1-\ell_2|, |\ell_1-\ell_2|+1, |\ell_1-\ell_2|+2, ..., (\ell_1+\ell_2). \tag{9.1.41}$$

Thus the second constraints for the Clebsch-Gordan coefficients is that, simultaneously with $m = m_1 + m_2$, we must also have $|j_1 - j_2| \le j \le (j_1 + j_2)$. Note that m takes values from $-j$ to $+j$.

9.2 Calculation of Clebsch–Gordan Coefficients

The computation of Clebsch-Gordan coefficients from first principles is somewhat cumbersome. For practical use, however, they have been tabulated in the literature and can be readily obtained. In some simple cases, these coefficients can be determined in a straightforward manner. For instance, it can be shown that the CG coefficients corresponding to two limiting cases $\{m_1 = \ell_1, m_2 = \ell_2, j = \ell_1 + \ell_2, m = (\ell_1 + \ell_2)\}$ and $\{m_1 = -\ell_1, m_2 = -\ell_2, j = \ell_1 + \ell_2, m = -(\ell_1 + \ell_2)\}$ are equal to 1. That is

$$\langle \ell_1, \ell_2, \ell_1, \ell_2 | (\ell_1 + \ell_2), (\ell_1 + \ell_2) \rangle = 1, \tag{9.2.1}$$

$$\langle \ell_1, \ell_2, -\ell_1, -\ell_2 | (\ell_1 + \ell_2), -(\ell_1 + \ell_2) \rangle = 1. \tag{9.2.2}$$

In general, to calculate CG coefficients, other than the aforementioned simple cases, one uses *either the recursion relations between the CG coefficients or the ladder operator method*. Below, we shall demonstrate these methods of calculation by taking up a concrete problem.

***Recursion relations between CG Coefficients*:** To determine the recursion relations it is required to evaluate the matrix elements,

$$\langle \ell_1, \ell_2; m_1, m_2 | \hat{J}_\pm | j, m \rangle,$$

in two different ways and equate the results. Let us do it. Firstly, using (9.1.17) and (9.1.18), we have

$$\langle \ell_1, \ell_2; m_1, m_2 | \hat{J}_\pm | j, m \rangle = \hbar \sqrt{(j \mp m)(j \pm m + 1)}\ \langle \ell_1, \ell_2; m_1, m_2 | j, m \pm 1 \rangle. \tag{9.2.3}$$

Secondly, replacing $\hat{J}_\pm$ by $\hat{J}_{1\pm} + \hat{J}_{2\pm}$, where

$$\hat{J}_{k\pm} = \frac{1}{2}(\hat{L}_{kx} \pm i\hat{L}_{ky}), \qquad k = 1, 2, \tag{9.2.4}$$

and acting on the bra $\langle \ell_1, \ell_2; m_1, m_2 |$, we obtain

$$\begin{aligned} \langle \ell_1, \ell_2; m_1, m_2 | \hat{J}_\pm | j, m \rangle &= \hbar \sqrt{(\ell_1 \pm m_1)(\ell_1 \mp m_1 + 1)}\ \langle \ell_1, \ell_2; m_1 \mp 1, m_2 | j, m \rangle \\ &\quad + \hbar \sqrt{(\ell_2 \pm m_2)(\ell_2 \mp m_2 + 1)}\ \langle \ell_1, \ell_2; m_1, m_2 \mp 1 | j, m \rangle. \end{aligned} \tag{9.2.5}$$

From (9.2.3) and (9.2.5), we arrive at the first recursion relation between CG coefficients

$$\sqrt{(j \mp m)(j \pm m + 1)}\ \langle \ell_1, \ell_2; m_1, m_2 | j, m \pm 1 \rangle$$

$$= \sqrt{(\ell_1 \pm m_1)(\ell_1 \mp m_1 + 1)}\ \langle \ell_1, \ell_2; m_1 \mp 1, m_2 | j, m\rangle$$
$$+ \sqrt{(\ell_2 \pm m_2)(\ell_2 \mp m_2 + 1)}\ \langle \ell_1, \ell_2; m_1, m_2 \mp 1 | j, m\rangle. \tag{9.2.6}$$

Now, considering the matrix elements

$$\langle \ell_1, \ell_2; m_1, m_2 | \hat{J}_\pm | j, m \mp 1\rangle,$$

and repeating the same steps that led to (9.2.6), we obtain the second recursion relation

$$\sqrt{(j \pm m)(j \mp m + 1)}\ \langle \ell_1, \ell_2; m_1, m_2 | j, m\rangle$$
$$= \sqrt{(\ell_1 \pm m_1)(\ell_1 \mp m_1 + 1)}\ \langle \ell_1, \ell_2; m_1 \mp 1, m_2 | j, m \mp 1\rangle$$
$$+ \sqrt{(\ell_2 \pm m_2)(\ell_2 \mp m_2 + 1)}\ \langle \ell_1, \ell_2; m_1, m_2 \mp 1 | j, m \mp 1\rangle. \tag{9.2.7}$$

The recursion relations (9.2.6) and (9.2.7) along with the orthonormality conditions (9.1.25) and (9.1.27) enable one to calculate all the CG coefficients, except for the sign, for given values of ℓ_1, ℓ_2 and j.

The sign is determined by the so-called phase convention (Wigner's convention), according to which the coefficient

$$\langle \ell_1, \ell_2, \ell_1, (j - \ell_1) | j, j\rangle, \tag{9.2.8}$$

is considered to be real and positive. Since all the CG coefficients are obtained from this single coefficient by repeated applications of the recursion relations, and since this coefficient is taken to be real, all other CG coefficients must be real.

Example 9.2.1: Find the ClebschGordan coefficients associated with the coupling of the spins of two spin $1/2$ particles with zero orbital angular momentum.

Solution: We wish to solve this problem by both the methods mentioned above.

I. *The recursion relation method*[1]: Since $\ell_1 = s_1 = \frac{1}{2}$, $\ell_2 = s_2 = \frac{1}{2}$, and $j = s = 1, 0$. When $s = 0$, we have $m = m_s = 0$ and there is a spin singlet state $|0,0\rangle$. When $s = 1$, $m = m_s$ can take three values -1, 0, and 1. Correspondingly, there is a triplet of spin states: $|1,-1\rangle$, $|1,0\rangle$ and $|1,1\rangle$.

Let us expand these states in terms of the states $|s_1, s_2; m_1, m_2\rangle$, where $m_1 = m_{s_1}$ and $m_2 = m_{s_2}$, such that $m = m_1 + m_2$:

[1] N. Zettili, Quantum Mechanics: Concepts and Applications, John Wiley, 2009.

$$|0,0\rangle = \left\langle \frac{1}{2},\frac{1}{2};\frac{1}{2},-\frac{1}{2}|0,0\right\rangle \left|\frac{1}{2},\frac{1}{2};\frac{1}{2},-\frac{1}{2}\right\rangle + \left\langle \frac{1}{2},\frac{1}{2};-\frac{1}{2},\frac{1}{2}|0,0\right\rangle \left|\frac{1}{2},\frac{1}{2};-\frac{1}{2},\frac{1}{2}\right\rangle, \tag{9.2.9}$$

$$|1,1\rangle = \left\langle \frac{1}{2},\frac{1}{2};\frac{1}{2},\frac{1}{2}|1,1\right\rangle \left|\frac{1}{2},\frac{1}{2};\frac{1}{2},\frac{1}{2}\right\rangle, \tag{9.2.10}$$

$$|1,0\rangle = \left\langle \frac{1}{2},\frac{1}{2};\frac{1}{2},-\frac{1}{2}|1,0\right\rangle \left|\frac{1}{2},\frac{1}{2};\frac{1}{2},-\frac{1}{2}\right\rangle + \left\langle \frac{1}{2},\frac{1}{2};-\frac{1}{2},\frac{1}{2}|1,0\right\rangle \left|\frac{1}{2},\frac{1}{2};-\frac{1}{2},\frac{1}{2}\right\rangle, \tag{9.2.11}$$

$$|1,-1\rangle = \left\langle \frac{1}{2},\frac{1}{2};-\frac{1}{2},-\frac{1}{2}|1,-1\right\rangle \left|\frac{1}{2},\frac{1}{2};-\frac{1}{2},-\frac{1}{2}\right\rangle. \tag{9.2.12}$$

Calculation of the coefficients $\langle \frac{1}{2},\frac{1}{2};\pm\frac{1}{2},\mp\frac{1}{2}|0,0\rangle$

In order to calculate these coefficients, let us go to the recursion relation (9.2.6) with upper signs and put $j = s = 0, m = 0$ and $m_1 = m_2 = \frac{1}{2}$ to obtain

$$\left\langle \frac{1}{2},\frac{1}{2};-\frac{1}{2},\frac{1}{2}|0,0\right\rangle = -\left\langle \frac{1}{2},\frac{1}{2};\frac{1}{2},-\frac{1}{2}|0,0\right\rangle. \tag{9.2.13}$$

Also, if we now put $j = s = 0$ and $m = 0$ in (9.1.26) and take into account the constraint that $m = m_1 + m_2$, we are led to

$$\left\langle \frac{1}{2},\frac{1}{2};-\frac{1}{2},\frac{1}{2}|0,0\right\rangle^2 + \left\langle \frac{1}{2},\frac{1}{2};\frac{1}{2},-\frac{1}{2}|0,0\right\rangle^2 = 1. \tag{9.2.14}$$

From (9.2.13) and (9.2.14), we obtain

$$\left\langle \frac{1}{2},\frac{1}{2};\frac{1}{2},-\frac{1}{2}|0,0\right\rangle = \pm\frac{1}{\sqrt{2}}. \tag{9.2.15}$$

Note that, according to the phase convention, the sign of $\langle \frac{1}{2},\frac{1}{2};\frac{1}{2},-\frac{1}{2}|0,0\rangle$ has to be positive. Thus, we have

$$\left\langle \frac{1}{2},\frac{1}{2};\frac{1}{2},-\frac{1}{2}|0,0\right\rangle = \frac{1}{\sqrt{2}}. \tag{9.2.16}$$

Then, in accordance with (9.2.13), we get that

$$\left\langle \frac{1}{2},\frac{1}{2};-\frac{1}{2},\frac{1}{2}|0,0\right\rangle = -\frac{1}{\sqrt{2}}. \tag{9.2.17}$$

Calculation of the coefficients $\langle \frac{1}{2}, \frac{1}{2}; \frac{1}{2}, \frac{1}{2}|1,1\rangle$ ***and*** $\langle \frac{1}{2}, \frac{1}{2}; -\frac{1}{2}, -\frac{1}{2}|1,-1\rangle$

For $(j = s = 1, m = 1)$ and $(j = s = 1, m = -1)$, we, after taking into account that $m = m_1 + m_2$, get from (9.1.26) that

$$\left\langle \frac{1}{2}, \frac{1}{2}; \frac{1}{2}, \frac{1}{2}|1,1\right\rangle^2 = 1, \tag{9.2.18}$$

$$\left\langle \frac{1}{2}, \frac{1}{2}; -\frac{1}{2}, -\frac{1}{2}|1,-1\right\rangle^2 = 1. \tag{9.2.19}$$

Since, according to the phase convention, both $\langle \frac{1}{2}, \frac{1}{2}; \frac{1}{2}, \frac{1}{2}|1,1\rangle$ and $\langle \frac{1}{2}, \frac{1}{2}; -\frac{1}{2}, -\frac{1}{2}|1,-1\rangle$ are real and positive, we obtain

$$\left\langle \frac{1}{2}, \frac{1}{2}; \frac{1}{2}, \frac{1}{2}|1,1\right\rangle = 1, \tag{9.2.20}$$

$$\left\langle \frac{1}{2}, \frac{1}{2}; -\frac{1}{2}, -\frac{1}{2}|1,-1\right\rangle = 1. \tag{9.2.21}$$

Calculation of the coefficients $\langle \frac{1}{2}, \frac{1}{2}; \frac{1}{2}, -\frac{1}{2}|1,0\rangle$ ***and*** $\langle \frac{1}{2}, \frac{1}{2}; -\frac{1}{2}, \frac{1}{2}|1,0\rangle$

These coefficients are obtained by putting $j = s = 1, m = 0, m_1 = \frac{1}{2}, m_2 = -\frac{1}{2}$ and $j = s = 1, m = 0, m_1 = -\frac{1}{2}, m_2 = \frac{1}{2}$ into the recursion relation (9.2.7) with lower sign, respectively. Using (9.2.21), we arrive at

$$\left\langle \frac{1}{2}, \frac{1}{2}; \frac{1}{2}, -\frac{1}{2}|1,0\right\rangle = \frac{1}{\sqrt{2}} \left\langle \frac{1}{2}, \frac{1}{2}; \frac{1}{2}, \frac{1}{2}|1,1\right\rangle = \frac{1}{\sqrt{2}}, \tag{9.2.22}$$

$$\left\langle \frac{1}{2}, \frac{1}{2}; -\frac{1}{2}, \frac{1}{2}|1,0\right\rangle = \frac{1}{\sqrt{2}} \left\langle \frac{1}{2}, \frac{1}{2}; \frac{1}{2}, \frac{1}{2}|1,1\right\rangle = \frac{1}{\sqrt{2}}. \tag{9.2.23}$$

II. *Ladder Operator Method*[2]: The calculations are done in the following steps:

***Step* 1.** We start with the state with the maximal total spin. It corresponds to the case when both the particles have spins parallel and "up". The normalized spin wave function of the system is given by

$$\chi_{s=1}^{m_s=1} = \chi\left(s_1 = \frac{1}{2}, m_{s_1} = \frac{1}{2}\right) \otimes \chi\left(s_2 = \frac{1}{2}, m_{s_2} = \frac{1}{2}\right) \equiv \chi_+^{(1)} \chi_+^{(2)}, \tag{9.2.24}$$

[2]David J. Griffiths, Introduction to Quantum Mechanics, Pearson Prentice Hall (2005).

where the superscript stands for the particle number. The subscript + denotes spin "up". This is also written as

$$|1,1\rangle = \left|\frac{1}{2},\frac{1}{2};\frac{1}{2},\frac{1}{2}\right\rangle. \tag{9.2.25}$$

***Step* 2.** We now subject the maximal spin state to the lowering operator, $\hat{S}_-$, corresponding to the total spin: $\hat{S}_- = \hat{S}_-^{(1)} + \hat{S}_-^{(2)}$, where $\hat{S}_-^{(1)}$ and $\hat{S}_-^{(2)}$ are the lowering operators corresponding to the first and the second particles, respectively.

On one hand, making use of the equation (8.1.7) with lower sign, we obtain

$$\hat{S}_-\,|1,1\rangle = \sqrt{2}\hbar\,|1,0\rangle. \tag{9.2.26}$$

On the other hand,

$$\hat{S}_-\,|1,1\rangle = (\hat{S}_-^{(1)} + \hat{S}_-^{(2)})\left|\frac{1}{2},\frac{1}{2};\frac{1}{2},\frac{1}{2}\right\rangle = \hbar\left(\left|\frac{1}{2},\frac{1}{2};-\frac{1}{2},\frac{1}{2}\right\rangle + \left|\frac{1}{2},\frac{1}{2};\frac{1}{2},-\frac{1}{2}\right\rangle\right). \tag{9.2.27}$$

From (9.2.26) and (9.2.27), we obtain

$$|1,0\rangle = \frac{1}{\sqrt{2}}\left(\left|\frac{1}{2},\frac{1}{2};-\frac{1}{2},\frac{1}{2}\right\rangle + \left|\frac{1}{2},\frac{1}{2};\frac{1}{2},-\frac{1}{2}\right\rangle\right). \tag{9.2.28}$$

***Step* 3.** Further, we act on $|1,0\rangle$ with $\hat{S}_- = \hat{S}_-^{(1)} + \hat{S}_-^{(2)}$.

On one hand, we have

$$\hat{S}_-\,|1,0\rangle = \sqrt{2}\hbar\,|1,-1\rangle. \tag{9.2.29}$$

On the other hand,

$$\begin{aligned}
\hat{S}_-\,|1,0\rangle &= \frac{1}{\sqrt{2}}\hat{S}_-\left(\left|\frac{1}{2},\frac{1}{2};\frac{1}{2},-\frac{1}{2}\right\rangle + \left|\frac{1}{2},\frac{1}{2};-\frac{1}{2},\frac{1}{2}\right\rangle\right) \\
&= \frac{1}{\sqrt{2}}\hat{S}_-^{(1)}\left(\left|\frac{1}{2},\frac{1}{2};\frac{1}{2},-\frac{1}{2}\right\rangle + \left|\frac{1}{2},\frac{1}{2};-\frac{1}{2},\frac{1}{2}\right\rangle\right) + \\
&+ \frac{1}{\sqrt{2}}\hat{S}_-^{(2)}\left(\left|\frac{1}{2},\frac{1}{2};\frac{1}{2},-\frac{1}{2}\right\rangle + \left|\frac{1}{2},\frac{1}{2};-\frac{1}{2},\frac{1}{2}\right\rangle\right) \\
&= \frac{1}{\sqrt{2}}\left[\hbar\left|\frac{1}{2},\frac{1}{2};-\frac{1}{2},-\frac{1}{2}\right\rangle + 0 + 0 + \hbar\left|\frac{1}{2},\frac{1}{2};-\frac{1}{2},-\frac{1}{2}\right\rangle\right] \\
&= \sqrt{2}\hbar\left|\frac{1}{2},\frac{1}{2};-\frac{1}{2},-\frac{1}{2}\right\rangle.
\end{aligned} \tag{9.2.30}$$

From (9.2.29) and (9.2.30), we obtain

$$|1,-1\rangle = \left|\frac{1}{2},\frac{1}{2};-\frac{1}{2},-\frac{1}{2}\right\rangle. \tag{9.2.31}$$

This completes the construction of the manifold of spin $s=1$ states-the spin triplet states:

$$s=1:\text{triplet}\rightarrow\begin{cases}|1\,1\rangle & = & |\uparrow_1\rangle|\uparrow_2\rangle\\ |1\,0\rangle & = & \frac{1}{\sqrt{2}}(|\downarrow_1\rangle|\uparrow_2\rangle+|\uparrow_1\rangle|\downarrow_2\rangle)\\ |1-1\rangle & = & |\downarrow_1\rangle|\downarrow_2\rangle\end{cases} \tag{9.2.32}$$

***Step* 4.** The state $|s=s_{\max}-1, m=s_{\max}-1\rangle = |0,0\rangle$ must be orthogonal to the state $|s=s_{\max}, m=s_{\max}-1\rangle$. Hence, we first write the state $|0,0\rangle$ as the linear combination

$$|0,0\rangle = a\left|\frac{1}{2},\frac{1}{2};-\frac{1}{2},\frac{1}{2}\right\rangle + b\left|\frac{1}{2},\frac{1}{2};\frac{1}{2},-\frac{1}{2}\right\rangle. \tag{9.2.33}$$

The orthonormality with $|1,0\rangle$, given by (9.2.28), leads to

$$\langle 1,0|0,0\rangle = \frac{a+b}{\sqrt{2}} = 0. \tag{9.2.34}$$

On the other hand, the normalization of $|0,0\rangle$ gives

$$\langle 0,0|0,0\rangle = a^2+b^2 = 1. \tag{9.2.35}$$

From the above two equations we obtain

$$b = \pm\frac{1}{\sqrt{2}}a. \tag{9.2.36}$$

In accordance with the phase convention (in the given context) the coefficient $\langle s_1,s_2, s_1,(s-s_1)|s,s\rangle = \langle s_1,s_2,s_1,(s-s_1)|0,0\rangle$ must be positive. Therefore $b=\langle 1/2,1/2,1/2, -1/2|0,0\rangle = 1/\sqrt{2}$. Then $a=-1/\sqrt{2}$. Thus, the final result is

$$|0,0\rangle = \frac{1}{\sqrt{2}}\left|\frac{1}{2},\frac{1}{2};\frac{1}{2},-\frac{1}{2}\right\rangle - \frac{1}{\sqrt{2}}\left|\frac{1}{2},\frac{1}{2};-\frac{1}{2},\frac{1}{2}\right\rangle. \tag{9.2.37}$$

If we apply the lowering operator ($\hat{S}_- = \hat{S}_-^{(1)}+\hat{S}_-^{(2)}$) or the raising operator ($\hat{S}_+ = \hat{S}_+^{(1)}+\hat{S}_+^{(2)}$) to $|0,0\rangle$, we obtain zero. This means that the state $|0,0\rangle$ is a singlet state.

If we now compare (9.3.25), (9.2.28), (9.2.31), and (9.2.37) with the respective expressions in the system (9.2.9)-(9.2.12), we obtain the following results for the CG coefficients

$$\left\langle \frac{1}{2},\frac{1}{2};\frac{1}{2},-\frac{1}{2}|0,0\right\rangle = \frac{1}{\sqrt{2}}, \tag{9.2.38}$$

$$\left\langle \frac{1}{2},\frac{1}{2};-\frac{1}{2},\frac{1}{2}|0,0\right\rangle = -\frac{1}{\sqrt{2}}, \tag{9.2.39}$$

$$\left\langle \frac{1}{2},\frac{1}{2};\frac{1}{2},-\frac{1}{2}|1,0\right\rangle = \left\langle \frac{1}{2},\frac{1}{2};-\frac{1}{2},\frac{1}{2}|1,0\right\rangle = \frac{1}{\sqrt{2}}, \tag{9.2.40}$$

$$\left\langle \frac{1}{2},\frac{1}{2};\frac{1}{2},\frac{1}{2}|1,1\right\rangle = \left\langle \frac{1}{2},\frac{1}{2};-\frac{1}{2},-\frac{1}{2}|1,-1\right\rangle = 1. \tag{9.2.41}$$

We see that the same values for the CG coefficients are the same as obtained earlier by recursion relation method. Thus, both the methods yield identical results as it should be.

Thus we conclude that *addition of the spins of two spin half particles, leads to two possible spin states of the composite system:* (i) $|1,1\rangle_{\text{triplet}} = \{|1,1\rangle, |1,0\rangle, |1,-1\rangle$, *which is symmetric with respect to the interchange of the spin "up" state with the spin "down" state and vice versa, and* (ii) $|0,0\rangle_{\text{singlet}}$, *which is anti-symmetric with respect to the interchange of the spin "up" state with the spin "down" state and vice versa.*

9.3 Algebraic Addition of the Orbital and the Spin Angular Momenta

Let us consider the addition of the orbital angular momentum and the spin angular momentum, i.e., $\hat{\vec{J}} = \hat{\vec{L}} + \hat{\vec{S}}$, of a spin half particle (say, of an electron). In the given case $\ell_1 = \ell$ (an integer) $m_1 = m_\ell$ (takes values from $-\ell$ to ℓ), $\ell_2 = s = \frac{1}{2}$, and $m_2 = m_s = \pm\frac{1}{2}$.

The value of j in this case is restricted in the interval

$$\left|\ell - \frac{1}{2}\right| \le j \le \left|\ell + \frac{1}{2}\right|. \tag{9.3.1}$$

Clearly, j can have two limiting values $j_{\max} = \ell + \frac{1}{2}$ and $j_{\min} = -\ell + \frac{1}{2}$.

The maximal set of commuting observables in this case is given by: $\{\hat{J}^2, \hat{L}^2, \hat{S}^2, \hat{J}_z\}$. The joint eigenvectors of these operators are:$|\ell, s, m_\ell, m_s\rangle$. The eigenvectors of $\hat{J}^2$ are: $|j,m\rangle \equiv |\ell, s; j, m\rangle$; ℓ and s being fixed. Obviously, the following hold:

$$\hat{J}^2\,|j,m\rangle = \hbar^2\, j(j+1)\,|j,m\rangle\,, \tag{9.3.2}$$

$$\hat{L}^2\,|j,m\rangle = \hbar^2\, \ell(\ell+1)\,|j,m\rangle\,, \tag{9.3.3}$$

$$\hat{S}^2 |j,m\rangle = \hbar^2 s(s+1) |j,m\rangle = \frac{3}{4}\hbar^2 |j,m\rangle, \tag{9.3.4}$$

$$\hat{J}_z |j,m\rangle = \hbar m |j,m\rangle. \tag{9.3.5}$$

The state with maximal total angular momentum $j = \ell + \frac{1}{2}$ and $m_{\max} = m_{1\max} + m_{s\max} = = (m_\ell)_{\max} + (m_s)_{\max} = \ell + \frac{1}{2}$ is

$$|j_{\max}, m_{\max}\rangle = \left|\ell + \frac{1}{2}, \ell + \frac{1}{2}\right\rangle = |\ell,\ell\rangle \otimes \left|\frac{1}{2}, \frac{1}{2}\right\rangle. \tag{9.3.6}$$

The corresponding CG coefficient

$$\left\langle \ell, \frac{1}{2}; \ell, \frac{1}{2} \middle| \ell + \frac{1}{2}, \ell + \frac{1}{2} \right\rangle = 1, \tag{9.3.7}$$

in accordance with our earlier discussions.

Let us act on the state $|j_{\max}, m_{\max}\rangle$ with the lowering operator $\hat{J}_- = \hat{L}_- + \hat{S}_-$ to generate states with $m = m_{\max} - 1, m = m_{\max} - 2$ and so on till we reach $m_{\max} = -j$. On one hand we have

$$\begin{aligned}\hat{J}_- \left|\ell + \frac{1}{2}, \ell + \frac{1}{2}\right\rangle &= \hbar \sqrt{\left[\left(\ell + \frac{1}{2}\right) + \left(\ell + \frac{1}{2}\right)\right]\left(\ell + \frac{1}{2} - \ell - \frac{1}{2} + 1\right)} \left|\ell + \frac{1}{2}, \ell - \frac{1}{2}\right\rangle \\ &= \hbar\sqrt{2\ell + 1} \left|\ell + \frac{1}{2}, \ell - \frac{1}{2}\right\rangle,\end{aligned} \tag{9.3.8}$$

while on the other

$$\begin{aligned}(\hat{L}_- + \hat{S}_-) \left|\ell + \frac{1}{2}, \ell + \frac{1}{2}\right\rangle &= \hat{L}_- |\ell,\ell\rangle \otimes \left|\frac{1}{2}, \frac{1}{2}\right\rangle + |\ell,\ell\rangle \otimes \hat{S}_- \left|\frac{1}{2}, \frac{1}{2}\right\rangle \\ &= \hbar\sqrt{2\ell} |\ell, \ell - 1\rangle \otimes \left|\frac{1}{2}, \frac{1}{2}\right\rangle + \hbar |\ell,\ell\rangle \otimes \left|\frac{1}{2}, -\frac{1}{2}\right\rangle.\end{aligned} \tag{9.3.9}$$

From (9.3.8) and (9.3.9) we get that

$$\left|\ell + \frac{1}{2}, \ell - \frac{1}{2}\right\rangle = \frac{1}{\sqrt{2\ell + 1}} \left[\sqrt{2\ell} |\ell, \ell - 1\rangle \otimes \left|\frac{1}{2}, \frac{1}{2}\right\rangle + |\ell,\ell\rangle \otimes \left|\frac{1}{2}, -\frac{1}{2}\right\rangle\right]. \tag{9.3.10}$$

We now act on $|\ell + \frac{1}{2}, \ell - \frac{1}{2}\rangle$ with $\hat{J}_-$ to find $|\ell + \frac{1}{2}, \ell - \frac{1}{2}\rangle$. We have

$$\hat{J}_- \left|\ell + \frac{1}{2}, \ell - \frac{1}{2}\right\rangle = \hbar\sqrt{2\ell \times 2} \left|\ell + \frac{1}{2}, \ell - \frac{3}{2}\right\rangle. \tag{9.3.11}$$

on the other hand

$$(\hat{L}_- + \hat{S}_-)\left|\ell + \frac{1}{2}, \ell - \frac{1}{2}\right\rangle = \sqrt{\frac{2\ell}{2\ell+1}}\,\hat{L}_-|\ell,\ell-1\rangle \otimes \left|\frac{1}{2},\frac{1}{2}\right\rangle$$
$$+\sqrt{\frac{2\ell}{2\ell+1}}|\ell,\ell-1\rangle \otimes \hat{S}_-\left|\frac{1}{2},\frac{1}{2}\right\rangle + \frac{1}{\sqrt{2\ell+1}}\hat{L}_-|\ell,\ell\rangle \otimes \left|\frac{1}{2},-\frac{1}{2}\right\rangle$$
$$+\frac{1}{\sqrt{2\ell+1}}|\ell,\ell\rangle \otimes \hat{S}_-\left|\frac{1}{2},-\frac{1}{2}\right\rangle. \tag{9.3.12}$$

Since $\hat{S}_-\left|\frac{1}{2},-\frac{1}{2}\right\rangle = 0$, we obtain

$$(\hat{L}_- + \hat{S}_-)\left|\ell + \frac{1}{2}, \ell - \frac{1}{2}\right\rangle = \hbar\sqrt{\frac{4\ell(2\ell-1)}{2\ell+1}}|\ell,\ell-2\rangle \otimes \left|\frac{1}{2},\frac{1}{2}\right\rangle$$
$$+\hbar\sqrt{\frac{2\ell}{2\ell+1}}|\ell,\ell-1\rangle \otimes \left|\frac{1}{2},-\frac{1}{2}\right\rangle + \hbar\sqrt{\frac{2\ell}{2\ell+1}}|\ell,\ell-1\rangle \otimes \left|\frac{1}{2},-\frac{1}{2}\right\rangle \tag{9.3.13}$$

From (9.3.11) and (9.3.13), we obtain

$$\left|\ell + \frac{1}{2}, \ell - \frac{3}{2}\right\rangle = \sqrt{\frac{2\ell-1}{2\ell+1}}|\ell,\ell-2\rangle \otimes \left|\frac{1}{2},\frac{1}{2}\right\rangle + \sqrt{\frac{2}{2\ell+1}}|\ell,\ell-1\rangle \otimes \left|\frac{1}{2},-\frac{1}{2}\right\rangle. \tag{9.3.14}$$

The other $\left|\ell+\frac{1}{2},m\right\rangle$ states are given by

$$\left|\ell + \frac{1}{2}, m\right\rangle = \sqrt{\frac{\ell+m+\frac{1}{2}}{2\ell+1}}\left|\ell, m-\frac{1}{2}\right\rangle \otimes \left|\frac{1}{2},\frac{1}{2}\right\rangle$$
$$+\sqrt{\frac{\ell-m+\frac{1}{2}}{2\ell+1}}\left|\ell, m+\frac{1}{2}\right\rangle \otimes \left|\frac{1}{2},-\frac{1}{2}\right\rangle, \tag{9.3.15}$$

where

$$m = \ell+\frac{1}{2}, \ell-\frac{1}{2}, \ell-\frac{3}{2}, \ldots -\ell+\frac{1}{2}, -\left(\ell+\frac{1}{2}\right). \tag{9.3.16}$$

All these states are with $j = \ell + \frac{1}{2}$. The states with $j = \ell - \frac{1}{2}$ are found in an identical manner starting with the state $\left|\ell-\frac{1}{2},\ell-\frac{1}{2}\right\rangle$ which must be orthogonal to the state

$$\left|\ell + \frac{1}{2}, \ell - \frac{1}{2}\right\rangle = \frac{1}{\sqrt{2\ell+1}}\left[\sqrt{2\ell}|\ell,\ell-1\rangle \otimes \left|\frac{1}{2},\frac{1}{2}\right\rangle + |\ell,\ell\rangle \otimes \left|\frac{1}{2},-\frac{1}{2}\right\rangle\right]. \tag{9.3.17}$$

If we do the calculations the same way as earlier and take into account the phase convention, we obtain

$$\left|\ell-\frac{1}{2},\ell-\frac{1}{2}\right\rangle=\frac{1}{\sqrt{2\ell+1}}\left[\sqrt{2\ell}\,|\ell,\ell\rangle\otimes\left|\frac{1}{2},-\frac{1}{2}\right\rangle-|\ell,\ell-1\rangle\otimes\left|\frac{1}{2},\frac{1}{2}\right\rangle\right]. \tag{9.3.18}$$

Calculations similar to those leading to (9.3.14) yield

$$\left|\ell-\frac{1}{2},\ell-\frac{3}{2}\right\rangle=\sqrt{\frac{2\ell-1}{2\ell+1}}\,|\ell,\ell-1\rangle\otimes\left|\frac{1}{2},-\frac{1}{2}\right\rangle-\sqrt{\frac{2}{2\ell+1}}\,|\ell,\ell-2\rangle\otimes\left|\frac{1}{2},\frac{1}{2}\right\rangle. \tag{9.3.19}$$

The other $|\ell-\frac{1}{2},m\rangle$ states are given by

$$\begin{aligned}\left|\ell-\frac{1}{2},m\right\rangle &= \sqrt{\frac{\ell+m+\frac{1}{2}}{2\ell+1}}\left|\ell,m+\frac{1}{2}\right\rangle\otimes\left|\frac{1}{2},-\frac{1}{2}\right\rangle\\ &\quad-\sqrt{\frac{\ell-m+\frac{1}{2}}{2\ell+1}}\left|\ell,m-\frac{1}{2}\right\rangle\otimes\left|\frac{1}{2},\frac{1}{2}\right\rangle,\end{aligned} \tag{9.3.20}$$

where

$$m=\ell-\frac{1}{2},\ell-\frac{3}{2},\ldots,-\ell+\frac{3}{2},-\left(\ell-\frac{1}{2}\right). \tag{9.3.21}$$

The required CG coefficients are readily read off from (9.3.14), (9.3.15) and (9.3.17)-(9.3.20).

Example 9.3.1: Consider the case of $\ell=1$ and $s=\frac{1}{2}$. Find all the states and the corresponding CG coefficients.

Solution: In this case, for the states $|\ell+\frac{1}{2},m\rangle$, the equation (9.3.16) shows that m can take four values $\frac{1}{2}$, $\frac{3}{2}$, $-\frac{1}{2}$, and $-\frac{3}{2}$. Therefore, from (9.3.15) we get

$$\begin{aligned}\left|\frac{3}{2},\frac{3}{2}\right\rangle &= \sqrt{\frac{1+\frac{3}{2}+\frac{1}{2}}{2+1}}\left|1,\frac{3}{2}-\frac{1}{2}\right\rangle\otimes\left|\frac{1}{2},\frac{1}{2}\right\rangle+\sqrt{\frac{1-\frac{3}{2}+\frac{1}{2}}{2+1}}\left|1,\frac{3}{2}+\frac{1}{2}\right\rangle\otimes\left|\frac{1}{2},-\frac{1}{2}\right\rangle\\ &= |1,1\rangle\otimes\left|\frac{1}{2},\frac{1}{2}\right\rangle\equiv\left|1,\frac{1}{2};1,\frac{1}{2}\right\rangle,\end{aligned} \tag{9.3.22}$$

$$\left|\frac{3}{2},\frac{1}{2}\right\rangle = \sqrt{\frac{1+\frac{1}{2}+\frac{1}{2}}{2+1}}\left|1,\frac{1}{2}-\frac{1}{2}\right\rangle\otimes\left|\frac{1}{2},\frac{1}{2}\right\rangle+\sqrt{\frac{1-\frac{1}{2}+\frac{1}{2}}{2+1}}\left|1,\frac{1}{2}+\frac{1}{2}\right\rangle\otimes\left|\frac{1}{2},-\frac{1}{2}\right\rangle$$

$$= \sqrt{\frac{2}{3}}\,|1,0\rangle \otimes \left|\frac{1}{2},\frac{1}{2}\right\rangle + \sqrt{\frac{1}{3}}\,|1,1\rangle \otimes \left|\frac{1}{2},-\frac{1}{2}\right\rangle$$

$$\equiv \sqrt{\frac{2}{3}}\left|1,\frac{1}{2};0,\frac{1}{2}\right\rangle + \sqrt{\frac{1}{3}}\left|1,\frac{1}{2};1,-\frac{1}{2}\right\rangle, \tag{9.3.23}$$

$$\left|\frac{3}{2},-\frac{1}{2}\right\rangle = \sqrt{\frac{1-\frac{1}{2}+\frac{1}{2}}{2+1}}\left|1,-\frac{1}{2}-\frac{1}{2}\right\rangle \otimes \left|\frac{1}{2},\frac{1}{2}\right\rangle + \sqrt{\frac{1+\frac{1}{2}+\frac{1}{2}}{2+1}}\left|1,\frac{1}{2}-\frac{1}{2}\right\rangle \otimes \left|\frac{1}{2},-\frac{1}{2}\right\rangle$$

$$= \sqrt{\frac{1}{3}}\,|1,-1\rangle \otimes \left|\frac{1}{2},\frac{1}{2}\right\rangle + \sqrt{\frac{2}{3}}\,|1,0\rangle \otimes \left|\frac{1}{2},-\frac{1}{2}\right\rangle$$

$$\equiv \sqrt{\frac{1}{3}}\left|1,\frac{1}{2};-1,\frac{1}{2}\right\rangle + \sqrt{\frac{2}{3}}\left|1,\frac{1}{2};0,-\frac{1}{2}\right\rangle, \tag{9.3.24}$$

$$\left|\frac{3}{2},-\frac{3}{2}\right\rangle = \sqrt{\frac{1-\frac{3}{2}+\frac{1}{2}}{2+1}}\left|1,-\frac{3}{2}-\frac{1}{2}\right\rangle \otimes \left|\frac{1}{2},\frac{1}{2}\right\rangle + \sqrt{\frac{1+\frac{3}{2}+\frac{1}{2}}{2+1}}\left|1,-\frac{3}{2}+\frac{1}{2}\right\rangle \otimes \left|\frac{1}{2},-\frac{1}{2}\right\rangle$$

$$= |1,-1\rangle \otimes \left|\frac{1}{2},-\frac{1}{2}\right\rangle \equiv \left|1,\frac{1}{2};-1,-\frac{1}{2}\right\rangle. \tag{9.3.25}$$

Similarly, for the states $\left|\ell-\frac{1}{2},m\right\rangle$ the equation (9.3.21) shows that m can take two values $\frac{1}{2}$ and $-\frac{1}{2}$. Therefore, from (9.3.15) we get

$$\left|\frac{1}{2},\frac{1}{2}\right\rangle = \sqrt{\frac{1+\frac{1}{2}+\frac{1}{2}}{2+1}}\left|1,\frac{1}{2}+\frac{1}{2}\right\rangle \otimes \left|\frac{1}{2},-\frac{1}{2}\right\rangle - \sqrt{\frac{1-\frac{1}{2}+\frac{1}{2}}{2+1}}\left|1,\frac{1}{2}-\frac{1}{2}\right\rangle \otimes \left|\frac{1}{2},\frac{1}{2}\right\rangle$$

$$= \sqrt{\frac{2}{3}}\,|1,1\rangle \otimes \left|\frac{1}{2},-\frac{1}{2}\right\rangle - \sqrt{\frac{1}{3}}\,|1,0\rangle \otimes \left|\frac{1}{2},\frac{1}{2}\right\rangle$$

$$\equiv \sqrt{\frac{2}{3}}\left|1,\frac{1}{2};1,-\frac{1}{2}\right\rangle - \sqrt{\frac{1}{3}}\left|1,\frac{1}{2};0,\frac{1}{2}\right\rangle, \tag{9.3.26}$$

$$\left|\frac{1}{2},-\frac{1}{2}\right\rangle = \sqrt{\frac{1-\frac{1}{2}+\frac{1}{2}}{2+1}}\left|1,-\frac{1}{2}+\frac{1}{2}\right\rangle \otimes \left|\frac{1}{2},-\frac{1}{2}\right\rangle - \sqrt{\frac{1+\frac{1}{2}+\frac{1}{2}}{2+1}}\left|1,-\frac{1}{2}-\frac{1}{2}\right\rangle \otimes \left|\frac{1}{2},\frac{1}{2}\right\rangle$$

$$= \sqrt{\frac{1}{3}}\,|1,0\rangle \otimes \left|\frac{1}{2},-\frac{1}{2}\right\rangle - \sqrt{\frac{2}{3}}\,|1,-1\rangle \otimes \left|\frac{1}{2},\frac{1}{2}\right\rangle$$

$$\equiv \sqrt{\frac{1}{3}}\left|1,\frac{1}{2};0,-\frac{1}{2}\right\rangle - \sqrt{\frac{2}{3}}\left|1,\frac{1}{2};-1,\frac{1}{2}\right\rangle. \tag{9.3.27}$$

By taking the appropriate scalar products in (9.3.22)-(9.3.27), we obtain the required CG coefficients.

Example 9.3.2: In an atom there are two valence electron; each in the state $3p$. (a) What are the possible values for the total spin quantum number, s, where $\vec{S} = \vec{S}_1 + \vec{S}_2$? (b) What are the possible values for the total angular momentum quantum number, ℓ, where $\vec{L} = \vec{L}_1 + \vec{L}_2$? (c) Given that under the exchange symmetry the parity of the spatial part of the wave function of the two-electron system is decided by the parity of the total orbital quantum number ℓ, determine which combinations of s and ℓ are allowed states for the two-electron system. (d) For each allowed combination, what are the possible values of the quantum number j, where $\vec{J} = \vec{L} + \vec{S}$?

Solution:

(a) As discussed s will take values from $s_{\max}$ to $s_{\min}$ in integer steps. Since $s_{\max} = s_1 + s_2 = 1$ and $s_{\min} = s_1 - s_2 = 0$, s can take only two values: $s = 0, 1$.

(b) Both the electrons are in the $3p$ state and, therefore, $\ell_1 = \ell_2 = 1$. Hence $\ell_{\max} = \ell_1 + \ell_2 = 2$ and $\ell_{\min} = \ell_1 - \ell_2 = 0$. As a consequence $\ell = 0, 1, 2$.

(c) The total wave function of the system must be antisymmetric (see Chapter 10). Now, for $s = 0$ the spin part of the wave function of the system will be antisymmetric and hence only those values of ℓ can be paired with $s = 0$ which make the total wave function antisymmetric. We know that the parity of the spatial part of the two-electron system is decided by whether $(-1)^\ell$ is $+1$ or -1: when $(-1)^\ell = +1$ it is symmetric, whereas when $(-1)^\ell = -1$ it is antisymmetric. Therefore, $\ell = 0$ and $\ell = 2$ will correspond to the symmetric spatial part, while $\ell = 1$ will give antisymmetric spatial part of the wave function. Hence, the possible pairs in this case are: $(\ell, s) = (0,0), (0,2)$.

   On the other hand, since the spin part of the wave function is symmetric under the exchange symmetry for $s = 1$ (see Chapter 10), the only allowed pair is $(s, \ell) = (1, 1)$.

(d) Since j varies from $|\ell - s|$ to $\ell + s$ in integer steps, for $(\ell, s) = (0,0)$, the only possible value of j is $j = 0$. For both the other two combinations $(\ell, s) = (1,1)$ and $(\ell, s) = (0,2)$, three values of j are possible: $j = 0, 1, 2$.

9.4 Vectorial Addition of the Orbital and the Spin Angular Momenta for an Electron

Consider an electron on its orbit in an atom with orbital angular momentum $\vec{L}$ and spin angular momentum $\vec{S}$. Since the components of the angular momentum operator, $\hat{L}_k, k = 1, 2, 3$, do not commute, no two components of $\vec{L}$ can be measured simultaneously accurately. Therefore, it is impossible to assign a vector (in the usual sense) to the orbital angular momentum in quantum mechanics. It is simply not defined. The same applies to the spin angular momentum vector $\vec{S}$. In spite of this difficulty, it turns out that one can use the so-called semi-classical model for adding two or more angular momenta vectorially. For instance, it can be used to add orbital and spin angular momenta. How it is done is explained in what follows.

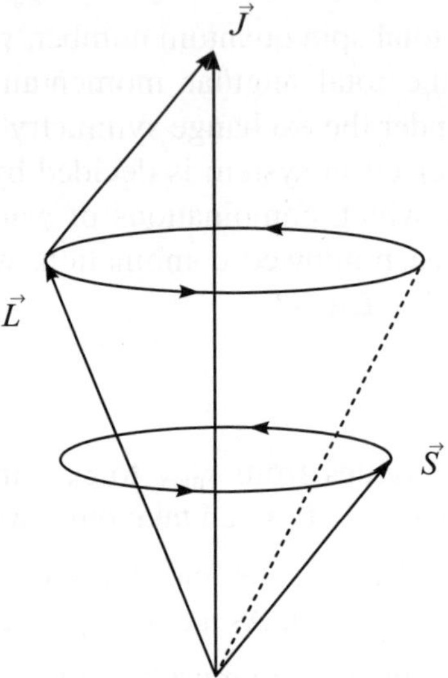

Figure 9.1 Semi-classical model for the vectorial addition of orbital and spin angular momenta.

According to this model, we can treat $\vec{L}$ and $\vec{S}$ as usual three-dimensional vectors keeping in mind their quantized nature. The resultant angular momentum vector, usually written as $\vec{J}$, is obtained by the familiar rule of parallelogram for the addition of vectors in vector algebra:

$$\vec{J} = \vec{L} + \vec{S}. \tag{9.4.1}$$

We know that $\vec{L}$ and $\vec{S}$ have their corresponding magnetic moments $\vec{\mu}_\ell$ and $\vec{\mu}_s$, respectively, through which they can interact. Classically, due to this interaction, both $\vec{L}$ and $\vec{S}$ precess around the direction of the total angular momentum $\vec{J}$ (see Fig. 9.1). This classical treatment, however, is supplemented by quantum conditions. In particular, $\vec{L}$, which is numerically equal to $\hbar\sqrt{\ell(\ell+1)}$, where ℓ is the orbital quantum number, cannot be arbitrarily oriented in space. According to the phenomenon of space quantization, $\vec{L}$ has to have only those orientations in space for which its projection on the vertical direction (z-axis), $L_z = m_\ell \hbar$, where m_ℓ is the magnetic quantum number. Similarly, $\vec{S}$, which is numerically equal to $\hbar\sqrt{s(s+1)}$, where s is the spin quantum number, cannot be arbitrarily oriented in space. It also has to have such orientations that its projection on the vertical direction (z-axis) $S_z = m_s \hbar$, where m_s is the spin magnetic quantum number.

Due to this reason, the angle between $\vec{L}$ and $\vec{S}$ cannot be arbitrary. It is bound to take only discrete set of values. Consequently, the magnitude of the total angular momentum $\vec{J}$ takes discrete set of numerical values given by

$$|\vec{J}| = \hbar\sqrt{j(j+1)}, \tag{9.4.2}$$

where

$$j = \ell + s = \ell \pm \frac{1}{2}. \tag{9.4.3}$$

The projection of $\vec{J}$ on the z-axis takes values according to

$$J_z = m_j \hbar, \tag{9.4.4}$$

where m_j, the magnetic quantum number corresponding to the total angular momentum, can take $(2j+1)$ values from $-j$ to $+j$.

This vector model can also be generalized to the case of atoms with more than one electrons. It allows one not only to explain the fine structure of atomic spectra but also the details of anomalous Zeeman splitting of spectral lines in the presence of a magnetic field. This vector model can also be generalized to the case of atoms with more than one electron.

Homework Problems

1. A particle is in the $j = 1$ state. The measurement of $\hat{J}_z$ in this state yields the value $\hbar$. If $\hat{J}_x$ is now measured, what values would result and with what probabilities?
2. Add angular momenta $\ell_1 = 1$ and $\ell_2 = 1$. Using the ladder operator method with the steps given in the chapter, express all the eigenvectors $|j,m\rangle$ in terms of the eigenvectors $|\ell_1, \ell_2, m_1, m_2\rangle$.
3. Let $\vec{S}_1$ and $\vec{S}_2$ denote the spins of a spin 1 and a spin 2 particles, respectively and let $\vec{S}$ be the total spin of the combined system. (a) If $\vec{S}^2$ is measured, what are the possible results? (b) If the system is in a spin state with $\vec{S}^2 = 2\hbar^2$ and the z component of spin of the spin 1 particle is measured, what is the probability of getting the result $\hbar$?
4. Consider the system of particles mentioned in Problem 3. The measurement of S_z yields a value $\hbar$ when the system is in a state with $\vec{S}^2 = 2\hbar^2$. If immediately afterward $\vec{S}^2$ is measured again, what is the probability of obtaining a value $12\hbar^2$?
5. Let $\vec{J} = \vec{L} + \vec{S}$. Using the method described in this chapter, identify and calculate all non-zero Clebsch–Gordan coefficients for the case when $\ell = 2$ and $s = \frac{1}{2}$.
6. A particle of spin $\frac{1}{2}$ is in a state of orbital angular momentum $\ell = 2$. What are its possible states of total angular momentum?

7. Consider a system of three non-identical particles, each with angular momentum $\frac{1}{2}$. Find the possible values of the total spin S of this system and specify the number of angular momentum eigenstates corresponding to each value of S.
8. Consider a system of three non-identical particles, each with angular momentum $\frac{3}{2}$. Find the possible values of the total spin S of this system and specify the number of angular momentum eigenstates corresponding to each value of S.

Chapter 10

Quantum Mechanics of Many-Particle Systems

10.1 General Theory

Consider a system consisting of N particles with masses $m_1, m_2, m_3, ..., m_N$. Let $\vec{r}_j, j = 1, 2, 3, ..., N$, be the position vector of the jth particle. The wave function of such a system will depend on the position vectors of all the particles and time:

$$\psi = \psi(\vec{r}_1, \vec{r}_2, \vec{r}_3, ..., \vec{r}_N, t). \tag{10.1.1}$$

The quantum mechanical formalism for a many-particle system is developed by generalizing the single-particle machinery to the N-particle system. The Schrödinger equation for this N-particle system is written as

$$i\hbar \frac{\partial \psi}{\partial t} = \hat{H} \psi(\vec{r}_1, \vec{r}_2, \vec{r}_3, ..., \vec{r}_N, t), \tag{10.1.2}$$

where the Hamiltonian $\hat{H}$ is given by

$$\hat{H} = -\sum_{j=1}^{N} \frac{\hbar^2}{2m_j} \vec{\nabla}_j^2 + V(\vec{r}_1, \vec{r}_2, \vec{r}_3, ..., \vec{r}_N). \tag{10.1.3}$$

Here, $V(\vec{r}_1, \vec{r}_2, \vec{r}_3, ..., \vec{r}_N)$ is the potential energy of the system, and $\vec{\nabla}_j^2$ is the Laplace operator with respect to the coordinates of the jth particle, that is,

$$\vec{\nabla}_j^2 = \frac{\partial^2}{\partial x_j^2} + \frac{\partial^2}{\partial y_j^2} + \frac{\partial^2}{\partial z_j^2}. \tag{10.1.4}$$

In analogy with the single-particle case, the quantity

$$|\psi(\vec{r}_1, \vec{r}_2, \vec{r}_3, ..., \vec{r}_N, t)|^2 \, d\tau_1 d\tau_2 d\tau_3 ... d\tau_N, \tag{10.1.5}$$

is interpreted as the probability, at a given instant t, of finding the particle 1 in the infinitesimal volume element $d\tau_1$ around $\vec{r}_1$, particle 2 in the infinitesimal volume element $d\tau_2$ around $\vec{r}_2$, particle 3 in the infinitesimal volume element $d\tau_3$ around $\vec{r}_3$, and so on, particle N in $d\tau_N$ around $\vec{r}_N$. Therefore, as earlier, the normalization for the wave function is written as

$$\int_{-\infty}^{+\infty} d\tau_1 \int_{-\infty}^{+\infty} d\tau_2 \int_{-\infty}^{+\infty} d\tau_3 ... \int_{-\infty}^{+\infty} d\tau_N |\psi(\vec{r}_1,\vec{r}_2,\vec{r}_3,...,\vec{r}_N,t)|^2 = 1. \tag{10.1.6}$$

If the potential, V, is time independent, the stationary states of an N-particle system are characterized by the wave functions of the form

$$\psi(\vec{r}_1,\vec{r}_2,\vec{r}_3,...,\vec{r}_N,t) = \phi(\vec{r}_1,\vec{r}_2,\vec{r}_3,...,\vec{r}_N)e^{-\frac{i}{\hbar}Et}, \tag{10.1.7}$$

where E is the total energy of the system and the function $\phi(\vec{r}_1,\vec{r}_2,\vec{r}_3,...,\vec{r}_N)$ satisfies the following time independent Schrödinger equation

$$-\sum_{j=1}^{N} \frac{\hbar^2}{2m_j}\vec{\nabla}_j^2\phi(\vec{r}_1,...,\vec{r}_N) + V(\vec{r}_1,...,\vec{r}_N)\phi(\vec{r}_1,...,\vec{r}_N) = E\phi(\vec{r}_1,...,\vec{r}_N). \tag{10.1.8}$$

As in the case of a single-particle system, the probability density, ρ,

$$\rho = |\psi|^2 = \phi^*(\vec{r}_1,...,\vec{r}_N)e^{\frac{i}{\hbar}Et}\phi(\vec{r}_1,...,\vec{r}_N)e^{-\frac{i}{\hbar}Et} = |\phi(\vec{r}_1,...,\vec{r}_N)|^2, \tag{10.1.9}$$

and the probability current density, $\vec{j}$,

$$\vec{j} = \frac{\hbar}{2i}\sum_{k=1}^{N}\frac{1}{m_k}\left[\left(\vec{\nabla}_k\phi^*(\vec{r}_1,...,\vec{r}_N)\right)\phi(\vec{r}_1,...,\vec{r}_N) - \phi^*(\vec{r}_1,...,\vec{r}_N)\left(\vec{\nabla}_k\phi(\vec{r}_1,...,\vec{r}_N)\right)\right], \tag{10.1.10}$$

do not depend on time in a stationary state. Also, the expectation value,

$$\langle\psi|\hat{A}|\psi\rangle = \int_{-\infty}^{+\infty} \phi^*(\vec{r}_1,...,\vec{r}_N)\,\hat{A}\,\phi(\vec{r}_1,...,\vec{r}_N)\,d\tau_1 d\tau_2 d\tau_3 ... d\tau_N, \tag{10.1.11}$$

of a time-independent observable $\hat{A}$ does not depend on time in a stationary state. Consequently, it is conserved. For instance, energy of a many-particle system in a stationary state is conserved.

So far as the commutation relations for the operators are concerned, the operators representing observables related to different (distinct) particles commute, while those related to a given (specific) particle satisfy the commutation relations valid for a single-particle system. For instance, the position and momentum operators satisfy the following commutation relations

$$[(\hat{r}_\alpha)_k, (\hat{p}_\beta)_\ell] = i\hbar\delta_{k\ell}\delta_{\alpha\beta}, \tag{10.1.12}$$

where the Roman indices, $k, \ell, \ldots$, stand for the particle's number $(1,2,3,...,N)$ in the system, while the Greek indices $\alpha, \beta, \ldots$, represent the Cartesian components of the position vector, $\vec{r}$, and momentum, $\vec{p}$. Note that if $k = \ell$, we are talking about the one and the same particle. For instance, if $k = \ell = 1$, we have the following commutation relations among the position and momentum operators of the first particle:

$$[(\hat{r}_\alpha)_1, (\hat{p}_\beta)_1] = i\hbar\delta_{11}\delta_{\alpha\beta} = i\hbar\delta_{\alpha\beta}. \tag{10.1.13}$$

If $\alpha = \beta = 1$ in (10.1.13), we get that $[(\hat{x}_1)_1, (\hat{p}_1)_1] = i\hbar\delta_{11}$ or $[\hat{x}_1, \hat{p}_{x1}] = i\hbar$. On the other hand, if $\alpha = 1$ and $\beta = 2$, we have $[\hat{x}_1, \hat{p}_{x2}] = i\hbar\delta_{12} = 0$. Similarly, we can calculate the aforementioned commutators for other values of α and β.

In summary, the coordinate and momentum operators of different particles commute, while the coordinate and momentum operators of the same particle satisfy the usual single-particle commutation relations.

It is quite clear now that in order to study the physical characteristics of a many-particle system in a stationary state, we have to solve the equation (10.1.8) for a given potential energy operator V. In the general case of arbitrary V, it is very difficult (almost impossible) to solve the equation (10.1.8). Our earlier experience tells us that it would be possible to find the solutions if (10.1.8) could somehow be split into a system of N single-particle time independent Schrödinger equations. It turns out that this can be achieved in a special case of systems consisting of the so-called, independent particles. We shall discuss these in the following subsections.

10.2 System of Independent and Distinguishable Particles

When the particles belonging to a quantum mechanical system do not interact among themselves and are subject solely to an externally applied potential, they are called independent. This is because of the fact that each of them experiences its own potential, independent of all other particles of the system. For such a system, the potential in (10.1.8) can be written as

$$V(\vec{r}_1, \vec{r}_2, \vec{r}_3, ..., \vec{r}_N) = \sum_{j=1}^{N} V_j(\vec{r}_j), \tag{10.2.1}$$

where $V_j(\vec{r}_j)$ is the potential experienced by the jth particle. If, in addition, the particles can be distinguished from each other in terms of one or several individual properties, they are called distinguishable. The system of particles is then said to be consisting of distinguishable independent particles. Let us assume that the particles of our system are distinguishable by their masses, that is, each of the particles has its own mass m_j, $j = 1,2,3,\ldots,N$, different from the masses of all other particles of the system. Under these conditions, the time-independent Schrödinger equation (10.1.8) permits separation of variables leading to N independent single-particle Schrödinger equations

$$-\frac{\hbar^2}{2m_j}\vec{\nabla}_j^2\phi(\vec{r}_j) + V(\vec{r}_j)\phi_j(\vec{r}_j) = E_j\phi_j(\vec{r}_j), \quad j = 1,2,3,...,N. \tag{10.2.2}$$

The solution of each of these equations yields a single-particle wave function ϕ_{n_j} corresponding to the energy eigenvalue $E_{n_j}, j = 1,2,3,...,N$, where n_j stands for the entire set of quantum numbers of the jth particle. The stationary state wave function of the system is then given by the product of the single-particle wave functions

$$\begin{aligned}\psi(\vec{r}_1,...,\vec{r}_N,t) &= \phi_1(\vec{r}_1)\phi_2(\vec{r}_2)\phi_3(\vec{r}_3)...\phi_N(\vec{r}_N)\ e^{-\frac{i}{\hbar}(E_1+E_2+E_3+...+E_N)t} \\ &= \left(\prod_{j=1}^{N}\phi_j(\vec{r}_j)\right)\ e^{-\frac{i}{\hbar}Et},\end{aligned} \tag{10.2.3}$$

with energy

$$E = E_1 + E_2 + E_3 + ... + E_N = \sum_{j=1}^{N} E_j. \tag{10.2.4}$$

Example 10.2.1: Three spinless non-interacting particles, with respective masses m_1, m_2, and m_3 in the ratio $m_1 : m_2 : m_3 = 1 : 2 : 3$, are subject to a common infinite square well potential of width L in one spatial dimension. Determine the energies and the corresponding wave functions in the three lowest lying states of the system.

Solution: In the given case, the stationary Schrödinger equation (10.1.8)) splits up into three independent single-particle equations (one each for the individual particles):

$$-\frac{\hbar^2}{2m_j}\frac{d^2\phi(x_j)}{dx_j^2} + V(x_j)\phi_j(x_j) = E_j\phi_j(x_j), \quad j = 1,2,3. \tag{10.2.5}$$

The corresponding single-particle wave functions and energies are:

$$\phi_{n_j}(x_j) = \sqrt{\frac{2}{L}}\sin\left(\frac{n_j\pi}{L}x_j\right), \quad j = 1,2,3, \tag{10.2.6}$$

$$E_j = \frac{n_j^2\pi^2\hbar^2}{2m_jL^2}, \quad j = 1,2,3. \tag{10.2.7}$$

Ground state: For the ground state, we have $n_1 = n_2 = n_3 = 1$, and the energy of the system will be

$$E_{111} = \frac{\pi^2\hbar^2}{2L^2}\left(\frac{1}{m_1} + \frac{1}{m_2} + \frac{1}{m_3}\right) = \frac{11\pi^2\hbar^2}{12m_1L^2}. \tag{10.2.8}$$

The corresponding ground state wave function is given by

$$\psi_{111}(x_1,x_2,x_3) = \sqrt{\frac{8}{L^3}}\sin\left(\frac{\pi}{L}x_1\right)\sin\left(\frac{\pi}{L}x_2\right)\sin\left(\frac{\pi}{L}x_3\right). \tag{10.2.9}$$

First excited state: Since $m_3 > m_2 > m_1$, the first excited state will correspond to $n_1 = n_2 = 1$ and $n_3 = 2$. This is because of the fact that this combination of the quantum numbers yields the minimum value of energy that must be given to the system to go from the ground state to the first excited state. Consequently, the energy of the first excited state, E_{112}, will be

$$E_{112} = \frac{\pi^2\hbar^2}{2L^2}\left(\frac{1}{m_1} + \frac{1}{m_2} + \frac{4}{m_3}\right) = \frac{17\pi^2\hbar^2}{12m_1L^2}. \tag{10.2.10}$$

The wave function of the first excited state reads

$$\psi_{112}(x_1,x_2,x_3) = \sqrt{\frac{8}{L^3}}\sin\left(\frac{\pi}{L}x_1\right)\sin\left(\frac{\pi}{L}x_2\right)\sin\left(\frac{2\pi}{L}x_3\right). \tag{10.2.11}$$

Second excited state: The second excited state corresponds to the case when $n_1 = 1$ and $n_2 = n_3 = 2$. Hence, its energy, E_{122}, equals:

$$E_{122} = \frac{\pi^2\hbar^2}{2L^2}\left(\frac{1}{m_1} + \frac{4}{m_2} + \frac{4}{m_3}\right) = \frac{13\pi^2\hbar^2}{6m_1L^2}. \tag{10.2.12}$$

The corresponding wave function is given by

$$\psi_{122}(x_1,x_2,x_3) = \sqrt{\frac{8}{L^3}}\sin\left(\frac{\pi}{L}x_1\right)\sin\left(\frac{2\pi}{L}x_2\right)\sin\left(\frac{2\pi}{L}x_3\right). \tag{10.2.13}$$

Similarly, one can determine the energies and the corresponding wave functions of all other excited states of this three-particle system.

Example 10.2.2: Consider two distinguishable non-interacting particles 1 and 2 with masses m_1 and m_2, respectively. If $m_1 > m_2$ and they are subject to a common three-dimensional potential

$$V(x,y,z) = \begin{cases} 0, & \text{for } 0 < x < a, 0 < y < b, 0 < z < c, \\ \infty & \text{for } x \geq a, y \geq b, z \geq c, \end{cases} \tag{10.2.14}$$

where $a < b < c$ are positive constants, determine the energies and the wave functions of the ground and the first excited states of the system. What is the energy that will be required to excite the system from the ground state to the first excited state?

Solution: In this case, the single-particle wave functions and energies are:

$$\phi_{n_{x_1}n_{y_1}n_{z_1}}(x_1,y_1,z_1) = \sqrt{\frac{8}{abc}}\sin\left(\frac{n_{x_1}\pi}{a}x_1\right), \sin\left(\frac{n_{y_1}\pi}{b}y_1\right)\sin\left(\frac{n_{z_1}\pi}{c}z_1\right), \tag{10.2.15}$$

$$\phi_{n_{x_2}n_{y_2}n_{z_2}}(x_2,y_2,z_2) = \sqrt{\frac{8}{abc}}\sin\left(\frac{n_{x_2}\pi}{a}x_2\right), \sin\left(\frac{n_{y_2}\pi}{b}y_2\right)\sin\left(\frac{n_{z_2}\pi}{c}z_2\right), \tag{10.2.16}$$

$$E_{n_{x_j}n_{y_j}n_{z_j}} = \frac{\pi^2\hbar^2}{2m_j}\left(\frac{n_{x_j}^2}{a^2}+\frac{n_{y_j}^2}{b^2}+\frac{n_{z_j}^2}{c^2}\right), \quad j=1,2. \tag{10.2.17}$$

Ground state: For the ground state of the system, both the particles will occupy the single-particle ground state with $n_{x_j} = n_{y_j} = n_{z_j} = 1, j = 1,2$. The energy of the system in the ground state will be

$$E^{(0)} = \frac{(m_1+m_2)\pi^2\hbar^2}{2m_1m_2}\left(\frac{1}{a^2}+\frac{1}{b^2}+\frac{1}{c^2}\right). \tag{10.2.18}$$

The ground state wave function of the system is simply the product of the single-particle wave functions, that is,

$$\psi^{(0)}(x_1,y_1,z_1;x_2,y_2,z_2) = \phi_{111}(x_1,y_1,z_1)\phi_{111}(x_2,y_2,z_2)$$

$$= \frac{8}{abc}\sin\left(\frac{\pi}{a}x_1\right)\sin\left(\frac{\pi}{a}x_2\right)\sin\left(\frac{\pi}{b}y_1\right)\sin\left(\frac{\pi}{b}y_2\right)\sin\left(\frac{\pi}{c}z_1\right)\sin\left(\frac{\pi}{c}z_1\right). \tag{10.2.19}$$

The first excited state: One of the particles is in the single-particle ground state, while the other is in the single-particle first excited state. Since $z = c$ is the largest side of the box, the particle in the first excited state will have $n_x = n_y = 1$ and $n_z = 2$. But $m_1 > m_2$, therefore the first particle with mass m_1 will be in the first excited state. Hence, the first excited state will have energy

$$E^{(1)} = \pi^2\hbar^2\left\{\left(\frac{(m_1+m_2)}{2m_1m_2}\left[\frac{1}{a^2}+\frac{1}{b^2}\right]\right)+\frac{(m_1+4m_2)}{2m_1m_2c^2}\right\}. \tag{10.2.20}$$

The wave function of the first excited state of the system will be

$$\psi^{(1)}(x_1,y_1,z_1;x_2,y_2,z_2) = \phi_{112}(x_1,y_1,z_1)\phi_{111}(x_2,y_2,z_2)$$

$$= \frac{8}{abc} \sin\left(\frac{\pi}{a}x_1\right) \sin\left(\frac{\pi}{a}x_2\right) \sin\left(\frac{\pi}{b}y_1\right) \sin\left(\frac{\pi}{b}y_2\right) \sin\left(\frac{2\pi}{c}z_1\right) \sin\left(\frac{\pi}{c}z_1\right). \quad (10.2.21)$$

The energy required to excite the system will be

$$E^{(1)} - E^{(0)} = \frac{3\pi^2\hbar^2}{2m_1c^2}. \quad (10.2.22)$$

10.3 System of Identical Particles

Let all the particles constituting the system be identical, that is, they all have the same physical characteristics. In classical mechanics, these particles, despite being identical, may be distinguished from each other. For instance, we can colour them differently at $t = 0$ and then keep track of their individual trajectories separately in time. This will enable us to distinguish them at any instant of time t. We thus conclude that in classical mechanics, identical particles are always distinguishable.

Let us see whether identical and classically distinguishable particles remain distinguishable in quantum mechanics or not. In quantum mechanics, colouring the particles means putting separate tags on them which we cannot do. This is because putting a tag on them means specifying some distinct physical characteristic for each of the particles of the system and this cannot be achieved in view of the fact that all of them have the same *maximal set of commuting observables*. Secondly, due to the uncertainty principle, even if the position of a particle is known at a given instant of time, its momentum is completely indeterminate. Therefore, the very concept of trajectory of a quantum particle loses its meaning and we cannot follow trajectories of the individual particles, the way we proposed to do in classical mechanics. Therefore, there is no way to distinguish between identical particles in quantum mechanics. Clearly, *identical particles are inevitably indistinguishable in quantum mechanics.* This indistinguishability of identical quantum particles has some interesting consequences, which we are going to discuss here.

It turns out that, due to indistinguishability, it is possible to deduce some important properties of the wave functions of a system of N identical particles without solving (10.1.8). For this purpose, let us define the so-called permutation operator $\hat{P}_{jk}$, which interchanges the particles that are at the positions $\vec{r}_j$ and $\vec{r}_k$. Its action on the wave function of the system will then read

$$\hat{P}_{jk}\phi(\vec{r}_1,\vec{r}_2,...,\underbrace{\vec{r}_j,...,\vec{r}_k}_{},...,\vec{r}_N) = \phi(\vec{r}_1,\vec{r}_2,...,\underbrace{\vec{r}_k,...,\vec{r}_j}_{},...,\vec{r}_N). \quad (10.3.1)$$

Since the particles are indistinguishable, no experiment can determine which of the particles of the system is at $\vec{r}_j$ and which one is at $\vec{r}_k$. The probability density, therefore, should remain unchanged, that is,

$$|\phi(\vec{r}_1,\vec{r}_2,...,\vec{r}_j,...,\vec{r}_k,...,\vec{r}_N)|^2 = |\phi(\vec{r}_1,\vec{r}_2,...,\vec{r}_k,...,\vec{r}_j,...,\vec{r}_N)|^2. \tag{10.3.2}$$

This, in turn, gives

$$\phi(\vec{r}_1,\vec{r}_2,...,\vec{r}_j,...,\vec{r}_k,...,\vec{r}_N) = \pm\phi(\vec{r}_1,\vec{r}_2,...,\vec{r}_k,...,\vec{r}_j,...,\vec{r}_N). \tag{10.3.3}$$

As a consequence, the wave function of a system of N identical particles can either be symmetric or anti-symmetric with respect to the interchange of any pair of particles of the system. In nature, as confirmed by experiments, particles with integer spin ($s = 0, 1\hbar, 2\hbar, 3\hbar, ...$) have symmetric wave functions, while the particles with half-odd integer spin ($\hbar/2, 3\hbar/2, 5\hbar/2, ...$) are characterized by the anti-symmetric wave functions. The former satisfy *Bose–Einstein* statistics and are called *bosons*, whereas the latter satisfy *Fermi–Dirac* statistics and are called *fermions*. Note that the relation between spin and statistics can be derived only in relativistic quantum mechanics. In our non-relativistic quantum mechanics, it is taken to be as an axiom.

Composite particles: The natural question arises: What are the symmetry properties of the wave functions of composite particles under the interchange transformation?

Particles that are not elementary but consist of several identical elementary particles (electrons, positrons, muons, etc) are called composite particles. They can also be classified as fermions and bosons. The thing is that the spin of a composite particle can be obtained by adding up the spins of its constituents. If the spins of the constituent particles add up to a half-odd integer (in the units of $\hbar$), the composite particle has a half-odd integer spin and it behaves like a fermion. Consequently, it obeys Fermi–Dirac statistics. If, on the other hand, the resultant spin has an integer value, the composite particle behaves like a boson and obeys Bose–Einstein statistics. For instance, nucleons are fermions because they consist of three quarks with half-odd integer spins, while mesons are bosons because they consist of two quarks only. Atoms can also be classified likewise. For instance, a hydrogen atom consisting of two fermions (an electron and a proton) is a boson, while the isotope ^{3}He of the helium atom is a fermion since it consists of three fermions: one neutron and two protons. *The wave functions of all such composite particles also abide by the symmetry properties discussed earlier.*

10.4 Exchange Degeneracy

The Hamiltonian of a system of N identical particles is a sum of the kinetic energy operators and the potential energy operators of all the particles

$$\hat{H}(\vec{r}_1,\vec{r}_2,...,\vec{r}_j,...,\vec{r}_k,...,\vec{r}_N) = \sum_{j=1}^{N}\frac{\hat{p}_j^2}{2m} + \hat{V}(\vec{r}_1,\vec{r}_2,...,\vec{r}_j,...,\vec{r}_k,...,\vec{r}_N) \tag{10.4.1}$$

If we exchange any pair of particles, say the jth and the kth, the potential energy must remain unchanged, that is,

$$\hat{V}(\vec{r}_1,...,\vec{r}_j,...,\vec{r}_k,...,\vec{r}_N) \rightarrow \hat{V}(\vec{r}_1,...,\vec{r}_k,...,\vec{r}_j,...,\vec{r}_N) = \hat{V}(\vec{r}_1,...,\vec{r}_j,...,\vec{r}_k,...,\vec{r}_N). \quad (10.4.2)$$

If it is not so, the particles will be distinguishable and that will contradict the quantum mechanical assertion that identical particles are indistinguishable. Since the kinetic energy part of $\hat{H}$ remains unchanged, if any two particles are interchanged, the total Hamiltonian of the system will be invariant under the exchange of any pair of particles of the system. In other words, $\hat{H}$ is completely symmetric with respect to the coordinates of the particles. This fact leads to a novel phenomenon called the *exchange degeneracy*.

Consider the eigenvalue problem

$$\hat{H}(\vec{r}_1,...,\vec{r}_j,...,\vec{r}_k,...,\vec{r}_N)\phi(\vec{r}_1,...,\vec{r}_j,...,\vec{r}_k,...,\vec{r}_N) = E\phi(\vec{r}_1,...,\vec{r}_j,...,\vec{r}_k,...,\vec{r}_N). \quad (10.4.3)$$

In view of the invariance of the Hamiltonian under the exchange of any pair of particles, the wave functions corresponding to all possible permutations of particles of the system will have one and the same energy E. *That is, the eigenstates of the Hamiltonian are degenerate*. This is called the *exchange degeneracy*.

Furthermore, we have

$$\begin{aligned}
&\hat{H}\hat{P}_{jk}\phi(\vec{r}_1,...,\vec{r}_j,...,\vec{r}_k,...,\vec{r}_N) = \hat{H}\phi(\vec{r}_1,...,\vec{r}_k,...,\vec{r}_j,...,\vec{r}_N)\\
&= E\phi(\vec{r}_1,...,\vec{r}_k,...,\vec{r}_j,...,\vec{r}_N) = E\hat{P}_{jk}\phi(\vec{r}_1,...,\vec{r}_j,...,\vec{r}_k,...,\vec{r}_N)\\
&= \hat{P}_{jk}E\phi(\vec{r}_1,...,\vec{r}_j,...,\vec{r}_k,...,\vec{r}_N) = \hat{P}_{jk}\hat{H}\phi(\vec{r}_1,...,\vec{r}_j,...,\vec{r}_k,...,\vec{r}_N). \qquad (10.4.4)
\end{aligned}$$

In other words,

$$(\hat{H}\hat{P}_{jk} - \hat{P}_{jk}\hat{H})\phi(\vec{r}_1,...,\vec{r}_j,...,\vec{r}_k,...,\vec{r}_N) = 0. \qquad (10.4.5)$$

The last equation shows that the operator $\hat{P}_{jk}$ commutes with the Hamiltonian

$$(\hat{H}\hat{P}_{jk} - \hat{P}_{jk}\hat{H}) \equiv [\hat{H},\hat{P}_{jk}] = 0. \qquad (10.4.6)$$

It means that the symmetry property of the wave function of a system of N identical particles is conserved in time, that is, if at $t = t_0$ the system starts out with a symmetric (anti-symmetric) wave function, the wave function remains symmetric (anti-symmetric) at any instant $t > t_0$. In addition, since $\hat{P}_{jk}$ and $\hat{H}$ are hermitian and commute, they possess a complete set of common eigenfunctions (see Chapter 3).

10.5 Symmetric and Anti-symmetric Wave Functions and the Pauli Exclusion Principle

Let us construct the wave functions for a system of identical particles. Let us for the sake of convenience, combine the spatial and the spin variables together and write them as ξ, that is, $\xi \equiv (\vec{r}, S)$. By doing so, we put a label on the particles using their position vector $\vec{r}$ and spin S. For instance, if we write $\psi_n(\xi_k)$, then what we mean is that this is the wave function of the nth energy state of the particle with spin variable S_k and the position $\vec{r}_k$. Sometimes, it is also written as $\psi_n(k)$, meaning thereby the wave function of the kth particle or of a particle with a label k. Note that the latter method of labelling identical particles is not quite acceptable because the particles are indistinguishable and we cannot identify which one is the ith and which one is the jth or which one is the kth etc.

Proceeding further, let $\phi_{n_1}(\xi_1)$, $\phi_{n_2}(\xi_2)$, $\phi_{n_3}(\xi_3)$,..., $\phi_{n_N}(\xi_N)$ be the normalized single-particle wave functions, where each of the indices n_1, n_2, n_3,...,n_N stands for the total set of quantum numbers relevant to the problem at hand. We shall assume, for now, that n_1, n_2, n_3,...,n_N are all different.

The first guess could be to write the wave function of the system as a product

$$\phi_{n_1 n_2 \dots n_N}(\xi_1, \xi_2, \xi_3, \dots, \xi_N) = \phi_{n_1}(\xi_1)\phi_{n_2}(\xi_2)\phi_{n_3}(\xi_3)\dots\phi_{n_N}(\xi_N). \tag{10.5.1}$$

However, this is incorrect because writing the wave function this way means that we can distinguish between the particles and that contradicts our earlier assertion about the indistinguishability of identical particles in quantum mechanics. Secondly, such a product function is neither symmetric nor anti-symmetric, whereas our wave function has to be either symmetric or anti-symmetric in view of the indistinguishability of the particles. The problem is overcome by taking the linear combination of the products of the single-particle wave functions corresponding to all possible permutations of the particles.

Thus, the symmetric and the anti-symmetric wave functions, $\phi_s(\xi_1, \xi_2, \xi_3, ..., \xi_N)$ and $\phi_a(\xi_1, \xi_2, \xi_3, ..., \xi_N)$ respectively, of the system of N identical and indistinguishable particles are written as

$$\phi_s(\xi_1, \xi_2, \xi_3, ..., \xi_N) = \frac{1}{\sqrt{N!}} \sum_P P\{\phi_{n_1}(\xi_1)\phi_{n_2}(\xi_2)\phi_{n_3}(\xi_3)...\phi_N(\xi_N)\}, \tag{10.5.2}$$

$$\phi_a(\xi_1, \xi_2, \xi_3, ..., \xi_N) = \frac{1}{\sqrt{N!}} \sum_P (-1)^P P\{\phi_{n_1}(\xi_1)\phi_{n_2}(\xi_2)\phi_{n_3}(\xi_3)...\phi_N(\xi_N)\} \tag{10.5.3}$$

where the sum stands for the summation over all possible permutations ($N!$ in all) of the particles. It is worth noting that in the case of the anti-symmetric wave function, $(-1)^P = +1$, if $(\xi_1, \xi_2, ..., \xi_j, ..., \xi_i, ..., \xi_N)$ (resulting from the interchange of the ith and the jth particles) is an even permutation of $(\xi_1, \xi_2, ..., \xi_i, ..., \xi_j, ..., \xi_N)$, while $(-1)^P = -1$, if $(\xi_1, \xi_2, ..., \xi_j, ..., \xi_i, ..., \xi_N)$ is an odd permutation of $(\xi_1, \xi_2, ..., \xi_i, ..., \xi_j, ..., \xi_N)$. Note that the factor $1/\sqrt{N!}$ comes from the normalization and, as stated earlier, all $n_j, j = 1, 2, 3, ..., N$ have been taken to be different.

According to this prescription, the symmetric wave function for a system of two indistinguishable particles assumes the form

$$\phi_s(\xi_1, \xi_2) = \frac{1}{\sqrt{2}}[\phi_{n_1}(\xi_1)\phi_{n_2}(\xi_2) + \phi_{n_1}(\xi_2)\phi_{n_2}(\xi_1)], \tag{10.5.4}$$

while the anti-symmetric wave function for the same system can be written as

$$\phi_a(\xi_1,\xi_2) = \frac{1}{\sqrt{2}}\left[\phi_{n_1}(\xi_1)\phi_{n_2}(\xi_2) - \phi_{n_1}(\xi_2)\phi_{n_2}(\xi_1)\right]. \tag{10.5.5}$$

The factor of $1/\sqrt{2}$, in these formulae, comes from normalization of the two-particle wave function. It is easy to check that if we interchange the particles, $\phi_s(\xi_2,\xi_1) = \phi_s(\xi_1,\xi_2)$ but $\phi_a(\xi_2,\xi_1) = -\phi_a(\xi_1,\xi_2)$, as it should be.

Similarly, for a three-particle system, the symmetric wave function has the form

$$\begin{aligned}\phi_s(\xi_1,\xi_2,\xi_3) = \frac{1}{\sqrt{3!}}\,&[\phi_{n_1}(\xi_1)\phi_{n_2}(\xi_2)\phi_{n_3}(\xi_3) + \phi_{n_1}(\xi_2)\phi_{n_2}(\xi_3)\phi_{n_3}(\xi_1)\\ &+ \phi_{n_1}(\xi_3)\phi_{n_2}(\xi_1)\phi_{n_3}(\xi_2) + \phi_{n_1}(\xi_1)\phi_{n_2}(\xi_3)\phi_{n_3}(\xi_2)\\ &+ \phi_{n_1}(\xi_3)\phi_{n_2}(\xi_2)\phi_{n_3}(\xi_1) + \phi_{n_1}(\xi_2)\phi_{n_2}(\xi_1)\phi_{n_3}(\xi_3)],\end{aligned} \tag{10.5.6}$$

while the anti-symmetric wave function for the three-particle system can be written as

$$\begin{aligned}\phi_a(\xi_1,\xi_2,\xi_3) = \frac{1}{\sqrt{3!}}\,&[\phi_{n_1}(\xi_1)\phi_{n_2}(\xi_2)\phi_{n_3}(\xi_3) + \phi_{n_1}(\xi_2)\phi_{n_2}(\xi_3)\phi_{n_3}(\xi_1)\\ &+ \phi_{n_1}(\xi_3)\phi_{n_2}(\xi_1)\phi_{n_3}(\xi_2) - \phi_{n_1}(\xi_1)\phi_{n_2}(\xi_3)\phi_{n_3}(\xi_2)\\ &- \phi_{n_1}(\xi_3)\phi_{n_2}(\xi_2)\phi_{n_3}(\xi_1) - \phi_{n_1}(\xi_2)\phi_{n_2}(\xi_1)\phi_{n_3}(\xi_3)].\end{aligned} \tag{10.5.7}$$

As in the previous case, the factor $1/\sqrt{3!}$ comes from normalization. Clearly, using the general formulae, we can write down the wave functions for a system of any given number of identical particles.

Slater determinant

The anti-symmetric wave functions (10.5.5) and (10.5.7) can also be written as determinants:

$$\phi_a(\xi_1,\xi_2) = \frac{1}{\sqrt{2}}\begin{vmatrix} \phi_{n_1}(\xi_1) & \phi_{n_1}(\xi_2) \\ \phi_{n_2}(\xi_1) & \phi_{n_2}(\xi_2) \end{vmatrix}, \tag{10.5.8}$$

$$\phi_a(\xi_1,\xi_2,\xi_3) = \frac{1}{\sqrt{3!}}\begin{vmatrix} \phi_{n_1}(\xi_1) & \phi_{n_1}(\xi_2) & \phi_{n_1}(\xi_3) \\ \phi_{n_2}(\xi_1) & \phi_{n_2}(\xi_2) & \phi_{n_2}(\xi_3) \\ \phi_{n_3}(\xi_1) & \phi_{n_3}(\xi_2) & \phi_{n_3}(\xi_3) \end{vmatrix}. \tag{10.5.9}$$

In general, the N-particle anti-symmetric wave function can be written as

$$\phi_a(\xi_1,\xi_2,\xi_3,...,\xi_N)=\frac{1}{\sqrt{N!}}\begin{vmatrix}\phi_{n_1}(\xi_1) & \phi_{n_1}(\xi_2) & \cdots & \phi_{n_1}(\xi_N)\\ \phi_{n_2}(\xi_1) & \phi_{n_2}(\xi_2) & \cdots & \phi_{n_2}(\xi_N)\\ \cdots & \cdots & \cdots & \cdots\\ \cdots & \cdots & \cdots & \cdots\\ \phi_{n_N}(\xi_1) & \phi_{n_N}(\xi_2) & \cdots & \phi_{n_N}(\xi_N)\end{vmatrix}. \tag{10.5.10}$$

This is known as the *Slater determinant.* The product of the diagonal elements of the determinant gives the original unsymmetrized product of the single-particle wave functions, and the rest of the terms correspond to all permutations of the particles with proper signs. Note that interchanging any two identical particles is equivalent to interchanging the corresponding columns of the Slater determinant. In other words, exchanging two columns exchanges the labels on two particles. From the properties of the determinants, we know that interchanging two columns of a determinant results in the multiplication of the determinant by (-1), that is, the sign of the determinant changes. Thus, writing down the wave function of a system, consisting of fermions, as a Slater determinant is consistent with the anti-symmetry of the wave function under the exchange of any pair of fermions.

Let us now ask the question: What will be the expressions for the functions ϕ_s and ϕ_a when some (or may be all) of $n_j, j=1,2,3,...,N$ coincide?

1. *Symmetric wave function*

   (a) In this case, if all $n_j, j=1,2,3,...,N$ coincide ($n_1=n_2=n_3=...=n_N\equiv n$), the symmetric wave function is given by

   $$\phi_s(\xi_1,...,\xi_N)=\phi_n(\xi_1)\phi_n(\xi_2)\phi_n(\xi_3)...\phi_n(\xi_N). \tag{10.5.11}$$

   (b) If some of these n_j coincide, then we have to avoid double counting. For instance, if n_1 occurs m_1 times, n_2 occurs m_2 times, ..., n_N occurs m_N times, then the total number of distinct permutation of N indices will be

   $$P=\frac{N!}{m_1!m_2!m_3!...m_N!}, \tag{10.5.12}$$

   and hence, the symmetric wave function of the system will be

   $$\phi_s(\xi_1,...,\xi_N)=\sqrt{\frac{m_1!m_2!m_3!...m_N!}{N!}}\sum_P \hat{P}\{\phi_{n_1}(\xi_1)\phi_{n_2}(\xi_2)...\phi_{n_N}(\xi_N)\}. \tag{10.5.13}$$

   For instance, consider a system of four independent identical bosons with $n_1=n_2=n_3=n$, and $n_4\neq n$. Since $m_1=3$, we get

$$\phi_s = \sqrt{\frac{3!}{4!}} \sum_P \hat{P}\{\phi_{n_1}(\xi_1)\phi_{n_2}(\xi_2)\phi_{n_3}(\xi_3)\phi_{n_4}(\xi_4)\}$$

$$= \frac{1}{\sqrt{4}}\left[\phi_n(\xi_1)\phi_n(\xi_2)\phi_n(\xi_3)\phi_{n_4}(\xi_4) + \phi_n(\xi_1)\phi_n(\xi_2)\phi_{n_4}(\xi_3)\phi_n(\xi_4)\right.$$

$$\left. + \phi_n(\xi_1)\phi_{n_4}(\xi_2)\phi_n(\xi_3)\phi_n(\xi_4) + \phi_{n_4}(\xi_1)\phi_n(\xi_2)\phi_n(\xi_3)\phi_n(\xi_4)\right]. \quad (10.5.14)$$

2. *Anti-symmetric wave function.*

   In this case, all n_j have to be different, otherwise the wave function will vanish. For instance, if the particles at positions x_j and x_k are in the same spatial and spin states, that is, $n_j = n_k$, then the jth and the kth rows of the Slater determinant will coincide and the determinant will vanish. Consequently, the wave function of the system will be identically equal to zero: $\phi_a(\xi_1, \xi_2, \xi_3, ..., \xi_N) \equiv 0$.

   Conclusion: *In a system of N identical fermions, no two fermions can occupy the same single-particle state at a time; every single-particle state can be occupied by (at most) one fermion only.* This is known as the *Pauli exclusion principle.*

   Note that the Pauli exclusion principle does not apply to a system of identical bosons. There is no restriction on the number of bosons that can occupy a single-particle state. On the contrary, it so happens that, under suitable conditions, bosons tend to occupy the same quantum state, the ground state. This phenomenon has been experimentally observed and is known as the *Bose–Einstein condensation.*

   Note that ξ includes spatial as well as spin variables. Since spin represents an internal degree of freedom (independent of the spatial degrees of freedom), the wave function of a particle is written as a product of the spatial and the spin parts (see Chapter 5), that is, $\phi(\xi) = \phi(\vec{r}, \vec{S}) = \phi(\vec{r})\chi(\vec{S})$. Generalizing it to the system of N identical particles, we have

$$\phi(\vec{r}_1, ..., \vec{r}_N, \vec{S}_1, ..., \vec{S}_N) = \phi(\vec{r}_1, ..., \vec{r}_N)\,\chi(\vec{S}_1, ..., \vec{S}_N). \quad (10.5.15)$$

Since this wave function, as discussed earlier, has to be either symmetric or anti-symmetric, the parities of the spatial part and that of the spin part of the wave function cannot be arbitrary. They must be such that their product gives, depending on the nature of the particles, the required parity of the total wave function. In the case of identical bosons, when the wave function must be symmetric, the spatial and the spin parts must have the same parity, that is, they are both either symmetric or anti-symmetric. Thus,

$$\phi_s(\vec{r}_1, ..., \vec{r}_N, \vec{S}_1, ..., \vec{S}_1) = \begin{cases} \phi_s(\vec{r}_1, ..., \vec{r}_N)\,\chi_s(\vec{S}_1, ..., \vec{S}_N) \\ \phi_a(\vec{r}_1, ..., \vec{r}_N)\,\chi_a(\vec{S}_1, ..., \vec{S}_N), \end{cases} \quad (10.5.16)$$

where the suffixes s and a stand for the symmetric and the anti-symmetric wave functions, respectively.

For a system of N identical fermions, the wave function must be overall anti-symmetric and, therefore, the spatial and the spin parts of the wave function must have opposite parities, that is, if one of them is symmetric, the other has to be anti-symmetric, and vice versa. Thus,

$$\phi_a(\vec{r}_1,...,\vec{r}_N,\vec{S}_1,...,\vec{S}_1) = \begin{cases} \phi_s(\vec{r}_1,...,\vec{r}_N)\,\chi_a(\vec{S}_1,...,\vec{S}_N) \\ \phi_a(\vec{r}_1,...,\vec{r}_N)\,\chi_s(\vec{S}_1,...,\vec{S}_N). \end{cases} \tag{10.5.17}$$

Example 10.5.1: Two identical non-interacting particles are in an isotropic harmonic oscillator potential. Find the degeneracy of the ground state and the first excited state of the system (a) if the particles are spin-1/2 fermions and (b) when they are spin-1 bosons.

As shown in Chapter 4, the nth stationary state of a single-particle in an isotropic harmonic oscillator potential can be characterized by a triplet of non-negative integers n_x, n_y and n_z. It has energy $E_{n_x n_y n_z} = \left(n_x + n_y + n_z + \frac{3}{2}\right)\hbar\omega$. The ground state corresponds to $n_x = n_y = n_z = 0$, while the first excited state corresponds to $n_x = 1, n_y = n_z = 0$ or $n_x = 0, n_y = 1, n_z = 0$ or $n_x = 0, n_y = 0, n_z = 1$.

(a) In the ground state of the system, both the particles are in the single-particle ground states ($n_{x_j} = n_{y_j} = n_{z_j} = 0$, $j = 1,2$) with opposite spins. Since the quantum numbers coincide, the anti-symmetric spatial part of the wave function,

$$\phi_a(\vec{r}_1,\vec{r}_2) = \frac{1}{\sqrt{2}}\Big[\phi_{n_{x_1} n_{y_1} n_{z_1}}(x_1,y_1,z_1)\,\phi_{n_{x_2} n_{y_2} n_{z_2}}(x_2,y_2,z_2) - \phi_{n_{x_1} n_{y_1} n_{z_1}}(x_2,y_2,z_2)\,\phi_{n_{x_2} n_{y_2} n_{z_2}}(x_1,y_1,z_1)\Big]. \tag{10.5.18}$$

vanishes. It means that the spatial part of the total wave function of the system will be symmetric:

$$\phi_s(\vec{r}_1,\vec{r}_2) = \frac{1}{\sqrt{2}}\Big[\phi_{n_{x_1} n_{y_1} n_{z_1}}(x_1,y_1,z_1)\,\phi_{n_{x_2} n_{y_2} n_{z_2}}(x_2,y_2,z_2) + \phi_{n_{x_1} n_{y_1} n_{z_1}}(x_2,y_2,z_2)\,\phi_{n_{x_2} n_{y_2} n_{z_2}}(x_1,y_1,z_1)\Big], \tag{10.5.19}$$

Therefore, in view of the fact that, for spin $\frac{1}{2}$ particles, the overall wave function must be anti-symmetric, the spin part of the wave function must be the anti-symmetric singlet spin function. Hence, the ground-state energy is non-degenerate, that is, its degeneracy equals 1.

The first excited state corresponds to one particle in the single-particle ground state, $n_x = n_y = n_z = 0$ and the other in the first excited state with $n_x = 1, n_y = n_z = 0$

or $n_x = 0, n_y = 1, n_z = 0$ or $n_x = 0, n_y = 0, n_z = 1$. Since the spatial parts of the single-particle wave function of the particles are different, both the symmetric and the anti-symmetric spatial parts of the total wave function will be non-zero. The former will have to be combined with the anti-symmetric singlet spin state, while the latter has to be combined with the symmetric triplet spin state. Hence, there are in all four spin configurations. In addition, as mentioned earlier, the first excited state of one of the particles can be realized in three different ways. Therefore, the total degeneracy of the first excited state of the system is $3 \times 4 = 12$.

(b) In the case of two spin-1 bosons, the overall wave function must be symmetric. The system's spin function is obtained by combining the spins of the two particles. As we know, there are six symmetric and three anti-symmetric spin functions for this system. For the ground state of the system, when both the bosons are in the single-particle ground state with $n_{x_j} = n_{y_j} = n_{z_j} = 0,\ j = 1,2$, there is a single symmetric spatial part of the wave function, which must be combined with one of the six symmetric spin functions to give an overall symmetric wave function. Hence, the degeneracy of the ground state in this case is six.

The first excited state, analogous to the case of fermions, will have a symmetric or an anti-symmetric spatial part of the wave function. Once again the symmetric spatial part is combined with the three anti-symmetric spin parts and the anti-symmetric spatial part is combined with six symmetric spin parts of the wave function. In this case too, the first excited state of one of the particles can be realized in three different ways. So, in this case of bosonic system, the total degeneracy of the first excited state will be $9 \times 3 = 27$.

Example 10.5.2: Two non-interacting particles, each of mass m, are confined to move in a one-dimensional potential well: $V(x) = 0$, for $0 < x < 2a$ and $V(x) = \infty$ elsewhere, where a is a positive constant. What are the energies and the corresponding degeneracies of the three lowest lying states of the system, if the particles are (i) indistinguishable spin-1/2 fermions?, and (ii) distinguishable spin-1/2 fermions?.

Solution: The single-particle spatial part of the stationary state wave functions, satisfying the standard boundary conditions at $x = 0$ and $x = 2a$, are

$$\psi_n(\xi,t) = \phi_n(x)\, e^{-\frac{i}{\hbar}\mathscr{E}_n t}, \tag{10.5.20}$$

where n is a positive integer,

$$\phi_n(x) = \sqrt{\frac{1}{a}}\, \sin\left(\frac{n\pi}{2a}x\right) \tag{10.5.21}$$

and $\mathscr{E}_n$, given by

$$\mathscr{E}_n = \frac{n^2\pi^2\hbar^2}{8ma^2} \tag{10.5.22}$$

is the single-particle energy in the nth state.

It is obvious that the nth energy state of the system will be characterized by two sets of quantum numbers n_1 and n_2. The corresponding stationary state wave function of the system will be

$$\psi_{n_1n_2}(\xi_1,\xi_2,t) = \phi_{n_1n_2}(x_1,x_2)\,\chi(S_1,S_2)\,e^{-\frac{i}{\hbar}E_{n_1n_2}t}, \tag{10.5.23}$$

where

$$E_{n_1n_2} = \mathscr{E}_{n_1} + \mathscr{E}_{n_2} = \left(n_1^2+n_2^2\right)\frac{\pi^2\hbar^2}{8ma^2} \tag{10.5.24}$$

gives the total energy of the system.

(i) Since the particles are indistinguishable fermions, the ground state of the system will have both the fermions in the single-particle states with $n_1 = n_2 = 1$ under the condition that they will have opposite spins. Hence, the ground state will have energy $E_{11} = \mathscr{E}_1 + \mathscr{E}_1 = \pi^2\hbar^2/4ma^2$.

We have $n_1 = n_2 = 1$, which means that the anti-symmetric spatial part of the wave function will be zero. Since for a fermionic system the overall wave function must be anti-symmetric, the ground state wave function will be given by

$$\psi_{11}(\xi_1,\xi_2) = \frac{1}{a}\sin\left(\frac{n_1\pi}{2a}x_1\right)\sin\left(\frac{n_2\pi}{2a}x_2\right)\chi_{\text{singlet}}(s_1,s_2), \tag{10.5.25}$$

where $\chi_{\text{singlet}}(s_1,s_2)$ given by

$$\chi_{\text{singlet}}(s_1,s_2) = \frac{1}{\sqrt{2}}\left[\chi_1^{(+)}\chi_2^{(-)} - \chi_1^{(-)}\chi_2^{(+)}\right], \tag{10.5.26}$$

is anti-symmetric with respect to the interchange of particles. The superscripts '(+)' and '(−)' stand for *spin up* and *spin down*, respectively. Evidently, the ground state will be non-degenerate (*degeneracy equal to one*).

The first excited state of the system will correspond to $n_1 = 2, n_2 = 1$ or $n_1 = 1, n_2 = 2$. This state will have energy

$$E_{12} = E_{21} = \frac{5\pi^2\hbar^2}{8ma^2}. \tag{10.5.27}$$

The wave function of the system will be given by

$$\psi_a(\xi_1,\xi_2) = \begin{cases} \phi_s(x_1,x_2)\,\chi_{\text{singlet}}(s_1,s_2) \\ \phi_a(x_1,x_2)\,\chi_{\text{triplet}}(s_1,s_2). \end{cases} \tag{10.5.28}$$

where $\chi_{\text{singlet}}(s_1,s_2)$ is given by (10.5.26), while $\chi_{\text{triplet}}(s_1,s_2)$ stands for three possible symmetric spin functions given by

$$\chi_{\text{triplet}}(s_1,s_2) = \begin{cases} \chi_1^{(+)}\chi_2^{(+)}, \\ \frac{1}{\sqrt{2}}\left[\chi_1^{(+)}\chi_2^{(-)} + \chi_1^{(-)}\chi_2^{(+)}\right] \\ \chi_1^{(+)}\chi_2^{(+)}. \end{cases} \tag{10.5.29}$$

The spatial parts of the wave function are

$$\phi_s(x_1,x_2) = \frac{1}{\sqrt{2a}}\left[\sin\left(\frac{\pi x_1}{2a}\right)\sin\left(\frac{\pi x_2}{a}\right) + \sin\left(\frac{\pi x_1}{a}\right)\sin\left(\frac{\pi x_2}{2a}\right)\right], \tag{10.5.30}$$

$$\phi_a(x_1,x_2) = \frac{1}{\sqrt{2a}}\left[\sin\left(\frac{\pi x_1}{2a}\right)\sin\left(\frac{\pi x_2}{a}\right) - \sin\left(\frac{\pi x_1}{a}\right)\sin\left(\frac{\pi x_2}{2a}\right)\right]. \tag{10.5.31}$$

Since there are four possible spin configurations, the first excited state of the system is 4-fold degenerate.

The second excited state of the system corresponds to $n_1 = n_2 = 2$ and the energy of the system in this state will be

$$E^{(2)} = E_{22} = \frac{\pi^2\hbar^2}{ma^2}. \tag{10.5.32}$$

Once again, since the overall wave function of the system must be anti-symmetric and $n_1 = n_2 = 2$, the anti-symmetric spatial part of the wave function will be zero. Hence, the total wave function of the second excited state will be

$$\psi_{22}(\xi_1,\xi_2) = \frac{1}{a}\sin\left(\frac{\pi x_1}{a}\right)\sin\left(\frac{\pi x_2}{a}\right)\chi_{\text{singlet}}(s_1,s_2). \tag{10.5.33}$$

Just like the ground state, the second excited state of the system will also be non-degenerate.

(ii) Since the particles are distinguishable fermions, there is no restriction on the symmetry of the wave functions: neither on the spatial part nor on the spin part.

The *ground state* of the system will correspond to $n_1 = n_2 = 1$ and its energy will be the same as in the previous part. The spatial part of the ground-state wave function is given by

$$\phi_{11}(x_1,x_2) = \frac{1}{a}\sin\left(\frac{\pi x_1}{2a}\right)\sin\left(\frac{\pi x_2}{2a}\right). \tag{10.5.34}$$

This can be combined with four possible spin functions

$$\chi_1^{(+)}\chi_2^{(+)},\ \chi_1^{(-)}\chi_2^{(+)},\ \chi_1^{(+)}\chi_2^{(-)},\ \chi_1^{(-)}\chi_2^{(-)},$$

where '+' and '−' stand for spin up and spin down, respectively. Therefore, the ground state in this case is 4-fold degenerate.

The first excited state: It will have $n_1 = 1, n_2 = 2$ or $n_1 = 2, n_2 = 1$ and its energy will be given by

$$E^{(1)} = \frac{5\pi^2\hbar^2}{8ma^2}. \tag{10.5.35}$$

So far as the spatial part of the wave function is concerned, there are two possibilities

$$\phi_{12}(x_1,x_2) = \frac{1}{a}\sin\left(\frac{\pi x_1}{2a}\right)\sin\left(\frac{\pi x_2}{a}\right) \tag{10.5.36}$$

or

$$\phi_{21}(x_1,x_2) = \frac{1}{a}\sin\left(\frac{\pi x_1}{a}\right)\sin\left(\frac{\pi x_2}{2a}\right). \tag{10.5.37}$$

Each one of these can be combined with the previously written four possible spin functions. Hence, the first excited state of the system will have 8-fold degeneracy.

The *second excited state* of the system will correspond to $n_1 = n_2 = 2$ with energy

$$E^{(2)} = E_{22} = \frac{\pi^2\hbar^2}{ma^2}. \tag{10.5.38}$$

The spatial part of the second excited state wave function is given by

$$\phi_{22}(x_1,x_2) = \frac{1}{a}\sin\left(\frac{\pi x_1}{2a}\right)\sin\left(\frac{\pi x_2}{2a}\right). \tag{10.5.39}$$

This can again be combined with four possible spin functions

$$\chi_1^{(+)}\chi_2^{(+)},\ \chi_1^{(-)}\chi_2^{(+)},\ \chi_1^{(+)}\chi_2^{(-)},\ \chi_1^{(-)}\chi_2^{(-)}.$$

Therefore, the second excited state is also 4-fold degenerate.

Example 10.5.3: Two identical and non-interacting spin-1 particles are moving in a common infinite square well potential of width a. Determine the energy and the wave functions of this two-particle system in the ground state and the first excited state.

Solution: The single-particle energy levels and the corresponding spatial parts of the wave functions are

$$\mathscr{E}_{n_1} = \frac{n_1^2\pi^2\hbar^2}{2ma^2}, \qquad \mathscr{E}_{n_2} = \frac{n_2^2\pi^2\hbar^2}{2ma^2}, \tag{10.5.40}$$

$$\phi_{n_1}(x_1) = \sqrt{\frac{2}{a}}\sin\left(\frac{n_1\pi}{a}x_1\right), \tag{10.5.41}$$

$$\phi_{n_2}(x_2) = \sqrt{\frac{2}{a}}\sin\left(\frac{n_2\pi}{a}x_2\right), \tag{10.5.42}$$

where the quantum numbers n_1 and n_2 take positive integer values.

The stationary state wave function of the system has to be symmetric. Therefore, it is given by the linear combination of the two symmetric wave functions that can be constructed with the help of the full (spatial plus the spin part) single-particle wave functions:

$$\psi_{n_1n_2}(\xi_1,\xi_2,t) = \frac{1}{\sqrt{2}}\left[\phi_s(x_1,x_2)\,\chi_s(s_1,s_2) + \phi_a(x_1,x_2)\,\chi_a(s_1,s_2)\right]\, e^{-\frac{i}{\hbar}E_{n_1n_2}t}. \tag{10.5.43}$$

Here, $E_{n_1n_2} = \mathscr{E}_{n_1} + \mathscr{E}_{n_2}$ and, as earlier, the subscripts s and a stand for the symmetric and the anti-symmetric wave functions, respectively. The symmetric and the anti-symmetric spatial parts of the wave function are given by

$$\phi_s(x_1,x_2) = \frac{1}{\sqrt{2}}\left[\phi_{n_1}(x_1)\phi_{n_2}(x_2) + \phi_{n_1}(x_2)\phi_{n_2}(x_1)\right], \tag{10.5.44}$$

$$\phi_a(x_1,x_2) = \frac{1}{\sqrt{2}}\left[\phi_{n_1}(x_1)\phi_{n_2}(x_2) - \phi_{n_1}(x_2)\phi_{n_2}(x_1)\right]. \tag{10.5.45}$$

The spin part of the wave function of the system, $\chi(s_1,s_2)$, is obtained by combining the spins of the two particles. We know that for a system of two spin-1 particles, there are in total, six states that are symmetric: *five* $|2,m_s\rangle$ *states, with* $m_s = 0,\pm1,\pm2$ *corresponding to the total spin of* 2 *and a singlet state corresponding to* $s = 0$. In addition, we have a triplet of states, $|1,m_s\rangle$ with $m_s = 0,\pm1$, which correspond to $s = 1$ and are anti-symmetric. Given these inputs, let us find the required energy states and the wave functions of the system.

Ground state: In this case, both the particles will be in the single-particle ground state with $n_1 = 1$ and $n_2 = 1$. The energy of the system will be given by

$$E^{(0)} = E_{11} = \mathscr{E}_1 + \mathscr{E}_2 = \frac{\pi^2\hbar^2}{ma^2}. \tag{10.5.46}$$

Since $n_1 = n_2 = 1$, the anti-symmetric spatial part $\phi_a(x_1,x_2)$ will vanish and the wave function of the system will be

$$\psi^{(0)}(x_1,s_1,x_2,s_2) = \phi_1(x_1)\phi_1(x_2)\,\chi_s(s_1,s_2) = \frac{2}{a}\sin\left(\frac{\pi}{a}x_1\right)\sin\left(\frac{\pi}{a}x_2\right)\chi_s(s_1,s_2), \tag{10.5.47}$$

where $\chi_s(s_1,s_2)$ can be any one of the six symmetric states. Clearly, the ground state of the system is 6-fold degenerate.

First excited state: One of the particles will be in the single-particle ground state, while the other will be in the first excited state, that is, either $n_1 = 1, n_2 = 2$ or $n_1 = 2, n_2 = 1$. The energy of the system in this state will be

$$E^{(1)} = E_{12} = E_{21} = \frac{5\pi^2\hbar^2}{2ma^2}. \tag{10.5.48}$$

The state will have either symmetric or anti-symmetric spatial part of the wave function. The former must be coupled with the six symmetric spin functions $\chi_s(s_1,s_2)$, while the latter should be coupled with the three anti-symmetric spin functions $\chi_a(s_1,s_2)$. Consequently, the wave function of the system will be given by

$$\psi^{(1)}(\xi_1,\xi_2) = \begin{cases} \phi_s(x_1,x_2)\,\chi_s(s_1,s_2), \\ \phi_a(x_1,x_2)\,\chi_a(s_1,s_2). \end{cases}$$

$$= \begin{cases} \frac{\sqrt{2}}{a}\left[\sin\left(\frac{\pi}{a}x_1\right)\sin\left(\frac{2\pi}{a}x_2\right) + \sin\left(\frac{2\pi}{a}x_1\right)\sin\left(\frac{\pi}{a}x_2\right)\right]\chi_s(s_1,s_2), \\ \frac{\sqrt{2}}{a}\left[\sin\left(\frac{\pi}{a}x_1\right)\sin\left(\frac{2\pi}{a}x_2\right) + \sin\left(\frac{2\pi}{a}x_1\right)\sin\left(\frac{\pi}{a}x_2\right)\right]\chi_a(s_1,s_2). \end{cases} \tag{10.5.49}$$

The first excited state will be 9-fold degenerate.

Second excited state: In the second excited state of the system, both the particles will be in the single-particle first excited state with $n_1 = n_2 = 2$. The energy of the system equals

$$E^{(2)} = E_{22} = \mathscr{E}_2 + \mathscr{E}_2 = \frac{4\pi^2\hbar^2}{ma^2}. \tag{10.5.50}$$

Since $n_1 = n_2 = 2$, $\phi_a(x_1,x_2)$ vanishes and the wave function of the system will be

$$\psi^{(2)}(x_1,s_1,x_2,s_2) = \phi_2(x_1)\phi_2(x_2)\chi_s(s_1,s_2) = \frac{2}{a}\sin\left(\frac{2\pi}{a}x_1\right)\sin\left(\frac{2\pi}{a}x_2\right)\chi_s(s_1,s_2). \tag{10.5.51}$$

The second excited state of the system is 6-fold degenerate.

Example 10.5.4: Consider three identical and non-interacting particles moving, in one spatial dimension, in a common infinite square well potential of width a: $V(x) = 0$ for $0 < x < a$ and ∞ otherwise. Determine the eigenfunctions and the corresponding energies of the system for the ground state, the first excited state and the second excited state, in the following cases: (i) the particles are spinless bosons and (ii) spin-1/2 fermions.

Solution: As discussed earlier, the total wave function is a product of the spatial part $\phi_{n_1n_2n_3}(x_1,x_2,x_3)$ and the spin part $\chi(S_1,S_2,S_3)$, which are to be constructed from the single-particle wave functions. The spatial parts of the single-particle wave functions vanish for $x < 0$ and $x > a$. In the region $0 < x < a$, they satisfy their individual time-independent Schrödinger equation:

$$\frac{\partial^2\phi_j(x_j)}{\partial x_j^2} + k_j^2\phi_j(x_j) = 0, \quad j = 1,2,3, \tag{10.5.52}$$

where $k_j^2 = 2m\mathscr{E}_j/\hbar^2$. The energy eigenvalues, $\mathscr{E}_{n_j}$ and the corresponding wave functions, $\phi_{n_j}(x_j)$ (satisfying the required boundary conditions: $\phi_{n_j}(x_j) = 0$ at $x_j = 0$ and $x_j = a$), are given by

$$\phi_{n_1}(x_1) = \sqrt{\frac{2}{a}}\sin\left(\frac{n\pi x_1}{a}\right), \quad \mathscr{E}_{n_1} = \frac{n_1^2\pi^2\hbar^2}{2ma^2}, \quad n_1 = 1,2,3,... \tag{10.5.53}$$

$$\phi_{n_2}(x_2) = \sqrt{\frac{2}{a}}\sin\left(\frac{n_2\pi x_2}{a}\right), \quad \mathscr{E}_{n_2} = \frac{n_2^2\pi^2\hbar^2}{2ma^2}, \quad n_2 = 1,2,3,... \tag{10.5.54}$$

$$\phi_{n_3}(x_3) = \sqrt{\frac{2}{a}}\sin\left(\frac{n_3\pi x_3}{a}\right), \quad \mathscr{E}_{n_3} = \frac{n_3^2\pi^2\hbar^2}{2ma^2}, \quad n_3 = 1,2,3,... . \tag{10.5.55}$$

(i) In this case, the particles are spinless bosons and, therefore, the ground state corresponds to $n_1 = n_2 = n_3 = 1$. The ground state energy is given by

$$E^{(0)} = E_{111} = \frac{\pi^2\hbar^2}{2ma^2}(1+1+1) = \frac{3\pi^2\hbar^2}{2ma^2}. \tag{10.5.56}$$

Since there is no spin part of the wave function to be taken into account, and $n_1 = n_2 = n_3 = 1$, the ground-state wave function, $\phi^{(0)}(x_1,x_2,x_3)$, is given by the simple product of the single-particle wave functions

$$\psi^{(0)}(x_1,x_2,x_3) = \phi_1(x_1)\phi_1(x_2)\phi_1(x_3) = \sqrt{\frac{8}{a^3}}\sin\left(\frac{\pi x_1}{a}\right)\sin\left(\frac{\pi x_2}{a}\right)\sin\left(\frac{\pi x_3}{a}\right). \quad (10.5.57)$$

The first excited state corresponds to the case when two of the bosons are in the ground state while one boson is in the first excited state. There are three possible particle configurations with $n_1 = 2, n_2 = n_3 = 1$, $n_2 = 2, n_1 = n_3 = 1$ and $n_3 = 2, n_1 = n_2 = 1$ with the same total energy which is equal to

$$E^{(1)} = E_{211} = E_{121} = E_{112} = \frac{\pi^2\hbar^2}{2ma^2}(1+1+4) = \frac{3\pi^2\hbar^2}{ma^2}. \quad (10.5.58)$$

Clearly, the state is 3-fold degenerate. The second excited state corresponds to the case when one of the bosons is in the ground state ($n_j = 1$), while the other two are in the first excited state ($n_j = 2$). The corresponding energy level is again 3-fold degenerate and the energy of the system in this state is

$$E^{(2)} = E_{221} = E_{122} = E_{212} = \frac{\pi^2\hbar^2}{2ma^2}(1+4+4) = \frac{9\pi^2\hbar^2}{2ma^2}. \quad (10.5.59)$$

To find the wave function of the first excited state, we notice that, since the particles are indistinguishable, we cannot say which particle is in the ground state (with $n = 1$) and which in the first excited state (with $n = 2$); all that we can say is that two of the particles are in the ground state and one is in the first excited state. In other words, two of the indices n_1, n_2 and n_3 coincide. The total number of distinct permutations is, therefore, equal to $3!/2!$. Consequently, the wave function for the first excited state of the system will be

$$\begin{aligned}\psi^{(1)}(x_1,x_2,x_3) = \phi_{111}(x_1,x_2,x_3) &= \sqrt{\frac{2!}{3!}}\,[\phi_1(x_1)\phi_1(x_2)\phi_2(x_3) + \phi_1(x_2)\phi_1(x_3)\phi_2(x_1) \\ &\quad + \phi_1(x_3)\phi_1(x_1)\phi_2(x_2)] \\ &= \sqrt{\frac{2!}{3!}}\left[\sin\left(\frac{\pi x_1}{a}\right)\sin\left(\frac{\pi x_2}{a}\right)\sin\left(\frac{2\pi x_3}{a}\right)\right. \\ &\quad + \sin\left(\frac{\pi x_2}{a}\right)\sin\left(\frac{\pi x_3}{a}\right)\sin\left(\frac{2\pi x_1}{a}\right) \\ &\quad \left. + \sin\left(\frac{\pi x_3}{a}\right)\sin\left(\frac{\pi x_1}{a}\right)\sin\left(\frac{2\pi x_2}{a}\right)\right].. \end{aligned} \quad (10.5.60)$$

The wave function corresponding to the second excited state of the system will be

$$\psi^{(2)}(x_1,x_2,x_3) = \sqrt{\frac{2!}{3!}}\,[\phi_1(x_1)\phi_2(x_2)\phi_2(x_3) + \phi_2(x_1)\phi_1(x_2)\phi_2(x_3)$$
$$+ \phi_2(x_1)\phi_2(x_2)\phi_1(x_3)]$$
$$= \sqrt{\frac{2!}{3!}}\left[\sin\left(\frac{\pi x_1}{a}\right)\sin\left(\frac{2\pi x_2}{a}\right)\sin\left(\frac{2\pi x_3}{a}\right)\right.$$
$$+\sin\left(\frac{2\pi x_1}{a}\right)\sin\left(\frac{\pi x_2}{a}\right)\sin\left(\frac{2\pi x_3}{a}\right)$$
$$\left.+\sin\left(\frac{2\pi x_1}{a}\right)\sin\left(\frac{2\pi x_2}{a}\right)\sin\left(\frac{\pi x_3}{a}\right)\right]. \tag{10.5.61}$$

(ii) Now the particles are spin-1/2 fermions. Therefore, we have to take the Pauli principle into account. Consequently, the ground state of the system is the one in which two of the particles are in the single-particle ground state, $n_j = 1$, with opposite spins and the third particle is in the first excited state, $n_j = 2$, with either positive (up) or negative (down) spin. The ground state energy is given by

$$E^{(0)} = E_{211} = E_{112} = E_{121} = \frac{\pi^2\hbar^2}{2ma^2}(1+1+4) = \frac{3\pi^2\hbar^2}{ma^2}. \tag{10.5.62}$$

In this case, *the ground state* wave function is given by the corresponding Slater determinant:

$$\psi_a^{(0)}(x_1,x_2,x_3,s_1,s_2,s_3) = \begin{vmatrix} \phi_1(x_1)\,\chi(s_1) & \phi_1(x_1)\,\chi(s_2) & \phi_2(x_1)\,\chi(s_3) \\ \phi_1(x_2)\,\chi(s_1) & \phi_1(x_2)\,\chi(s_2) & \phi_2(x_2)\,\chi(s_3) \\ \phi_1(x_3)\,\chi(s_1) & \phi_1(x_3)\,\chi(s_2) & \phi_2(x_3)\,\chi(s_3) \end{vmatrix}. \tag{10.5.63}$$

Note that since there are four ways in which we can configure the spins of three fermions, the ground state is 4-fold degenerate.

Let us write down the wave function of the ground state for one of the four configurations shown in Figure 10.1. The wave function corresponding to this configuration will be

$$\psi_a^{(0)} = \begin{vmatrix} \phi_1(x_1)\,\chi^{(+)}(s_1) & \phi_1(x_1)\,\chi^{(-)}(s_2) & \phi_2(x_1)\,\chi^{(+)}(s_3) \\ \phi_1(x_2)\,\chi^{(+)}(s_1) & \phi_1(x_2)\,\chi^{(-)}(s_2) & \phi_2(x_2)\,\chi^{(+)}(s_3) \\ \phi_1(x_3)\,\chi^{(+)}(s_1) & \phi_1(x_3)\,\chi^{(-)}(s_2) & \phi_2(x_3)\,\chi^{(+)}(s_3) \end{vmatrix}, \tag{10.5.64}$$

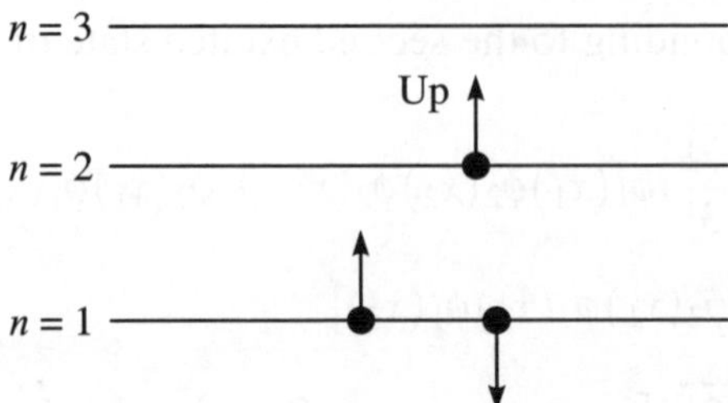

Figure 10.1 One of the four possible configurations of the system with two of the fermions in the single-particle ground state with $n = 1$ and the third fermion in the single-particle first excited state with $n = 2$.

where

$$\chi^{(+)} = \left|\frac{1}{2},\frac{1}{2}\right\rangle, \; \chi^{(-)} = \left|\frac{1}{2},-\frac{1}{2}\right\rangle, \; \chi^{(\pm)} = \left|\frac{1}{2},\pm\frac{1}{2}\right\rangle. \tag{10.5.65}$$

Simplifying we arrive at

$$\begin{aligned}\phi_a^{(0)} &= \frac{1}{\sqrt{3}}\left[\phi_1(x_1)\phi_1(x_2)\phi_2(x_3) - \phi_1(x_1)\phi_2(x_2)\phi_1(x_3) - \phi_1(x_1)\phi_1(x_2)\phi_2(x_1)\right] \\ &\times \frac{1}{\sqrt{2}}\left[\chi^{(+)}(s_1)\chi^{(-)}(s_2) - \chi^{(-)}(s_1)\chi^{(+)}(s_2)\right]\chi^{(+)}(s_3).\end{aligned} \tag{10.5.66}$$

The *first excited state* of the system will be the one in which two of the fermions, with opposite spins, will be in the first excited state, $n_j = 2$, while the third will be in the ground state, $n_j = 1$, with either up or down spin. Therefore, the first excited state energy of the system will be given by

$$E^{(1)} = E_{221} = E_{122} = E_{212} = \frac{\pi^2\hbar^2}{2ma^2}(1+4+4) = \frac{9\pi^2\hbar^2}{ma^2}. \tag{10.5.67}$$

The corresponding first excited state wave function will be

$$\psi^{(1)}(x_1,x_2,x,s_1,s_2,s_3) = \begin{vmatrix} \phi_1(x_1)\,\chi(s_1) & \phi_2(x_1)\,\chi(s_2) & \phi_2(x_1)\,\chi(s_3) \\ \phi_1(x_2)\,\chi(s_1) & \phi_2(x_2)\,\chi(s_2) & \phi_2(x_2)\,\chi(s_3) \\ \phi_1(x_3)\,\chi(s_1) & \phi_2(x_3)\,\chi(s_2) & \phi_2(x_3)\,\chi(s_3) \end{vmatrix}. \tag{10.5.68}$$

Once again, there are four ways in which we can configure the spins of three fermions. Consequently, there are four states corresponding to the same energy $E^{(1)}$. Hence, the first excited state is 4-fold degenerate. The situation is depicted in Figure 10.2.

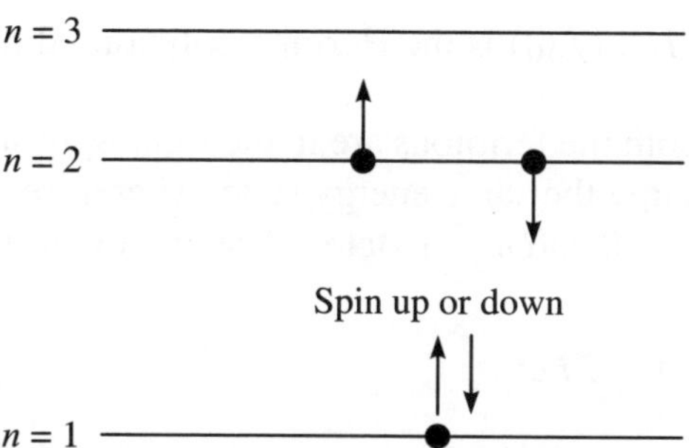

Figure 10.2 One of the four possible configurations of the system with two of the fermions in the single-particle first excited state with $n = 2$ and opposite spins and the third fermion in the single-particle ground state with $n = 1$ and spin up.

In the *second excited state*, two of the fermions will be in the single-particle ground state with $n_j = 1$ while the third will be in the single-particle second excited state with $n_j = 3$. Hence, the second excited state of the system will have energy

$$E^{(2)} = E_{131} = E_{113} = E_{131} = \frac{\pi^2\hbar^2}{2ma^2}(1+1+9) = \frac{11\pi^2\hbar^2}{ma^2}. \tag{10.5.69}$$

The corresponding second excited state wave function of the system will be

$$\psi^{(2)}(x_1,x_2,x,s_1,s_2,s_3) = \begin{vmatrix} \phi_1(x_1)\,\chi(s_1) & \phi_1(x_1)\,\chi(s_2) & \phi_3(x_1)\,\chi(s_3) \\ \phi_1(x_2)\,\chi(s_1) & \phi_1(x_2)\,\chi(s_2) & \phi_3(x_2)\,\chi(s_3) \\ \phi_1(x_3)\,\chi(s_1) & \phi_1(x_3)\,\chi(s_2) & \phi_3(x_3)\,\chi(s_3) \end{vmatrix}. \tag{10.5.70}$$

This state too is 4-fold degenerate.

Example 10.5.5: A system of two independent identical spin-1/2 particles are subject to a common one-dimensional harmonic oscillator potential of frequency ω. Both the particles are in the spin-down state characterized by the spin function $\left|s = \frac{1}{2}, m_s = -\frac{1}{2}\right\rangle \equiv \left|\frac{1}{2}, -\frac{1}{2}\right\rangle$. Find the energies and the wave functions of the ground and the second excited states.

Solution: We know from Chapter 3 that, for the one-dimensional harmonic oscillator potential, the single-particle energies, $\mathscr{E}_n$, and the spatial part of the corresponding wave functions, $\phi_n(x)$, are

$$\mathscr{E}_n = \left(n + \frac{1}{2}\right)\hbar\omega, n = 0,1,2,3,\ldots \tag{10.5.71}$$

$$\phi_n(x) = \frac{1}{\sqrt{\sqrt{\pi}2^n n! x_0}} e^{-(x^2/2x_0^2)} H_n(x/x_0), \tag{10.5.72}$$

where $x_0 = \sqrt{\hbar/m\omega}$ and $H_n(x/x_0)$ is the Hermite polynomial of degree n.

The ground state: Since both the fermions are in the same spin state, according to the Pauli principle, they cannot occupy the same energy state. Therefore, the ground state will have one fermion each in the $n = 0$ and $n = 1$ states. The energy of the system will be

$$E^{(0)} = \left(\frac{1}{2} + \frac{3}{2}\right)\hbar\omega = 2\hbar\omega. \tag{10.5.73}$$

Now, the wave function of our system of spin-1/2 particles must be anti-symmetric. Therefore, because both particles are in the same spin state, the spatial part of the wave function has to be anti-symmetric. The ground state wave function is therefore given by

$$\psi^{(0)} = \frac{1}{\sqrt{2}}\left(\phi_0(x_1)\phi_1(x_2) - \phi_0(x_2)\phi_1(x_1)\right)\left|\frac{1}{2}, -\frac{1}{2}\right\rangle. \tag{10.5.74}$$

The first excited state: It will correspond to one particle in the $n = 0$ energy state and the other in the $n = 2$ energy state. The state with $n = 1$ will be empty. As a result, the second excited state will have energy

$$E^{(2)} = \left(\frac{1}{2} + \frac{5}{2}\right)\hbar\omega = 3\hbar\omega, \tag{10.5.75}$$

and its wave function will be

$$\psi^{(2)} = \frac{1}{\sqrt{2}}\left(\phi_0(x_1)\phi_2(x_2) - \phi_0(x_2)\phi_2(x_1)\right)\left|\frac{1}{2}, -\frac{1}{2}\right\rangle. \tag{10.5.76}$$

The particle distribution among the energy states for both the cases are depicted in Figure 10.3.

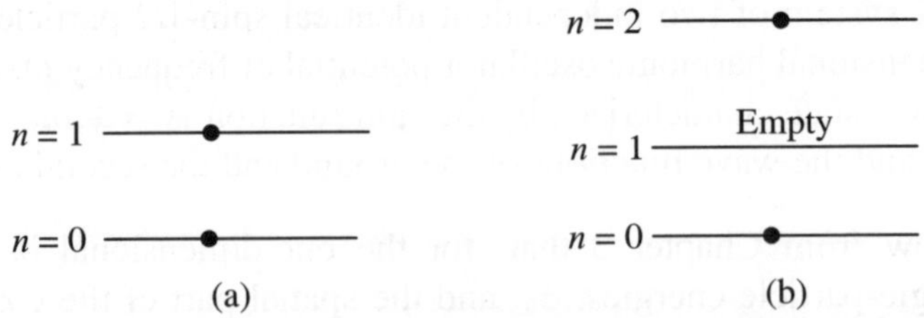

Figure 10.3 Distribution of particles among the energy states for the system of two spin-1/2 fermions in a common one-dimensional harmonic oscillator potential: (a) in the ground state and (b) in the first excited state.

Homework Problems

1. Consider a system of *four* non-interacting distinguishable bosons that are confined to move in a one-dimensional infinite potential well of length a with walls at $x = 0$ and $x = a$. Determine the energies and wave functions of the ground state, the first excited state and the second excited state, if their respective masses satisfy the relation: $m_1 = 2m_2 = 4m_3 = 8m_4$.
2. Three non-interacting distinguishable particles move in a common external one-dimensional harmonic oscillator potential. Find the energies and the wave functions of the ground state, the first excited state, and the second excited state of the system, if their respective masses satisfy the relation: $m_1 = 4m_2 = 8m_3$.
3. Prove that the exchange operator $\hat{P}_{ij}$ that interchanges the particles at $\vec{r}_i$ and $\vec{r}_j$ is hermitian.
4. Prove that two interchangeable operators $\hat{P}_{ij}$ and $\hat{P}_{k\ell}$ commute if the sets of indices (i, j) and (k, ℓ) refer to different pairs of particles.
5. Check whether the following functions are symmetric or anti-symmetric under the exchange of particles:

$$(a)\ \phi(x_1,x_2) = \frac{(x_1 - x_2)^2}{5x_1x_2} + \frac{3x_1 + x_2}{x_1^2 + x_2^2},$$

$$(b)\ \phi(x_1,x_2) = \frac{2}{a}\left[\sin\left(\frac{2\pi x_1}{a}\right)\sin\left(\frac{5\pi x_2}{a}\right) - \sin\left(\frac{5\pi x_1}{a}\right)\sin\left(\frac{2\pi x_2}{a}\right)\right].$$

$$(c)\ \phi(x_1,x_2) = \frac{e^{-(x_1^2+x_2^2)}}{x_1^5 + 10}$$

6. Symmetrize the following wave function and normalize it to find the constant A:

$$\phi(x_1,x_2) = A\sin\left(\frac{5\pi x_1}{a}\right)\sin\left(\frac{10\pi x_2}{a}\right).$$

7. Anti-symmetrize the wave function:

$$\phi(x_1,x_2) = A\sin\left(\frac{2\pi x_1}{a}\right)\sin\left(\frac{5\pi x_2}{a}\right).$$

   and normalize it to find the constant A.
8. Three non-interacting identical bosons (each of mass m) are subject to a common external one-dimensional harmonic oscillator potential. Find the energy levels and wave functions of the ground state, the first excited state and the second excited state of the system.

9. Consider a system of *four* non-interacting identical spin-1/2 particles (each of mass m) that are confined to move in a one-dimensional infinite potential well of length a with walls at $x = 0$ and $x = a$. Determine the energies and wave functions of the ground state and the first three excited states. Draw a figure showing how the particles are distributed among the levels.
10. Two identical particles of spin-1/2 are enclosed in a one-dimensional infinite square well potential of length a with rigid walls at $x = 0$ and $x = a$. Assuming that the two-particle system is in a *singlet* spin state, find the energy levels, the wave functions and the degeneracies corresponding to the three lowest states.
11. Two identical spin-1/2 particles are moving under the influence of a one-dimensional harmonic oscillator potential. Assuming that the two-particle system is in a triplet spin state, find the energy levels, the wave functions and the degeneracies corresponding to the three lowest states.
12. A system of three independent identical spin-1/2 particles are subject to a common one-dimensional harmonic oscillator potential of frequency ω. Both the particles are in the spin-up state characterized by the spin function $\left|s = \frac{1}{2}, m_s = \frac{1}{2}\right\rangle \equiv \left|\frac{1}{2}, \frac{1}{2}\right\rangle$. Find the energies and the wave functions of the first three lowest-lying states.

Chapter 11

Symmetry and Conservation Laws

According to Herman Weyl, by symmetry of an object (or a physical system) we mean the property of the object to appear unchanged after some operation has been done on it. We then say that the object is symmetrical under the given operation. For instance, consider a square. It is indistinguishable after rotations by $\frac{1}{2}\pi$, π and $\frac{3}{2}\pi$ about the axis passing through its geometrical center and perpendicular to its plane (Shown by the dot in the figure). This axis is said to be the axis of symmetry of the square. Note that the angle of rotation, for which the square possesses symmetry, takes on only discrete values. Consequently, it has, as we say, a discrete symmetry. On the other hand, a sphere looks unchanged after all rotations (infinitesimal or finite) about its axis of symmetry. Since the angle of rotation can take continuous values, the rotational symmetry of the sphere is a continuous symmetry.

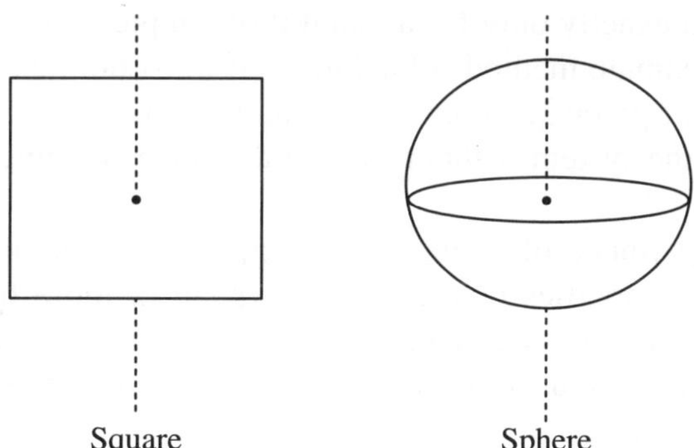

Figure 11.1 Invariance of a square under discrete rotations and that of a sphere under continuous rotations, about the respective axes of symmetry.

It turns out that, for each continuous symmetry of a physical system, there exists a conserved quantity, i.e., a physical characteristic that remains constant as the system evolves in time according to a given dynamical equation. This result is known as the celebrated Nöther theorem. For example, if we place a system of particles in empty space, far from anything that might affect it, it does not make a difference where exactly we put it. There are no preferred locations in empty space; all locations are equivalent. As a consequence, there is a symmetry for a system of particles with respect to translations in empty space. This translational symmetry leads to the law of conservation of the total linear momentum of the system. Similarly, there exists a symmetry for a system of particles in empty space with respect to rotations of the system as a whole because there are no preferred directions in empty space. This rotational symmetry leads to conservation of the total angular momentum of the system. Another important symmetry is the symmetry with respect to shift in time. It turns out that it does not matter when we perform an experiment on an isolated system. The results will be the same. This symmetry with respect to shift in the origin of time gives rise to the law of conservation of energy.

The aforementioned conservation laws hold good in classical as well as in quantum mechanics. However, in quantum mechanics, there appear some new laws of conservation related to the invariance of the system with respect to a change in its quantum mechanical characteristics. For instance, the invariance of the probability density $\psi^*\psi$ under a change of the phase of the wave function by a constant quantity leads to the conservation of charge. Another example is the exchange symmetry (discussed in Chapter 10), related to a strictly quantum mechanical phenomenon of indistinguishability of identical particles, which leads to novel physical consequences that do not have any classical analogue.

The relationship between the symmetries and the conservation laws is important. This is because, besides allowing us to formulate the known conservation laws, it also enables us to discover new laws of conservation that play crucial roles in physics at large. Apart from that, there is another important aspect that requires a serious study of the invariance properties of a quantum mechanical system. It is related to the fact that the Schrödinger equation can be solved exactly only for a handful of simple cases; in all other cases, one has to adhere to approximate methods of solution. It turns out that a detailed study of the symmetry properties of a given quantum system allows us to deduce a number of important physical properties of the system without solving the corresponding Schrödinger equation explicitly.

In view of the importance of symmetry principles in quantum mechanics, we shall discuss the invariance properties of a quantum system in more detail and derive some important results. In doing so, we shall use the language of group theory, which is the appropriate mathematical language for dealing with such properties.

11.1 Transformation of the Wave Function under Coordinate Transformations

In general a coordinate transformation may be either *active* or *passive*. By active transformation we mean the one in which the position vector of the point is changed while the coordinate system remains unchanged. For instance, the position vector $\vec{r}$ of the point P can be rotated. On the other hand, in a passive transformation the position vector of the point remains unchanged, while the basis vectors, that define the coordinate axes of the coordinate system, undergo a change. In our case, we consider the wave function, $\psi(\vec{r})$, of a particle at the point P with the position vector $\vec{r}$ and aim is to deduce its transformation property, when we go over from one system of coordinates to another.

Let $\hat{g}$ stand for the operation with the help of which we move over to a new coordinate system in which the point P is represented by the position vector $\vec{r}\,'$. That is,

$$\vec{r}\,' = \hat{g}\vec{r}. \tag{11.1.1}$$

The inverse transformation is given by

$$\vec{r} = \hat{g}^{-1}\vec{r}\,'. \tag{11.1.2}$$

Since the value of the wave function at P has not changed as result of the coordinate transformation, we must find the same value of ψ for $\vec{r}\,'$ which was there for $\vec{r}$. Hence, we obtain

$$\psi'(\vec{r}\,') = \psi(\vec{r}). \tag{11.1.3}$$

Taking into account (11.1.2), we land up with

$$\psi'(\vec{r}\,') = \psi(\hat{g}^{-1}\vec{r}\,'). \tag{11.1.4}$$

Note, however, that the functional form of the wave function might change due to the coordinate transformation. This fact is indicated by putting *prime* on the wave function for $\vec{r}\,'$. For instance, consider the case of one spatial dimension and let $\psi = A_0 \exp(-ax)$ and $x' = \exp(bx)$, where A_0, a and b are constants, then

$$\psi'(x') = \psi(x) = \psi\left(S^{-1}(x')\right) = A_0 \exp\left(-\frac{a}{b}\, ln(x')\right) = \frac{A_0}{[x']^{(a/b)}}. \tag{11.1.5}$$

We now argue that the wave function $\psi'(\vec{r}\,')$ can be obtained from $\psi(\vec{r}\,')$ by acting on $\psi(\vec{r}\,')$ with a suitable operator $\hat{R}_g$, where the subscript g shows that this operation is induced by the coordinate transformation (11.1.1). Thus, we can write

$$\psi'(\vec{r}\,') = \hat{R}_g\psi(\vec{r}\,'). \tag{11.1.6}$$

Comparing (11.1.4) and (11.1.6) we get the rule according to which $\hat{R}_g$ acts on $\psi(\vec{r}\,')$:

$$\hat{R}_g\psi(\vec{r}\,') = \psi(\hat{g}^{-1}\vec{r}\,'). \tag{11.1.7}$$

Removing the prime without affecting the generality we obtain

$$\hat{R}_g\psi(\vec{r}) = \psi(\hat{g}^{-1}\vec{r}), \tag{11.1.8}$$

which determines the rule according to which the wave function should transform under the coordinate transformation (11.1.1).

Since the physical properties of a system should not depend on the coordinate system chosen to describe the system, the normalization of the wave function must be preserved. Hence,

$$\langle \psi'(\vec{r}\,') \,|\, \psi'(\vec{r}\,') \rangle = \left\langle \psi(\vec{r}\,') \left| \hat{R}_g^\dagger \hat{R}_g \right| \psi(\vec{r}\,') \right\rangle = \langle \psi(\vec{r}) \,|\, \psi(\vec{r}) \rangle = 1. \tag{11.1.9}$$

It follows from (11.1.9) that the operator $\hat{R}_g$ must be a unitary operator:

$$\hat{R}_g^\dagger \hat{R}_g = \hat{R}_g \hat{R}_g^\dagger = \hat{I}, \tag{11.1.10}$$

where $\hat{I}$ is the unit operator.

Consider now the effect of two successive transformations of the coordinates: first by applying the operator $\hat{g}_1$ and then by applying the operator $\hat{g}_2$:

$$\vec{r} \to \vec{r}\,' = \hat{g}_1 \vec{r}, \qquad \vec{r}\,' \to \vec{r}\,'' = \hat{g}_2 \vec{r}\,'. \tag{11.1.11}$$

Let $\hat{R}_{g_1}$ and $\hat{R}_{g_2}$ be the operators, corresponding to $\hat{g}_1$ and $\hat{g}_2$, respectively, which act on the wave function $\psi(\vec{r})$. Then, on one hand

$$\psi''(\vec{r}\,'') = \hat{R}_{g_2} \psi'(\vec{r}\,'') = \hat{R}_{g_2} \hat{R}_{g_1} \psi(\vec{r}\,''). \tag{11.1.12}$$

On the other hand, since the value of the wave function should not change as a result of the coordinate transformations, we must have

$$\psi''(\vec{r}\,'') = \psi(\vec{r}) = \psi(\hat{g}_1^{-1} \hat{g}_2^{-1} \vec{r}\,''). \tag{11.1.13}$$

It follows from (11.1.12) and (11.1.13) that, after the aforementioned two successive coordinate transformations, we shall have

$$\hat{R}_{g_2} \hat{R}_{g_1} \psi(\vec{r}) = \psi(\hat{g}_1^{-1} \hat{g}_2^{-1} \vec{r}). \tag{11.1.14}$$

Equation (11.1.14) gives the transformation of the wave function under two successive coordinate transformations. The point to be noted here is that the order of the g-inverse operators acting on the argument of the wave function is reversed in comparison with the order in which the operators $\hat{R}_{g_1}$ and $\hat{R}_{g_2}$ act on the wave function.

It is obvious now that, if the set of coordinate transformations, represented by the operators $\{\hat{g}\}$, form a group, say g, then the group $G(\hat{R})$, formed by the operators $\{\hat{R}_g\}$, is said to be the representation of g in the linear vector space whose elements are nothing but the set of all possible wave functions of the system.

11.2 Group of Symmetry of the Schrödinger Equation and the Conservation Laws

Consider the Schrödinger equation for a system of n particles

$$i\hbar\frac{\partial\psi}{\partial t} = \hat{H}\psi, \tag{11.2.1}$$

where $\psi = \psi(\vec{r}_1,\vec{r}_2,\ldots,\vec{r}_n,t)$ and $\hat{H}(\vec{r}_1,\vec{r}_2,\ldots,\vec{r}_n)$ are the wave function and the Hamiltonian operator of the system, respectively. Mathematically, the values of the wave function can either be numbers or elements of a multi-dimensional linear vector space. If the wave function belongs to a m-dimensional linear vector space, then we choose an appropriate orthonormal basis, $\{|f_j(\vec{r})\rangle\}, j = 1,2,3,\ldots,m$

$$\langle f_j\,|\,f_k\rangle = \delta_{jk} = \begin{cases} 0, & \text{if } j \neq k, \\ 1, & \text{if } j = k. \end{cases} \tag{11.2.2}$$

in this space and characterize the vector wave function, $|\psi(\vec{r}_1,\vec{r}_2,\ldots,\vec{r}_n,t)\rangle$, by its set of components

$$\psi_j(\vec{r}_1,\vec{r}_2,\ldots,\vec{r}_n,t) = \langle f_j\,|\,\psi\rangle, \quad j = 1,2,3,\ldots,n. \tag{11.2.3}$$

In this case, the Schrödinger equation (11.2.1) is rewritten as

$$i\hbar\frac{\partial\psi_j}{\partial t} = \sum_{k=1}^{m} H_{jk}\psi_k, \tag{11.2.4}$$

where $H_{jk} = \langle f_j|\hat{H}|f_k\rangle$ are the matrix elements of the Hamiltonian operator, $\hat{H}$, in this basis.

By definition, the Schrödinger equation (11.2.1) is said to be invariant with respect to a certain operation, represented by an operator $\hat{R}$, if every solution ψ of this equation is transformed by $\hat{R}$ into a new wave function, $\psi' = \hat{R}\psi$, which also satisfies the Schrödinger equation (11.2.1) and all the required boundary conditions. That is

$$i\hbar\frac{\partial\psi'}{\partial t} = \hat{H}\psi', \implies i\hbar\frac{\partial(\hat{R}\psi)}{\partial t} = \hat{H}\hat{R}\psi \tag{11.2.5}$$

If the set of all such operators $\{\hat{R}\}$ forms a group, say $G(\hat{R})$, then G or any of its subgroups is called the group of symmetry of the Schrödinger equation (11.2.1).

It is usually assumed that the operators $\{\hat{R}\}$, besides being linear, commute with the operation of differentiation with respect to time, that is,

$$\frac{\partial}{\partial t}(\hat{R}\psi) = \hat{R}\frac{\partial\psi}{\partial t}. \tag{11.2.6}$$

Multiplying (12.2.1) by $\hat{R}$ from left and using (12.2.6), we get that

$$i\hbar \frac{\partial(\hat{R}\psi)}{\partial t} = \hat{R}\,(\hat{H}\,\psi). \tag{11.2.7}$$

From (12.1.5) and (12.1.7) we arrive at

$$(\hat{H}\hat{R} - \hat{R}\hat{H})\,\psi = 0. \tag{11.2.8}$$

Since ψ is an arbitrary solution of the Schrödinger equation, it follows from (11.2.8) that

$$\hat{R}\hat{H} - \hat{H}\hat{R} = [\hat{R},\hat{H}] = 0. \tag{11.2.9}$$

Hence, the elements of the group of the Schrödinger equation commute with the Hamiltonian. Clearly, the operator, $\hat{R}^\dagger$, hermitian conjugate to $\hat{R}$, also commute with $\hat{H}$:

$$[\hat{R}^\dagger,\hat{H}] = 0. \tag{11.2.10}$$

As a consequence, the hermitian operators

$$\hat{F}_1 = \frac{1}{2}\,(\hat{R}+\hat{R}^\dagger), \quad \textit{and} \quad \hat{F}_2 = \frac{1}{2i}\,(\hat{R}-\hat{R}^\dagger), \tag{11.2.11}$$

also commute with the Hamiltonian.

The commutativity of the operators $\hat{F}_1$ and $\hat{F}_2$ leads to the following important consequences. In particular, it turns out that $\hat{F}_1$ and $\hat{F}_2$ are conserved in time. In order to establish that, let us prove the following simple theorems.

Theorem 11.2.1: If at a given instant t_0 the wave function $\psi(\vec{r}_1,\vec{r}_2,\ldots,\vec{r}_n,t)$ is an eigenfunction of the operators $\hat{F}_1$ and $\hat{F}_2$,

$$\hat{F}_j\,\psi(\vec{r}_1,\vec{r}_2,\ldots,\vec{r}_n,t_0) = \lambda_j\,\psi(\vec{r}_1,\vec{r}_2,\ldots,\vec{r}_n,t_0), \;\; j = 1,2, \tag{11.2.12}$$

then at any other instant, t, it is again the eigenfunction of these operators with the same eigenvalues, λ_j.

Proof: Consider the function

$$\phi_j(\vec{r}_1,\vec{r}_2,\ldots,\vec{r}_n,t) = \hat{F}_j\,\psi(\vec{r}_1,\vec{r}_2,\ldots,\vec{r}_n,t) - \lambda_j\,\psi(\vec{r}_1,\vec{r}_2,\ldots,\vec{r}_n,t), \;\; j = 1,2. \tag{11.2.13}$$

The operators $\hat{F}_1$ and $\hat{F}_2$, being the symmetry operators, commute both with $\partial/\partial t$ and the Hamiltonian. Hence $\hat{F}_j\,\psi(\vec{r}_1,\vec{r}_2,\ldots,\vec{r}_n,t)$ satisfies the Schrödinger equation. It means that both the terms on the right hand-side of (11.2.13) satisfy the Schrödinger equation.

Therefore, the function $\phi(\vec{r}_1,\vec{r}_2,\ldots,\vec{r}_n,t)$ also satisfies the the Schrödinger equation. Since the Hamiltonian does not depend on time explicitly, the solution to the Schrödinger equation at any $t > t_0$ is uniquely given by

$$\phi_j(\vec{r}_1,\vec{r}_2,\ldots,\vec{r}_n,t) = e^{-\frac{i}{\hbar}(t-t_0)\hat{H}}\phi_j(\vec{r}_1,\vec{r}_2,\ldots,\vec{r}_n,t_0) \quad j=1,2. \tag{11.2.14}$$

Because of the fact that at $t = t_0$ we have $\phi_j(\vec{r}_1,\vec{r}_2,\ldots,\vec{r}_n,t_0) \equiv 0$, the equation (11.2.4) tells us that $\phi_j(\vec{r}_1,\vec{r}_2,\ldots,\vec{r}_n,t) = 0$ at any $t > t_0$. Hence,

$$\hat{F}_j\,\psi(\vec{r}_1,\vec{r}_2,\ldots,\vec{r}_n,t) = \lambda_j\,\psi(\vec{r}_1,\vec{r}_2,\ldots,\vec{r}_n,t), \quad j=1,2 \quad \forall\ t > t_0. \tag{11.2.15}$$

Therefore, $\psi_j(\vec{r}_1,\vec{r}_2,\ldots,\vec{r}_n,t)$ are eigenfunctions of $\hat{F}_j, j=1,2$ at any $t > t_0$ corresponding to the same eigenvalues λ_j.

Theorem 11.2.2: The average values of the operators $\hat{F}_1$ and $\hat{F}_2$, in a given state of a quantum system with a time-independent Hamiltonian, are conserved.

Proof: The time evolution of the average values (in a given state $\psi(t)$) of these operators is governed by the following Ehrenfest's equations:

$$i\hbar\frac{d\langle F_1\rangle}{dt} = \langle[\hat{F}_1,\hat{H}]\rangle, \tag{11.2.16}$$

$$i\hbar\frac{d\langle F_2\rangle}{dt} = \langle[\hat{F}_2,\hat{H}]\rangle, \tag{11.2.17}$$

where

$$\langle F_j\rangle = \frac{\int \psi^*(\hat{F}_j\psi)d^3x}{\int \psi^*\psi d^3x}, \quad j=1,2, \tag{11.2.18}$$

$$\langle[\hat{F}_j,\hat{H}]\rangle = \frac{\int \psi^*([\hat{F}_j,\hat{H}]\psi)d^3x}{\int \psi^*\psi d^3x}, \quad j=1,2. \tag{11.2.19}$$

Since $\hat{F}_1$ and $\hat{F}_2$ commute with the Hamiltonian, we conclude that the average values of the operators $\hat{F}_1$ and $\hat{F}_2$ are conserved in time.

Theorem 11.2.1 and Theorem 11.2.2 show that the hermitian operators $\hat{F}_1$ and $\hat{F}_2$ represent conserved physical quantities. This, in turn means that to every element of the group of symmetry of the Schrödinger equation, there corresponds a law of conservation. However, all these conservation laws are not independent. Therefore, the question arises: How can we, out of the totality of conservation laws provided by a group of symmetry of the Schrödinger equation, find the smallest number of independent conservation laws from which all other conservation laws can be derived? Note that this is very important in the context of continuous groups of symmetry which lead to a continuum of conservation laws.

The following theorem, which is proved using the theory of Lie groups[1], provides the answer to this question.

Theorem 11.2.3: Let the symmetry group, $G(\hat{R})$, of the Schrödinger equation be an m dimensional Lie group with m parameters $\alpha_1, \alpha_2, \alpha_3, \ldots, \alpha_m$ and the corresponding generators $I_1, I_2, I_3, \ldots, I_m$. Then the following hold good: (a) every generator $I_j, j = 1,2,3,\ldots,m$ corresponds to a conserved physical quantity, and (b) all conservation laws related to the elements of the group G follow from the m conservation laws corresponding to the generators.

Proof: Consider a group element $\hat{R}(\alpha), \alpha = \{\alpha_1, \ldots, \alpha_m\}$. From the theory of Lie groups we know that $\hat{R}$ can be represented as

$$\hat{R}(\alpha_1, \ldots, \alpha_m) = \exp\left(\sum_{j=1}^{m} i\alpha_j \hat{I}_j\right), \tag{11.2.20}$$

where the generator $\hat{I}_j$, corresponding to the parameter α_j is calculated as

$$\hat{I}_j = -i \left. \frac{\partial \hat{R}(0,0,\ldots,\alpha_j,\ldots,0,0\}}{\partial \alpha_j} \right|_{\alpha_j = 0}. \tag{11.2.21}$$

Since $\hat{R}$ is an element of the symmetry group of the Schrödinger equation, it commutes with the Hamiltonian. Hence,

$$\begin{aligned} [\hat{R},\hat{H}] &= \left[\sum_{n=0}^{\infty} \frac{1}{n!}\left(\sum_{j=1}^{m} i\alpha_j \hat{I}_j\right)^n, \hat{H}\right] = \left[\left(1 + i\sum_{j=1}^{m} \alpha_j \hat{I}_j + \frac{1}{2!} i^2 \sum_{j,k=1}^{m} \alpha_j \alpha_k \hat{I}_j \hat{I}_k \right.\right. \\ &\left.\left. + \frac{1}{3!} i^3 \sum_{j,k,\ell=1}^{m} \alpha_j \alpha_k \alpha_\ell \hat{I}_j \hat{I}_k \hat{I}_\ell + \ldots \right), \hat{H}\right] = 0. \end{aligned} \tag{11.2.22}$$

For the above relation to hold good, each of the m generators, $\hat{I}_j, j = 1,2,3,\ldots,m$, must individually commute with the Hamiltonian. That is, we must have

$$[\hat{I}_j, \hat{H}] = 0 \qquad j = 1,2,3,\ldots,m. \tag{11.2.23}$$

Equation (11.2.23) shows that each generator of the group of symmetry of the Schrödinger equation, G, leads to a conservation law. Evidently, in view of (11.2.20), all conservation laws related to the elements of the group G follow from the m conservation laws corresponding to the generators.

[1] M. Tinkham, *Group Theory and Quantum Mechanics*, New York: McGraw-Hill Book, 1964.

11.3 Homogeneity of Time and Space: Conservation of Energy and Momentum

(a) *Homogeneity of time*: Consider a quantum mechanical system with constant external conditions. For such a system, all the instants of time are equivalent. This is known as *homogeneity of time*. It means that if $\psi(\vec{r},t)$ is a solution of the Schrödinger equation at a given instant t, and the system is displaced in time by an infinitesimal amount τ, that is, t goes to $t' = \hat{g}_\tau t = t + \tau$, where we have $-\infty < \tau < +\infty$, then the wave functions

$$\psi'(\vec{r},t') = \hat{R}_g\,\psi(\vec{r},t) = \psi(\vec{r},\hat{g}_\tau^{-1} t) = \psi(\vec{r},t-\tau) \tag{11.3.1}$$

will also be the solutions of the Schrödinger equation. In this case, the group of symmetry of the Schrödinger equation is a *one-parameter Lie group*, $G = \{\hat{R}(\alpha = \tau)\}$, of time translations whose generator is calculated as

$$\hat{I}(t)\,\psi(\vec{r},t) = -i\frac{\partial}{\partial\tau}\left(\hat{R}(\tau)\,\psi(\vec{r},t)\right)_{\tau=0} = -i\left(\frac{\partial\psi(\vec{r},t-\tau)}{\partial\tau}\right)_{\tau=0} = i\frac{\partial}{\partial t}\psi(\vec{r},t). \tag{11.3.2}$$

Hence,

$$\hat{I}(t) = i\frac{\partial}{\partial t}. \tag{11.3.3}$$

Since $G = \{\hat{R}(\tau)\}$ is the group of symmetry, the generator $\hat{I} = i(\partial/\partial t)$ must commute with the Hamiltonian and hence, it is conserved.

Let us recollect that in quantum mechanics, the operator for the total energy of a system under constant external conditions is given by $\hat{E} = i\hbar\,(\partial/\partial t)$. If we now look at $\hat{I}$, we come to two important conclusions: (i) The quantum mechanical operator for energy is proportional to the generator of the group of time translations and the constant of proportionality is given by $\hbar$; and (ii) the physical quantity that is conserved due to the invariance of the Schrödinger equation with respect to translations in time of the system, as a whole, is the total energy of the system.

(b) *Homogeneity of space*: Consider now a quantum mechanical system (for simplicity consider a single-particle system with position vector $\vec{r}$) which is subject to spatially homogeneous external conditions. This means that all the points in space are indistinguishable for the system. As a consequence, the potential energy does not change if the system as a whole is translated in space by a constant vector $\vec{a}$, say from $\vec{r}$ to $\vec{r}+\vec{a}$. Then, the set of operators $\{\hat{R}(\vec{a})\}$, whose action on the wave function is defined through

$$\hat{R}(\vec{a})\,\psi(\vec{r},t) = \psi(\hat{g}_{\vec{a}}^{-1}\vec{r},t) = \psi(\vec{r}-\vec{a},t), \tag{11.3.4}$$

forms the symmetry group, G, for the Schrödinger equation. It is a three-parameter Lie group of spatial translations whose parameters are taken to be the Cartesian components ($a_1 = a_x, a_2 = a_y$, and $a_3 = a_z$) of the vector $\vec{a}$. Correspondingly, there are three generators $\hat{I}_j, j = 1,2,3$, that are computed as

$$\begin{aligned}\hat{I}_j(a_j)\psi(\vec{r},t) &= -i\frac{\partial}{\partial a_j}\left(\hat{R}(\vec{a})\psi(\vec{r},t)\right)\big|_{a_j=0} = -i\frac{\partial}{\partial a_j}\left(\psi(\vec{r}-\vec{a},t)\right)\big|_{a_j=0} \\ &= -i\frac{\partial}{\partial a_j}\left(\exp\left(-\vec{a}\cdot\vec{\nabla}\psi(\vec{r},t)\right)\right)\Big|_{a_j=0} = i\frac{\partial}{\partial x_j}\psi(\vec{r},t). \qquad (11.3.5)\end{aligned}$$

Hence,

$$\hat{I}_j(a_j) = i\frac{\partial}{\partial x_j}, \quad j = 1,2,3. \qquad (11.3.6)$$

Since the group, G, of space translations is the symmetry group for the Schrödinger equation, these generators must commute with the Hamiltonian and be conserved.

As we know, in quantum mechanics, $\hat{p}_j = -i\hbar\hat{I}_j, j = 1,2,3$ is the operator for the jth component of the linear momentum. Taking this into account, we come to the conclusions: (i) The quantum mechanical operator for linear momentum is proportional to the generator of the group of space translations and the constant of proportionality is given by $-\hbar$; and (ii) the physical quantity, which is conserved due to the invariance of the Schrödinger equation with respect to spatial translations of the system as a whole in a given direction, is the linear momentum of the system along that direction.

11.4 Isotropy of Space: Conservation of Angular Momentum

Isotropy of space means that there is no preferred direction in space: all directions are equivalent. Consider a one-particle system on which no external force is acting. Because of the isotropy of space, the properties of this system will remain unchanged under arbitrary rotations of the system as a whole. Let us see what are the consequences, if the system is subject to an infinitesimal rotation about a given direction in space.

Consider an infinitesimal rotation by an angle $\delta\phi$, about an axis passing through a point in 3-dimensional space. Such a rotation is characterized by a vector $\vec{\delta\phi}$, whose magnitude coincides with the angle of rotation and whose direction is along the direction of the axis of rotation. If we perform such a rotation, the change in the position vector, $\vec{r}$, of a particle (system) is given by $\delta\vec{r} = [\vec{\delta\phi}\times\vec{r}]$. In other words,

$$\vec{r}\to\vec{r} + [\vec{\delta\phi}\times\vec{r}]. \qquad (11.4.1)$$

In this case, the set of operators, $\hat{R}(\vec{\delta\phi})$, defined by

$$\hat{R}(\vec{\delta\phi})\psi(\vec{r},t) = \psi(\vec{r} - [\vec{\delta\phi}\times\vec{r}],t), \qquad (11.4.2)$$

forms the symmetry group, $G(\vec{\delta\phi})$, of the Schrödinger equation. It is a 3-parameter Lie group of rotations whose parameters are taken to be the Cartesian components ($\delta\phi_1 = \delta\phi_x, \delta\phi_2 = \delta\phi_y$, and $\delta\phi_3 = \delta\phi_z$) of the vector $\vec{\delta\phi}$.

Since

$$-(\vec{\delta\phi} \times \vec{r}) \cdot \vec{\nabla} = - \sum_{j,k,\ell=1}^{3} \varepsilon_{\ell jk} \, \delta\phi_j \, x_k \frac{\partial}{\partial x_\ell}, \tag{11.4.3}$$

where $\varepsilon_{\ell jk}$ are the components of the Levi-Civita tensor density:

$$\varepsilon_{\ell jk} = \begin{cases} +1, & \text{if } (\ell jk) \text{ are even permutations of } (123), \\ -1, & \text{if } (\ell jk) \text{ are odd permutations of } (123), \\ 0, & \text{otherwise,} \end{cases} \tag{11.4.4}$$

the three generators, $\hat{I}_m(\delta\phi)$ ($m = 1,2,3$), of this group are given by

$$\begin{aligned} \hat{I}_m(\delta\phi)\psi(\vec{r},t) &= -i \frac{\partial}{\partial(\delta\phi_m)} \left(\hat{R}(\vec{\delta\phi})\psi(\vec{r},t) \right)\Big|_{\delta\phi_m=0} \\ &= -i \frac{\partial}{\partial(\delta\phi_m)} \left(e^{-(\vec{\delta\phi}\times\vec{r})\cdot\vec{\nabla}} \psi(\vec{r},t) \right)\Big|_{\delta\phi_m=0} \\ &= i \sum_{k,\ell=1}^{3} \varepsilon_{\ell mk} \, x_k \frac{\partial}{\partial x_\ell} \psi(\vec{r},t) = i \sum_{k,\ell=1}^{3} \varepsilon_{mk\ell} \, x_k \frac{\partial}{\partial x_\ell} \psi(\vec{r},t). \end{aligned} \tag{11.4.5}$$

Hence,

$$\hat{I}_m(\delta\phi_m) = i \sum_{k,\ell=1}^{3} \varepsilon_{mk\ell} \, x_k \frac{\partial}{\partial x_\ell}, \quad m = 1,2,3. \tag{11.4.6}$$

Recollecting that the quantum mechanical operator for the angular momentum, $\hat{\vec{L}}$, is given by

$$\hat{\vec{L}} = \hat{\vec{r}} \times \hat{\vec{p}}. \tag{11.4.7}$$

Or,

$$\hat{L}_m = \sum_{k,\ell=1}^{3} \varepsilon_{mk\ell} \, x_k \, \hat{p}_\ell, \tag{11.4.8}$$

we conclude that: (i) the components of the angular momentum operator are also proportional to the corresponding generators of the group of rotations in three dimensional space, and the constant of proportionality is given by $-\hbar$, that is, $\hat{L}_m = -\hbar \hat{I}_m$, $m = 1,2,3$, and (ii) the physical quantity, which is conserved due to the invariance of the Schrödinger equation with respect to the rotation of the system as a whole about a given direction in space, is the component of the angular momentum along that direction.

Comment: *Note that Quantum Mechanics does not give a general prescription for constructing operators corresponding to an arbitrary function of classical variables. In this context, the above analysis shows that the relevant operators in quantum mechanics have their origin in the invariance of the classical analogue of the quantum mechanical system, under the continuous groups of time translations, space translations and rotations which, in turn represent homogeneity of time, homogeneity of space and the isotropy of space, respectively. The individual operators are proportional to the corresponding generators with position operator being proportional to the generator of the group of translations in the momentum space.*

11.5 Symmetry of the Hamiltonian and Degeneracy

We have seen that in many of the problems in two or three spatial dimensions the energy eigenvalues are degenerate. For instance, all the stationary states of the hydrogen atom, barring the ground state with $n = 1$, are degenerate. While talking about degeneracy we did mention that it was the result of some underlying symmetry of the Hamiltonian. Our main aim in this section will be to understand the cause of degeneracy of energy states of quantum systems.

Consider the time-independent Schrödinger equation

$$\hat{H}|\phi_n\rangle = E_n|\phi_n\rangle \tag{11.5.1}$$

where $\hat{H}$ is the Hamiltonian operator. Let $G(\hat{R}_g)$ be the m-dimensional representation of some symmetry group of the Hamiltonian, where $\hat{R}_g$ corresponds to the element g of the symmetry group. Then, starting with a given eigenfunction $|\phi_n\rangle$ of $\hat{H}$ with the eigenvalue E_n, we can generate m more linearly independent eigenfunctions $\{|\phi_n^1\rangle, |\phi_n^2\rangle, |\phi_n^3\rangle, \ldots, |\phi_n^m\rangle\}$ of the Hamiltonian by acting on $|\phi_n\rangle$ with all the elements of the group G. All these $m+1$ eigenfunctions will correspond to the same energy eigenvalue E_n. It is therefore clear that if $\hat{H}$ has a group of symmetry of order m, each of the energy states of the system has the potential to be $(m+1)$-fold degenerate. Thus degeneracy has its origin in the symmetry properties of the Hamiltonian.

The wave functions corresponding to a degenerate energy state can be used to form an irreducible representation of the symmetry group G in which they serve as basis functions. In general, basis functions belonging to different irreducible representations of G must correspond to different energy eigenvalues. Thus, for each energy eigenvalue E_n, there is a corresponding irreducible representation of the symmetry group G with dimension equal to the degree of degeneracy of E_n. Evidently, an energy level cannot be degenerate unless the symmetry group has an irreducible representation of dimension two or more.

In many cases, it turns out that, for certain values of physical parameters, two or more eigenfunctions, belonging to different irreducible representations, have the same energy eigenvalue. This type of degeneracy, which is not a consequence of any symmetry of the system, is called accidental degeneracy. The degeneracy related to the symmetry properties of the system is usually called the *essential degeneracy.*

To clarify the difference between these two types of degeneracy, let us take up an example[2]. Consider a two level (states *I* and *II*) finite potential well with electrons, in the absence of any magnetic field. It is obvious that each of these levels is two-fold degenerate in the sense that each of them can accommodate two electrons, one with spin 'up' and the other with spin 'down'. *This degeneracy of the energy states is an essential one* because the electronic states with spin up and spin down are indistinguishable in the absence of any external perturbation. If we now switch on an external perturbation in the form of a magnetic field, B, the degeneracy is removed and each of the energy states *I* and *II* split up into two. The separation between the sub-levels (I_1, I_2) and (II_1, II_2) is proportional to the strength of the magnetic field B. As a consequence, for a particular value of the magnetic field B, the energies of the sub-levels I_2 and II_1 will coincide (point P in the Fig. 11.2). The two-fold degeneracy at the point P is an accidental one because it does not have its origin in any symmetry property of the original Hamiltonian.

In order to throw some more light on accidental degeneracy, let us take up another example[3]. Consider an infinite rectangular well potential with sides a, b, and c (also referred to as rectangular box). We know that the stationary state wave functions and the energies are given by

$$\phi_{n_x n_y n_z}(x,y,z) = \sqrt{\frac{1}{8abc}} \sin\left(\frac{n_x \pi x}{a}\right) \sin\left(\frac{n_y \pi y}{b}\right) \sin\left(\frac{n_z \pi z}{c}\right), \tag{11.5.2}$$

$$E_{n_x n_y n_z} = \frac{\pi^2 \hbar^2}{2m}\left(\frac{n_x^2}{a^2} + \frac{n_y^2}{b^2} + \frac{n_z^2}{c^2}\right), \tag{11.5.3}$$

where n_x, n_y and n_z are nonzero integers.

Unless at least two of the sides a, b, and c have integer values, there is no degeneracy. In the opposite case, however, there is degeneracy. For example, if $a = p$ and $b = q$, where p and q are integers, then

$$E_{(5p)(5q)n_z} = E_{(7p)qn_z} \tag{11.5.4}$$

Obviously, this type of parameter-dependent degeneracy is accidental and not essential.

So far we have talked about continuous symmetries of the Hamiltonian and the degeneracy related to them. In such symmetries finite transformations can be realized by

[2] A.W. Joshi, Elements of Group Theory for Physicists, John Wiley and Sons Ltd., p. 177, 1988.

[3] C. A. Hollingsworth, Accidental Degeneracies of the Particle in a Box, J. Chem. Edu., 67, No. 12, p. 999, 1990.

successive infinitesimal transformations. There are, however, symmetry operations that cannot be obtained in this way, for instance, space inversion, lattice translation and time reversal. They fall into the category of the so-called discrete transformations. In what follows, we shall discuss the degeneracies related to such discrete symmetries of the Hamiltonian, if it does exist.

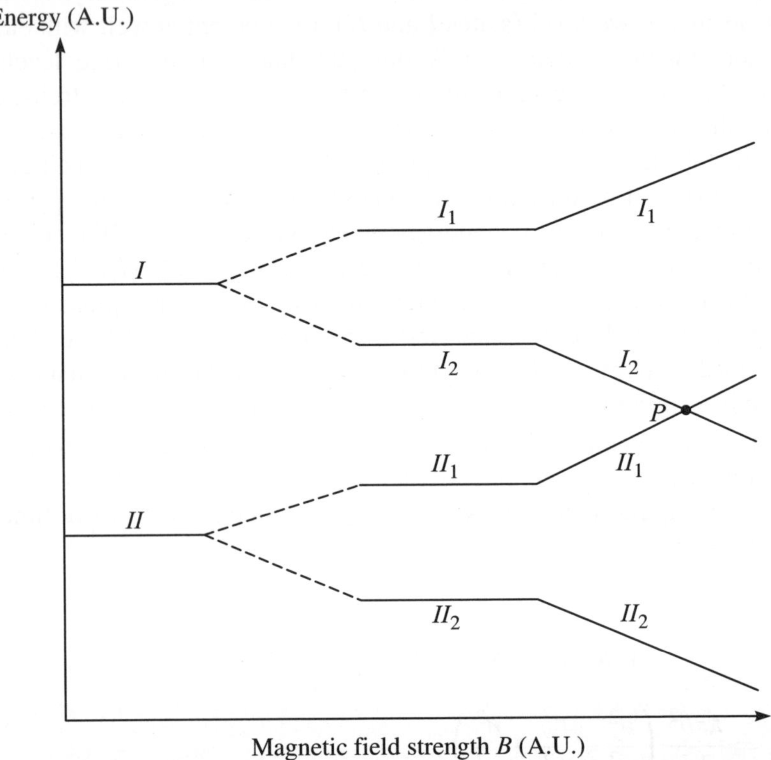

Figure 11.2 The schematic illustration of double degeneracy of energy states: The two-fold degeneracy of the states I and II is essential, while the double degeneracy of energy states at the point P is accidental.

11.6 Space Inversion Symmetry

We have seen in Chapter 2 that the space inversion transformation consists of reflecting the coordinate system with respect to the origin

$$x \to -x, \quad y \to -y, \quad z \to -z. \tag{11.6.1}$$

This is a discrete operation, i.e. the operation which cannot be composed of infinitesimal operations. As a consequence, it does not possess any generators. It is accomplished by the so-called parity operator $\hat{\mathscr{P}}$ which is hermitian as well as unitary and it can have only two eigenvalues $\lambda = \pm 1$. There we also discussed about the properties of the eigenfunctions of the parity operator. Let us, following *Sakurai*, reformulate the the entire thing in terms of the eigenkets of the position and momentum operators.

Let us keep the coordinate system fixed and assume the parity operator to act on the elements of the Hilbert space of states of a quantum system. In other words, given a state $|\psi\rangle$, we introduce a space-inverted state obtained by acting on $|\psi\rangle$ with the parity operator $\hat{\mathscr{P}}$:

$$|\psi\rangle \to \hat{\mathscr{P}}|\psi\rangle. \tag{11.6.2}$$

We require $\hat{\mathscr{P}}$ to be norm preserving. So, it is unitary: $\hat{\mathscr{P}}^\dagger = \hat{\mathscr{P}}^{-1}$. We also require the average value of the position operator $\hat{\vec{r}}$ to satisfy

$$\left\langle \psi \middle| \hat{\mathscr{P}}^\dagger \hat{\vec{r}} \hat{\mathscr{P}} \middle| \psi \right\rangle = -\langle \psi|\hat{\vec{r}}|\psi\rangle \quad \forall |\psi\rangle. \tag{11.6.3}$$

It then follows from (11.6.3) that

$$\hat{\mathscr{P}}^\dagger \hat{\vec{r}} \hat{\mathscr{P}} = -\hat{\vec{r}}. \tag{11.6.4}$$

Taking into account that the parity operator is unitary, we conclude that the position operator anticommutes with the parity operator:

$$[\hat{\mathscr{P}}, \hat{\vec{r}}]_+ = 0. \tag{11.6.5}$$

We now wish to find the result of action of the parity operator on the eigenvectors of the position operator. Let $|\vec{r}\,'\rangle$ be the position eigenket:

$$\hat{\vec{r}}|\vec{r}\,'\rangle = r'|\vec{r}\,'\rangle. \tag{11.6.6}$$

Then, we have

$$\hat{\vec{r}} \hat{\mathscr{P}} |\vec{r}\,'\rangle = -\hat{\mathscr{P}} \hat{\vec{r}} |\vec{r}\,'\rangle = -r' \hat{\mathscr{P}} |\vec{r}\,'\rangle. \tag{11.6.7}$$

It follows from (11.6.7) that $\hat{\mathscr{P}}|\vec{r}\,'\rangle$ is an eigenket of the position operator with eigenvalue $-r'$. Hence, $\hat{\mathscr{P}}|\vec{r}\,'\rangle$ must be proportional to $|-\vec{r}\,'\rangle$:

$$\hat{\mathscr{P}}|\vec{r}\,'\rangle = e^{i\alpha}|\vec{r}\,'\rangle, \tag{11.6.8}$$

where $e^{i\alpha}$ is the phase factor and α a real constant. The phase factor is usually taken to be unity. We thus have the rule according to which the parity operator acts on the position eigenket:

$$\hat{\mathscr{P}}|\vec{r}\,'\rangle = |-\vec{r}\,'\rangle. \tag{11.6.9}$$

Applying the parity operator once more on the position eigenket, we regain the initial state

$$\hat{\mathscr{P}}^2 |\vec{r}'\rangle = \hat{\mathscr{P}} |-\vec{r}'\rangle = |\vec{r}'\rangle. \tag{11.6.10}$$

As a consequence, we have

$$\hat{\mathscr{P}}^2 = \hat{I}. \tag{11.6.11}$$

Thus, the parity operator is not only unitary, it is also hermitian

$$\hat{\mathscr{P}}^\dagger = \hat{\mathscr{P}}^{-1} = \hat{\mathscr{P}}. \tag{11.6.12}$$

Since $\hat{\mathscr{P}}^2 = \hat{I}$, the parity operator can have only two eigenvalues $\lambda = \pm 1$.

Let us now deduce the result of the action of the parity operator on the momentum eigenkets. For that we require the operation of infinitesimal space translation followed by space inversion to be equivalent to the operation of space inversion followed by infinitesimal space translation in the opposite direction. So, if $\hat{D}(\vec{d}r)$ stands for the generator of infinitesimal translation, then our requirement boils down to the following equality

$$\hat{\mathscr{P}}\hat{D}(\vec{d}r) = \hat{D}(-\vec{d}r)\hat{\mathscr{P}}. \tag{11.6.13}$$

We know that the momentum operator in quantum mechanics is proportional to the generator of spatial translation. Therefore, using the substitution

$$\hat{D}(\vec{d}r) = 1 - \frac{i}{\hbar}\hat{\vec{p}}, \tag{11.6.14}$$

in the aforementioned equation, we obtain

$$\hat{\mathscr{P}}\left(1 - \frac{i}{\hbar}\hat{\vec{p}}\right) = \left(1 + \frac{i}{\hbar}\hat{\vec{p}}\right)\hat{\mathscr{P}}. \tag{11.6.15}$$

It follows from (11.6.15) that the momentum operator anticommutes with the parity operator

$$[\hat{\mathscr{P}}, \hat{\vec{p}}]_+ = 0. \tag{11.6.16}$$

Also,

$$\hat{\mathscr{P}}^\dagger \hat{\vec{p}} \hat{\mathscr{P}} = -\hat{\vec{p}}. \tag{11.6.17}$$

So far as the operator of orbital angular momentum, $\hat{\vec{L}} = \hat{\vec{r}} \times \hat{\vec{p}}$ is concerned it should commute with the parity operator because both, the position and the momentum operators are odd with respect to inversion. Thus, we have

$$[\hat{\mathscr{P}}, \hat{\vec{L}}] = 0, \text{ or } \hat{\mathscr{P}}^\dagger \hat{\vec{L}} \hat{\mathscr{P}} = \hat{\vec{L}}. \tag{11.6.18}$$

The spin operator, $\hat{\vec{S}}$ also transforms as $\hat{\vec{L}}$:

$$[\hat{\mathscr{P}}, \hat{\vec{S}}] = 0, \text{ or } \hat{\mathscr{P}}^\dagger \hat{\vec{S}} \hat{\mathscr{P}} = \hat{\vec{S}}. \tag{11.6.19}$$

***Wave functions under space inversion*:** Now we wish to look at the parity property of wave functions. Consider a spinless particle in the state $|\psi\rangle$. Let $\psi(\vec{r}\,')$ be the wave function of particle in the state $|\psi\rangle$. We know from Chapter 4 that

$$\psi(\vec{r}\,') = \langle \vec{r}\,' | \psi \rangle, \tag{11.6.20}$$

where we have chosen the basis in the Hilbert space consisting of the eigenkets, $\{|\vec{r}\,'\rangle\}$, of the position operator. The wave function of the space-inverted state, represented by the ket $\hat{\mathscr{P}}|\psi\rangle$ will be

$$\langle \vec{r}\,' | \hat{\mathscr{P}} | \psi \rangle = \langle -\vec{r}\,' | \psi \rangle = \psi(-\vec{r}\,'), \tag{11.6.21}$$

where we have taken into account (11.6.9) and the fact that $\hat{\mathscr{P}}$ is hermitian. Equivalently, we have

$$\hat{\mathscr{P}} \psi(\vec{r}\,') = \psi(-\vec{r}\,'). \tag{11.6.22}$$

Suppose that $|\psi\rangle$ is a parity eigenstate, i.e.,

$$\hat{\mathscr{P}} |\psi\rangle = \pm |\psi\rangle. \tag{11.6.23}$$

Then the space-inverted wave function satisfies

$$\psi(-\vec{r}\,') = \langle \vec{r}\,' | \hat{\mathscr{P}} | \psi \rangle = \pm \langle \vec{r}\,' | \psi \rangle = \pm \psi(\vec{r}\,'). \tag{11.6.24}$$

If $\psi(-\vec{r}\,') = \psi(\vec{r}\,')$, the wave function is said to have even parity, while, if $\psi(-\vec{r}\,') = -\psi(\vec{r}\,')$, it is said to have odd parity. Thus the eigenfunctions of the parity operator have definite parity: they are either symmetric or antisymmetric functions of coordinates.

It is worth mentioning here that not all wave functions have parity properties in the aforementioned sense, i.e., not all wave functions are eigenfunctions of the parity operator. For instance, the eigenfunctions of the momentum operator, given by $\exp[i(\vec{p} \cdot \vec{r})/\hbar]$, is neither symmetric nor antisymmetric under space inversion.

So far as the wave functions of the angular momentum operator $\hat{\vec{L}}$ (given by the spherical harmonics $Y_\ell^m(\theta, \varphi)$) are concerned, under space inversion transformation

$$\begin{aligned} r &\to r, \\ \theta &\to (\pi - \theta), \\ \varphi &\to (\varphi + \pi), \end{aligned} \tag{11.6.25}$$

they transform as

$$\hat{\mathscr{P}} Y_\ell^m(\theta,\varphi) = (-1)^\ell Y_\ell^m(\theta,\varphi). \tag{11.6.26}$$

Let us now have a look at the parity properties of energy eigenfunctions. Consider the case when the potential energy operator is invariant under inversion. Then, although $\vec{p}$ changes its sign, the Hamiltonian operator, being quadratic in momentum operator, will be invariant under inversion. It then follows that the parity operator, being a symmetry operator for the Hamiltonian, must commute with it

$$[\hat{\mathscr{P}}, \hat{H}] = 0. \tag{11.6.27}$$

This means that, if $\psi(\vec{r})$ is an eigenfunction of the Hamiltonian with a nondegenerate eigenvalue E, $\hat{\mathscr{P}}\psi$ is also an eigenfunction of the Hamiltonian with the same energy E.

Note that the theorem fails when the non-degeneracy condition for E is not met. For instance, although the Hamiltonian for a free particle is invariant under space inversion and commutes with the parity operator, the eigenfunctions of $\hat{\vec{p}}$ (which are also the eigenfunctions of the Hamiltonian) are not the eigenfunctions of the parity operator. However, since these energy wave functions are two-fold degenerate (with $\exp\left[i(\vec{p}\cdot\vec{r})/\hbar\right]$ and $\exp\left[-i(\vec{p}\cdot\vec{r})/\hbar\right]$ corresponding to the same energy eigenvalue), it is possible to form two linear combinations resulting into $\sin(\vec{p}\cdot\vec{r}/\hbar)$ and $\cos(\vec{p}\cdot\vec{r}/\hbar)$, which are eigenfunctions of the parity operator with eigenvalues -1 and $+1$, respectively.

In general, if the Hamiltonian is invariant under parity transformation and the solution, ψ, of the eigenvalue equation for the Hamiltonian is neither symmetric nor anti-symmetric function of coordinates, one construct the linear combinations

$$\psi_s(\vec{r}) = \frac{1}{2}\left(\psi(\vec{r}) + \hat{\mathscr{P}}\psi(\vec{r})\right) \tag{11.6.28}$$

$$\psi_a(\vec{r}) = \frac{1}{2}\left(\psi(\vec{r}) - \hat{\mathscr{P}}\psi(\vec{r})\right), \tag{11.6.29}$$

which will be eigenfunctions of the parity operator as well as the eigenfunctions of the Hamiltonian with the same energy as that of the state ψ.

11.7 Time Reversal Symmetry and Time Reversal Operator

Many physical systems contain an invariance under the reversal of the direction of flow of time. This is true of classical systems as well as quantum mechanical systems. If for a motion picture of a mechanical system it is not possible to decide whether it is shown in the forward or reverse direction, the system is said to have time-reversal symmetry.

In order to clarify the meaning, for concreteness consider a classical particle in three spatial dimensions described by a Hamiltonian $H(\vec{r},\vec{p})$, where $\vec{r}$ and $\vec{p}$ are the position vector and momentum, respectively. The equations of motion are given by Hamilton's equations

$$\frac{d\vec{r}}{dt} = \frac{\partial H}{\partial \vec{p}}, \quad \frac{d\vec{p}}{dt} = -\frac{\partial H}{\partial \vec{r}}. \tag{11.7.1}$$

The classical notion of time-reversal symmetry, as stated above, is directly related to a symmetry property of the Hamiltonian. Namely, if the Hamiltonian is invariant under the $\vec{p} \to -\vec{p}$,then the equations of motion (11.7.1) are invariant under the transformation

$$(\vec{r},\vec{p},t) \to (\vec{r},-\vec{p},-t). \tag{11.7.2}$$

What this actually means is that, if the pair $(\vec{q}(t),\vec{p}(t))$ describes the trajectory of a possible motion of the particle in the phase space with the initial conditions $(\vec{q}_0,\vec{p}_0)$, then the pair $(\vec{q}(-t),-\vec{p}(-t))$ also does the same but with the initial conditions $(\vec{q}_0,-\vec{p}_0)$. For the configuration space, it will mean that if $\vec{q}(t)$ is a solution of the equations of motion, then $\vec{q}(-t)$ is also a solution of the same equations of motion.

To check whether quantum dynamics is invariant under time reversal or not, consider the time-dependent Schrödinger equation

$$i\hbar\frac{\partial \psi(\vec{r},t)}{\partial t} = -\frac{\hbar^2}{2m}\vec{\nabla}^2\psi(\vec{r},t) + V(\vec{r})\psi(\vec{r},t). \tag{11.7.3}$$

If we replace t by $-t$, the Schrödinger equation goes into

$$-i\hbar\frac{\partial \psi(\vec{r},-t)}{\partial t} = -\frac{\hbar^2}{2m}\vec{\nabla}^2\psi(\vec{r},-t) + V(\vec{r})\psi(\vec{r},-t), \tag{11.7.4}$$

which does not coincide with the previous equation. That means that Schrödinger equation is not invariant under time reversal. If, however, we perform complex conjugate on both sides of (11.7.4), we get

$$i\hbar\frac{\partial \psi^*(\vec{r},-t)}{\partial t} = -\frac{\hbar^2}{2m}\vec{\nabla}^2\psi^*(\vec{r},-t) + V(\vec{r})\psi^*(\vec{r},-t). \tag{11.7.5}$$

Equation (11.7.5) for $\psi^*(\vec{r},-t)$ coincides with the Schrödinger equation (11.7.3) for $\psi(\vec{r},t)$. Thus, the quantum dynamics is invariant under time reversal transformation followed by complex conjugation: $\Theta : (t \to -t) \times Complex\ conjugation$. This fact tells us that even if $|\psi(\vec{r},-t)\rangle$ is not a solution it is possible to find an anti-linear operator that transforms $|\psi(\vec{r},-t)\rangle$ into the time-reversed solution $|\psi_R(\vec{r},t)\rangle = |\psi^*(\vec{r},-t)\rangle$ of the Schrödinger equation.

Furthermore, we want that if the system is in a time-reversed state $|\psi_R(\vec{r},t)\rangle$, the probability of finding it in a state $|\phi_R(\vec{r},t)\rangle$ is equal to the probability of finding it, at the time $-t$, in the state $|\phi(\vec{r},t)\rangle$ when the system is known to be the original state $|\psi(\vec{r},t)\rangle$. In fact, since $\langle\phi_R(\vec{r},t)|\psi_R(\vec{r},t)\rangle = \langle\phi(\vec{r},-t)|\psi(\vec{r},-t)\rangle^*$,

$$|\langle\phi_R(\vec{r},t)|\psi_R(\vec{r},t)\rangle|^2 = |\langle\phi(\vec{r},-t)|\psi(\vec{r},-t)\rangle|^2. \tag{11.7.6}$$

Keeping all this in mind, let us define the time reversal operator $\hat{\Theta}$ by

$$|\psi_R(\vec{r},t)\rangle = \hat{U}|\psi(\vec{r},-t)\rangle^* = \hat{U}\hat{K}|\psi(\vec{r},-t)\rangle = \hat{\Theta}|\psi(\vec{r},-t)\rangle, \tag{11.7.7}$$

where $\hat{U}^\dagger\hat{U} = \hat{U}\hat{U}^\dagger = \hat{I}$ and $\hat{K}$ is the complex conjugation operator. The operator $\hat{\Theta} = \hat{U}\hat{K}$ is antiunitary, i.e., it is anti-linear and preserves the norm:

$$\left(\langle\psi\hat{\Theta}^\dagger\right)\left(\hat{\Theta}\psi\rangle\right) = \langle\psi|\psi\rangle. \tag{11.7.8}$$

Note that the effect of the operator $\hat{K}$ depends on the representation used.This is because, in a chosen basis for a particular representation, the state vector is represented by its coefficients. Then, by definition $|\psi\rangle^*$ is the vector obtained by taking, in the same basis, the complex conjugate of these coefficients. Thus, if

$$|\psi\rangle = \sum_i |\phi_i\rangle\langle\phi_i|\psi\rangle, \tag{11.7.9}$$

in an orthonormal basis $\{|\phi_i\rangle\}$, then

$$|\psi\rangle^* = \sum_i |\phi_i\rangle\langle\phi_i|\psi\rangle^*. \tag{11.7.10}$$

Also, The operator, $\hat{O}^*$ complex conjugate to the operator $\hat{O}$ is defined in this basis as

$$\hat{O}^* = \sum_{i,j} |\phi_i\rangle\langle\phi_i|\hat{K}|\phi_j\rangle^*\langle\phi_j|. \tag{11.7.11}$$

Further, since the operation of complex conjugation carried out twice is equivalent to the identity operation, $\hat{K}^2 = \hat{I}$, and therefore $\hat{K}^{-1} = \hat{K}$. As a consequence, we have $\hat{\Theta}^{-1} = \hat{K}\hat{U}^\dagger$. Note that the unitary operator $\hat{U}$ depends on the nature of the Hamiltonian and like $\hat{K}$ depends on the representation used for the wave function.

Recollecting the effect of time reversal in classical mechanics, we require the transformation properties of the operators $\hat{\vec{r}}$ and $\hat{\vec{p}}$, under time reversal, to be such that

$$\langle \psi(\vec{r},-t)|\hat{\vec{r}}|\psi(\vec{r},-t)\rangle = \langle \psi_R(\vec{r},t)|\hat{\vec{r}}|\psi_R(\vec{r},t)\rangle, \tag{11.7.12}$$

$$\langle \psi(\vec{r},-t)|\hat{\vec{p}}|\psi(\vec{r},-t)\rangle = -\langle \psi_R(\vec{r},t)|\hat{\vec{p}}|\psi_R(\vec{r},t)\rangle. \tag{11.7.13}$$

These in turn require that

$$\hat{\Theta}\hat{\vec{r}}\hat{\Theta}^{-1} = \hat{\vec{r}}, \quad \hat{\Theta}\hat{\vec{p}}\hat{\Theta}^{-1} = -\hat{\vec{p}}. \tag{11.7.14}$$

Let us show that the relations given by (11.7.14) do preserve (11.7.12) and (11.7.13). We have

$$\langle \psi(\vec{r},-t)|\hat{\vec{r}}|\psi(\vec{r},-t)\rangle = (\langle \psi(\vec{r},-t)|\hat{\Theta}^{-1})\,\hat{\vec{r}}\,(\hat{\Theta}|\psi(\vec{r},-t)\rangle) = \langle \psi_R(\vec{r},t)|\hat{\vec{r}}|\psi_R(\vec{r},t)\rangle, \tag{11.7.15}$$

$$\langle \psi(\vec{r},-t)|\hat{\vec{p}}|\psi(\vec{r},-t)\rangle = -(\langle \psi(\vec{r},-t)|\hat{\Theta}^{-1})\,\hat{\vec{p}}\,(\hat{\Theta}|\psi(\vec{r},-t)\rangle) = -\langle \psi_R(\vec{p},t)|\hat{\vec{p}}|\psi_R(\vec{r},t)\rangle. \tag{11.7.16}$$

It is easy to check that the fundamental commutation relation $[\hat{x}_i, \hat{p}_j] = i\hbar\,\delta_{ij}$ remains invariant under the transformation (11.7.14).

So far as the operator for the orbital angular momentum is concerned, it transforms as

$$\hat{\Theta}\hat{\vec{L}}\hat{\Theta}^{-1} = -\hat{\vec{L}}, \tag{11.7.17}$$

and anticommutes with the time reversal operator: $[\hat{\Theta}, \hat{\vec{L}}]_+ = 0$. Also, this transformation law preserves the commutation relation for the components of the angular momentum operator $[\hat{L}_i, \hat{L}_j] = i\hbar\,\varepsilon_{ijk}\hat{L}_k$.

The spin operator, $\hat{\vec{S}}$, being intrinsic angular momentum, transforms the same way as $\hat{\vec{L}}$ transforms

$$\hat{\Theta}\hat{\vec{S}}\hat{\Theta}^{-1} = -\hat{\vec{S}}, \tag{11.7.18}$$

and anticommutes with the time reversal operator: $[\hat{\Theta}, \hat{\vec{S}}]_+ = 0$. The commutation relation between the components of the spin operator is also preserved: $[\hat{S}_i, \hat{S}_j] = i\hbar\,\varepsilon_{ijk}\hat{S}_k$.

Time-reversal operator for spinless particles: Consider a particle moving in a time-independent potential $V(\vec{r})$. In the coordinate representation, in view of the fact that $\hat{\vec{r}}$ is real, we have

$$\hat{\Theta}\hat{\vec{r}}\psi = \hat{U}(\hat{K}\hat{\vec{r}}\hat{K}^{-1})\hat{K}\,\psi = \hat{U}\,\hat{\vec{r}}\psi^*. \tag{11.7.19}$$

Since $\hat{\Theta}\hat{\vec{r}}\psi = \hat{\vec{r}}\hat{\Theta}\psi = \hat{\vec{r}}\hat{\Theta}\psi^*$, we conclude that $\hat{U}$ commutes with $\hat{\vec{r}}$.

Furthermore, in the coordinate representation $\hat{\vec{p}} = -i\hbar\vec{\nabla}$ and we have

$$\hat{\Theta}(-i\hbar\vec{\nabla})\psi = i\hbar\hat{U}\vec{\nabla}\,\psi^* = i\hbar\,(\hat{U}\,\vec{\nabla}\hat{U}^{-1})\hat{U}\,\psi^* = i\hbar\vec{\nabla}\hat{U}\,\psi^*. \tag{11.7.20}$$

It then follows that $\hat{U}$ commutes with $\hat{\vec{r}}$ and $\vec{\nabla}$ and so it can neither be a function of the coordinates nor a differential operator of the coordinates. Hence, in the coordinate representation, $\hat{U}$ has to be a multiplicative constant of modulus unity. Since the phase factor can always be consumed in the wave function, we assume $\hat{U}$ to be unity. We thus see that, in the coordinate representation, the time-reversal operator for spinless particles, $\hat{\Theta}$, coincides with the complex conjugation operator, $\hat{K}$, except for a phase factor which can always be taken to be unity. Thus, in coordinate representation we have

$$\hat{\Theta} = \hat{K}. \tag{11.7.21}$$

and

$$\psi_R(\vec{r},t) = \psi^*(\vec{r},-t). \tag{11.7.22}$$

The invariance of the Hamiltonian under time reversal, i.e., under $\vec{p} \to -\vec{p}$ is equivalent to saying that $\hat{H}$ commutes with $\hat{\Theta}$, i.e.,

$$[\hat{\Theta},\hat{H}] = 0. \tag{11.7.23}$$

Remembering that $\hat{\vec{p}} = -i\hbar\vec{\nabla}$, we see that (11.7.23) will hold if $\hat{H}$ does not contain any odd powers of $\hat{\vec{p}}$, i.e., if $\hat{H}$ is real.

***Time-reversal operator for particles with spin*:** In order to determine the time-reversal operator for particles with spin, we start with the transformation properties of angular momentum operator under time-reversal. Since $\hat{\vec{L}}$ and $\vec{S}$ both anticommute with the time reversal operator, the total angular momentum $\vec{J}$ will also anti-commute with $\hat{\Theta}$. So, we have

$$\hat{\Theta}\hat{\vec{J}}\hat{\Theta}^{-1} = -\hat{\vec{J}}. \tag{11.7.24}$$

We have seen that in the standard representation (S_z-representation in which the z-axis is taken to be the axis of projection), the matrices corresponding to $\hat{S}_z$ and $\hat{S}_x$ are real, while the matrix corresponding to $\hat{S}_y$ is purely imaginary. Under the action of the complex conjugation operator, we therefore have

$$\hat{K}\hat{S}_x\hat{K}^\dagger = \hat{S}_x, \qquad \hat{K}\,\hat{S}_z\hat{K}^\dagger = \hat{S}_z, \qquad \hat{K}\hat{S}_y\hat{K}^\dagger = -\hat{S}_y. \tag{11.7.25}$$

Now, $\hat{\Theta} = \hat{U}\hat{K}$ and $\hat{K}^{-1} = \hat{K}$ and therefore we can write $\hat{U}$ as $\hat{U} = \hat{\Theta}\hat{K}$. Using this and the equations (17.7.14), (11.7.18) and (11.7.21), we arrive at

$$\hat{U}\vec{r}\hat{U}^{\dagger} = \hat{\Theta}(\hat{K}\vec{r}\hat{K}^{\dagger})\hat{\Theta}^{-1} = \hat{\Theta}\vec{r}\hat{\Theta}^{-1} = \vec{r}, \tag{11.7.26}$$

$$\hat{U}\vec{p}\hat{U}^{\dagger} = \hat{\Theta}(\hat{K}\vec{p}\hat{K}^{\dagger})\hat{\Theta}^{-1} = -\hat{\Theta}\vec{p}\hat{\Theta}^{-1} = \vec{p}, \tag{11.7.27}$$

$$\hat{U}\hat{S}_x\hat{U}^{\dagger} = \hat{\Theta}(\hat{K}\hat{S}_x\hat{K}^{\dagger})\hat{\Theta}^{-1} = \hat{\Theta}\hat{S}_x\hat{\Theta}^{-1} = -\hat{S}_x, \tag{11.7.28}$$

$$\hat{U}\hat{S}_y\hat{U}^{\dagger} = \hat{\Theta}(\hat{K}\hat{S}_y\hat{K}^{\dagger})\hat{\Theta}^{-1} = -\hat{\Theta}\hat{S}_y\hat{\Theta}^{-1} = \hat{S}_y, \tag{11.7.29}$$

$$\hat{U}\hat{S}_z\hat{U}^{\dagger} = \hat{\Theta}(\hat{K}\hat{S}_z\hat{K}^{\dagger})\hat{\Theta}^{-1} = \hat{\Theta}\hat{S}_z\hat{\Theta}^{-1} = -\hat{S}_z. \tag{11.7.30}$$

Since $\hat{U}$ commutes with both $\hat{\vec{r}}$ and $\hat{\vec{p}}$, $\hat{U}$ has an effect only on the spin variables of the particle. The equations (11.7.28)-(11.7.30) in fact show that $\hat{U}$ corresponds to a rotation through π about the S_y axis in the spin space of the particle. As a consequence, we obtain

$$\hat{U} = \exp\left(-\frac{i}{\hbar}\pi\hat{S}_y\right), \tag{11.7.31}$$

which leads to

$$\hat{\Theta} = \exp\left(-\frac{i}{\hbar}\pi\hat{S}_y\right)\hat{K}. \tag{11.7.32}$$

For a spin-half particle, $\hat{S}_y = \frac{1}{2}\hbar\hat{\sigma}_y$ and we get that $\hat{U} = -i\hat{\sigma}_y$. Therefore, for a spin-half particle

$$\hat{\Theta} = -i\hat{\sigma}_y\hat{K}. \tag{11.7.33}$$

The above result is easily generalized to a system of n particles

$$\hat{\Theta} = \exp\left(-\frac{i}{\hbar}\pi\hat{S}_{1y}\right)\exp\left(-\frac{i}{\hbar}\pi\hat{S}_{2y}\right)\exp\left(-\frac{i}{\hbar}\pi\hat{S}_{3y}\right)\ldots\exp\left(-\frac{i}{\hbar}\pi\hat{S}_{ny}\right)\hat{K}, \tag{11.7.34}$$

where $\hat{\vec{S}}_i$ is the spin operator of the $i-th$ particle and $\hat{S}_{iy}$ is the y-th component of the spin operator of the $i-th$ particle. Since $\hat{S}_{iy}$ acts on the spin variable of the $i-th$ particle alone, all these operators commute with each other and the order of the factors in (11.7.37) is thus immaterial.

11.8 Kramers' Degeneracy and Kramers' Theorem

According to Kramers, in the absence of any magnetic field, the energy states of a system with odd number of electrons is at least doubly degenerate even in the presence of an external electric field. Since it is a quite general statement of degeneracy, one suspects that this may be related to some symmetry of the Hamiltonian of the quantum system. Wigner showed that this underlying symmetry is time-reversal invariance.

In fact, from (11.7.37), we get

$$\hat{\Theta}^2 = \exp\left(-2\frac{i}{\hbar}\pi\hat{S}_{1y}\right)\exp\left(-2\frac{i}{\hbar}\pi\hat{S}_{2y}\right)\exp\left(-2\frac{i}{\hbar}\pi\hat{S}_{3y}\right)\dots\exp\left(-2\frac{i}{\hbar}\pi\hat{S}_{ny}\right), \tag{11.8.1}$$

where we have taken into account that $\hat{K}^2 = \hat{I}$. Each factor on the right-hand side of (11.8.1) denotes a rotation through 2π about the S_y-axis in the spin space. The ith factor will be equal to $+1$ or -1 depending on whether the spin of the ith particle is an integral multiple or a half-odd-integral multiple of $\hbar$. It then follows that $\hat{\Theta}^2 = \pm 1$, where $+1$ corresponds to the case when the number of particles with half-odd-integral spin is even and -1 corresponds to the case when it is odd.

Now, if ψ is is an eigenfunction of the system, whose Hamiltonian is invariant under time-reversal operation,, then $\hat{\Theta}\psi$ is also an eigenfunction of the system. If ψ is a non-degenerate eigenfunction, $\hat{\Theta}\psi$ must be proportional to ψ. Let $\hat{\Theta}\psi = \alpha\,\psi$, where α is a complex constant. Operating once again by $\hat{\Theta}$, we have

$$\hat{\Theta}^2\psi = \hat{\Theta}(\alpha\,\psi) = \alpha^*\,\hat{\Theta}\,\psi = \alpha^*\,\alpha\,\psi = |\alpha|^2\,\psi. \tag{11.8.2}$$

Thus, if $\hat{\Theta}^2 = +1$, then $|\alpha|^2 = 1$ and α is just a phase factor. But if $\hat{\Theta}^2 = -1$, there is no number α for which $|\alpha|^2 = -1$, so that the eigenfunction $\hat{\Theta}\psi$ and ψ must be linearly independent. Since both ψ and $\hat{\Theta}\psi$ correspond to the same energy eigenvalue, the energy states of the system is at least two-fold degenerate. This is known as *Kramers' degeneracy*. Further, since $\hat{\Theta}^2\,\psi = -\,\psi$ is a multiple of the original wave function ψ, the degeneracy must be *even-fold*. We thus arrive at the following theorem[4]:

Kramers' Theorem: *Every energy level of a system with an odd number of electrons in the presence of any electric field but no magnetic field is even-fold degenerate.*

There is another theorem closely related to Kramer's theorem which states that *the expectation value of the magnetic moment is zero in any non-degenerate state*. It has important consequences in the theory of paramagnetic susceptibilities[5]. Before we end this section, let us note that Kramers' degeneracy is removed by applying an external

[4]H.A. Kramers, Proc. Amsterdam Acad., v 33, p. 959, 1930.

[5]Martin J. Klein, Am. J. Phys., vol. 20, p. 65, 1952.

magnetic field because the latter introduces additional terms in the Hamiltonian which are not invariant under time-reversal.

Homework Problems

1. Which components or combinations of components of the linear momentum $\vec{p}$ and the angular momentum $\vec{L}$ are conserved when a particle is moving in the external potential field of a homogeneous cone?
2. Which components or combinations of components of the linear momentum $\vec{p}$ and the angular momentum $\vec{L}$ are conserved when a particle is moving in the external potential field of an infinite homogeneous helix?
3. Show that the Runge–Lenz vector

$$\vec{M} = \frac{1}{2\mu}\left(\vec{p}\times\vec{L}-\vec{L}\times\vec{p}\right) - \frac{Ze^2}{4\pi\varepsilon_0}\frac{\vec{r}}{r}, \tag{11.8.3}$$

   is conserved for a charged particle moving in a Coulombic potential
4. Which components or combinations of components of the linear momentum $\vec{p}$ and the angular momentum $\vec{L}$ are conserved when a particle is moving in the external potential field of an infinite homogeneous prism?
5. Derive the quantum mechanical operator that generates translational symmetry for a charged particle moving in external homogeneous electric and magnetic fields.

Chapter 12

Relativistic Generalization

According to Einstein's special principle of relativity, the laws of physics must be formulated in a form which is Lorentz invariant, that is, the description should not allow one to differentiate between frames of reference that are moving relative to each other with a constant velocity $\vec{V}$. Therefore, the equations governing the physical laws of the micro-world must also be formulated in the so-called Lorentz invariant/covariant form. In view of this requirement, we shall discuss the relativistic generalization of quantum mechanics. We shall, however, confine ourselves to the discussions of the Klein–Gordon and Dirac equations, which happen to be the starting point of quantum field theory or any other related advanced physical theories. We start with the revision of the basic principles and consequences of the special theory of relativity. We then proceed to derive the aforementioned relativistic equations, find their plane wave solutions and discuss the properties of the latter.

12.1 Lorentz Transformations

Einstein's special theory of relativity (STR) is the generalization of Galilean invariance of Newtonian mechanics to include electromagnetism in its framework. It was based on the basic properties of space and time, namely, homogeneity of space and time, the isotropy of space and the constancy of the speed of light, c, in vacuum in all inertial frames. The result yielded the so-called Lorentz transformations that relate the space and time coordinates, (x,y,z,t) of a point in an inertial frame K with the coordinates, (x',y',z',t'), of the same point in the frame K', which is in rectilinear motion with respect to the frame K along the x direction at a constant speed V. It is assumed, for convenience that the origins O and O' of the two systems coincide at $t = t' = 0$ and that the respective coordinate axes of the two inertial frames are parallel. Such an arrangement of the two inertial frames of interest is known as the *standard configuration*, which is shown in Figure 12.1.

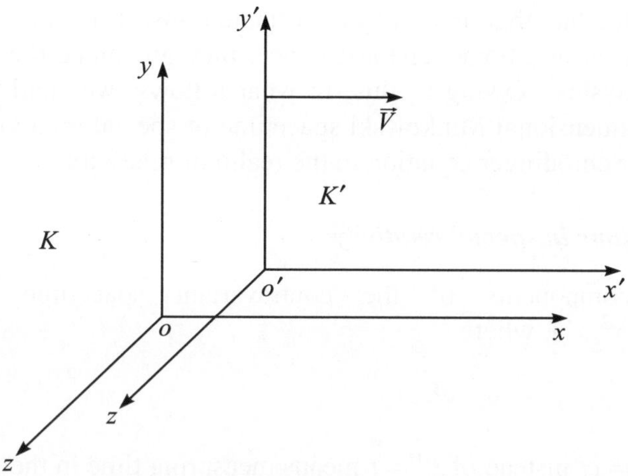

Figure 12.1 The standard configuration of two inertial frames with K' in rectilinear motion with respect to the frame K along the positive x direction at a constant speed V.

The resulting Lorentz transformations are

$$x' = \frac{x - Vt}{\sqrt{1 - \frac{V^2}{c^2}}}, \tag{12.1.1}$$

$$y' = y, \tag{12.1.2}$$

$$z' = z, \tag{12.1.3}$$

$$t' = \frac{t - (V/c^2)x}{\sqrt{1 - \frac{V^2}{c^2}}}. \tag{12.1.4}$$

It is worth mentioning here that, while using the aforementioned Lorentz transformations, one should make sure that the layout of the inertial frames is consistent with the standard configuration. The factor

$$\gamma = \frac{1}{\sqrt{1 - \frac{V^2}{c^2}}} \equiv \frac{1}{\sqrt{1 - \beta^2}}, \tag{12.1.5}$$

with $\beta = V/c$, known as the Lorentz factor (or, relativistic factor), is a measure of the importance of the relativistic effect: *larger the value of γ, greater is the necessity for taking into account the relativistic effect.*

We know from the mathematical theory of special relativity that the Lorentz covariant form of the equations of physics can be obtained if we could write them in tensorial form.

This follows from the fact that under any coordinate transformation, tensors transform in a manner that is linear and homogeneous; hence, they guarantee the form invariance of the equations of physics. Owing to this, in what follows, we shall give a geometrical structure to the 4-dimensional Minkowski spacetime of special relativity that will help us in generalizing the Schrödinger equation to the realm of relativity.

Geometry of spacetime in special relativity

We write the components of the contravariant spacetime 4-vector, x, as $x = (x^\mu) = (x^0, x^1, x^2, x^3)$, where

$$x^0 = ct, \quad x^1 = x, \quad x^2 = y, \quad x^3 = z. \tag{12.1.6}$$

Note that using $x^0 = ct$ instead of $x^0 = t$ means measuring time in the units of length. For instance, $x^0 = 1$ m means the time taken by light in traversing a distance of 1 m in vacuum. The Lorentz transformations can now be written as

$$x'^0 = \gamma(x^0 - \beta x^1), \tag{12.1.7}$$

$$x'^1 = \gamma(x^1 - \beta x^0), \tag{12.1.8}$$

$$x'^2 = x^2, \tag{12.1.9}$$

$$x'^3 = x^3, \tag{12.1.10}$$

Since tensors are usually represented in the matrix form, let us rewrite the aforementioned set of equations in the matrix form as

$$x' = \Lambda x, \tag{12.1.11}$$

where

$$x = \begin{pmatrix} x^0 \\ x^1 \\ x^2 \\ x^3 \end{pmatrix}, \quad x' = \begin{pmatrix} x'^0 \\ x'^1 \\ x'^2 \\ x'^3 \end{pmatrix}, \tag{12.1.12}$$

are the column matrices representing the spacetime 4-vectors in the unprimed and the primed systems of coordinates, respectively, and the Lorentz matrix Λ is given by

$$\Lambda = \begin{pmatrix} \gamma & -\beta\gamma & 0 & 0 \\ -\beta\gamma & \gamma & 0 & 0 \\ 0 & 0 & 1 & 0 \\ 0 & 0 & 0 & 1 \end{pmatrix}. \tag{12.1.13}$$

In the component form, the transformation equations are written as

$$x'^{\mu} = \sum_{\nu=0}^{3} \Lambda^{\mu}{}_{\nu} x^{\nu}. \tag{12.1.14}$$

Note that in the matrix element Λ^{μ}_{ν}, μ enumerates the rows, while ν represents the number of columns.

Four vectors: An arbitrary four vector a^{μ} is defined as a set of four components (a^0, a^1, a^2, a^3), which, under Lorentz transformation, transforms in the same way as the spacetime 4-vector x:

$$a'^0 = \gamma(a^0 - \beta a^1), \tag{12.1.15}$$

$$a'^1 = \gamma(a^1 - \beta a^0), \tag{12.1.16}$$

$$a'^2 = a^2, \tag{12.1.17}$$

$$a'^3 = a^3. \tag{12.1.18}$$

Scalar product of** 4-**vectors: Clearly, the scalar product of any two 4-vectors, say, $a = (a^0, a^1, a^2, a^3)$ and $b = (b^0, b^1, b^2, b^3)$, has to be defined in such a manner that it remains invariant under Lorentz transformations. Accordingly, it is defined as

$$a \cdot b = a^0 b^0 - a^1 b^1 - a^2 b^2 - a^3 b^3. \tag{12.1.19}$$

It is easy to check that it is indeed invariant under Lorentz transformations:

$$\begin{aligned} a'^0 \cdot b'^0 &= \gamma^2 (a^0 - \beta a^1)(b^0 - \beta b^1) - \gamma^2 (a^1 - \beta a^0)(b^1 - \beta b^0) - a^2 b^2 - a^3 b^3 \\ &= \gamma^2 [(1-\beta^2) a^0 b^0 + (1-\beta^2) a^1 b^1] - a^2 b^2 - a^3 b^3 \\ &= a^0 b^0 - a^1 b^1 - a^2 b^2 - a^3 b^3 = a \cdot b. \end{aligned} \tag{12.1.20}$$

In order to keep track of the minus sign in the scalar product, we introduce another type of 4-vector with lower indices:

$$a_{\mu} = (a_0, a_1, a_2, a_3) = (a^0, -a^1, -a^2, -a^3). \tag{12.1.21}$$

The 4-vector with upper indices is called a *contravariant vector*, whereas the one with the lower indices is called a *covariant vector*. The scalar product can now be written as

$$a \cdot b = \sum_{\mu=0}^{3} a^{\mu} b_{\mu} = \sum_{\mu=0}^{3} a_{\mu} b^{\mu}. \tag{12.1.22}$$

In literature, it has been agreed that the scalar product in (12.1.22) is to be written as

$$a \cdot b = a^{\mu} b_{\mu} = a_{\mu} b^{\mu}, \tag{12.1.23}$$

in which one adheres to Einstein's summation convention. According to this convention, if an index is repeated, once as a covariant index and once as a contravariant index, it is summed over from 0 to 3.

As evident, raising and lowering of the indices costs a *minus sign for the spatial parts*, while it *costs nothing for the temporal part*. It is usually done by introducing the so-called metric tensor $g_{\mu\nu}$ or $g^{\mu\nu}$,

$$g_{\mu\nu} = g^{\mu\nu} = \begin{pmatrix} 1 & 0 & 0 & 0 \\ 0 & -1 & 0 & 0 \\ 0 & 0 & -1 & 0 \\ 0 & 0 & 0 & -1 \end{pmatrix}, \tag{12.1.24}$$

which is a tensor of rank 2.

It is easy to check that

$$a_{\mu} = \sum_{\nu=0}^{3} g_{\mu\nu} a^{\nu} \quad \text{and} \quad a^{\mu} = \sum_{\nu=0}^{3} g^{\mu\nu} a_{\nu}. \tag{12.1.25}$$

Note that, using the metric tensor, the scalar product in (12.1.19) can be re-written as

$$a \cdot b = g_{\mu\nu} a^{\mu} b^{\nu} = g^{\mu\nu} a_{\mu} b_{\nu}. \tag{12.1.26}$$

Interval: The most important contribution of Einstein's special theory of relativity to physics is the unification of space and time into one entity called the spacetime. Because of this, the concept of distance between physical events taking place at two spatially separated points in usual Newtonian physics had to be changed accordingly. In special relativity, distance is replaced by the concept called *interval*. It is constructed as follows. Suppose two events, 1 and 2, take place at spacetime points x_1^{μ} and $x_2^{\mu} = x_1^{\mu} + dx^{\mu}$, respectively. Then dx with components dx^{μ}, $\mu = 0,1,2,3$, is the displacement 4-vector between these two points, where $dx^{\mu} = x_2^{\mu} - x_1^{\mu}$. The scalar product of this displacement vector with itself is given by

$$dx \cdot dx = g_{\mu\nu}\, dx^{\mu}\, dx^{\nu} = (dx^0)^2 - (dx^1)^2 - (dx^2)^2 - (dx^3)^2. \tag{12.1.27}$$

Or, in terms of the time difference dt between the events and the spatial separation $|d\vec{r}|$,

$$dx \cdot dx = c^2 dt^2 - dx^2 - dy^2 - dz^2 = c^2 dt^2 - (d\vec{r})^2. \tag{12.1.28}$$

The interval between the given two events is defined as

$$ds = \sqrt{dx_\mu dx^\mu} = c\,dt\sqrt{1 - \frac{\vec{u}^2}{c^2}}, \tag{12.1.29}$$

where $\vec{u} = d\vec{r}/dt$ is the usual three-dimensional velocity of the particle. Note that in spite of the fact that time as well as spatial coordinates change when we go over from one inertial frame to the other, the interval between two events does not change, that is, ds is invariant under Lorentz transformations.

Proper and improper Lorentz transformations: Invariance of the interval under Lorentz transformations yields

$$\Lambda^\mu_\alpha\, g_{\mu\nu}\, \Lambda^\nu_\beta = g_{\alpha\beta}. \tag{12.1.30}$$

Further, we can write (12.1.30) in matrix form as

$$\Lambda^T\, g\, \Lambda = g. \tag{12.1.31}$$

Taking the determinant on both sides of (12.1.31), we get

$$(\det\Lambda)\,(\det\Lambda^T) = (\det\Lambda)^2 = 1, \tag{12.1.32}$$

from where it follows that

$$\det\Lambda = \pm 1. \tag{12.1.33}$$

Transformations for which $\det\Lambda = +1$ are called proper Lorentz transformations. They include boost with a constant velocity along a given direction, three-dimensional rotations and three-dimensional translations. They are continuous transformations in the sense that the parameters of transformations take on continuous values. As a consequence, any finite proper Lorentz transformation can be obtained from the *identity* transformation by successively applying infinitesimal transformations.

Transformations for which $\det(\Lambda) = -1$ are called improper Lorentz transformations. They are discrete transformations and contain spatial and temporal reflections. Since the parameters of improper Lorentz transformations take discrete values, no improper Lorentz transformation can be obtained from the identity transformation by successive infinitesimal transformations.

Elements of relativistic mechanics: We now want to present the basic concepts and important formulae of relativistic mechanics of particles, which will be very useful for making the transition to relativistic quantum mechanics.

Proper time: For the generalization of Newtonian mechanics to the realm of relativity, it is advantageous to introduce a time interval that will be invariant under Lorentz transformation. Since the interval ds and the speed of light c are invariant quantities, we define the proper time interval, $d\tau$, as

$$d\tau = \frac{ds}{c} = dt\sqrt{\left(1 - \frac{d\vec{r}}{c^2 dt}\right)} = dt\sqrt{\left(1 - \frac{\vec{u}^2}{c^2}\right)}. \tag{12.1.34}$$

τ is called the proper time. It is the time measured by the clocks moving with the particle, that is, by the clock that is at rest with respect to the particle. Using this proper time interval, one defines a velocity 4-vector U^μ

$$U^\mu = \frac{dx^\mu}{d\tau} \tag{12.1.35}$$

with components

$$U^0 = \frac{dx^0}{d\tau} = \frac{c}{\sqrt{\left(1 - \frac{\vec{u}^2}{c^2}\right)}}, \quad U_1 = \frac{dx^1}{d\tau} = \frac{u_x}{\sqrt{\left(1 - \frac{\vec{u}^2}{c^2}\right)}},$$

$$U_2 = \frac{dx^2}{d\tau} = \frac{u_y}{\sqrt{\left(1 - \frac{\vec{u}^2}{c^2}\right)}}, \quad U_3 = \frac{dx^3}{d\tau} = \frac{u_z}{\sqrt{\left(1 - \frac{\vec{u}^2}{c^2}\right)}}. \tag{12.1.36}$$

Note that

$$U_\mu U^\mu = U_0 U^0 + \vec{U}^2 = \frac{c^2}{\left(1 - \frac{\vec{u}^2}{c^2}\right)} - \frac{\vec{u}^2}{\left(1 - \frac{\vec{u}^2}{c^2}\right)} = c^2. \tag{12.1.37}$$

The spatial part, $\vec{U}$, of U^μ is called the *proper velocity*. Since the U^μ transforms like a 4-vector under Lorentz transformations, it is advantageous to work with U^μ while dealing with mechanical problems.

Now we are in a position to define the momentum of a particle in relativistic mechanics. The 4-vector of momentum of a particle is defined as

$$p^\mu = m_0 U^\mu = \left(\frac{m_0 c}{\sqrt{(1 - \vec{u}^2/c^2)}}, \frac{m_0 \vec{u}}{\sqrt{(1 - \vec{u}^2/c^2)}}\right), \tag{12.1.38}$$

where m_0 is a numerical quantity, which characterizes the inertial properties of the particle and is called the *rest mass*. Later, we shall see why it is called the rest mass.

Having defined the 4-vector of velocity, we now define the 4-vector of acceleration as

$$a^\mu = \frac{dU^\mu}{d\tau} = \frac{d^2x^\mu}{d\tau^2}. \tag{12.1.39}$$

Newton's equations of motion can now be written as

$$m_0\frac{dU^\mu}{d\tau} = m_0\frac{d^2x^\mu}{d\tau} = F^\mu, \tag{12.1.40}$$

where F^μ is the 4-dimensional generalization of force whose spatial components are the components of the three-dimensional force $\vec{F}$, which appears in Newton's equations of motion. This form of Newton's equation is relativistically covariant in the sense that both sides of the equation transform, under Lorentz transformations, as 4-vectors. What is left now is to determine the physical meaning of various components of this equation and relate them to their non-relativistic counterparts.

Since $U_\mu U^\mu = c^2$, differentiating this expression with respect to τ, we obtain

$$\frac{d(U_\mu U^\mu)}{d\tau} = U_\mu\frac{dU^\mu}{d\tau} = 0. \tag{12.1.41}$$

Multiplying Newton's equations of motion, (12.1.40), with U_μ and summing over μ, we arrive at

$$U_\mu F^\mu = U_0F^0 - \vec{U}\cdot\vec{F} = 0. \tag{12.1.42}$$

Or,

$$F^0 = \frac{\vec{U}\cdot\vec{F}}{U_0} \tag{12.1.43}$$

If we now go back to (12.1.40) and consider its zeroth component, we get

$$m_0\frac{dU^0}{d\tau} = F^0 = \frac{\vec{U}\cdot\vec{F}}{U_0} \tag{12.1.44}$$

Using (12.1.35) and the expression for U^0, we arrive at

$$\frac{d\left(m_0c^2/\sqrt{(1-\vec{u}^2/c^2)}\right)}{dt} = \vec{u}\cdot\vec{F}\sqrt{(1-\vec{u}^2/c^2)}. \tag{12.1.45}$$

In order to see whether we can assign any physical meaning to this equation, let us introduce

$$\vec{f} = \sqrt{(1-\vec{u}^2/c^2)}\vec{F}, \tag{12.1.46}$$

$$m = \frac{m_0}{\sqrt{(1-\vec{u}^2/c^2)}}, \tag{12.1.47}$$

where $\vec{f}$ is the relativistic generalization of the three-dimensional force and the velocity dependent quantity, m, is called relativistic mass. It is easy to check that in the non-relativistic limit $u \ll c$ $(u \approx 0)$, m goes into m_0 and that is why m_0 is called the rest mass.

We now see that the right hand-side of (12.1.45) represents the rate of work done on the particle by the force acting on it. Therefore, invoking the work–energy theorem of classical mechanics, we conclude that the quantity $E = mc^2$ under the differential sign in the numerator on the left hand-side of (12.1.45), must be the total energy of the particle. The fact that this is actually so is confirmed by the non-relativistic limit, $u \ll c$, of the quantities mc^2:

$$mc^2 = m_0c^2\left(1 + \frac{\vec{u}^2}{2c^2} + ...\right) \approx m_0c^2 + \frac{1}{2}m_0\vec{u}^2 \tag{12.1.48}$$

Since the second term on the right-hand side in (12.1.48) is the non-relativistic expression for the kinetic energy of the particle and the first term, m_0c^2, has dimensions of energy; $m_0c^2$is called the rest energy of the particle. Thus, the total energy of the particle is equal to the sum of its rest energy and the kinetic energy.

The spatial part of (12.1.40) reads:

$$\frac{d\left(m_0\vec{u}/\sqrt{(1-\vec{u}^2/c^2)}\right)}{dt} = \vec{F}\sqrt{(1-\vec{u}^2/c^2)}. \tag{12.1.49}$$

Since

$$\vec{F}\sqrt{1 - \frac{\vec{u}^2}{c^2}} = \vec{f}, \tag{12.1.50}$$

we arrive at the following relativistic form of Newton's second law

$$\frac{d(m\vec{u})}{dt} = \vec{f}, \tag{12.1.51}$$

which goes into $d(m_0\vec{u})/dt = \vec{F}$ in the non-relativistic limit.

Thus we see that the zeroth component of (12.1.40) represents the work–energy theorem in relativistic form and establishes the equivalence of energy and mass through the famous Einstein's formula $E = mc^2$, where m is the relativistic mass. On the other hand, the spatial part of (12.1.40) gives the relativistic generalization of Newton's equations of motion.

***Relativistic relation between energy and momentum*:** We have

$$p_\mu p^\mu = (p^0)^2 - (\vec{p})^2 = \frac{m_0^2 c^2}{(1-\vec{u}^2/c^2)} - \frac{m_0^2 \vec{u}^2}{(1-\vec{u}^2/c^2)} = \frac{c^2-\vec{u}^2}{c^2-\vec{u}^2} m_0^2 c^2 = m_0^2 c^2. \quad (12.1.52)$$

On the other hand,

$$p_\mu p^\mu = \frac{E^2}{c^2} - (\vec{p})^2. \quad (12.1.53)$$

$$E^2 - p^2 c^2 = m_0^2 c^4. \quad (12.1.54)$$

This formula allows one to calculate E when $\vec{p}$ is known and vice versa, without ever having to determine the velocity.

12.2 Klein–Gordon Equation

The Klein–Gordon equation was the first relativistic (Lorentz covariant) quantum mechanical model. It serves as an excellent pedagogical tool for the introduction of basic concepts related to relativistic generalization of quantum mechanics. As we shall see later in this Chapter, the analysis of this equation in the general framework of quantum mechanics led to the contradiction with the probabilistic interpretation of the wave function. This fact forced researchers to look for other possible relativistic generalizations of the Schrödinger equation and ultimately led to the discovery of the Dirac equation.

The Hamiltonian of a free particle in nonrelativistic mechanics is given by

$$H = \frac{\vec{p}^2}{2m_0}, \quad (12.2.1)$$

where m_0 is the nonrelativistic mass (rest mass) of the particle and $\vec{p}$ is the 3-momentum of the particle. The usual Schrödinger equation for a single-particle, with no force acting on it, is obtained from the correspondence principle by replacing the energy of the particle, E, and its momentum, $\vec{p}$, by their respective operators in the equation

$$E = H = \frac{\vec{p}^2}{2m_0}. \quad (12.2.2)$$

Recollecting that

$$E \to i\hbar \frac{\partial}{\partial t}, \quad \vec{p} \to -i\hbar \vec{\nabla}, \quad (12.2.3)$$

where $\vec{\nabla}$ is the gradient operator, we obtain the required Schrödinger equation for the free particle as

$$i\hbar\frac{\partial\psi}{\partial t} = -\frac{\hbar^2}{2m_0}\vec{\nabla}^2\psi, \tag{12.2.4}$$

where $\psi(\vec{r},t)$ is its wave function. One can argue in the same heuristic way for a particle of mass m_0 moving in an external field described by the potential $V(\vec{r})$. In this case the Hamiltonian equals the total energy of the particle

$$E = H = \frac{\vec{p}^2}{2m_0} + V(\vec{r}), \tag{12.2.5}$$

and the resulting Schrödinger equation is

$$i\hbar\frac{\partial\psi}{\partial t} = \left(-\frac{\hbar^2}{2m_0}\vec{\nabla}^2 + V(\vec{r})\right)\psi. \tag{12.2.6}$$

We have used the correspondence $\hat{\vec{r}} = \vec{r}$ and the fact that $\hat{V}(\hat{\vec{r}}) = V(\vec{r})$.

If we use the relativistic relation between energy and momentum of a particle of rest mass m_0, given by (12.1.54), we obtain the following second order partial differential equation for the wave function $\psi(\vec{r},t)$:

$$\frac{1}{c^2}\frac{\partial^2\psi}{\partial t^2} - \vec{\nabla}^2\psi + \frac{m_0^2c^2}{\hbar^2}\psi = 0. \tag{12.2.7}$$

It was derived by Schrödinger in 1926. It was also independently proposed by Gordon in 1926 and Klein in 1927. However, in literature, it is known as the Klein–Gordon (KG) equation.

The standard probabilistic interpretation of the wave function is also assumed to be valid for the Klein–Gordon equation. Keeping this in mind, let us derive the continuity equation for the probability density, ρ, and the probability current density, $\vec{j}$, analogous to the case of the non-relativistic Schrödinger equation.

The KG equation for the complex conjugate wave function $\psi^*(\vec{r},t)$ will read:

$$\frac{1}{c^2}\frac{\partial^2\psi^*}{\partial t^2} - \vec{\nabla}^2\psi^* + \frac{m_0^2c^2}{\hbar^2}\psi^* = 0. \tag{12.2.8}$$

If we multiply (12.2.7) by ψ^* from the left, we get

$$\frac{1}{c^2}\psi^*\frac{\partial^2\psi}{\partial t^2} - \psi^*\vec{\nabla}^2\psi + \frac{m_0^2c^2}{\hbar^2}\psi^*\psi = 0. \tag{12.2.9}$$

Similarly, by multiplying (12.2.8) by ψ from the right, we obtain

$$\frac{1}{c^2}\frac{\partial^2\psi^*}{\partial t^2}\,\psi - \vec{\nabla}^2\psi^*\,\psi + \frac{m_0^2c^2}{\hbar^2}\,\psi\,\psi^* = 0. \tag{12.2.10}$$

Subtracting (12.2.9) from (12.2.10), we get:

$$-\frac{1}{c^2}\frac{\partial}{\partial t}\left(\psi^*\frac{\partial\psi}{\partial t} - \psi\frac{\partial\psi^*}{\partial t}\right) + \vec{\nabla}\cdot\left(\psi^*\,\vec{\nabla}\psi - \vec{\nabla}\psi^*\,\psi\right) = 0. \tag{12.2.11}$$

Note that, if we decide to have the expression for the current density, $\vec{j}$, the same as in the case of the nonrelativistic Schrödinger equation, then we shall have to multiply (12.2.11) throughout by $\frac{\hbar}{2m_0 i}$ so that we obtain

$$\frac{\partial\rho}{\partial t} + \vec{\nabla}\cdot\vec{j} = 0, \tag{12.2.12}$$

with

$$\rho(\vec{r},t) = \frac{i\hbar}{2m_0c^2}\left(\psi^*\frac{\partial\psi}{\partial t} - \frac{\partial\psi^*}{\partial t}\psi\right) = \frac{i\hbar}{2m_0c}\left(\psi^*\frac{\partial\psi}{\partial x^0} - \frac{\partial\psi^*}{\partial x^0}\psi\right) \tag{12.2.13}$$

and

$$\vec{j}(\vec{r},t) = \frac{\hbar}{2\,m_0\,i}\left(\psi^*\vec{\nabla}\,\psi - \vec{\nabla}\psi^*\,\psi\right). \tag{12.2.14}$$

Note that in the non-relativistic limit, when we put $i\hbar\partial_t\psi = E\psi$ (E being the total energy of the particle) and take into account that $E \approx mc^2$, ρ reduces to its non-relativistic expression: $\rho = |\psi|^2$.

Since the wave function in the Klein–Gordon equation has only one component, it transforms like a scalar under Lorentz transformations:

$$\psi(x) = \psi(x'), \tag{12.2.15}$$

where $x = (x^0, \vec{r})$ and $x' = \left(x'^0, \vec{r}'\right)$ represent the spacetime coordinates of a point in the inertial frames K and K', respectively. Therefore, the particles described by $\psi(x)$ cannot have any other degrees of freedom except translations in spacetime. Also, since ψ transforms like a scalar under Lorentz transformation, it follows from the group theoretic analysis[1] that it must describe a particle with zero spin.

The Klein–Gordon equation can be written in the relativistically covariant form as

$$\left(\Box + \frac{m_0^2c^2}{\hbar^2}\right)\psi(x) = 0, \tag{12.2.16}$$

[1]M. Tinkham, *Group Theory and Quantum Mechanics*, New York: McGraw-Hill Book, 1964.

where,

$$\Box = \partial_\mu \partial^\mu = \frac{1}{c^2}\frac{\partial^2}{\partial t^2} - \vec{\nabla}^2, \tag{12.2.17}$$

is the D'Alembertian operator and the summation from 0 to 3 over the repeated index is implied. Note that, in (12.2.17), we have $\partial_\mu = \partial/\partial x^\mu, \partial^\mu = g^{\mu\nu}\partial_\nu$.

If we introduce current density 4-vector $j^\mu = (c\rho, \vec{j})$, where c is the speed of light in vacuum, the continuity equation takes the following Lorentz invariant form

$$\partial_\mu j^\mu = 0. \tag{12.2.18}$$

12.3 Properties and Physical Interpretation

I. We have seen above that, similar to the one for the nonrelativistic Schrödinger equation, it is possible to derive the continuity equation for the Klein–Gordon equation. However, there is a problem here with ρ. Since, Klein–Gordon equation is second-order in time derivative, irrespective of the choice of the probability current density, ρ has to contain the first-order time derivative of the wave function. On the other hand, the solution of KG equation requires ψ and $\partial_t\psi$ to be prescribed at some initial moment $t = t_0$. These initial conditions, however, can be prescribed arbitrarily and depend on space coordinates $\vec{r}$. Hence, as ψ evolves in time according to the KG equation, ρ can assume positive as well as negative values. Therefore, due to the fact that the probability density must always be positive definite, ρ for the KG equation, cannot be interpreted as probability density. Because of this difficulty related to the probabilistic interpretation, Klein–Gordon equation was abandoned for many years before Pauli and Weisskopf[2] interpreted KG equation as a classical field equation and formulated its quantized theory. In this formulation, ρ and $\vec{j}$ are interpreted as charge and current densities, respectively, of the particles of the field.

II. *Plane wave solutions of the Klein–Gordon equation*: It can be directly verified by substitution that the Klein–Gordon equation admits plane wave solutions of the form

$$\psi(\vec{r},t) = A_0\, e^{-\frac{i}{\hbar}p_\mu x^\mu} = A_0\, e^{-\frac{i}{\hbar}(p_0 x_0 - \vec{p}\cdot\vec{r})}, \tag{12.3.1}$$

provided

$$p_0 = \frac{E}{c} = \pm\sqrt{\vec{p}^2 + m_0^2 c^2}. \tag{12.3.2}$$

Here, in these equations, A_0 is an arbitrary constant and $p^\mu = \left(\frac{E}{c}, \vec{p}\right)$ is the 4-momentum of the particle. Note that either sign on the right-hand side of (12.3.2) gives a solution. It

[2]W. Pauli and V. Weisskopf, Helv. Phys. Acta, Vol. 7, p. 709, 1934.

means that Klein–Gordon equation allows for solutions with both positive and negative energies. This is a characteristic property of relativistic quantum mechanics.

The occurrence of negative energy solutions should not present any problem for free particles. If initially the particle is in a positive energy state with energy, $E = c\sqrt{\vec{p}^2 + m_0^2 c^2}$, then it will always remain in that state due to the absence of any interaction. In such a state, the probability density is given by

$$\rho = \frac{E}{mc^2} |\psi|^2, \tag{12.3.3}$$

and is clearly positive definite, $\rho \geq 0$, and remains so for all times by virtue of the equations of motion.

In the presence of interaction with external fields, a particle, initially in the positive energy state may make transitions to the negative energy states with $E = -c\sqrt{\vec{p}^2 + m_0^2 c^2}$. According to the quantized theory of Pauli and Weisskopf, such states should be interpreted as particles of negative charge (if positive E corresponds to positive charge) but of positive energy. Transition from a state of positive to one of negative E is interpreted as the production (or annihilation) of a pair of particles of opposite charge.

12.4 Electrically Charged Spin Zero Particle and Interaction with the Electromagnetic Field

The basic description of a charged spin zero particle is carried out in exactly the same manner except that the Hamiltonian is modified by the so-called minimal coupling formalism with the following gauge invariant replacements for the momentum $\vec{p}$ and the total energy E of the particle in SI units (see Chapter 7):

$$\vec{p} \to \vec{p} - e\vec{A}, \tag{12.4.1}$$

$$E \to E - e\Phi, \tag{12.4.2}$$

where $\Phi(\vec{r},t)$ and $\vec{A}(\vec{r},t)$ are the scalar and the vector potentials of the electromagnetic field, respectively. The particle has been assumed to be negatively charged with charge e. The Klein–Gordon equation is now modified to

$$\left(i\hbar\frac{\partial}{\partial t} - e\Phi\right)^2 \psi = c^2\left(-i\hbar\vec{\nabla} - e\vec{A}\right)^2 \psi + m_0^2 c^4 \psi \tag{12.4.3}$$

Using (12.4.3) and the same procedure as before, it can be shown that the continuity equation still holds in the presence of the electromagnetic field with

$$\rho(\vec{r},t) = \frac{i\hbar}{2mc^2}\left(\psi^*\frac{\partial\psi}{\partial t} - \frac{\partial\psi^*}{\partial t}\psi\right) - \frac{e\Phi}{mc^2}\psi^*\psi, \tag{12.4.4}$$

and

$$\vec{j}(\vec{r},t) = \frac{\hbar}{2im}\left(\psi^*\vec{\nabla}\psi - \vec{\nabla}\psi^*\psi\right) - \frac{e}{im}\vec{A}(\psi^*\psi). \tag{12.4.5}$$

The interpretation of the solutions of the Klein–Gordon equation in the presence of an external electromagnetic field is no longer simple. For instance, consider the scattering of a spin zero particle by an external potential that is non-zero for a small time interval ΔT. Clearly, since we are dealing with a real incident particle, the wave function of the incident particle is a superposition of the positive energy solutions of the Klein–Gordon equation. Now, after time ΔT has elapsed, it is possible for the wave function of the scattered particle to have negative energy components due to interaction with the external potential. This means that after scattering, the probability of finding the particle in the negative energy state becomes non-zero to which, a priori, it is not possible to give any physical explanation[3].

It is natural to ask: What will happen, if the external potential is time independent? In this case, the variables will separate to allow for the stationary state solutions of the form

$$\psi(t,\vec{r}) = \phi(\vec{r})\, e^{-\frac{i}{\hbar}Et}, \tag{12.4.6}$$

to exist. For such solutions, the probability density takes the form

$$\rho = \frac{E - e\Phi}{mc^2}\,|\phi|^2. \tag{12.4.7}$$

For instance, as we know, in the case of Coulomb potential when

$$e\Phi(\vec{r}) = -\frac{Ze^2}{4\pi\varepsilon_0\, r}, \tag{12.4.8}$$

the Klein–Gordon equation allows for positive as well as negative energy solutions to exist[4]. Although, in the given case, a particle initially in the positive energy case will remain in that for all times to come, the difficulty arises because of the fact that the probability density (12.4.7) becomes negative for sufficiently small r for which the motion is essentially relativistic and the one-particle interpretation breaks down. In spite of the fact that it is not possible to give an acceptable physical interpretation of the solutions of the Klein–Gordon equation in the presence of an external field, its solutions are physically relevant in the field theoretic interpretation as shown by Pauli and Weisskopf.

12.5 The Dirac Equation

Dirac started with an aim to derive a relativistically covariant equation free from the problem of negative probability encountered by the Klein–Gordon equation. The reason

[3]Silvan S. Schweber, An Introduction to Relativistic Quantum Field Theory, Row, Peterson and Company, New York, 1961.
[4]L.I. Schiff, Quantum Mechanics, McGraw-Hill, 1949.

behind this negative probability density was the appearance of the first order time derivative in the expression for ρ. This was, in turn, related to the fact that the Klein–Gordon equation contained second order time derivatives. Therefore, Dirac's main idea was to avoid the second order time derivative in the differential equation describing a relativistic particle. But then, since in the theory of relativity x, y, z and ct are treated symmetrically, the wave function ψ in the required equation must satisfy a first order differential equation in all four coordinates. Furthermore, the equation must be linear so that the superposition principle of quantum mechanics holds. Apart from that, the correspondence principle also requires that the Klein–Gordon equation be satisfied so that in the limit of large quantum numbers, classical relativity holds and we have the correct relativistic relation between energy and momentum: $E^2 = c^2\vec{p}^2 + m_0^2c^4$.

In general, the first order differential equation (with appropriate dimensional factors) for an N-component wave function

$$\psi = \begin{pmatrix} \psi_1 \\ \psi_2 \\ \psi_3 \\ \cdot \\ \cdot \\ \cdot \\ \psi_N \end{pmatrix}, \tag{12.5.1}$$

of a free particle, where N is yet unspecified, can be written as[5]

$$\frac{1}{c}\frac{\partial\psi}{\partial t} + \sum_{j=1}^{3} \alpha^j \frac{\partial\psi}{\partial x^j} + \frac{imc}{\hbar}\beta\,\psi = 0, \tag{12.5.2}$$

where $\alpha_i, i = 1,2,3$ and β are dimensionless square matrices of order N. Note that homogeneity of spacetime for a free particle requires $\alpha_i, i = 1,2,3$ and β to be independent of the spacetime coordinates x^0, x^1, x^2, x^3. For convenience, introducing the vector matrix

$$\vec{\alpha} = \alpha^1\,\hat{\imath} + \alpha^2\,\hat{\jmath} + \alpha^3\,\hat{k}, \tag{12.5.3}$$

where $\hat{\imath}, \hat{\jmath}$ and $\hat{k}$ are the unit vectors along the coordinate axes, x, y and z, respectively, we rewrite (12.5.2) as

$$\frac{1}{c}\frac{\partial\psi}{\partial t} + \vec{\alpha}\cdot(\vec{\nabla}\,\psi) + \frac{imc}{\hbar}\beta\,\psi = 0. \tag{12.5.4}$$

[5]For convenience in writing the equations, we have dropped the subscript 0 from the symbol m_0 for the rest mass of the particle.

As in the case of the Klein–Gordon equation, we wish to find an expression for the probability density ρ and the probability current density $\vec{j}$ that satisfy the continuity equation (12.2.12). We, therefore define, analogous to the Klein–Gordon equation,

$$\rho = c\psi^\dagger \psi, \tag{12.5.5}$$

where $\psi^\dagger$ is the matrix hermitian conjugate to ψ. Thus, $\psi^\dagger$ is a row matrix, with one row and N columns, whose elements are complex conjugate of the corresponding components of ψ:

$$\psi^\dagger = \begin{pmatrix} \psi_1^* & \psi_2^* & \psi_3^* & \cdot & \cdot & \cdot & \psi_N^* \end{pmatrix}, \tag{12.5.6}$$

where *asterisk* stands for complex conjugation. Taking the hermitian conjugate of (12.5.4), we get the equation satisfied by $\psi^\dagger$:

$$\frac{1}{c}\frac{\partial \psi^\dagger}{\partial t} + (\vec{\nabla}\,\psi^\dagger)\cdot\vec{\alpha}^\dagger - \frac{imc}{\hbar}\,\psi^\dagger\,\beta^\dagger = 0, \tag{12.5.7}$$

where we have used the well-known formula $(AB)^\dagger = B^\dagger A^\dagger$ for any two matrices A and B.

Multiplying (12.5.4) on the left by $\psi^\dagger$, (12.5.7) on the right by ψ and adding, we obtain

$$\frac{1}{c}\left(\psi^\dagger\frac{\partial\psi}{\partial t} + \frac{\partial\psi^\dagger}{\partial t}\psi\right) + \psi^\dagger\vec{\alpha}\cdot(\vec{\nabla}\psi) + (\vec{\nabla}\psi^\dagger)\cdot\vec{\alpha}^\dagger\psi + \frac{imc}{\hbar}\psi^\dagger(\beta-\beta^\dagger)\psi = 0, \tag{12.5.8}$$

If

$$\vec{\alpha} = \vec{\alpha}^\dagger, \quad \beta = \beta^\dagger, \tag{12.5.9}$$

then (12.5.8) takes the form of the continuity equation with the probability density, ρ, defined by (12.5.5) and the probability current, $\vec{j}$, given by

$$\vec{j} = c\,\psi^\dagger\,\vec{\alpha}\,\psi. \tag{12.5.10}$$

Note that the matrices α^k, $k = 1,2,3$ and β must be hermitian; this follows from the fact that the (12.5.4) can be written as

$$i\hbar\frac{\partial\psi}{\partial t} = \hat{H}_D\,\psi, \tag{12.5.11}$$

where the Dirac Hamiltonian $\hat{H}_D$ is given by

$$\hat{H}_D = \left(-i\hbar c\,\vec{\alpha}\cdot\vec{\nabla} + mc^2\,\beta\right) = \left(c\,\vec{\alpha}\cdot(\hat{\vec{p}}) + mc^2\,\beta\right). \tag{12.5.12}$$

Thus, for the Hamiltonian to be hermitian, the matrices $\vec{\alpha} = \{\alpha^k\}, k = 1,2,3$ and β have to be hermitian.

It is quite obvious that we need to know more properties of the $\vec{\alpha}$ and β matrices. They follow by imposing the condition that ψ must also satisfy the Klein–Gordon equation.

Let us act on the equation (12.5.2) with the operator

$$\frac{1}{c}\frac{\partial}{\partial t} - \sum_{k=1}^{3} \alpha^k \frac{\partial}{\partial x^k} - \frac{imc}{\hbar}\beta. \tag{12.5.13}$$

It is easy to check that the result is

$$\frac{1}{c^2}\frac{\partial^2 \psi}{\partial t^2} = \frac{1}{2}\sum_{j=1}^{3}\sum_{k=1}^{3}\left(\alpha^j\alpha^k + \alpha^k\alpha^j\right)\frac{\partial^2 \psi}{\partial x^j \partial x^k} - \frac{m^2c^2}{\hbar^2}\beta^2\,\psi$$
$$+\frac{imc}{\hbar}\sum_{k=1}^{3}\left(\alpha^k\beta + \beta\,\alpha^k\right)\psi = 0. \tag{12.5.14}$$

In deriving this result, we have used the fact that $\partial/\partial x^j$ and $\partial/\partial x^k$ commute and hence, we can symmetrize the product $(\alpha^j\,\partial/\partial x^j)\,(\alpha^k\,\partial/\partial x^k)$. Now, for (12.5.14) to coincide with the Klein–Gordon equation,

$$\frac{1}{c^2}\frac{\partial^2 \psi}{\partial t^2} - \vec{\nabla}^2\,\psi + \frac{m^2c^2}{\hbar^2}\,\psi = 0, \tag{12.5.15}$$

the following relations must hold good:

$$\alpha^j\alpha^k + \alpha^k\alpha^j = 2\delta_{jk} = 2\left\{\begin{array}{ll} 0 & \text{if}, j \neq k \\ 1 & \text{if}, j = k \end{array}\right., \tag{12.5.16}$$

$$\alpha^k\,\beta + \beta\,\alpha^k = 0 \tag{12.5.17}$$

$$\beta^2 = I, \tag{12.5.18}$$

where I is the unit matrix. It follows from (12.5.16) that

$$(\alpha^j)^2 = I, \quad j = 1,2,3. \tag{12.5.19}$$

Theorem 12.5.1: The order (dimension) of Dirac matrices $\alpha^i, i = 1,2,3$ and β must be even.

Proof: From (12.5.9) and the equations (12.5.16)–(12.5.18), it follows that all the 4-Dirac matrices are hermitian, anti-commute with each other and their squares equal the unit matrix. Now, from hermiticity, it follows that the eigenvalues of all the Dirac matrices are real. Since $(\alpha^j)^2 = I, j = 1,2,3$, and $\beta^2 = I$, the eigenvalues of these matrices can be either $+1$ or -1.

Furthermore, since α^k and β anti-commute, we have

$$\alpha^k \beta = (-I)\, \beta\, \alpha^k. \tag{12.5.20}$$

Taking determinants on both sides of (12.5.20), we obtain

$$\det(\alpha^k)\det(\beta) = \det(-I)\det(\beta)\det(\alpha^k) = (-1)^N \det(\beta)\det(\alpha^k). \tag{12.5.21}$$

Since none of the matrices is singular and all of them can have inverses, none of the determinants in (12.6.2) vanishes. This leads to

$$(-1)^N = 1. \tag{12.5.22}$$

The last equation shows that the order (dimension) of the Dirac matrices, $\alpha^j, j = 1,2,3,$ and β, must be even. Q.E.D.

Theorem 12.5.2: Dirac matrices, $\alpha^j, j = 1,2,3,$ and β, are traceless.

Proof: Once more we have from (12.5.17),

$$-\beta = (\alpha^k)^{-1}\, \beta\, \alpha^k. \tag{12.5.23}$$

Taking the trace of both sides of (12.5.23) we obtain

$$-\mathrm{tr}(\beta) = \mathrm{tr}((\alpha^k)^{-1}\, \beta\, \alpha^k) = \mathrm{tr}(\alpha^k\, (\alpha^k)^{-1}\, \beta) = \mathrm{tr}(\beta), \tag{12.5.24}$$

where we have used the property of the trace of the product of matrices, according to which $\mathrm{tr}(ABC) = \mathrm{tr}(BAC) = \mathrm{tr}(CAB)$ for any three matrices A, B and C. Therefore,

$$\mathrm{tr}(\beta) = 0. \tag{12.5.25}$$

Similarly, we can show that

$$\mathrm{tr}(\alpha^k) = 0, \quad k = 1,2,3. \tag{12.5.26}$$

Thus, all Dirac matrices are traceless. Q.E.D.

A possible representation of Dirac matrices: We have shown that the dimension of Dirac matrices must be even. The lowest even dimension is 2. Since we have four distinct Dirac matrices, $\alpha^k, k = 1,2,3,$ and β, while for $N = 2$, we have only three 2×2 linearly independent and anti-commuting matrices in the form of Pauli matrices $\hat{\sigma}^k, k = 1,2,3,$ $N = 2$ cannot be acceptable. Thus, we conclude that the minimum value of N for which the Dirac matrices will satisfy the requirements given by the equations (12.5.16)–(12.5.18) must be 4.

Assuming $N = 4$, we can now construct one of the possible representations of Dirac matrices. Thus, Dirac matrices must have at least four rows and four columns; we shall restrict ourselves to 4×4 matrices. We have seen that the trace of α^k and β must be zero. In addition, since β is hermitian and can always be diagonalized, it is convenient to represent β by a diagonal matrix. This, together with $\text{tr}\beta = 0$ and $\beta^2 = I$, leads to the choice

$$\beta = \begin{pmatrix} I & 0 \\ 0 & -I \end{pmatrix}, \tag{12.5.27}$$

where I is the 2×2 identity matrix.

The rest of the three matrices $\alpha^k, k = 1,2,3$, must anti-commute with β and also be hermitian. These requirements can be fulfilled if we make use of the three Pauli matrices, $\hat{\sigma}^j, j = 1,2,3$ (which anti-commute among themselves) and put

$$\alpha^k = \begin{pmatrix} 0 & \hat{\sigma}^k \\ \hat{\sigma}^k & 0 \end{pmatrix}. \tag{12.5.28}$$

12.6 Relativistically Covariant Form of Dirac Equation

In order to have a more symmetrical and covariant look of the Dirac equation, let us go back to (12.5.2) and multiply it throughout by the matrix β from the left to obtain

$$i\beta \frac{\partial \psi}{\partial x^0} + i \sum_{j=1}^{3} \beta\, \alpha^j \frac{\partial \psi}{\partial x^j} - \frac{mc}{\hbar}\, \psi = 0, \tag{12.6.1}$$

where we have used $\beta^2 = I$. Let us now introduce new matrices by

$$\gamma^0 = \beta = \begin{pmatrix} I & 0 \\ 0 & -I \end{pmatrix}, \tag{12.6.2}$$

$$\gamma^j = \beta\, \alpha^j = \begin{pmatrix} 0 & \hat{\sigma}^j \\ -\sigma^j & 0 \end{pmatrix}, \quad (j = 1,2,3). \tag{12.6.3}$$

In the aforementioned formulae, I is a 2×2 unit matrix and 0 stands for a 2×2 null matrix. Thus, the explicit forms of 4×4 Dirac matrices are

$$\gamma^0 = \begin{pmatrix} 1 & 0 & 0 & 0 \\ 0 & 1 & 0 & 0 \\ 0 & 0 & -1 & 0 \\ 0 & 0 & 0 & -1 \end{pmatrix}, \quad \gamma^1 = \begin{pmatrix} 0 & 0 & 0 & 1 \\ 0 & 0 & 1 & 0 \\ 0 & -1 & 0 & 0 \\ -1 & 0 & 0 & 0 \end{pmatrix}, \tag{12.6.4}$$

$$\gamma^2 = \begin{pmatrix} 0 & 0 & 0 & -i \\ 0 & 0 & i & 0 \\ 0 & i & 0 & 0 \\ -i & 0 & 0 & 0 \end{pmatrix}, \quad \gamma^3 = \begin{pmatrix} 0 & 0 & 1 & 0 \\ 0 & 0 & 0 & -1 \\ -1 & 0 & 0 & 0 \\ 0 & 1 & 0 & 0 \end{pmatrix}, \tag{12.6.5}$$

Usually these matrices are grouped together and written as

$$\gamma^\mu = \{\gamma^0, \gamma^j\} \equiv \{\gamma^0, \vec{\gamma}\}. \tag{12.6.6}$$

The indices on the γs are raised and lowered with the help of the metric tensor ($g_{\mu\nu}$ and $g^{\mu\nu}$):

$$\gamma_\mu = g_{\mu\nu}\gamma^\mu, \quad \gamma^\mu = g^{\mu\nu}\gamma_\mu, \tag{12.6.7}$$

even though they are not components of a 4-vector. With the help of these γ matrices, we can rewrite the Dirac equation (12.5.4) in the following covariant form

$$\left(i\gamma^\mu \partial_\mu - \frac{mc}{\hbar}\right)\psi = 0, \tag{12.6.8}$$

where summation from 0 to 3 over the repeated index μ is understood. This is the relativistically covariant form of the Dirac equation.

Using (12.6.6), we can also write the continuity equation in the following invariant form

$$\partial_\mu j^\mu = 0, \tag{12.6.9}$$

where the current density 4-vector j^μ is given by

$$j^\mu = c\bar{\psi}\gamma^\mu\psi. \tag{12.6.10}$$

Here, in the expression for j^μ, $\bar{\psi}$ is the so-called Dirac conjugate wave function, which is defined through

$$\bar{\psi} = \psi^\dagger(\gamma^0)^\dagger = \psi^\dagger\gamma^0, \tag{12.6.11}$$

where $\psi^\dagger$ is the hermitian conjugate of ψ.

12.7 Properties of γ Matrices

1. Since β is hermitian, γ^0 is hermitian. But, γ^j is anti-hermitian:

$$(\gamma^j)^\dagger = (\beta\,\alpha^j)^\dagger = (\alpha^j)^\dagger\,\beta^\dagger = \alpha^j\,\beta = -\beta\,\alpha^j = -\gamma^j, \tag{12.7.1}$$

where we have made use of (12.5.17). These hermiticity relations can be summarized compactly by

$$(\gamma^\mu)^\dagger = \gamma^0 \gamma^\mu \gamma^0. \tag{12.7.2}$$

2. Using the commutation relations for the α and β matrices along with the equations (12.5.18) and (12.5.19), it is easy to show that the γ matrices satisfy the following commutation relation

$$\gamma^\mu \gamma^\nu + \gamma^\nu \gamma^\mu = 2I\, g^{\mu\nu}, \tag{12.7.3}$$

where I is the unit matrix and $g^{\mu\nu}$ is the Minkowski metric tensor given by (12.1.24).

We may form new matrices from the four γ^μ matrices by multiplying two or more of them together. Since the square of each of the γ^μ matrices equals ± 1, we need to consider only products whose factors are different. The order of the factors in the product is irrelevant since different γ matrices either commute or anti-commute. Since the number of ways in which one can make distinct combinations out of n objects is $(2^n - 1)$, we shall have altogether $2^4 - 1 = 15$ different products of γ matrices. If we also include the unit matrix I, we can enumerate 16 different matrices, which have been tabulated as follows:

$$I, \ \gamma^0, \ i\gamma^1, \ i\gamma^2 \ i\gamma^3,$$

$$\gamma^0\gamma^1, \ \gamma^0\gamma^2, \ \gamma^0\gamma^3, \ i\gamma^2\gamma^3, \ i\gamma^3\gamma^1, \ i\gamma^1\gamma^2, \tag{12.7.4}$$

$$i\gamma^0\gamma^2\gamma^3, \ i\gamma^0\gamma^3\gamma^1, \ i\gamma^0\gamma^1\gamma^2, \ \gamma^1\gamma^2\gamma^3,$$

$$i\gamma^0\gamma^1\gamma^2\gamma^3 \equiv i\gamma_5, \tag{12.7.5}$$

where the factor i has been inserted so that the square of each element is $+I$.

Let us denote the elements in the aforementioned array by $\Gamma_m, m = 1, 2, 3, \ldots, 16$. The following properties can be verified (no summation over repeated Latin indices implied):

$$\Gamma_m \Gamma_n = a_{mn} \Gamma_\ell, \ \text{where } a_{mn} = \pm 1 \text{ or} \pm i, \tag{12.7.6}$$

$$\Gamma_m \Gamma_n = I, \ \text{iff } m = n, \tag{12.7.7}$$

where Γ_ℓ in (12.7.6) is one of the elements of the array, other than Γ_m and Γ_n. It can also be checked that γ_5 anti-commutes with all γ^μ and that $(\gamma_5)^2 = -I$. Furthermore, we have the following important theorems.

Theorem 12.7.1: For each Γ_i, except for $\Gamma_1 = I$, we can always find a Γ_j such that

$$\Gamma_j \Gamma_i \Gamma_j = -\Gamma_i. \tag{12.7.8}$$

The *proof* consists in showing that there does exit a Γ_j for each Γ_i as required by (12.7.8). For instance, it is easy to check that for $i = 2,3,\ldots,5$, that is, for the last four elements in the first line of (12.7.4), $\Gamma_j = i\gamma^0\gamma^1\gamma^2\gamma^3$, for the elements of the *second* line of (12.7.4), Γ_j equals one of the last four elements of the *first* line, for the elements of the third line of (12.7.4), $\Gamma_j = i\gamma^0\gamma^1\gamma^2\gamma^3$ and for the last element $\Gamma_j = \gamma^0$.

Theorem 12.7.2: For $i \neq 1$, $\mathrm{tr}\Gamma_i = 0$.

Proof: Consider any Γ_i. Then, using the property (12.7.8), we have $\Gamma_j\Gamma_i\Gamma_j = -\Gamma_i$. Using now the property of the trace of the products of matrices, we obtain

$$\mathrm{tr}(\Gamma_j\Gamma_i\Gamma_j) = \mathrm{tr}(\Gamma_j\Gamma_j\Gamma_i) = \mathrm{tr}(\Gamma_j^2\Gamma_i) = \mathrm{tr}(\Gamma_i) = -\mathrm{tr}(\Gamma_i). \tag{12.7.9}$$

It thus follows that $\mathrm{tr}(\Gamma_i) = 0$.

Theorem 12.7.3: The matrices $\Gamma_j, j = 1,2,3,\ldots,16$ are linearly independent, that is, the equality $\sum_{j=1}^{16} c_j\Gamma_j = 0$ holds only if the constants $c_j = 0, j = 1,2,3,\ldots,16$.

Proof: Multiplying $\sum_{j=1}^{16} c_j\Gamma_j = 0$ by Γ_k, with $k \neq j$, and using the properties (12.7.6) and (12.7.7), we obtain

$$\sum_{j=1}^{16} c_j\Gamma_k\Gamma_j = c_jI + \sum_{k\neq j} c_j a_{jk}\Gamma_\ell = 0. \tag{12.7.10}$$

Taking the trace of (12.7.10), we get $c_j = 0, j = 1,2,3,\ldots,16$ because $tr(\Gamma_\ell) = 0$, whereas $tr(I) \neq 0$. Q.E.D.

***Corollary*:** It follows from Theorem 12.7.3 that any 4×4 matrix M can be written uniquely as a linear combination of the Γ_js:

$$M = \sum_{k=1}^{16} c_k\Gamma_k. \tag{12.7.11}$$

Multiplying (12.7.11) by Γ_ℓ and taking the trace, we obtain

$$tr(\Gamma_\ell M) = \sum_{k=1}^{16} c_k tr(\Gamma_\ell\Gamma_k) = c_\ell tr(\Gamma_\ell\Gamma_\ell) + \sum_{k\neq\ell}^{16} c_k tr(\Gamma_\ell\Gamma_k) = c_\ell tr(I) = 4c_\ell. \tag{12.7.12}$$

It thus follows that $c_\ell = \frac{1}{4} tr(\Gamma_\ell M)$ and hence

$$M = \sum_{k=1}^{16} \frac{1}{4} tr(\Gamma_k M)\,\Gamma_k. \tag{12.7.13}$$

Theorem 12.7.4: Any matrix M that commutes with $\gamma^\mu, \mu = 0,1,2,3$, is a multiple of the identity matrix I.

Proof: Assume M is not a multiple of the identity. If M commutes with all the γ^μ, it commutes with all the Γ_js and, therefore, we have $M = \Gamma_j M \Gamma_j$. From (12.7.10), we can write

$$M = c_j \Gamma_j + \sum_{k \neq j} c_k \Gamma_k. \tag{12.7.14}$$

Since, by assumption, M commutes with all the γ^μ, it commutes with all Γ_j. Hence, we have $M = \Gamma_\ell M \Gamma_\ell$. Therefore, if Γ_ℓ is the element for which $\Gamma_\ell \Gamma_j \Gamma_\ell = -\Gamma_j$, by multiplying (12.7.16) by Γ_ℓ from left and from the right, we obtain

$$M = \Gamma_\ell M \Gamma_\ell = c_j \Gamma_\ell \Gamma_j \Gamma_\ell + \sum_{k \neq j} c_k \Gamma_\ell \Gamma_k \Gamma_\ell = -c_j \Gamma_j + \sum_{k \neq j} c_k\,(\pm \Gamma_k), \tag{12.7.15}$$

where we have used the fact that Γ_k and Γ_ℓ either commute or anti-commute and hence, $\Gamma_\ell \Gamma_k \Gamma_\ell = \pm \Gamma_\ell^2 \Gamma_k = \pm \Gamma_k$.

Multiplying (12.7.14) and (12.7.15) by Γ_j and taking the trace we get that $c_j = -c_j$ and, therefore, $c_j = 0$. Since Γ_J is arbitrary except that it is not equal to I, it follows that all c_j in the expansion (12.7.11) are zero except c_1. That is,

$$M = c_1 \Gamma_1 = c_1 I. \tag{12.7.16}$$

This means that M is a multiple of the unit matrix I. Q.E.D.

Theorem 12.7.5: Consider two sets of 4×4 matrices γ^μ and γ'^μ, which satisfy the commutation relations

$$\gamma^\mu \gamma^\nu + \gamma^\nu \gamma^\mu = 2I\, g^{\mu\nu}, \tag{12.7.17}$$

$$\gamma'^\mu \gamma'^\nu + \gamma'^\nu \gamma'^\mu = 2I\, g^{\mu\nu}, \tag{12.7.18}$$

respectively. If Γ_j and Γ'_j are sets of sixteen products matrices formed in exactly the same manner from γ^μs and γ'^μs, respectively, then there exists a non-singular matrix S such that

$$\gamma'^\mu = S \gamma^\mu S^{-1}. \tag{12.7.19}$$

Proof: The proof lies in showing that we can indeed construct the required matrix S and that this matrix is nonsingular[6]. If Γ_j and Γ'_j are sets of sixteen products matrices formed in exactly the same manner from γ^μs and γ'^μs, respectively, then we set

$$S = \sum_{j=1}^{16} \Gamma'_j F \Gamma_j, \tag{12.7.20}$$

where F is an arbitrary 4×4 matrix, and show that it can be chosen such that S is nonsingular.

We have from (12.7.6)

$$\Gamma_i \Gamma_j \Gamma_i \Gamma_j = a_{ij}^2 \Gamma_k^2 = a_{ij}^2. \tag{12.7.21}$$

Multiplying (12.7.21) by Γ_i from the left and by Γ_j from the right we obtain

$$\Gamma_j \Gamma_i = a_{ij}^2 \Gamma_i \Gamma_j = a_{ij}^3 \Gamma_k. \tag{12.7.22}$$

Since, in the primed system, the γ'^μ's are constructed in the same manner as in the unprimed system, we also have

$$\Gamma'_i \Gamma'_j = a_{ij} \Gamma'_k. \tag{12.7.23}$$

Then, for any i,

$$\Gamma'_i S \Gamma_i = \sum_{j=1}^{16} \Gamma'_i \Gamma'_j F \Gamma_j \Gamma_i = \sum_{j=1}^{16} a_{ij}^4 \Gamma'_k F \Gamma_k. \tag{12.7.24}$$

Since $a_{ij}^4 = 1$, if we take into account that for fixed i the matrix $\Gamma_i \Gamma_j$ runs over all the sixteen elements of the algebra, the summation over j can be replaced by a sum over k. Then it follows from (12.7.24) that

$$\Gamma'_i S \Gamma_i = \sum_{k=1}^{16} \Gamma'_k F \Gamma_k = S. \tag{12.7.25}$$

By multiplying (12.7.25) by Γ_i from right and taking into account that $\Gamma_i^2 = I$, we get

$$\Gamma'_i S = S \Gamma_i. \tag{12.7.26}$$

From (12.7.26) we get

$$\Gamma'_i = S \Gamma_i S^{-1}. \tag{12.7.27}$$

[6] H. Bethe and R. Jackiw, Intermediate Quantum Mechanics, Westview Press, 1997

Now two things must be noted. Firstly, the matrix F can be chosen so that S is not zero. In the opposite case, by assuming S to be zero for all F, one can prove that all the matrices $\Gamma_j, j = 1,2,3,\ldots,16$ are linearly dependent which is a contradiction. Secondly, the matrix F can be chosen so that S is nonsingular. To prove this statement let us introduce a matrix S' defined by

$$S' = \sum_{j=1}^{16} \Gamma'_j G \Gamma_j. \tag{12.7.28}$$

where G is a 4×4 matrix to be chosen. For the moment, the matrix F in the definition of S,

$$S = \sum_{j=1}^{16} \Gamma'_j F \Gamma_j, \tag{12.7.29}$$

is also assumed to be arbitrary. By symmetry, with the primed and unprimed matrices interchanged, we can write

$$\Gamma_j S' = S' \Gamma'_j, \tag{12.7.30}$$

for any j. From (12.7.27) and (12.7.30), we obtain

$$\Gamma_j S'S = \Gamma_j S' \Gamma'_j S = \Gamma_j S' \Gamma'_j (\Gamma'_j S \Gamma_j) = \Gamma_j S' S \Gamma_j = S' S \Gamma_j. \tag{12.7.31}$$

Then according to Theorem 12.7.4,

$$S'S = aI, \tag{12.7.32}$$

that is, $S'S$ is a multiple of the unit matrix. Since F and G occurring (12.7.20) and (12.7.28)are arbitrary, they can be chosen so that a is not zero. Hence, S is nonsingular. Therefore, from (12.7.27), it follows that

$$\gamma'^{\mu} = S \gamma^{\mu} S^{-1}. \tag{12.7.33}$$

The above assertion is due to the fact that (12.7.27) holds for all $\Gamma_j, j = 1,2,3,\ldots,16$, including $\Gamma_1 = \gamma^0, \Gamma_2 = i\gamma^1, \Gamma_3 = i\gamma^2, \Gamma_4 = i\gamma^3$. Q.E.D.

Note that S is unique up to a constant, for suppose $\gamma'^{\mu} = S_1 \gamma^{\mu} S_1^{-1} = \gamma'^{\mu} = S_2 \gamma^{\mu} S_2^{-1}$. Then $S_2^{-1} S_1 \gamma^{\mu} = \gamma^{\mu} S_2^{-1} S_1$. Since $S_2^{-1} S_1$ commutes with all γ^{μ}, it follows from Theorem 12.7.4 that $S_2^{-1} S_1 = aI$. whence $S_1 = cS_2$, where c is a constant. Further, suppose that we are given four γ'^{μ} that satisfy (12.7.17). If we now define $\gamma'^{\mu} = S \gamma^{\mu} S^{-1}$, it is clear that the γ'^{μ} will also satisfy (12.7.17).

12.8 Form Invariance of Dirac Equation under Lorentz Transformations

Before we discuss the physical consequences of Dirac theory, we must be convinced that the Dirac equation (12.6.8) is consistent with STR in the sense that it is indeed covariant under Lorentz transformations. That is, it is form invariant under Lorentz transformations.

Consider the standard configuration of two inertial frames K and K' of section 12.1. In that context, the covariance of the Dirac equation requires two things:

(i) There must exist concrete rules to enable the observer in K' to construct his wave function $\psi'(x'^0,\vec{x}')$ on the basis of the information about the wave function $\psi(x^0,\vec{x})$, provided to him by the observer in K, so that both the Dirac wave functions, ($\psi(x^0,\vec{x}) \in K$ and $\psi'(x'^0,\vec{x}') \in K'$), describe the same quantum state of a given system.

(ii) To be consistent with the principle of special relativity, the wave function $\psi'(x'^0,\vec{x}')$ must also satisfy the Dirac equation

$$\left(i\hbar\gamma'^{\nu}\frac{\partial}{\partial x'^{\nu}} - mc\right)\psi' = 0, \tag{12.8.1}$$

written in the reference frame K'.

The transformation of the Dirac wave function under transition from K to K', via Lorentz transformations, must be linear because of the simple fact that both the Dirac equation and the Lorentz transformations are linear:

$$\psi'(x') = \psi'(\Lambda x) = S(\Lambda)\psi(x) = S(\Lambda)\psi(\Lambda^{-1}x'), \tag{12.8.2}$$

where Λ is the Lorentz matrix and $S(\Lambda)$ is a 4×4 matrix that depends on the parameters of Lorentz transformations and acts on the original Dirac wave function ψ to yield the transformed Dirac wave function ψ'. Note that through Λ, S depends on the relative velocity and the orientation of the two observers in K and K'.

According to the principle of relativity, the inverse transformation, represented by the matrix $S^{-1}(\Lambda)$ must exist so that the observer in K could also construct his wave function $\psi(x)$ on the basis of the knowledge of $\psi'(x')$. Now, this can be accomplished in two ways: (a) By acting on $\psi'(x')$ with the inverse of the matrix $S(\Lambda)$

$$\psi(x) = S^{-1}(\Lambda)\psi'(x') = S^{-1}(\Lambda)\psi'(\Lambda x), \tag{12.8.3}$$

and (b) By acting on $\psi'(x')$ with the matrix $S(\Lambda^{-1})$

$$\psi(x) = S(\Lambda^{-1})\psi'(x') = S(\Lambda^{-1})\psi'(\Lambda x). \tag{12.8.4}$$

The last two equations yield

$$S(\Lambda^{-1}) = S^{-1}(\Lambda). \tag{12.8.5}$$

Thus, if we construct $S(\Lambda)$, we are through. Let us proceed to achieve that.

For the observer in K, the Dirac equation reads

$$(i\hbar\gamma^{\mu}\,\partial_{\mu} - mc)\,\psi = 0. \tag{12.8.6}$$

Inserting here $\psi(x) = S(\Lambda^{-1})\psi'(x')$, we get

$$\left(i\hbar\gamma^{\mu}\,\frac{\partial}{\partial x_{\mu}}S(\Lambda^{-1}) - mc\,S(\Lambda^{-1})\right)\,\psi'(x') = 0. \tag{12.8.7}$$

Let us multiply (12.8.6) by $S(\Lambda)$ from the left, take into account (12.8.5) and the fact that $SS^{-1} = I$. Then we obtain

$$\left(i\hbar\left[S(\Lambda)\Lambda^{\nu}_{\mu}\gamma^{\mu}S^{-1}(\Lambda)\right]\frac{\partial}{\partial x'^{\nu}} - mc\right)\psi'(x') = 0. \tag{12.8.8}$$

Note that in the above equation (12.8.8), we have taken into account that

$$\frac{\partial}{\partial x^{\mu}} = \frac{\partial x'^{\nu}}{\partial x^{\mu}}\,\frac{\partial}{\partial x'^{\nu}} = \Lambda^{\nu}_{\mu}\,\frac{\partial}{\partial x'^{\nu}}. \tag{12.8.9}$$

Let us define $\Lambda^{\nu}_{\mu}\gamma^{\mu} = \gamma'^{\nu}$. Using the relation (12.1.31), it is easy to check that γ'^{ν} satisfy the commutation relation (12.7.18). Then, according Theorem 12.7.5, Which is also known as Pauli's fundamental theorem, there exists a (unique up to multiplicative constant) matrix S such that

$$\Lambda^{\nu}_{\mu}\gamma^{\mu} = S^{-1}\gamma^{\nu}S. \tag{12.8.10}$$

In the given case, it is $S(\Lambda)$ which must satisfy the constraint (12.8.10). If we use the above relation (12.8.10) in (12.8.9), we obtain

$$\left(i\hbar\gamma^{\nu}\,\frac{\partial}{\partial x'^{\nu}} - mc\right)\,\psi'(x') = 0. \tag{12.8.11}$$

The equation (12.8.11) shows that the Dirac equation will be form invariant under Lorentz transformations, if we succeed in constructing the required matrix $S(\Lambda)$.

It is clear from this analysis that constructing $S(\Lambda)$ is equivalent to solving (12.8.10), which holds good for both the proper and improper Lorentz transformations. In what follows, we shall discuss covariance of the Dirac equation only under proper Lorentz transformations. Also, time reversal will not be discussed here, nor will the related transformation of charge conjugation, which interchanges particles with anti-particles. While both of these operations are symmetries of the Dirac theory, they are discussed most easily in the field theoretical context, which, unfortunately, we do not have here. So far as parity transformation is concerned, the correct choice for S is γ^0.

Let us start with the proper Lorentz transformations, which are continuous, that is, they are parametrized by a continuously varying parameter. For instance, the relative velocity between two inertial frames is one of such parameters. It turns out that proper Lorentz transformations corresponding to a parameter form a Lie group. Therefore, any finite Lorentz transformation can be achieved from the identity transformation by repeatedly applying the corresponding infinitesimal transformation. As a consequence, it is sufficient to deal only with infinitesimal transformations.

Consider an infinitesimal proper Lorentz transformation by representing the Lorentz parameter Λ^{μ}_{ν} as

$$\Lambda^{\mu}_{\nu} = \delta^{\mu}_{\nu} + \Delta\omega^{\mu}_{\nu}, \tag{12.8.12}$$

where $|\Delta\omega| \ll 1$. From (12.1.31), we get

$$g_{\alpha\beta} = \Lambda^{\mu}_{\alpha} g_{\mu\nu} \Lambda^{\nu}_{\beta} = (\delta^{\mu}_{\alpha} + \Delta\omega^{\mu}_{\alpha}) g_{\mu\nu} (\delta^{\nu}_{\beta} + \Delta\omega^{\nu}_{\beta}) = g_{\alpha\beta} + (\Delta\omega_{\alpha\beta} + \Delta\omega_{\beta\alpha}) + O(\Delta\omega)^2. \tag{12.8.13}$$

As a consequence, we get

$$\Delta\omega_{\alpha\beta} + \Delta\omega_{\beta\alpha} = 0. \quad \Rightarrow \quad \Delta\omega_{\alpha\beta} = -\Delta\omega_{\beta\alpha}. \tag{12.8.14}$$

Consequently, there are six non-vanishing parameters each of which generates an infinitesimal Lorentz transformation. Let now consider $S(\Lambda) = S(I + \Delta\omega^{\mu\nu})$ and expand it in Taylor series around the identity transformation I:

$$S(\Lambda) = S(I + \Delta\omega^{\mu\nu}) = S(I) + \left(\left[\frac{\partial S}{\partial \Lambda} \right]_{\Lambda = I} \right)_{\mu\nu} \Delta\omega^{\mu\nu} + O(\Delta\omega)^2 \approx I - \frac{i}{4} \sigma_{\mu\nu} \Delta\omega^{\mu\nu}, \tag{12.8.15}$$

where we have taken into account that $S(I) = I$ and introduced the notation

$$\left(\left[\frac{\partial S}{\partial \Lambda} \right]_{\Lambda = I} \right)_{\mu\nu} = -\frac{i}{4} \sigma_{\mu\nu}. \tag{12.8.16}$$

Here, $\sigma_{\mu\nu}$ is a 4×4 matrix and the factor $-(i/4)$ has been introduced for convenience. Using $SS^{-1} = I$, it is easy to prove that $\sigma_{\mu\nu}$ is anti-symmetric

$$\sigma_{\mu\nu} = -\sigma_{\nu\mu}. \tag{12.8.17}$$

Let us now go back to (12.8.10). Up to the first order in $\Delta\omega$, we get

$$(g^{\nu}_{\mu} + \Delta\omega^{\nu}_{\mu}) \gamma^{\mu} = \left(I - \frac{i}{4} \sigma_{\alpha\beta} \Delta\omega^{\alpha\beta} \right) \gamma^{\nu} \left(I + \frac{i}{4} \sigma_{\alpha\beta} \Delta\omega^{\alpha\beta} \right). \tag{12.8.18}$$

Or,

$$\gamma^\nu + \Delta\omega^\nu_\mu \gamma^\mu = \gamma^\nu - \frac{i}{4}\gamma^\nu \sigma_{\alpha\beta}\Delta\omega^{\alpha\beta} + \frac{i}{4}\Delta\omega^{\alpha\beta}\sigma_{\alpha\beta}\gamma^\nu. \tag{12.8.19}$$

Cancelling γ^ν, we obtain

$$\Delta\omega^\nu_\mu \gamma^\mu = \frac{i}{4}(\sigma_{\alpha\beta}\gamma^\nu - \gamma^\nu\sigma_{\alpha\beta})\Delta\omega^{\alpha\beta} \tag{12.8.20}$$

Changing the dummy index μ to β, the left hand-side of (12.8.20) can be written as

$$\Delta\omega^\nu_\beta \gamma^\beta = g^\nu_\alpha \gamma^\beta \Delta\omega^\alpha_\beta = g^\nu_\alpha g_{\sigma\beta}\gamma^\beta \Delta\omega^{\alpha\sigma} = g^\nu_\alpha \Delta\omega^{\alpha\sigma}\gamma_\sigma = g^\nu_\alpha \Delta\omega^{\alpha\beta}\gamma_\beta. \tag{12.8.21}$$

Using now the anti-symmetry of $\Delta\omega^{\alpha\beta}$, we get

$$\Delta\omega^\nu_\beta \gamma^\beta = \frac{1}{2}\left(g^\nu_\alpha \Delta\omega^{\alpha\beta}\gamma_\beta + g^\nu_\beta \Delta\omega^{\beta\alpha}\gamma_\alpha\right) = \frac{1}{2}\left(g^\nu_\alpha \gamma_\beta - g^\nu_\beta \gamma_\alpha\right)\Delta\omega^{\alpha\beta}. \tag{12.8.22}$$

From (12.8.20)–(12.8.22), we get

$$\frac{i}{4}(\sigma_{\alpha\beta}\gamma^\nu - \gamma^\nu\sigma_{\alpha\beta})\Delta\omega^{\alpha\beta} = \frac{1}{2}\left(g^\nu_\alpha \gamma_\beta - g^\nu_\beta \gamma_\alpha\right)\Delta\omega^{\alpha\beta}. \tag{12.8.23}$$

Or,

$$2i\left(g^\nu_\beta \gamma_\alpha - g^\nu_\alpha \gamma_\beta\right) = [\sigma_{\alpha\beta}, \gamma^\nu]. \tag{12.8.24}$$

Thus, the problem of finding $S(\Lambda)$ has now reduced to finding the solution of the above equation (12.8.24) for $\sigma_{\alpha\beta}$. The required $\sigma_{\alpha\beta}$ is given by

$$\sigma_{\alpha\beta} = \frac{i}{2}[\gamma_\alpha, \gamma_\beta]. \tag{12.8.25}$$

Let us check whether it satisfies the aforementioned equation or not. We have

$$\begin{aligned}[\sigma_{\alpha\beta}, \gamma^\nu] &= \frac{i}{2}[[\gamma_\alpha, \gamma_\beta], \gamma^\nu] = \frac{i}{2}\{[\gamma_\alpha\gamma_\beta - \gamma_\beta\gamma_\alpha, \gamma^\nu]\} = \frac{i}{2}\{[2\gamma_\alpha\gamma_\beta, \gamma^\nu] - 2[g_{\alpha\beta}, \gamma^\nu]\} \\ &= i[\gamma_\alpha\gamma_\beta, \gamma^\nu] = i(\gamma_\alpha\gamma_\beta\gamma^\nu - \gamma^\nu\gamma_\alpha\gamma_\beta) = 2i\left(g^\nu_\beta\gamma_\alpha - g^\nu_\alpha\gamma_\beta\right),\end{aligned} \tag{12.8.26}$$

where we have used the anti-commutation relation for the γ matrices. Hence, the ansatz (12.8.25) does fulfill the condition (12.8.24). As a result, we finally have

$$S(\Lambda) = I - \frac{i}{4}\sigma_{\mu\nu}\Delta\omega^{\mu\nu} = I + \frac{1}{8}[\gamma_\mu, \gamma_\nu]\Delta\omega^{\mu\nu}. \tag{12.8.27}$$

Thus, we have shown that if $\gamma^\mu, \mu = 0,1,2,3$ do not change during Lorentz transformations and the Dirac wave function $\psi(x)$ transforms with $S(\Lambda)$, given by (12.8.27), the Dirac equation is covariant (that is, form invariant) under Lorentz transformations.

12.9 Free-Particle Solutions of Dirac Equation

The Dirac equation for a free particle is given by

$$i\hbar \frac{\partial \psi}{\partial t} = \left(c\,\vec{\alpha} \cdot \hat{\vec{p}} + mc^2\,\beta\right) \psi. \tag{12.9.1}$$

We look for the solution in the form

$$\psi(\vec{r},t) = \psi(\vec{r})\, e^{-\frac{i}{\hbar}E\,t}, \tag{12.9.2}$$

which, when substituted into the Dirac equation (12.9.1) yields the eigenvalue equation for the Dirac Hamiltonian $\hat{H}_D$:

$$\left(c\,\vec{\alpha} \cdot \hat{\vec{p}} + mc^2\,\beta\right) \psi(\vec{r}) = E\,\psi(\vec{r}), \tag{12.9.3}$$

where E is the energy eigenvalue. Since the operator on the left-hand side is a 4×4 matrix, the wave function $\psi(\vec{r},t)$ must have four components that are usually written as a four-component column vector and it is called a Dirac bi-spinor. Note that, since $\hat{H}_D$ is only a function of $\hat{\vec{p}}$, then $[\hat{\vec{p}}, \hat{H}_D] = 0$, so that the eigenvalues of $\hat{\vec{p}}$ can be used to characterize the states. In particular, we look for free-particle (plane-wave) solutions of the form

$$\psi(\vec{r}) = u(p)\, e^{\frac{i}{\hbar}(\vec{p}\cdot\vec{r})}, \tag{12.9.4}$$

where

$$u(p) = \begin{pmatrix} u_1(p) \\ u_2(p) \\ u_3(p) \\ u_4(p) \end{pmatrix}, \tag{12.9.5}$$

is a Dirac bi-spinor that satisfies, according to (12.9.3),

$$\left(c\,\vec{\alpha} \cdot \hat{\vec{p}} + mc^2\,\beta\right) u(p) = E\,u(p). \tag{12.9.6}$$

For convenience, we rewrite the Dirac bi-spinor in the following form

$$u(p) = \begin{pmatrix} \phi(p) \\ \chi(p) \end{pmatrix}, \tag{12.9.7}$$

so that (12.9.6) can be rewritten as

$$\begin{pmatrix} mc^2 & c\hat{\vec{\sigma}}\cdot\hat{\vec{p}} \\ c\hat{\vec{\sigma}}\cdot\hat{\vec{p}} & -mc^2 \end{pmatrix}\begin{pmatrix} \phi(p) \\ \chi(p) \end{pmatrix} = \begin{pmatrix} E & 0 \\ 0 & E \end{pmatrix}\begin{pmatrix} \phi(p) \\ \chi(p) \end{pmatrix}. \tag{12.9.8}$$

This matrix equation gives a system of two algebraic equations

$$(E - mc^2)\phi_p - (c\hat{\vec{\sigma}}\cdot\vec{p})\chi_p = 0, \tag{12.9.9}$$

$$-(c\hat{\vec{\sigma}}\cdot\vec{p})\phi_p + (E + mc^2)\chi_p = 0. \tag{12.9.10}$$

Taking into account that

$$\hat{p}_j\begin{pmatrix} \phi(p) \\ \chi(p) \end{pmatrix} = p_j\begin{pmatrix} \phi(p) \\ \chi(p) \end{pmatrix}, \quad j = 1,2,3, \tag{12.9.11}$$

we get from these equations

$$\phi_p = \frac{(c\hat{\vec{\sigma}}\cdot\vec{p})}{(E - mc^2)}\chi_p, \tag{12.9.12}$$

$$\chi_p = \frac{(c\hat{\vec{\sigma}}\cdot\vec{p})}{(E + mc^2)}\phi_p. \tag{12.9.13}$$

From the aforementioned two equations, we obtain a single equation for ϕ_p:

$$\left[(E - mc^2)(E + mc^2) - c^2(\hat{\vec{\sigma}}\cdot\vec{p})^2\right]\phi_p = 0 \tag{12.9.14}$$

Using the formula,

$$(\hat{\vec{\sigma}}\cdot\vec{a})(\hat{\vec{\sigma}}\cdot\vec{b}) = (\vec{a}\cdot\vec{b})I + i\sigma\cdot(\vec{a}\times\vec{b})I, \tag{12.9.15}$$

we obtain

$$(E^2 - c^2\vec{p}^2 - m^2c^4)\,\phi_p = 0. \tag{12.9.16}$$

Note that the same result is obtained by equating the determinant of the aforementioned homogeneous system of algebraic equations to zero. It is worth noting here that (12.9.16) represents the relativistic relation between energy and momentum of a free particle, which must be satisfied in any relativistic generalization of quantum mechanics. The Dirac equation, as we see, does preserve it.

For non-trivial solutions, we must have

$$(E^2 - c^2\vec{p}^2 - m^2c^4) = 0. \tag{12.9.17}$$

For a given value of the momentum, there are two solutions for the equation (12.9.17):

$$E_+ = +\sqrt{c^2\vec{p}^2 + m^2c^4}, \tag{12.9.18}$$

and

$$E_- = -\sqrt{c^2\vec{p}^2 + m^2c^4}, \tag{12.9.19}$$

which are nothing but the energy eigenvalues of the Hamiltonian. Thus, we see that, just like the Klein–Gordon equation, the Dirac equation too have solutions with positive as well as negative total energy.

For the positive energy eigenvalue, E_+, an appropriate solution is to take

$$\phi_p = \begin{pmatrix} 1 \\ 0 \end{pmatrix} \text{ or } \begin{pmatrix} 0 \\ 1 \end{pmatrix}. \tag{12.9.20}$$

Then,

$$\chi_p = \frac{(c\hat{\vec{\sigma}}\cdot\vec{p})}{(E+mc^2)}\begin{pmatrix} 1 \\ 0 \end{pmatrix} \text{ or } \frac{(c\hat{\vec{\sigma}}\cdot\vec{p})}{(E+mc^2)}\begin{pmatrix} 0 \\ 1 \end{pmatrix}. \tag{12.9.21}$$

But,

$$\hat{\vec{\sigma}}\cdot\vec{p} = \begin{pmatrix} p_z & p_x - ip_y \\ p_x + ip_y & -p_z \end{pmatrix}, \tag{12.9.22}$$

and, therefore, we have

$$\chi_p = \begin{pmatrix} \dfrac{cp_z}{E_+ + mc^2} \\ \dfrac{c(p_x + ip_y)}{E_+ + mc^2} \end{pmatrix} \text{ or } \begin{pmatrix} \dfrac{c(p_x - ip_y)}{E_+ + mc^2} \\ \dfrac{-cp_z}{E_+ + mc^2} \end{pmatrix}. \tag{12.9.23}$$

So, the full positive energy solutions are

$$\psi_+(\vec{r},t) = \begin{pmatrix} 1 \\ 0 \\ \dfrac{cp_z}{E_+ + mc^2} \\ \dfrac{c(p_x + ip_y)}{E_+ + mc^2} \end{pmatrix} e^{-i\frac{mc^2}{\hbar}t + \frac{i}{\hbar}(\vec{p}\cdot\vec{r})} \text{ or } \begin{pmatrix} 0 \\ 1 \\ \dfrac{c(p_x - ip_y)}{E_+ + mc^2} \\ \dfrac{-cp_z}{E_+ + mc^2} \end{pmatrix} e^{-i\frac{mc^2}{\hbar}t + \frac{i}{\hbar}(\vec{p}\cdot\vec{r})}, \tag{12.9.24}$$

both of which represent forward propagating solutions. These correspond to particles propagating forward in time with an energy $E = E_+$.

For the negative energy eigenvalue, E_-, an appropriate solution is to take

$$\chi_p = \begin{pmatrix} 1 \\ 0 \end{pmatrix} \quad \text{or} \quad \begin{pmatrix} 0 \\ 1 \end{pmatrix}. \tag{12.9.25}$$

so that the full negative energy solutions are given by

$$\psi_-(\vec{r},t) = \begin{pmatrix} \frac{-cp_z}{|E_-|+mc^2} \\ \frac{-c(p_x+ip_y)}{|E_-|+mc^2} \\ 1 \\ 0 \end{pmatrix} e^{i\frac{mc^2}{\hbar}t+\frac{i}{\hbar}(\vec{p}\cdot\vec{r})} \quad \text{or} \quad \begin{pmatrix} \frac{-c(p_x-ip_y)}{|E_-|+mc^2} \\ \frac{cp_z}{|E_-|+mc^2} \\ 0 \\ 1 \end{pmatrix} e^{i\frac{mc^2}{\hbar}t+\frac{i}{\hbar}(\vec{p}\cdot\vec{r})}. \tag{12.9.26}$$

These correspond to particles propagating backwards in time with an energy $E = |E_-|$.

Note that to have the normalized solution satisfying

$$(\phi_p)^*\phi_p + (\chi_p)^*\chi_p = 1, \tag{12.9.27}$$

we must multiply each of the components, ϕ_p and χ_p, by

$$N = \sqrt{\frac{E_+ + mc^2}{2E_+}}. \tag{12.9.28}$$

If we consider the non-relativistic limit in which $E = mc^2$, we notice that each of the aforementioned solutions have one of the two spinor components, ϕ_p and χ_p, of the order of $(v/c) \ll 1$. These components are called *small components*, while the other two are called *large components*. For instance, for the positive energy solutions, we have

$$E \approx mc^2 + \frac{\vec{p}^2}{2m}, \tag{12.9.29}$$

so that

$$\chi_p = \frac{c\,(\hat{\vec{\sigma}}\cdot\vec{p})}{2mc^2 + \frac{\vec{p}^2}{2m}}\,\phi_p. \tag{12.9.30}$$

Since $mc^2 \gg \frac{(\vec{p})^2}{2m}$, we conclude that $\chi_p \ll \phi_p$. Hence, for the positive energy solutions, ϕ_p is the large component, while χ_p is the small component. In the non-relativistic limit, we expect the large components to correspond to solutions of the Schrödinger equation for a free-particle.

12.10 Spin. Interpretation of the Negative Energy Solutions

It turns out that the Dirac equation describes a particle with spin equal to $\hbar/2$. This was one of the important features of Dirac's theory. To show this, let us consider the orbital angular momentum operator, $\hat{\vec{L}}$, and check whether it commutes with the Dirac's Hamiltonian given by (12.5.12). We have

$$[\hat{L}_x, \hat{H}_D] = [y\hat{p}_z - z\hat{p}_y, \left(\vec{\alpha}\cdot(\hat{\vec{p}}) + mc^2\beta\right)]. \tag{12.10.1}$$

Using the distributive property and the fundamental commutation relations (2.4.20), we get

$$[\hat{L}_x, \hat{H}_D] = i\hbar c\,(\alpha_y\,\hat{p}_z - \alpha_z\,\hat{p}_y)\,. \tag{12.10.2}$$

Similarly, we obtain

$$[\hat{L}_y, \hat{H}_D] = i\hbar c\,(\alpha_z\,\hat{p}_x - \alpha_x\,\hat{p}_z)\,, \tag{12.10.3}$$

$$[\hat{L}_z, \hat{H}_D] = i\hbar c\,(\alpha_x\,\hat{p}_y - \alpha_y\,\hat{p}_x)\,. \tag{12.10.4}$$

Combining the aforementioned three equations together, we arrive at

$$[\hat{\vec{L}}, \hat{H}_D] = i\hbar c\left(\vec{\alpha}\times\hat{\vec{p}}\right). \tag{12.10.5}$$

Hence, the orbital angular momentum operator does not commute with the Hamiltonian, and therefore, it is no longer a constant of motion. On the other hand, the existence of two linearly independent solutions corresponding to a given value of the energy indicates that there is some inherent symmetry in the Dirac Hamiltonian because of which this degeneracy occurs. This in turn means that there must exist an operator that commutes with the Hamiltonian.

If we go back to Section 8.2, we recognize that in the 4×4 representation, the spin operator takes the form

$$\hat{\vec{S}} = \frac{\hbar}{2}\hat{\vec{\sigma}} = \frac{\hbar}{2}\begin{pmatrix} \hat{\vec{\sigma}} & 0 \\ 0 & \hat{\vec{\sigma}} \end{pmatrix}. \tag{12.10.6}$$

Now, we consider the operator of total angular momentum

$$\hat{\vec{J}} = \hat{\vec{L}} + \hat{\vec{S}} \tag{12.10.7}$$

and calculate its commutator with the Dirac Hamiltonian.

Using the commutation relations for the Pauli matrices, we obtain

$$[\hat{S}_x, \hat{H}_D] = -i\hbar c\,(\alpha_y\,\hat{p}_z - \alpha_z\,\hat{p}_y)\,. \tag{12.10.8}$$

$$[\hat{S}_y, \hat{H}_D] = -i\hbar c\,(\alpha_z\,\hat{p}_x - \alpha_x\,\hat{p}_z)\,, \tag{12.10.9}$$

$$[\hat{S}_z, \hat{H}_D] = -i\hbar c\,(\alpha_x\,\hat{p}_y - \alpha_y\,\hat{p}_x)\,. \tag{12.10.10}$$

Therefore,

$$[\hat{J}_x, \hat{H}_D] = [\hat{L}_x, \hat{H}_D] + [\hat{S}_x, \hat{H}_D] = 0, \tag{12.10.11}$$

$$[\hat{J}_y, \hat{H}_D] = [\hat{L}_y, \hat{H}_D] + [\hat{S}_y, \hat{H}_D] = 0, \tag{12.10.12}$$

$$[\hat{J}_z, \hat{H}_D] = [\hat{L}_z, \hat{H}_D] + [\hat{S}_z, \hat{H}_D] = 0. \tag{12.10.13}$$

Or,

$$[\hat{\vec{J}}, \hat{H}_D] = 0. \tag{12.10.14}$$

On the basis of these results, we conclude that it is the the total angular momentum, $\vec{J}$, which is the integral of motion for the Dirac Hamiltonian and not the orbital angular momentum $\vec{L}$ or the spin angular momentum $\vec{S}$ individually.

Furthermore, for the positive energy Dirac particle at rest (that is, when $\vec{p} = 0$ and $E_0 = mc^2$), the small component, χ_p, vanishes and the full time-dependent wave function becomes

$$\psi_+ = \begin{pmatrix} 1 \\ 0 \\ 0 \\ 0 \end{pmatrix} e^{-i\frac{mc^2}{\hbar}t}, \tag{12.10.15}$$

or,

$$\psi_+ = \begin{pmatrix} 0 \\ 1 \\ 0 \\ 0 \end{pmatrix} e^{-i\frac{mc^2}{\hbar}t}. \tag{12.10.16}$$

For the negative energy Dirac particle at rest (that is, when $\vec{p} = 0$ and $E_0 = mc^2$), the small component, ϕ_p, vanishes and the full time-dependent wave function becomes

$$\psi_- = \begin{pmatrix} 0 \\ 0 \\ 1 \\ 0 \end{pmatrix} e^{i\frac{mc^2}{\hbar}t} \tag{12.10.17}$$

or,

$$\psi_- = \begin{pmatrix} 0 \\ 0 \\ 0 \\ 1 \end{pmatrix} e^{i\frac{mc^2}{\hbar}t}. \tag{12.10.18}$$

It is easy to check that the positive energy solution (12.9.15) and the negative energy solution (12.9.17) are eigenfunctions of the operator $\hat{S}_z$ with eigenvalue $\frac{\hbar}{2}$, while the positive energy solution (12.9.16) and the negative energy solution (12.9.18) are the eigenfunctions of $\hat{S}_z$ with eigenvalue $-\frac{\hbar}{2}$. *Thus, the particles that obey Dirac equation have half-integer spin in the units of* $\hbar$.

***Interpretation of the negative energy solution*:** We have seen that the Dirac equation has free-particle solutions for positive as well as negative energies. A plot of the energy levels is shown in Figure 12.1. As depicted, there is a continuum of energy levels for both $E > 0$ and $E < 0$. The gap between the edges of these regions is $\Delta E = 2mc^2$.

The existence of negative energy continuum of states for the Dirac electron, raises doubts about the stability of matter in Dirac's theory. Since particles can jump to lower energy states by emitting photons, an electron in the lowest positive energy state could emit a photon with energy equal to $2mc^2$ and jump to the highest negative energy state! Since it has lost energy, it would speed up, lose more energy by emitting photons and continue to fall deeper and deeper into the negative energy continuum states. Thus, the entire matter will be annihilated within a fraction of a second. As we know, it does not happen in reality. So, how do we understand the existence of these negative energy continuum states for the electron?

Note that in classical theory also negative energy solutions do exist. They are, however, excluded by imposing the initial condition that, *in the beginning*, all particles had positive energy. This initial condition is based on the requirement that a classical particle cannot make a transition from a positive energy state to the negative energy state without going through the intermediate energy levels[7]

A completely free, single quantum mechanical particle will not make transitions to negative energy states by itself. However, no particle is completely free and transitions can always be induced by external perturbations, for example, by radiation, if not by any

[7]H. Bethe and R. Jackiw, Intermediate Quantum Mechanics, Westview Press, 1997.

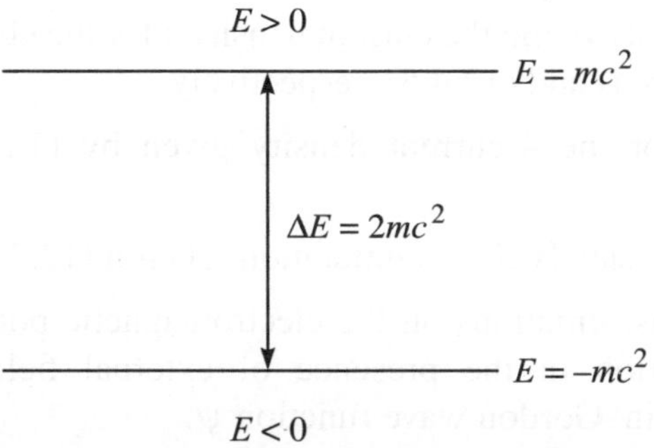

Figure 12.2 Schematic plot of the energy levels for free particles described by the Dirac equation.

other means. Therefore, to get rid of the instability of matter predicted by his theory, Dirac proposed the existence of the so-called 'Dirac sea' by assuming that all negative energy states are already filled with negative energy electrons. Since the negative energy electrons uniformly fill the Dirac sea, they cannot be directly observed. Each of these negative energy electrons have the following properties: *negative mass, negative energy, negative charge*. Because electrons are fermions and obey the Pauli exclusion principle, the transition of positive energy electrons to any of the negative energy state in the Dirac sea is not possible because that state is already occupied by an electron.

It is, however, possible that a negative energy electron, in one of the negative energy states in the Dirac sea, is excited to one of the positive energy states by an extremely high energy photon (such as gamma rays) with energy greater than or equal to $2mc^2$. That electron would now exist as a normal, positive energy electron. There will now be a hole (created by the absence of the negative energy electron) in the Dirac sea. This hole would behave as a particle as well. Since the hole is the absence of negative energy electron, it would have exactly the opposite properties of a negative energy electron. It would have *positive mass, positive energy and positive charge*. These particles would be anti-electrons (or positrons). Furthermore, if a normal electron ever encountered a positron (which was actually a hole in the Dirac sea), the electron would emit $2mc^2$ energy, and fall into the place of the positron in the Dirac sea.

Using the concept of the 'Dirac sea' of negative energy electrons, one can calculate the probability of pair production in the electric field of a nucleus by considering the probability of raising an electron from a negative to a positive state. Concluding this section, let us mention that similar techniques cannot be applied to the Klein–Gordon equation, because particles with integer spin do not obey any exclusion principle.

Homework Problems

1. Determine the infinitesimal Lorentz transformation when $\Delta\omega^{10} = -\Delta\omega^{01} = \Delta\beta$ and all other $\Delta\omega^{\mu\nu} = 0$. Here $\beta = V/c$ and $\Delta\beta = \Delta V/c$.
2. Determine the infinitesimal Lorentz transformation when $\Delta\omega^{12} = -\Delta\omega^{21} = \Delta\theta = constant$ and all other $\Delta\omega^{\mu\nu} = 0$.

3. Show that the charge density ρ and the current density $\vec{j}$ for the charged Klein–Gordon particle are given by (12.4.4) and (12.4.5), respectively.
4. Derive the expression for the 4-current density given by (12.6.10) for the Dirac equation.
5. Show that the γ^μ matrices satisfy the commutation relation (12.7.3).
6. Show that the gauge transformations on the electromagnetic potentials, occurring in the Klein–Gordon equation in the presence of external fields, induce a phase transformation of the Klein–Gordon wave function ψ.
7. Find the solutions of the Klein–Gordon for the case when the external field is a static Coulomb field for which
$$e\Phi = -\frac{Ze^2}{4\pi\varepsilon_0}\frac{1}{r}, \quad \vec{A}(\vec{r}) = 0.$$
8. Consider the Dirac equation in $1+1$ dimensions (that is, one space and one time dimension):
$$\left(\gamma^0\frac{\partial}{\partial x^0} + \gamma^1\frac{\partial}{\partial x^1} - \frac{mc}{\hbar}\right)\psi(x^0, x^1) = 0.$$
(a) Find a 2×2 matrix representation of γ^0 and γ^1 that satisfies $\{\gamma^\mu, \gamma^\nu\} = 2g^{\mu\nu}$ and has correct hermiticity. What is the physical reason that ψ can have only two components in $1+1$ dimensions?
(b) Find the representation of $\gamma^5 = \gamma^0\gamma^1$, $\gamma_5\gamma^\mu$ and $\sigma^{\mu\nu} = \frac{i}{2}[\gamma^\mu, \gamma^\nu]$. Are they independent? Define a minimal set of matrices that form a complete basis.
9. Consider the following matrices in the so-called Weyl representation
$$\vec{\alpha} = \begin{pmatrix} -\vec{\sigma} & 0 \\ 0 & \vec{\sigma} \end{pmatrix}, \qquad \beta = \begin{pmatrix} 0 & I \\ I & 0 \end{pmatrix}.$$
They are called Weyl matrices. Show that the Weyl matrices satisfy all the properties of the Dirac matrices $\alpha^1, \alpha^2, \alpha^3$ and β. Show that the Dirac matrices in the Weyl representation are
$$\vec{\gamma} = \begin{pmatrix} 0 & \vec{\sigma} \\ -\vec{\sigma} & 0 \end{pmatrix}, \qquad \beta = \begin{pmatrix} 0 & I \\ I & 0 \end{pmatrix}.$$
10. Solve the Dirac equation for an electron in a constant external magnetic field directed along the positive z-axis.

11. Derive the relativistic relation between the total energy, E, and momentum, $\vec{p}$, of a Dirac particle of mass m by computing the necessary and sufficient condition for the non-trivial solutions of the coupled system of algebraic equations (12.9.9) and (12.9.10).
12. Derive the normalization coefficient N given by (12.9.28).

Appendix A: Fundamental Constants

Quantity	Symbol/Equation	Value
Planck's constant	$\hbar$	$1.05457 \times 10^{-34}\ J\,s$
Speed of light	c	$2.9979 \times 10^{8}\ m\,s^{-1}$
Charge of proton	e	$1.602 \times 10^{-19}\ C$
Charge of electron	$-e$	$-1.602 \times 10^{-19}\ C$
Mass of electron	m_e	$9.109 \times 10^{-31}\ kg$
Mass of proton	m_p	$1.673 \times 10^{-27}\ kg$
Permittivity of free space	ε_0	$8.854 \times 10^{-12}\ F\,m^{-1}$
Permeability of free space	μ_0	$4\pi \times 10^{-7}\ N A^{-2}$
Bohr radius	$a = 4\pi\varepsilon_0\hbar^2/m_e c^2$	$0.529 \times 10^{-10}\ m$
Boltzmann constant	k	$1.381 \times 10^{-23}\ J K^{-1}$
Fine structure constant	$e^2/\hbar c$	$1/137.036$
Electron Compton wavelength	$\lambda = 2\pi\hbar/m_e c$	$2.426 \times 10^{-12}\ m$

Appendix B: Useful Integrals

- The integral

$$I_n = \int_0^{\infty} x^n \, e^{-\alpha x} \, dx$$

is computed by differentiating the integral

$$I_n = \int_0^{\infty} e^{-\alpha x} \, dx = \frac{1}{\alpha}$$

with respect to α. Straightforward calculations yield

$$I_n = \frac{1 \cdot 2 \cdot 3 \cdots n}{\alpha^{n+1}} = \frac{n!}{\alpha^{n+1}}.$$

- Consider the integral

$$I_{2n}^{(1)} = \int_0^{\infty} x^{2n} \, e^{-\alpha x^2} \, dx.$$

The value of such an integral is calculated by differentiating the integral

$$I_0 = \int_0^{\infty} e^{-\alpha x^2} \, dx = \frac{1}{2}\sqrt{\frac{\pi}{\alpha}}$$

with respect to the parameter α. The result is

$$I_{2n}^{(1)} = \frac{1 \times 3 \times 5 \times \ldots \times (2n-1)}{2^{n+1}} \sqrt{\frac{\pi}{\alpha^{2n+1}}}.$$

- Similarly,

$$I_{2n}^{(2)} = \int_0^\infty x^{2n}\, e^{-x^2/x_0^2}\, dx = \sqrt{\pi}\, \frac{(2n)!}{n!} \left(\frac{x_0}{2}\right)^{2n+1}.$$

$$I_{2n+1}^{(2)} = \int_0^\infty x^{2n+1}\, e^{-x^2/x_0^2}\, dx = \frac{n!}{2}\, (x_0)^{2n+2}$$

- The Gaussian integral:

$$\int_{-\infty}^{\infty} e^{-\alpha x^2 - 2\beta x}\, dx = \sqrt{\frac{\pi}{\alpha}}\, e^{\beta^2/\alpha}.$$

- For $a \neq b$,

$$\int \cos(ax)\, \sin(bx)\, dx = \frac{\cos[(a-b)x]}{2(a-b)} - \frac{\cos[(a+b)x]}{2(a+b)}.$$

- Using integration by parts for the functions $U(x)$ and $V(x)$

$$\int U dV = UV - \int V dU,$$

one computes the following integrals:

$$\int x \sin(ax)\, dx = \frac{1}{a^2} \sin(ax) - \frac{x}{a} \cos(ax),$$

$$\int x \cos(ax)\, dx = \frac{1}{a^2} \cos(ax) + \frac{x}{a} \cos(ax).$$

$$\int x \sin^2(ax)\, dx = \frac{x^2}{4} - \frac{x \sin(2ax)}{4a} - \frac{\cos(2ax)}{8a^2}.$$

$$\int x^2 \sin^2(ax)\, dx = \frac{x^3}{6} - \left(\frac{x^2}{4a} - \frac{1}{8a^3}\right) \sin(2ax) - \frac{x \cos(2ax)}{4a^2}.$$

Appendix C: Dirac Delta Function

If $f(x)$ is defined at the point $x = x_0$, then

$$\int_{-\infty}^{+\infty} f(x)\,\delta(x-x_0)\,dx = f(x_0).$$

Thus, Dirac delta function can formally be defined as

$$\delta(x-x_0) = \begin{cases} 0 & \text{if} \quad x \neq 0, \\ +\infty & \text{if} \quad x = x_0, \end{cases}$$

and

$$\int_{-\infty}^{+\infty} \delta(x-x_0)\,dx = 1.$$

Note that the delta function can also be defined as the limiting form of a function, $F(x)$, which is non-zero only in an infinitesimal interval in the vicinity of the point x_0 where it abruptly attains its positive maximal value such that

$$\int_{-\infty}^{+\infty} F(x)\,dx = 1.$$

For instance,

$$\delta(x-x_0) = \frac{1}{\pi}\lim_{a\to\infty}\frac{sin[a(x-x_0)]}{(x-x_0)} = \frac{1}{\pi}\lim_{\kappa\to\infty}\frac{1-cos[\kappa(x-x_0)]}{(x-x_0)^2}.$$

Or,

$$\delta(x-x_0) = \frac{1}{\pi}\lim_{\varepsilon\to 0}\frac{\varepsilon}{(x-x_0)^2+\varepsilon^2} = \lim_{\kappa\to 0}\frac{\Theta(x-x_0+\kappa)-\Theta(x-x_0)}{\kappa},$$

where $\Theta(x)$ is the Heaviside function

$$\Theta(x) = \begin{cases} 1 & \text{if} \quad x>0, \\ 0 & \text{if} \quad x<0, \end{cases} .$$

Properties of delta function

- $\delta(x) = \delta(-x)$.
- $\delta(ax) = \frac{1}{|a|}\delta(-x),\, a \neq 0$.
- $\delta[(x-a)(x-b)] = \frac{1}{|a-b|}[\delta(x-a)+\delta(x-b)],\, a \neq b$.
- $\delta(x^2-a^2) = \frac{1}{2|a|}[\delta(x+a)+\delta(x-a)],\, a \neq 0$.
- $x\,\delta(x) = 0$.
- $f(x)\,\delta(x-a) = f(a)\,\delta(x-a)$.
- $\int_{-\infty}^{+\infty}\delta(x-y)\,\delta(y-a)\,dy = \delta(x-a)$.
- $\delta'(x) = -\delta'(x)$.
- $\int_{-\infty}^{+\infty}\frac{d\delta(x)}{dx}f(x)\,dx = -\frac{df}{dx}(0) \equiv -f'(0)$.
- Repeated integrations by parts lead to the following general relation:
- $\int_{-\infty}^{+\infty}\frac{d^n\delta(x)}{dx^n}f(x)\,dx = (-1)^n\frac{d^nf}{dx^n}(0) \equiv (-1)^n - f^n(0)$.
- $x\delta'(x) = -\delta(x)$.
- $x^2\delta'(x) = 0$.
- $\delta(g(x)) = \sum_n\frac{\delta(x-x_n)}{|g'(x_n)|}$, where $g(x_n) = 0$ and $g'(x_n) \neq 0$.
- The Fourier transform of $\delta(x)$ is

$$\delta(x) = \frac{1}{2\pi}\int_{-\infty}^{+\infty}e^{ikx}\,dk.$$

- For the derivative of $\delta(x)$, we have

$$\delta'(x) = \frac{i}{2\pi}\int_{-\infty}^{+\infty}k\,e^{ikx}\,dk.$$

- The three-dimensional form of the delta function is given in Cartesian coordinates by

$$\delta(\vec{x}-\vec{x}') = \delta(x-x')\delta(y-y')\delta(z-z').$$

- The Fourier transform of the three-dimensional delta function is

$$\delta(\vec{x}-\vec{x}') = \frac{1}{(2\pi)^3}\int_{-\infty}^{+\infty} e^{i\vec{k}\cdot(\vec{x}-\vec{x}')}\, d^3k.$$

Appendix D: Important Formulae and Equations

The wave function: $\psi(\vec{r},t)$.

The position operator: $\hat{\vec{r}} = \{\hat{x},\hat{y},\hat{z}\} = \{x,y,z\}$.

The momentum operator: $\hat{\vec{p}} = -i\hbar\vec{\nabla}$.

Time-dependent Schrödinger Equation:

$$i\hbar\frac{\partial\psi(\vec{r},t)}{\partial t} = \hat{H}\,\psi(\vec{r},t),\, \hat{H} = \frac{\hat{\vec{p}}^2}{2m} + V(\vec{r}).$$

Stationary state wave function: $\psi(\vec{r},t) = \phi(\vec{r})\,e^{-\frac{i}{\hbar}Et}$.

Time-independent Schrödinger Equation:

$$\hat{H}\phi(\vec{r}) = E\,\phi(\vec{r}). \quad \text{Or,} \quad -\frac{\hbar^2}{2m}\vec{\nabla}^2\phi(\vec{r}) + V(\vec{r})\phi(\vec{r}) = E\,\phi(\vec{r}).$$

Canonical commutator: $[\hat{x}_i,\hat{p}_j] = i\hbar\,\delta_{ij}$, where $i,j = 1,2,3$.

Generalized uncertainty relation for two Hermitian but non-commuting operators $\hat{A}$ and $\hat{B}$:

$$\Delta A\,\Delta B \geq \tfrac{1}{2}|\langle[\hat{A},\hat{B}]\rangle|, \quad \text{where} \quad \Delta A = \sqrt{\langle\hat{A}^2\rangle - \langle\hat{A}\rangle^2}.$$

Heisenberg uncertainty relations: $\Delta x\,\Delta p \geq \frac{\hbar}{2}$, $\quad \Delta E\,\Delta t \geq \frac{\hbar}{2}$.

Measurement of an observable, $\hat{A}$, in a state $|\psi\rangle$ and probability:

$$\hat{A}|\phi_n\rangle = a_n\,|\phi_n\rangle,\quad P(a_n) = \frac{|\langle\phi_n|\psi\rangle|^2}{\langle\psi|\psi\rangle}.$$

The expectation value of an operator, $\hat{A}$, in a given state $|\psi\rangle$:

$$\langle\hat{A}\rangle = \frac{\langle\psi|\hat{A}|\psi\rangle}{\langle\psi|\psi\rangle}.\quad \text{Or,}\quad \langle\hat{A}\rangle = \int_{-\infty}^{+\infty}\psi^*(\vec{r})\,[\hat{A}\psi(\vec{r})]\,d^3x \Big/ \int_{-\infty}^{+\infty}\psi^*(\vec{r})\psi(\vec{r})\,d^3x.$$

Time evolution of expectation values (Ehrenfest's theorem):

$$\frac{d\langle\hat{A}\rangle}{dt} = \frac{\partial\langle A\rangle}{\partial t} + \frac{1}{i\hbar}\langle[\hat{A},\hat{H}]\rangle.$$

Probability density: $\rho(\vec{r},t) = \psi^*(\vec{r},t)\,\psi(\vec{r},t).$

Probability current density:

$$\vec{j}(\vec{r},t) = \frac{\hbar}{2im}\left(\psi^*(\vec{r},t)\,\vec{\nabla}\psi(\vec{r},t) - \vec{\nabla}\psi^*(\vec{r},t)\,\psi(\vec{r},t)\right).$$

Local conservation of probability (continuity equation):

$$\frac{\partial\rho}{\partial t} + \vec{\nabla}\cdot\vec{j} = 0.$$

Angular momentum operators in Cartesian coordinates:

$$\hat{L}_x == -i\hbar\left(y\frac{\partial}{\partial z} - z\frac{\partial}{\partial y}\right),\quad \hat{L}_y = -i\hbar\left(z\frac{\partial}{\partial x} - x\frac{\partial}{\partial y}\right),\quad \hat{L}_z = -i\hbar\left(x\frac{\partial}{\partial y} - y\frac{\partial}{\partial x}\right).$$

$$[\hat{L}_x,\hat{L}_y] = i\hbar\hat{L}_z,\quad [\hat{L}_y,\hat{L}_z] = i\hbar\hat{L}_x,\quad [\hat{L}_z,\hat{L}_x] = i\hbar\hat{L}_y.$$

$$\hat{L}^2|\ell,m\rangle = \hbar^2\ell(\ell+1)|\ell,m\rangle,\quad \hat{L}_z|\ell,m\rangle = \hbar m|\ell,m\rangle.$$

Spin angular momentum:

$$[\hat{S}_x,\hat{S}_y] = i\hbar\hat{S}_z,\quad [\hat{S}_y,\hat{S}_z] = i\hbar\hat{S}_x,\quad [\hat{S}_z,\hat{S}_x] = i\hbar\hat{S}_y.$$

$$\hat{S^2}|s,m_s\rangle = \hbar^2\,s(s+1)|s,m_s\rangle,\quad \hat{S}_z|s,m_s\rangle = \hbar m_s|s,m_s\rangle.$$

Pauli Matrices:

$$\sigma_x = \begin{pmatrix} 0 & 1 \\ 1 & 0 \end{pmatrix}, \quad \sigma_y = \begin{pmatrix} 0 & -i \\ i & 0 \end{pmatrix}, \quad \sigma_z = \begin{pmatrix} 1 & 0 \\ 0 & \text{-1} \end{pmatrix}.$$

Relativistic relation between energy and momentum:

$$E^2 - p^2c^2 = m_0^2c^4. \quad \Rightarrow \quad E = \pm\sqrt{c^2p^2 + m_0^2c^4}.$$

The Klein-Gordon equation:

$$\left(\Box + \frac{m^2c^2}{\hbar^2}\right)\psi(x) = 0, \quad \Box = \partial_\mu\partial^\mu = \frac{1}{c^2}\frac{\partial^2}{\partial t^2} - \vec{\nabla}^2.$$

Dirac Equation:

$$i\hbar\frac{\partial\psi}{\partial t} = \hat{H}_D\,\psi, \quad \hat{H}_D = \left(c\,\vec{\alpha}\cdot(\hat{\vec{p}}) + m\,c^2\,\beta\right).$$

$$\beta = \begin{pmatrix} I & 0 \\ 0 & -I \end{pmatrix}, \quad \alpha^k = \begin{pmatrix} 0 & \sigma^k \\ \sigma^k & 0 \end{pmatrix}.$$

Relativistically invariant form of Dirac equation:

$$\left(i\,\gamma^\mu\,\partial_\mu - \frac{mc}{\hbar}\right)\psi = 0,$$

where $m \equiv m_0$ is the rest mass of the particle.

Gamma Matrices: $\gamma^0 = \beta, \gamma^j = \beta\alpha^j$.

$$\gamma^\mu\gamma^\nu + \gamma^\nu\gamma^\mu = 2I\,g^{\mu\nu}.$$

$$\gamma^0 = \begin{pmatrix} 1 & 0 & 0 & 0 \\ 0 & 1 & 0 & 0 \\ 0 & 0 & -1 & 0 \\ 0 & 0 & 0 & -1 \end{pmatrix}, \quad \gamma^1 = \begin{pmatrix} 0 & 0 & 0 & 1 \\ 0 & 0 & 1 & 0 \\ 0 & -1 & 0 & 0 \\ -1 & 0 & 0 & 0 \end{pmatrix},$$

$$\gamma^2 = \begin{pmatrix} 0 & 0 & 0 & -i \\ 0 & 0 & i & 0 \\ 0 & i & 0 & 0 \\ -i & 0 & 0 & 0 \end{pmatrix}, \quad \gamma^3 = \begin{pmatrix} 0 & 0 & 1 & 0 \\ 0 & 0 & 0 & -1 \\ -1 & 0 & 0 & 0 \\ 0 & 1 & 0 & 0 \end{pmatrix}.$$

References

1. Bethe, H. and R. Jackiw. 1997. *Intermediate Quantum Mechanics*. Reading, Mass: Addison-Wesley
2. Blokhintsev, D. I. 1964. *Quantum Mechanics*. Dordrecht-Holland: D. Reidel Publishing.
3. Davydov, A. S. 1965. *Quantum Mechanics*. Oxford: Pergamon Press Limited.
4. Greiner W. 2001. *Quantum Mechanics: An Introduction.* Berlin Heidelberg: Springer-Verlag.
5. Griffiths, David J. 2005. *Introduction to Quantum Mechanics.* Upper Saddle River, NJ: Pearson Prentice Hall.
6. Haken, H. and H. C. Wolf. 1980. *The Physics of Atoms and Quanta*. Berlin Heidelberg: Springer-Verlag.
7. Hollingsworth, C. A. 1990. 'Accidental Degeneracies of the Particle in a Box.' *J. Chem. Edu.* 67 (12): 999.
8. Joshi, A. W. 1988. *Elements of Group Theory for Physicists*. Hoboken, New Jersey: John Wiley and Sons Ltd.
9. Klein, Martin J. 1952. 'On a Degeneracy Theorem of Kramer'. *Am. J. Physics* 20: 65–71.
10. Kramers, H. A. 1930. "General theory of paramagnetic rotation in crystals." In *Proc. Acad. Sci. Amsterdam,* vol. 33, p. 959.
11. L. I. Schiff. 1949. *Quantum Mechanics*. New York: McGraw-Hill.
12. Landau L. D., and E. M. Lifshitz. 1977. *Quantum MechanicsNonrelativistic Theory*. Oxford: Pergamon Press.
13. Messiah, Albert. 1967. *Quantum Mechanics: Volume I.* Amsterdam: North-Holland Publishing Company.
14. Pauli, W. and V. Weisskoff. 1934. *Helv. Phys. Acta* 7: 709.

15. Sakurai, J. J. 1994. *Modern Quantum Mechanics*. Reading, Massachusetts: Addison-Wesley Publishing Company.

16. Schwabl, F. 2007. *Quantum Mechanics*. Berlin: Springer-Verlag.

17. Schweber, Silvan S. 1961. *An Introduction to Relativistic Quantum Field Theory*. New York: Row, Peterson and Company.

18. Shankar, R. 1994. *Principles of Quantum Mechanics*. New York: Kluwer Academics/Plenum Publishers.

19. Tinkham, M. 1964. *Group Theory and Quantum Mechanics*. New York: McGraw-Hill Book.

20. Wigner, E. P. 1959. *Group Theory and its Application to the Quantum Mechanics of Atomic Spectra.* New York: Academic Press.

21. Zettili, Nuredine. 2001. *Quantum Mechanics: Concepts and Applications*. Chichester: John Wiley and Sons Ltd.

Index